W9-BIQ-515

FASHION!

by

Mary G. Wolfe
St. Michaels, Maryland

Publisher
The Goodheart-Willcox Company, Inc.
Tinley Park, Illinois

Manufactured in the United States of America.

Library of Congress Catalog Card Number 2005052518

ISBN-13: 978-1-59070-628-2
ISBN-10: 1-59070-628-5

3 4 5 6 7 8 9 10 – 06 – 11 10 09 08 07

Library of Congress Cataloging-in-Publication Data

Wolfe, Mary Gorgen
Fashion! / by Mary G. Wolfe.
p. cm.
ISBN 1-59070-628-5
1. Fashion. 2. Clothing trade. 3. Clothing trade--Vocational guidance. I. Title.

TT518 .W59 2005
746.9'2'023--dc22

2005052518

Introduction

Fashion! brings the exciting world of fashion to life through an in-depth look at how the apparel industries work. It opens your eyes to the many ways that you might be a better consumer and coordinate your lifestyle with fashion-related industries.

Fashion! is full of features to pique your interest. It

- discusses designer ready-to-wear, private label lines, and electronic marketing of fashion goods.
- looks at the future directions and trends that textile companies, apparel manufacturers, and retailers are likely to take.
- describes the many career opportunities related to fashion and apparel, including entrepreneurship.
- helps you become more fashionable by explaining how to use the elements and principles of design to your advantage, telling how to plan the best wardrobe for your needs, and teaching how to shop and care for your clothes.

Fashion! is also a great reference guide. Fashion terms are explained in Chapter 2, garment styles and parts are illustrated in Chapter 3, and popular apparel fabrics are described in Chapter 9. The extensive glossary at the end of the text defines other fashion and apparel terms that are used by industry professionals.

This information is presented in an easy-to-understand format. Simple, direct language is supplemented with hundreds of color photographs and illustrations. Each chapter begins with learning objectives and ends with review materials to make your learning more meaningful and enjoyable. The goal of this book is to help you gain knowledge that will be valuable to you whether you become an integral part of the fashion world or an informed consumer!

About the Author

Mary Wolfe has worked in all segments of the fashion industry, from textile research to retail sales. She designed for a national sportswear firm before opening her own apparel business. She gained recognition as the personal fashion designer for the wife of a U.S. Vice President. As a consultant to several garment manufacturers, she has assisted with collection designs and pattern specifications.

Mary received the Outstanding Faculty Member Award for her teaching of fashion-related courses at the University of Delaware. She has been a New Jersey Woman of the Year and has been listed in Outstanding Young Women of America and National Dean's List. Mary is also the author of books on fashion merchandising and pattern making. She received her bachelor's degree in Textiles and Clothing from Iowa State University and her Master's of Business Administration degree from West Chester University of Pennsylvania.

Contents in Brief

Contents

OREGON
N.Y.A.

PART ONE

Clothes and Fashion

Chapter 1
The Why of Clothes

Chapter 2
Knowing About Fashion

Chapter 3
Garment Styles and Parts

1 The Why of Clothes

Fashion Terms

protective clothing	status
adornment	values
beauty	attitudes
identification	conformity
dress code	individuality
modesty	personality

After studying this chapter, you will be able to

- explain the various reasons people wear clothes.
- analyze how clothing satisfies certain physical, psychological, and social needs.
- discuss how values, attitudes, conformity, individuality, and personality affect clothing selections.

Clothing does much more than just cover the body. Your appreciation of clothing will be broadened as you become more aware of the influences clothing has on you. Throughout history, clothing has had great meaning. It has indicated people's handicraft skills, artistic imagination, and cultural rituals. It has also reflected advances in technology.

In ancient times, clothing was made from items found in nature. The first clothes were probably made from animal skins. Today's technology provides us with many different fibers, fabrics, finishes, and manufacturing processes. Regardless of these advances, however, the reasons we wear clothes are the same as they have been throughout history.

Why People Wear Clothes

Prehistoric people clothed their bodies over 75,000 years ago. This has been shown by the discoveries of ancient cave drawings, statues, and remains of materials used for making clothing. From the beginning, clothing has served the same basic human needs. Those needs are protection (a physical need), adornment and identification (psychological needs), and modesty and status (social needs).

Protection

Unlike other animals, the human body needs protection, or physical safeguards. Clothing can prevent harm caused by the climate and the environment. It supplements our natural body covering like a "second skin." ***Protective clothing*** gives physical protection to the body. For instance, protection from drowning is provided by life jackets or other flotation clothing. As we will see, some clothing that offers protection is also stylish.

Protection from Weather

To preserve good physical health, clothing protects us from cold temperatures. It can protect us from sunshine or high winds.

It can also protect us from the wetness of rain, as shown in 1-1.

Warm sweaters, coats, gloves, and long underwear can protect us from frigid weather and snow. Wide-brimmed hats can keep hot sunshine off our faces and heads. We wear windbreakers and water-repellent jackets for protection against weather.

Sunglasses and hats are sometimes worn as fashion items as well as for protection. Fur pelts were essential for warmth and protection long ago. Now fur garments might provide controversy as well as high fashion and warmth.

Protection from Environmental Dangers

People need physical protection from dirt, insects, and other harmful agents in the environment. Shoes protect our feet from soil, hard objects, and hot and cold surfaces. Astronauts must have protective clothing to provide them with the correct atmospheric conditions to keep them alive in outer space. In medical professions, special clothing items, such as sterile gloves and face masks, help reduce the transfer of germs and maintain sanitation.

The Fashion Association/Lands' End

1-1 Water-repellent garments protect people from getting wet when it rains.

Natives of some geographic regions need protective clothing against insects, worms, and leeches. The swish of a grass skirt is meant to keep insects away. Some tropical natives grease their bodies. Some plaster mud on themselves for physical protection. These body coverings are substitutes for clothing and accomplish similar protective results.

Protection from Occupational Hazards

Some garments protect workers from the specific dangers of their jobs. Unlike turtles that can use their shells for protection, humans must use clothing to guard against bruises, cuts, burns, and other injuries.

As industry and technology have developed, special protective garments have evolved. Hard hats and safety goggles have been standard equipment at many job sites for years. More recently, special garments to guard against acids, static, lint, fumes, and other potential dangers have been developed. Some specialized clothing can protect against contamination, chemicals, radiation, and fires. These clothes were devised for firefighters, factory workers, miners, and others as in 1-2.

Athletes often wear protective clothing. The helmets, gloves, and pads pictured in 1-3 provide protection while skateboarding. The

1-2 Occupational clothing is able to provide protection against possible unsafe conditions of certain jobs.

Tultex Corporation, Swann-Neimann Photography

1-3 Sports and recreational activities can be more fun if the proper physical protection is used to prevent injury.

special shoes worn by tennis players and basketball players are designed to protect them from slipping.

Some garments were originally designed for protection while working. Later they became fashion items. Examples include blue jeans and tall leather boots that were first worn by ranchers and outdoor workers.

Protection from Enemies

Throughout history, clothing has protected people from attackers. Physical protection against human enemies centuries ago was provided by body shields or suits of armor. Today's Army helmet protects the head. Pockets and belts can hold military weapons. Camouflage fabric helps the wearer hide by blending in with the environment.

Police officers often wear bulletproof vests, as shown in 1-4, to protect their bodies from gunfire. Other clothing that protects against enemies may be less obvious. The helmets and pads worn by football players protect against opposing players. The reflective vests and jackets worn by joggers and bikers after dark offer protection from the drivers of motor vehicles.

Some cultures believe items of clothing can give them protection against such enemies as evil spirits, illness, or a bad harvest. Items such as good luck charms and medallions are sometimes used to make people feel luckier, healthier, safer, or braver.

Adornment

Clothing can affect a person's mental attitude or morale in a good way. This is done through ***adornment*** or decoration. Adornment provides a psychological feeling of well-being through beauty.

Beauty is a quality that gives pleasure to the senses. It creates a positive emotional reaction in the viewer. Most psychologists believe beauty is essential to human life. People have a need to make themselves look more attractive. Body adornment enhances self-concept and personality. The winter wear in 1-5 would

Dupont Kevlar®

1-4 Police officers can help protect their bodies from gunfire by wearing bulletproof vests under their uniforms.

Pendleton, Men's Fashion Association of America

1-5 The artistic pattern and colors of this sweater and the true red color of the jacket make a statement of adornment.

be just as efficient if it were all black. However, the combination of bright colors adds beauty to it.

Decorative clothing makes us more attractive. We wear clothing that is artistically designed, and we combine garments in artistic ways. Then we further adorn ourselves with earrings, bracelets, and neck chains. Makeup and nail polish add more decoration. The large sums of money spent each year on jewelry and beauty aids attest to the emphasis people place on personal appearance.

Adornment has been found in various cultures throughout history. Primitive people used colored clay or paint to decorate their bodies. Jewelry was carved from animal bones or horns. The body was decorated with what was available.

The decorations people use still depend on their native culture. In some regions, natives decorate their bodies with paints and ornaments. Some bodies are intentionally scarred or bound for adornment. People may wear necklaces made of animal teeth, shells, or seeds. Individuals in our culture imitate this type of decoration with tattoos and rings through the ears and nose.

The way one culture views beauty may be different from how other cultures view it. The desirability of certain decorations is determined by the standards, values, and traditions of each society. For instance, many cultures have popular, but different, hairstyles. Fabrics of different cultures vary in textures, patterns, and colors, as in 1-6. Often the traditional fabrics of a culture used to create its folk costumes have great importance.

Individuals also have different thoughts about beauty and adornment because of personal experiences. A professional athlete may consider casual attire to be more becoming than formal wear. Someone who enjoys ballet and opera might have dressier adornment preferences.

Ideas also change with time. What is beautiful to us one year may not look attractive at all to us a few years later.

Being well adorned with our clothes gives us a good psychological feeling of self-expression. We can express our creativity and individuality. Clothing can contribute to feelings of self-respect, self-acceptance, and self-esteem. When we improve our looks, we attract favorable attention. That satisfies our personal artistic nature.

BASF Corporation

1-6 Apparel fabrics used for body adornment in Nigeria are somewhat different from the textiles used in most American fashions.

Sometimes we adorn ourselves in a way that is different from usual for change or adventure. This gives us relief from boredom. It adds psychological zest to our life. A business person who wears dark suits all week might choose a bright outfit for a fun event on the weekend, as in 1-7.

Identification

Identification is the process of establishing or describing who someone is or what something does. Your clothing can identify you. Group identity is shown by dressing like those with whom you associate. Clothing can satisfy the psychological need to belong, such as to a profession, social group, association, or country of heritage. It indicates what "role" you play or what skills you have. It says who you are and what you do.

Uniforms are one way of identifying roles, as in 1-8. Uniforms are outfits or articles of clothing that are alike and specific to everyone in a certain group of people. They act as symbols of group identity. Besides giving a sense of belonging, they can indicate a position of authority. People who provide protection to the public, such as police officers, need easily recognized uniforms in order to carry out their duties. Uniforms indicate that someone actually is a mail carrier, a military officer, a hospital volunteer, a waiter, or a member of the clergy.

Uniforms can decrease racial, religious, and other barriers. Some schools require their students to wear uniforms so individual differences and tensions are minimized. Identifying with that school can then shift students' focus to academics. Uniforms give a unified appearance, or public image, among all individuals in a particular group.

The regular clothing of many people can be considered a type of "psychological uniform," because everyone dresses almost alike. Look around at your classmates. Are most of them dressed with the same general kinds of pants

The Fashion Association/Pronto Uomo

1-7 Adorning ourselves in ways that are different from our everyday attire can give us a psychological lift.

1-8 This uniform identifies the wearer as a cheerleader for a particular school.

or skirts, shirts, sweaters, and shoes? People who are close in age and who share similar interests often dress alike. By doing so, they gain confidence, acceptance, and psychological approval. They feel comfortable and secure. Such "rules of dress" are not formally written; silent understanding sets them.

Adults sometimes have the same habit of dressing almost alike. Men of a particular group might all wear dark suits and ties to work. To casual gatherings, they may all wear jeans and sweaters. This conformity, or unofficial uniform, makes them feel secure about being appropriately dressed.

Identification can also be accomplished with emblems, colors, badges, patches, and specific pieces of jewelry. Some students wear their class rings. This identification gives them a psychological feel of unity with others in the group. Athletic uniforms, nuns' habits, and nurses' outfits tell the world about those people's special identities. Scottish Highlanders wear particular tartan plaids to show they belong to certain clans.

Ceremonial garments can provide identification. The caps and gowns worn by students indicate they are graduating. A white gown and veil indicates a woman is a bride, as in 1-9. Some ceremonies are identified by the use of elaborate robes, costumes, masks, and other decorations.

Many businesses and schools have ***dress codes.*** Dress codes are written or unwritten rules of what should or should not be worn by a group of people. Although the garments worn are not uniforms, they must fall within a certain range of options. Besides achieving group identity, the clothes help the group members maintain a certain discipline of behavior. Sometimes, this results from the symbolic meaning of the clothing. At other times, it is because of the way the clothes look or feel. For instance, a business suit helps a person act in a businesslike manner.

Perry Ellis Formalwear

1-9 A wedding gown clearly identifies the woman wearing it as a bride.

Modesty

Human beings wear clothing to satisfy their social need for modesty. ***Modesty*** is the covering of a person's body according to the code of decency of that person's society. Our society has caused us to be embarrassed to go without clothes. Modesty dictates the proper way to cover the body for social acceptance. Our standards of decency are molded by our culture and social system. Each society has its own accepted standards of modesty.

In the 1800s, it was immodest for American women to let their ankles show. In the 1920s, older people were appalled at the short skirts worn by young women. However, short skirts soon became acceptable. Ladies' swimsuits were once made of thick fabric. They covered most of the body. Now fashionable swimwear exposes lots of skin! See 1-10. During the early part of the 1900s, the body was almost completely covered, even for sports. The standards of our society about modesty have changed a great deal.

The event you are attending also influences your amount of modesty. A man may wear a kilt to play a bagpipe at a traditional Scottish gathering. However, wearing a kilt anywhere else would probably embarrass him. By wearing appropriate or inappropriate clothing, people show their acceptance or rejection of their social environment.

JC Penney

1-10 Standards of modesty have changed over the years. Today's fashionable swimwear is much skimpier than in past eras.

JC Penney

1-11 This man achieves status through both the desirable trade name of the manufacturer and the revered name of the football team.

Status

A person's ***status*** is his or her position or rank in comparison to others. "Good" or "high" status is usually associated with recognition, prestige, and social acceptance. Clothing is sometimes used to gain a higher rank in society, along with social acceptance and peer approval. Thus, many people are willing to pay extra for garments with the most desired labels or logos, as in 1-11.

Adults may try to achieve a higher status by wearing furs, diamond jewelry, or expensive clothing items. Examples might be a cashmere sweater or a beaded gown, as in 1-12.

Some items have important social meanings and make the people wearing them feel important. The items enable individuals to show others what they have achieved. Service stripes on a military sleeve, merit badges on a Boy Scout's shirt, and a school letter on an athletic jacket all tend to raise the status of the wearer. In history, hunters adorned themselves with the pelts of their prey to impress others with their achievements.

Why People Select Certain Clothes

You have just read that people wear clothes to fulfill certain physical, psychological, and social needs. There are additional factors that influence the particular clothing choices people make. Some of the most important are people's values and attitudes, their tendencies toward conformity or individuality, and their personalities.

Fashion is a mirror of our times. It reflects our culture at a given point in time. Historic clothing has revealed many details about the lifestyles of people from various past cultures.

Some experts have commented that if only one book could be left from today for people to read in hundreds of years, it should be a fashion magazine. Many believe it would tell more than volumes written by philosophers, novelists, prophets, and scholars!

David's Bridal

1-12 A gown with decorative beading, such as this one, gives much more status than a plain long dress made of inexpensive fabric.

JC Penney

1-13 This young woman has chosen clothing to suit her values and attitudes.

Values and Attitudes

Values are the ideas and beliefs that are important to an individual. They are the underlying motivations for a person's actions. They are the basis of a person's decisions, lifestyle, and personal code of ethics. ***Attitudes*** are formed from values. They are an individual's feelings or reactions to people, things, or ideas.

Values and attitudes are learned concepts. They are influenced by cultural customs and traditions. Economic and social conditions of the time also affect them. Values and attitudes are passed from one generation to another. Family members, friends, and the community are important in forming them.

Some people select comfortable clothing because they value their own comfort, 1-13. Others always choose bargains because they value economy. Some people value easy care. Others must have the latest look in clothes, or expensive items, because they value prestige and want to be noticed. People who are in the businesses of making or selling clothing items try to identify the values and attitudes of their customers. Then they can provide the items that will be preferred, and ultimately bought, by that group of consumers.

Where people put their money shows their personal values. Some like to spend money on many clothes and accessories. Others have few clothes, preferring to spend their money on concerts, movies, ski trips, or other forms of recreation. Still others save their money for a car or other large purchases in the future.

Advertising can influence people's values, attitudes, and purchase decisions. Television commercials try to create a stronger desire for particular products. Fashion ads in newspapers and magazines play to desires for economy, status, easy care, adventure, and comfort.

Shopping malls attract many customers who value convenience.

Age influences people's clothing selections, too. As people go through life, their needs and values change. Students in their middle school years may consider conformity as the most important factor. High school and college students may have to count on economy, but they probably also want sex appeal. Business attire, as shown in 1-14, might be selected to rise through a professional career. Prestige and status may be desired during middle age. Comfort is important again with the slower years of old age.

Conformity versus Individuality

Pressure from other people has a great influence on how people dress. ***Conformity*** means obeying or agreeing with some given standard or authority. Humans learn early in life what others expect them to wear. Parents, teachers, and other authority figures set some of the rules of dress. Peer group pressure is another major force that contributes to conformity.

As discussed earlier, conformity can satisfy the needs for identification. By doing this, a safe feeling of belonging is achieved through approval. However, too much conformity can mean a loss of personal individuality.

Individuality is self-expression. It is the quality that distinguishes one person from another. It is the characteristic that makes each person unique. When people choose styles and colors of clothes that are different from those of their friends, they are communicating their individuality. They are satisfying their needs for adornment and possibly modesty. They are rejecting peer pressure and conformity. The girls in 1-15 have shown some individuality in their outfits.

The Fashion Association/Lands' End

1-14 The values and attitudes of this man are determined in part by his age and occupation. An attractive suit is valued as an asset to help his career advance.

JC Penney

1-15 Many individuals have fun showing their individuality through the way they put outfits together.

Most people balance the influences of conformity and individuality in their clothing. Their clothing choices depend on their moods as well as different settings and situations.

Personality

Personality can be defined as the total characteristics that distinguish an individual, especially his or her behavioral and emotional tendencies. Studies have found that certain ways of dressing give clues about specific personality traits. For instance, tests have shown that people who desire very decorative wearing apparel tend to be very sociable. People who mainly like comfort from their clothes tend to have self-control and confidence. They are often outgoing and secure, as in 1-16. People who are shown in the personality tests to prefer economy, rather than spending lots of money on their wardrobes, are usually responsible, alert, efficient, and precise.

JC Penney

1-16 These smiles confirm the results of studies concluding that people who choose comfortable clothes tend to be confident and outgoing.

Chapter Review

Summary

People wear clothes today for the same reasons they have worn them throughout history. One reason is for physical protection of the body from weather, environmental dangers, occupational hazards, and enemies. Another reason for wearing clothes is for adornment or decoration that gives a sense of beauty. Clothing is also worn for identification through uniforms, emblems and badges, ceremonial garments, and according to dress codes. Modesty, which follows the code of decency of society, is another reason people wear clothes. Finally, clothing is often worn for status to raise a person's recognition, prestige, and social acceptance.

People select certain clothes to wear because of their values and attitudes, their tendencies toward conformity or individuality, and their personalities. Values are the basis of people's decisions, lifestyles, and codes of ethics. Attitudes affect people's feelings or reactions to other people, things, or ideas. Conformity and individuality are usually balanced in people's clothing choices. People's clothing choices also give clues about their personality traits.

Fashion in Review

Write your answers on a separate sheet of paper.

Matching: Write the appropriate letter for the best reason for wearing each of the items listed below.

1. Designer clothes.	A. protection
2. Swimsuit.	B. adornment
3. Space suit.	C. identification
4. Colorful necklace.	D. modesty
5. Hard hat.	E. status
6. Uniform.	

True/False: Write true or false for each of the following statements.

7. The reasons we wear clothes are different today from those in historic times.
8. Clothes can offer protection from weather, environmental dangers, occupational hazards, and enemies.
9. Folk costumes provide information about how various cultures view beauty and adornment.
10. Uniforms eliminate group identity and take away the feeling of belonging.
11. All societies of the world have the same standards of modesty.
12. Clothing is sometimes used to achieve a higher status.
13. Clothing selection is influenced by people's values and attitudes.
14. Peer group pressure encourages individuality.
15. Personality tests have shown that people who wear very decorative clothes tend to be very sociable.

Fashion in Action

1. Cut at least two pictures from magazines, catalogs, or newspapers to illustrate protective clothing. Mount them on paper and be ready to explain in class why such clothing is necessary. Also try to describe the technological advances that have made the clothing possible.
2. Find at least two pictures of people wearing uniforms and mount them on paper. Write down your impressions of what each uniform shows or tells to those who see it being worn. Also explain how the uniform helps the person wearing it.
3. Ask two or more older people about the clothing they wore when they were your age. Did their clothes reflect traditions of their nationalities? Did they follow different codes of modesty? Did garments provide the physical protection needed for their jobs? What forms of decorative adornment were popular? Write a short report about what you learned.
4. Working in small groups, list clothes and accessories that people wear for protection, adornment, identification, modesty, and status.
5. As a class, discuss how clothing choices are affected by values, attitudes, desire for conformity or individuality, and personality.

2 Knowing About Fashion

Fashion Terms

style	fit
fashion	draped garments
apparel	tailored garments
garment	composite garments
silhouette	haute couture
fad	ready-to-wear
classic	fashion cycle

After studying this chapter, you will be able to

- use correct vocabulary to discuss fashion and clothing.
- identify and describe the three main methods of clothing construction.
- discuss the concept of fashion cycles that occur over time.
- analyze the influence that social and economic factors have on fashions.

Understanding fashion includes knowing many specific clothing terms. Some are already familiar to you. Many may have been written on the labels of clothing you have purchased. Several may have been mentioned by retail store personnel. The subject of fashion cannot be fully understood until these terms become a part of your vocabulary.

Fashion Terms

There are many styles of clothing. A ***style*** is a particular design, shape, or type of apparel item. The style of a garment is determined by the distinct features that create its overall appearance. Various styles that have been repeated in the history of clothing are recognizable. They have been given names such as A-line skirts, Bermuda shorts, Western shirts, and crewneck sweaters, such as in 2-1. Many distinct styles are described in the next chapter of this book.

Fashion is the display of the currently popular style of clothing. A ***fashion*** is the prevailing type of clothing that is favored by a large segment of the public at any given time. It is the clothing that is most accepted or up-to-date. The styles that are fashionable this year may seem very unfashionable in a few years.

Styles come and go. Fashion is always here in some form. Fashion reflects a continuing process of change in the styles of apparel that are accepted. If you are *stylish*, you are "in vogue" or are wearing the styles that are currently popular and fashionable.

The term ***apparel***, or wearing apparel, applies to any or all men's, women's, and children's clothing. A ***garment*** is any article of apparel, such as a dress, suit, coat, evening gown, or sweater. It is any particular clothing item. *Garment parts* are the sleeves, cuffs, collar, waistband, and other components that make up the complete garment.

The Fashion Association/Pronto Uomo, Div. of Mondo, Inc.

2-1 Crewneck sweaters are a particular style because they are pulled over the head and they have a high rounded neckline. The basic crewneck style is not affected by size, color, or fabric design (such as this geometric design).

The McCall Pattern Company, Vogue® Patterns

2-2 This sweater and skirt outfit presents a very tall, slender silhouette.

The ***silhouette*** is the shape of a clothing style. It is formed by the width and length of the neckline, sleeves, waistline, and pants or skirt. If you were to squint your eyes and look at a suit, dress, or coat, the outer lines (shape) of the garment would show its silhouette. Look at the straight silhouette in 2-2.

Silhouettes are always changing in fashion. The general direction that a silhouette takes (becomes wider, narrower, longer, shorter) shows a *fashion trend*. That is the direction in which fashion is moving. Exciting changes take place constantly in the world of fashion.

High fashion or *high style* items are the very latest or newest fashions. They are usually of top quality, with fine workmanship and beautiful fabrics, as in 2-3. Because of the quality, they are expensive. High fashion garments sometimes seem extreme and unusual. They originate from name designers in leading fashion cities. They are worn by wealthy or famous people who are fashion "pacesetters." Some details of high fashion garments filter down into generally accepted fashions.

Avant-garde clothes are the most daring and wild designs. They are unconventional and startling. They are too "far out" to be considered fashions of the times. Most features of these garments disappear completely after a few years. Avant-garde clothes are used to draw attention to the wearer. They do not appeal to very many people. They are often worn by rock groups on stage. Sometimes teenagers wear versions of avant-garde clothing and hairstyles. A "spiked" hairdo with purple and orange streaks is considered avant-garde.

A clothing ***fad*** is a temporary, passing fashion. It is an item or look that has great appeal to many people for a short period of

The Fashion Association/Marithe & Francois Girbaud

2-3 This crushed velvet duster and vest is combined with motorcycle pants and a sheer, crisp button-front shirt.

Veronica Smith Morrison, N.G.S. (C), National Geographic Society

2-4 Bright or patterned suspenders have been fad items worn by both males and females. At one time it was a fad to wear them hanging down at the sides of the body instead of over the shoulders.

time. It is usually out of the ordinary. A fad becomes popular fast and then dies out quickly. It is a "passing fancy" that is very well liked for a while. Soon after it reaches its height of popularity, its extreme design causes its popularity to wane. Often it is an accessory, such as the brightly colored suspenders shown in 2-4. It sometimes includes a particular fabric or decoration.

Many teenagers enjoy wearing the latest fads. Fads provide a feeling of adventure as well as a sense of belonging to the group. Ankle socks with lace ruffles are a fad of the past. Other examples are "hot pants" and tall, colorful plastic boots. Argyle kneesocks, various types of hats, and patchwork garments, such as in 2-5, are sometimes fad items.

Fads can be fun as long as they are not too expensive and the money being spent for them is not needed for other, more practical items. If

Men's Fashion Association of America/Dunnington

2-5 A patchwork appearance in apparel is sometimes a fad.

a faddish item is well designed and meets a clothing need, it can become a style or influence fashion. Fashions have more lasting aesthetic value than fads.

A *craze* is like a fad because it is a passing love for a new fashion. However, this has a display of emotion or crowd excitement with it. It is a mania! Sometimes very popular entertainers will start a craze. Advertising on television and in magazines and newspapers can heighten the craze. Stores have a hard time keeping such items in stock because people are so eager to buy them.

A ***classic*** item of clothing is one that continues to be popular even though fashions change. It is always acceptable. Classics were originally fashion items, but their general appeal and simple, stylish lines have kept them popular. They can be worn year after year. Some examples of classic garments are white dress shirts, dark business suits, navy blazers, pleated skirts, Chanel suits, shirtwaist dresses, and loafer shoes. Blue jeans are now also a classic. The trench coat shown in 2-6 is a classic garment.

A *wardrobe* consists of all the apparel a person owns. Your wardrobe includes all of your garments and accessories. Your *accessories* are the articles added to complete or enhance your outfits. Examples include belts, hats, jewelry, shoes, gloves, and scarves. These are the secondary items that you add to dress up, or set off, your garments. You need accessories in your wardrobe to achieve a total, completed look.

The Fashion Association/Newport Harbor

2-6 Trench coats have remained basically the same in style and use over many decades.

Clothing Construction Terms

The ***fit*** of a garment refers to how tight or loose it is on the person who is wearing it. A good fit means the garment is the right size and does not pull tightly or sag loosely when worn. Illustration 2-7 shows a well-fitting dress. The design of a garment can, however, give it intentional looseness. Garments can be designed to be loose, semifitted, or fitted (quite tight) to achieve different fashion looks. A *fitted garment* is shaped to follow the lines of the body.

Seams are the lines of stitches that join two garment pieces together. *Darts* are short, tapered, stitched areas that enable the garment to fit the figure. Seams and darts give shape to flat pieces of fabric so they can fit a three-dimensional body.

The *bodice* of a garment is the area above the waist, such as the upper part of a dress or jumpsuit. The bodice is usually closely fitted and is distinguished by a seam at the waistline of the garment. Most bodices are garments designed for females. Notice the seams, darts, and bodice in 2-8.

Throughout clothing history, there have been three main methods of making or constructing clothing. They are the draped, tailored, and composite methods. These three basic ways of putting fabric together do not change. Yet the results are styles of clothes that vary from year to year and from culture to culture.

Draped garments are those that are wrapped or hung on the human body. At first, animal skins were thrown over the body. Later, weaving produced pieces of fabric. An uncut square or rectangular piece of woven material was draped either tightly or loosely around the body. The same piece of fabric could be wrapped to achieve different looks or to serve several purposes. For instance, it could be used as a garment, as a bed covering, or for carrying things.

JC Penney

2-7 A well-fitting garment looks good and feels good when worn.

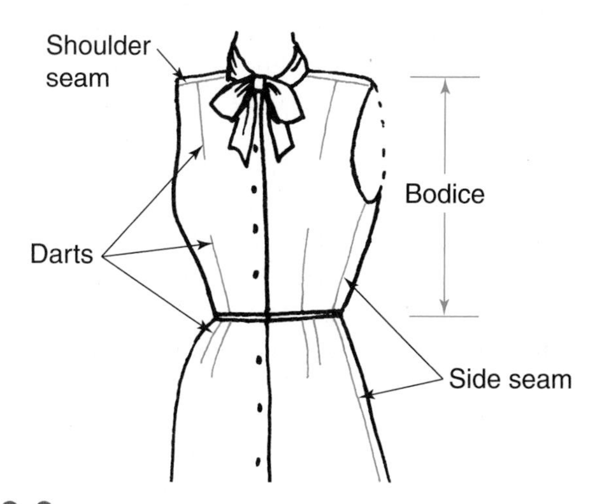

2-8 Seams are located at the shoulders, sides, and other places in garments. Darts stitch fabric together where the garment would be too big if left unstructured.

Examples of draped garments are the Roman toga and the Indian sari. There are also ponchos, as in 2-9, and draped skirts and gowns. They have characteristic folds of soft fabric. Draped garments are usually held in place with pins, buttons, toggles, or a sash or belt.

Tailored garments are made by first cutting garment pieces and then sewing them together to fit the shape of the body. The first tailored garments were made when the eyed needle was invented. Seams could then be sewn to join garment parts together.

Tailoring provided jackets and pants that were warmer than furs thrown loosely over the body. Animal skins were used at first since they do not ravel and are well-suited to this method. Garments were later cut and sewn from fabric. Examples of tailored garments of today include suits, pants, and fitted jackets, as in 2-10.

Composite garments are made by a combination of the tailored and draped methods.

Ecko Unlimited

2-9 Just as in historic times, flat fabrics can be draped to create garment designs.

Garment parts are cut and sewn. Some parts may fit close to the body, and some are draped. Folds of fabric often hang loosely from the shoulders, waist, or hips. Examples of composite garments are the Japanese kimono, tunics, bathrobes, caftans, and capes. The evening gown in 2-11 is a composite garment.

Clothing Business Terms

Haute couture (pronounced oat koo-tur') literally means "finest dressmaking" in French. However, it has come to mean the high fashion industry. It refers to a group of firms or "fashion houses," each with a designer who creates original, individually designed fashions. Designers of exclusive high fashion garments are called *couturiers.*

Haute couture clothes originate in Paris or other fashion centers. They are very expensive and are made in limited numbers. They are constructed of luxury fabrics with a great deal of handwork. They have intricate cut and details, and contain the exclusive label of the designer.

Haute couture garments are sometimes *custom-designed* specifically for a particular person. Custom-designed garments have special fit, design, and fabric for the one person who ordered them. *Made-to-order* or *custom-made* garments are not designed for a particular person although they are made for that person. He or she places an order after seeing a sample garment, sketch, or picture.

Copies of some haute couture garments are made in quantity by high-priced manufacturers. They look like the original haute couture garments, but are produced in factories. The

Kellwood Company

2-10 Although wearing ease allows for comfort, this jacket and pants are made with shaped garment pieces that have been cut and sewn together.

Vogue Patterns/Bellville Sassoon

2-11 In this composite dress, some garment parts are tailored and others are draped to achieve the overall design and fit.

quality of construction and fabric may be good, but not as high as a haute couture version. Good copies might sell in better fashion boutiques. Cheap copies sell at bargain shops.

Knock-offs are lower-priced copies of garments. They are produced in great volume with lower quality materials and construction. Copies and knock-offs are a result of *fashion piracy.* That is the stealing of design ideas without the consent of the originator, such as from Internet coverage of designer showings.

Ready-to-wear garments, as in 2-12, are those that are mass-produced in factories. They are manufactured in quantity according to standard sizes. Each garment design has thousands made. They are all alike and are for sale in many sizes. They are available in great numbers on the racks of stores throughout the country and world. They contain the manufacturing company's label rather than the name of the particular designer.

Consumers are the people who buy and wear the garments. Consumers purchase and use apparel as well as other economic goods. You, your family, and your friends are all consumers. Consumers are very important in determining what fashions will or will not become popular.

Retail stores sell to consumers. They advertise and sell their items directly to the general public. Retail stores include department stores, chain and discount stores, and small shops. Mail-order businesses and TV shopping channels will be discussed in Chapter 7.

The *wholesale* business, on the other hand, sells goods in large lots to retailers. Wholesalers usually distribute their goods from large warehouses. Each item costs less from a wholesaler, but usually dozens of each must be purchased.

Sometimes retail stores sell the extra clothes that were produced by manufacturers but were not ordered for regular selling. These items are called *overruns*. They are in perfect condition but are left over at the manufacturer at the end of the season.

Stores also may sell irregulars. *Irregulars* are articles with slight imperfections. The irregularities in merchandise are often not noticeable, but the goods must be labeled as such and priced lower than their perfect counterparts. *Seconds* are items that are soiled or

2-12 The clothes worn by almost all consumers are ready-to-wear garments offered in quantity on retail racks and shelves.

have flaws. For instance, they might have missing trim or mended runs and tears. They must also be priced lower than perfect goods.

Promotions to sell particular fashions are done nationally by the manufacturer of the goods, as well as locally by retail stores. The promotional activities are the advertising and merchandising efforts to increase sales.

Price Markets

The apparel industry offers garments at all prices along a sliding scale from high to moderate to low.

High-priced apparel is sold to the "class market" that is made up of the few people who buy high fashion clothing. These people accept more unusual styles and colors than most consumers. They are willing to spend more money on their wardrobe than most others. They often go to many important social events and like being seen in high fashion items. The garments are sold only in exclusive salons. Each item has the designer's name on the label.

The high-priced market has only a tiny percentage of the total sales of garments. However, the high-priced designers and their creations receive the most publicity. These designers also tend to be the most creative since they do not have restraints imposed by keeping production costs down. Their designs, and the publicity they receive, keep the general public aware of fashion trends.

The moderately priced market has almost one-third of all clothing sales. These garments are factory-produced in relatively small numbers. They have dependable brand names and are of good fabrics, such as the suit shown in 2-13. They are sold in small specialty stores or better department stores. Most fashion designers work in this price market. Usually the name of the manufacturing company is on the label instead of the name of the designer. The designer is most often a hired, anonymous talent.

Low-priced apparel is sold to the "mass market," which is made up of the bulk of people who are "average folk." The low-priced market has about two-thirds of all apparel sales. These garments are mass-produced in great volume in common styles and colors. They are sold on racks of discount stores, in basements of department stores, and through inexpensive catalog sales outlets. Not many fashion designers are employed for this market since very little original designing is done. Only the name of the manufacturing company is on the label.

Talbots

2-13 This casual suit is a moderately priced garment of good quality that has been produced in relatively small numbers.

Fashion Cycles

A ***fashion cycle*** is the periodic popularity, disappearance, and later reappearance of specific styles or general shapes. It is a rotation of particular styles. The fashion cycle is a regular round of different styles that are fashionable over time and then eventually repeat themselves.

How Cycles Occur

Our desire for new fashions causes garment silhouettes and details to constantly change. Fashions always change with the same series of events: the new style is introduced; it is eventually worn by many people; and finally it is discarded for a newer style. In other words, new fashions eventually peak, become old fashions, and disappear. New fashions are always being created because people want to own the newest and latest items.

In fashion cycles, high fashion is first introduced by the *fashion leaders* of the time. These are men, women, and young people with enough status and credibility to start new styles. They are not afraid to wear something before everyone else does. Every community has fashion leaders. They are the ones who are first to adopt and display new styles within their social groups.

In past centuries, the fashion leaders or trendsetters were royalty. Now they tend to be public celebrities, such as television and movie stars, rock musicians, sports heroes, and political figures. Some are members of the most prestigious social classes, while others are members of specific subcultures or simply people who are especially responsive to change.

While fashion leaders are wearing the new styles, other people are watching them. The others are forming opinions about the new fashions and deciding how they would look on them. Their eyes and minds are getting used to the new colors, shapes, and proportions. Soon adaptations of the fashions are worn by people in many social groups. The styles receive greater publicity, promotion, and availability in retail stores. Finally, the fashions become well established. They reach mass acceptance and are worn by the general public.

Following a fashion's highest level of social acceptance, it reaches "social saturation." This is a condition in which the fashion is worn by almost everyone. It is overused. The fashion becomes dull and boring. It is no longer novel or exclusive. It starts to decline in popularity and eventually becomes obsolete.

Meanwhile, a different high fashion look is emerging. For an existing fashion to die, there must be a more desirable one to replace it. The attention of consumers turns away from the old styles and focuses on the new. The new style eventually gains popularity with the average person. Consumers acquire, use, and eventually discard clothing items. This process is outlined in 2-14.

The life of each fashion look can be quite different. The period might range from several months to several years. Clothes from the past look strange to us today because they are no longer in fashion. However, each style is considered attractive when it is popularly worn.

As fashions come and go, they seem to be extreme and daring when first introduced, smart and stylish when they are popular, and dowdy and out-of-date after their peak. Chart 2-15, developed by James Laver, illustrates the feelings toward styles as they pass into and out of fashion. The years listed are approximate.

A fashion item

1. is created
2. is introduced to the public
3. is worn by certain fashion leaders
4. spreads to masses of consumers and gains a very high level of acceptance for a period of time
5. diminishes in popularity and use
6. is ended as an accepted fashion

2-14 Fashions always come into and go out of popularity with the same series of events. While some fashions are at the end of this cycle, other fashions are at the middle of it, and still others are just beginning the process.

The Swing of Fashion Popularity

10 years before its time—vulgar and indecent
5 years before its time—bold and shameless
1 year before its time—flashy and daring
When it is in fashion—elegant and smart
1 year after its time—tacky and dowdy
5 years after its time—hideous
10 years after its time—outrageous
20 years after its time—funny
50 years after its time—odd
100 years after its time—charming

2-15 Fashions are disliked before they become popular, then strongly liked when they are at their peak. After their popularity, they are strongly disliked. Eventually old fashions are considered to be charming and gorgeous.

Silhouettes of Fashion Cycles

This book will not go into depth in describing historic clothing. However, each period has had a characteristic style that set it apart from other times. Some styles became fashionable even though they were uncomfortable or unsafe. For instance, at times women were cinched in so tightly they could hardly breathe. Skirts were sometimes so long and tight that walking was difficult. Fashionable shoes have pinched toes or caused people to trip.

In past centuries, fashion cycles moved slowly. Three specific silhouettes would separately rise, peak, and fall in popularity. This always happened in the same order within about a 100 year span. The three basic silhouettes were bell, back fullness, and tubular. There were variations of dress and coat lengths, positions of the waistline, and styles within each silhouette category. The silhouettes would repeat, but with distinctly new fashion features. The old silhouettes were repeated in new ways. The changes took place gradually. Notice in 2-16 how the same silhouettes have cycled in and out of popularity over the years.

2-16 The bell, back fullness, and tubular silhouettes used to repeat themselves about every 100 years.

The *bell silhouette* dress has a fitted waist and full skirt. It is wide at the bottom. Capes and pants with legs that are wide and full at the bottom are also examples of this silhouette.

The *back fullness* silhouette has a skirt that puffs out in back but not in front. For this, there has been draping of fabric at the back hipline or a "bustle" to give back fullness. Sometimes suit jackets also flare out in back.

The *tubular silhouette* has a slim skirt all around. It sometimes has a high or low waistline. It often hangs from the shoulders to the hem without being belted. It has mostly vertical lines.

Fashion cycles in men's clothing can be seen in the different widths of neckties, lapels, and trouser legs. They fluctuate over time by moving from narrow, to medium, to wide, and then to narrow again. Also, the tops of trousers are sometimes full from the waist with pleats. At other times, they are tight to the hips. Trouser bottoms sometimes have cuffs and sometimes have no cuffs.

The chart in 2-17 summarizes the American fashion trends of the 1900s. Recently, cycles have not been as distinct as in past centuries. The pace of fashion change is tied to the overall pace of the culture. Rapid change takes place

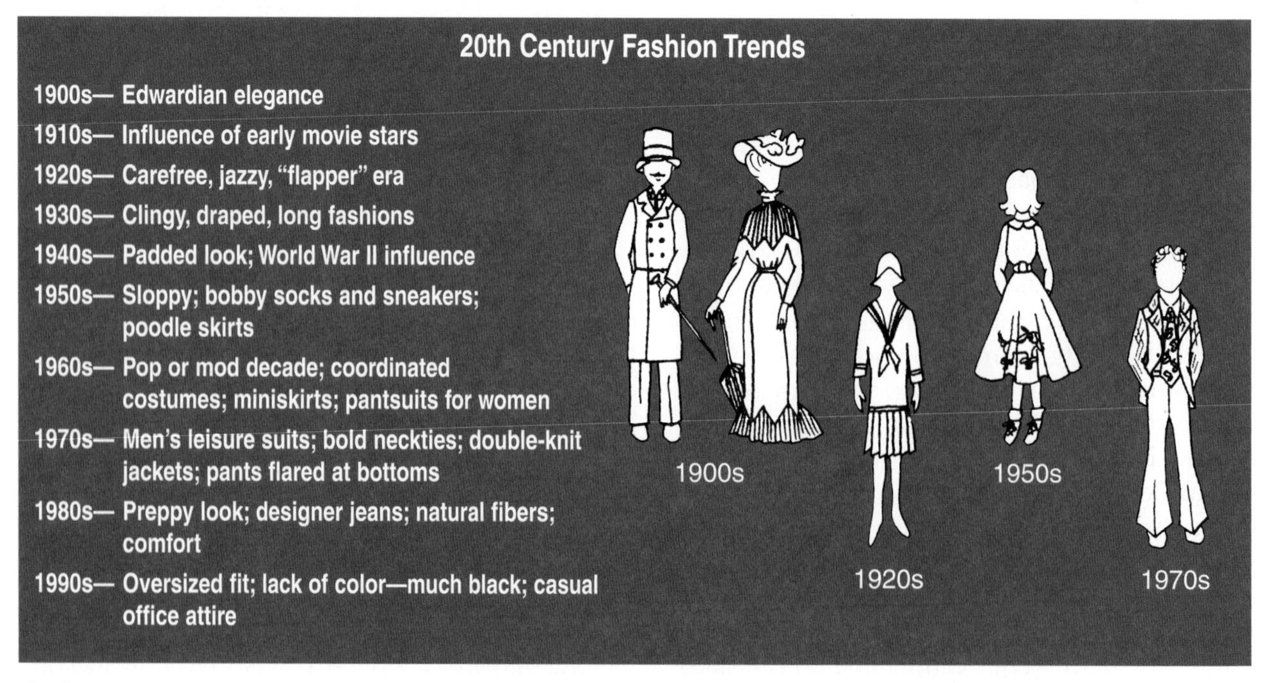

2-17 American fashion trends in the 20th Century were fun and interesting, but silhouettes were not as distinct as in other centuries.

now because of new technology in communications and manufacturing. Current fashion trends seem to occur in 20- to 30-year cycles, or the span of a generation. For instance, currently, many styles are recurring from the 1970s and 1980s.

Today most of our clothes are practical and comfortable. They have freedom of movement and are functional. Fashions seem unpredictable and individualistic. Consumers now think more freely. They buy what they like, what they need, and what they can afford. However, these fashions will also become part of clothing history.

Styles must sell to stay in fashion. Designers try to appeal to the public, so as many clothes as possible will be purchased. Because of our varied lifestyles, more clothing styles are popular at a time. Mass production allows a variety of fashion looks to be available to all. Also, worldwide communication, trade, and travel spread fashion ideas quickly. They permit many fashion looks to be popular everywhere. Thus, we now seem to have several basic styles at a time with minor trends of fashion changing quickly.

Social and Economic Influences on Fashion

Fashion has always reflected social and economic conditions, current events, technology, popular entertainment, and people's values and attitudes. The decoration on the clothes of the past related to the general thinking and art forms of the time. In the Victorian period, for instance, small, intricate decoration was used frequently. In architecture, it was achieved with endless varieties of jigsawed wood trims. In clothing, ribbons, laces, and braids were used.

Historic clothing shows that social and economic factors have always had a great deal of influence on fashion. Long ago, people dressed according to their "social class." Members of royalty were the only people who could afford to wear silks, pearls, embroidery, and certain colors. In fact, they were the only ones who were permitted to wear these items. For instance, purple dye was rare and expensive. Thus, it was a status symbol worn only by the wealthy.

Economic factors still affect fashion. For instance, in an era of hard times, clothing usually gives a serious, conservative image. In better times, the styles are brighter and more adventurous, since people are more apt to try new, different fashions. Clothing looks "perkier" during prosperous times.

The "Hemline Index" is a theory that was developed by a research director of a stock brokerage firm. He noticed that when hemlines rose (as in the early 1920s and 1960s), the stock market indexes also went up. When hemlines started to fall (as in the late 1920s and 1950s), the stock market indexes also went down. Hemlines were mixed in the early 2000s. Popular skirt lengths were simultaneously various levels of short and long, as in 2-18. This coincides with the "volatile" stock market that had sharp ups and downs!

QVC; John Romeo, Photographer

2-18 Although the "Hemline Index" is not considered scientific theory, it is still interesting to see if stock market trends are related to skirt lengths.

Many fashion and stock market experts think the Hemline Index is amusing and should be taken lightly. Even so, it reinforces the basic idea that people's moods are reflected in the way they dress. When they are down in the dumps, they tend to lose interest in their appearance. When their spirits are high, they dress in styles that are more fun and provocative. If the standards of dress change quickly, the basic social structure of the society has probably changed.

During wartimes, there is a military influence on apparel. During World War II, the government restricted the amount of fabric to be used for civilian clothing. This was necessary because of the demand for supplies needed for the armed forces. Styles became tight. The widths of hems and seam allowances were skimpy. Men's trousers no longer had cuffs. The country's economic needs made necessary limitations on the fashions of the time.

Chapter Review

Summary

To know about fashion, specific fashion terms must be understood. Clothing construction terms refer to how garments fit and are made. Clothing business terms deal with haute couture and ready-to-wear aspects of fashion.

Price markets differ. Price markets of high-priced, moderately-priced, and low-priced apparel have different characteristics.

Fashion cycles occur because of consumers' desire for new garment silhouettes and details. Fashion leaders introduce new looks that are created. The designs gain in popularity to finally achieve very high acceptance for a period of time. Eventually, they lose popularity, diminish in use, and disappear while other fashions are gaining popularity. At some later time, they reappear. Until recently, various silhouettes have recurred in the same order within 100-year spans.

Social and economic influences on fashion are shown historically by people dressing according to their social class and related to the general thinking and art forms of certain times. Hard times, stock market levels, and wars have influenced society's fashions during various periods.

Fashion in Review

Write your answers on a separate sheet of paper.

Matching: Write the letter for the general term that describes each specific term.

1. Stylish.
2. Retail.
3. Classic.
4. Draped.
5. Overruns.
6. Couturier.
7. Seams.
8. Fit.
9. Silhouette.
10. Darts.
11. Avant-garde.
12. Fashion piracy.

A. fashion term
B. clothing construction term
C. clothing business term

Short Answer: Write the correct answer to each of the following questions.

13. What is the difference between a fad and a craze?
14. Describe a fitted garment.
15. Explain the difference between a custom-designed garment and a custom-made garment.
16. What is fashion piracy?
17. What are the three silhouettes that were the basis for fashion cycles in past centuries?

True/False: Write true or false for each of the following statements.

18. In recent times, fashion cycles have not been as distinct as in past centuries.
19. During a recession, clothes tend to be fun and provocative to try to help people feel better.
20. According to the "Hemline Index," when hemlines rise, the stock market indexes also go up.

Fashion in Action

1. Bring pictures, detailed descriptions, or articles of clothing to class that show a fad of the past. Why do you think it is no longer in fashion? Show a current fad by clipping fashion ads that picture it. Use your imagination to draw or describe a possible fad for the future.
2. Visit a discount store and an expensive clothing boutique. In a written report, compare the kinds of clothes being promoted in each. What are the differences between the apparel for the mass market and the class market? What silhouettes and features are most common for each category? Which store sells more fad items? Which sells more classic styles?
3. Collect pictures of five different classic items of apparel from catalogs, newspapers, or magazines. Mount them on paper and label them with the names of the styles.
4. Clip or photocopy pictures of draped, tailored, and composite garments. Briefly tell how each is constructed.
5. Write an essay about one or more people whom you consider to be current fashion leaders. Describe the new styles that they are wearing. How and where do they display their fashion choices?

3 Garment Styles and Parts

Fashion Terms

asymmetrical	inseam
lapel	single-breasted
set-in sleeve	double-breasted
kimono sleeve	cardigan
raglan sleeve	pullover
cuff	closure
waistband	yoke

After studying this chapter, you will be able to

- describe the many styles of dresses.
- identify neckline and collar styles for men's and women's apparel.
- describe sleeve, skirt, pants, coat, and jacket styles.
- discuss how garment parts can be combined in different ways to achieve new and different fashions.

All garments are comprised of a combination of garment parts, such as sleeves and collars. As different garment parts are put together, new designs are created.

As you study this chapter, you will learn to identify basic garment styles and parts. You will develop a fashion awareness. You will understand how basic clothing styles are constantly being revived and modified to become current fashions.

Basic Dress Styles

Some basic dress styles are illustrated in 3-1.

Sheath dresses have no waistline seam. They hang from the shoulders and have inward shaping at the waist.

Shift or *chemise* dresses also have no waistline seam. They are straight and loose fitting with no inward shaping at the waist.

A-line dresses are narrow (fitted) at the shoulders. They have no waistline seam and become wider at the hemline. They are named after the "A" shape of their silhouette.

Tent dresses are large and billowy. They hang loosely from the shoulders. They are often worn in hot weather because they do not hug the body.

An *empire* (pronounced om-peer from French fashions) dress has a high waistline. Its opposite, the *lowered waistline* style, has a long torso. Its waistline seam is actually down toward the hips.

Other Dress Styles

Other dress styles are formed by using a variety of seams, fullness, or other characteristics. They modify basic dress styles, such as the examples shown in 3-2.

Princess dresses have seamlines going up and down their entire length. The vertical seaming is from the shoulder or the armhole. "Princess seams" provide fit as well as fashion. There is no horizontal waistline seam.

The *blouson* dress style has blousy fullness above the waist. It is usually belted. It most often has a fitted skirt.

A *shirtwaist* dress is like a long, semifitted, tailored shirt. It is as long as a regular dress and usually has a belt or sash at the waist.

A *coatdress* is a heavy dress that usually closes down the front like a coat. However, it is worn as the main garment rather than over another garment.

In an ***asymmetrical*** dress (or garment) design, the right side is different from the left side. If divided by a center line, the two halves are not the same. The asymmetrical dress pictured in 3-2 has a typical *surplice* closing, which has a diagonal overlap to one side.

Jumpers and *sundresses* have a brief bodice, often with shoulder straps and a low neckline. They may or may not have a waistline seam, and may be fitted or loose at the waist. A jumper is made of heavier fabric and is worn over a blouse or sweater. A sundress is worn in hot weather alone as a dress.

3-1 These dress styles may be made to look different by using various sleeves, collars, fabrics, and trims. However, the basic silhouette of each category remains the same.

3-2 Many dress styles are possible. Each one has a descriptive name.

Neckline Styles

There are many neckline styles. As you read the following descriptions of some of them, look at the drawings in 3-3.

Décolleté is the French term for a low neckline. It is usually used with bare shoulders, such as in an evening gown or sundress. Sometimes the garment is strapless.

A *jewel* or *round* neckline encircles the base of the neck. It is plain and rounded.

A *boat* or *bateau* neckline goes straight across from shoulder to shoulder. It is high at the front (and usually back) and is wide on the sides.

A *scoop* neckline is lowered and round. It is usually lower in front than it is in back.

A *horseshoe* neckline is up high at the neck in back but goes down like a "U" in front.

A *cowl* neckline is draped with flowing folds. It gets its name from a medieval monk's hood.

Necklines can be raised, normal, or low. Sometimes they are high in front and low in back. Sometimes they are just the opposite. They can be round, square, "V," or off the shoulder. They can have scalloped edges or special shapes such as the "sweetheart" or "keyhole" versions.

Collar Styles

Collars can be designed to have long or short points. Sometimes the corners are rounded. Study the different collars and their names in 3-4.

Collar styles range from narrow to wide. Some collars lie down flat on the garment, while others stand up around the neck. Most go up somewhat in back, fold over, and then fall back onto themselves and a small amount of the garment. A popular collar for shirts is the *button-down collar* with points that button to the shirt.

Collars on garments that fold open at the neckline often have lapels. A ***lapel*** is a pointed part of the garment below the collar. It turns back at the front neckline. It looks like a continuation of the collar going down from a "V" notch along the outer edge. It is shown in 3-4 on a *convertible* collar.

A *stock* collar is an imitation of an ascot. An ascot is an accessory added at the neck like a necktie. It is a broad scarf that is looped over itself. A true ascot was originally worn by men and is usually tucked into the front neckline of the shirt.

A *jabot* (pronounced ja-bow) was originally of lace or a ruffle worn on a man's shirt. Now it is worn on the collar front or going down from the neckline of women's garments, as well as on men's tuxedo shirts.

Sleeve Styles

The three basic types of sleeves are set-in, kimono, and raglan, as shown in 3-5. Any of these can be short, just above the elbow,

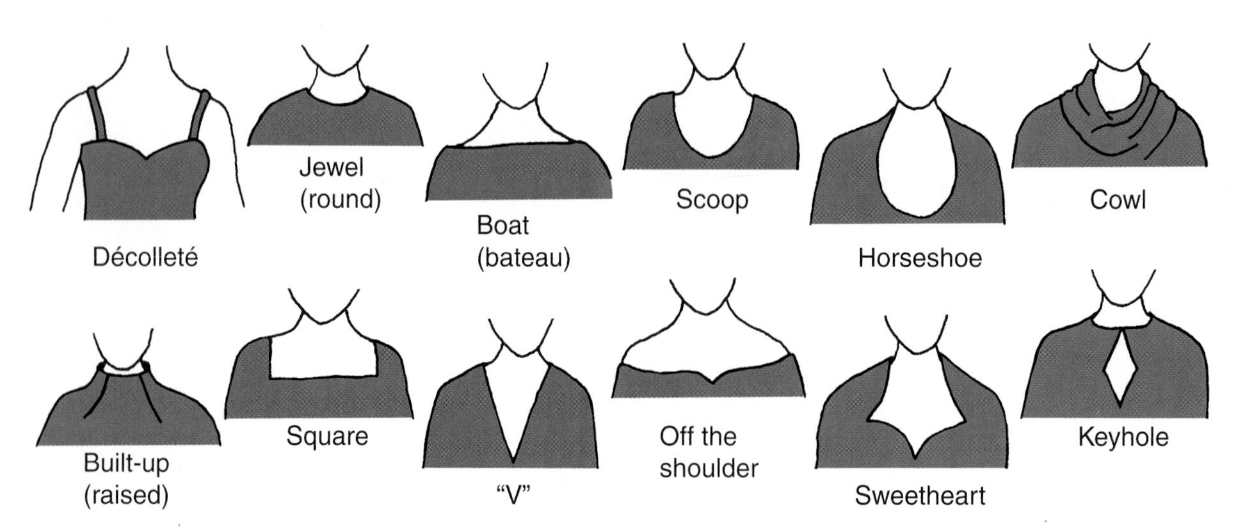

3-3 This is only a sampling of the many neckline options that are available for apparel.

3-4 Collars may be wide or narrow. They may lie down on the garment or stand up around the neck.

Basic Sleeves

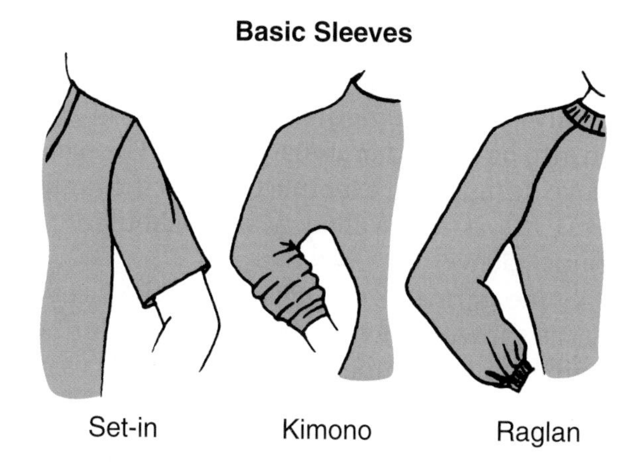

3-5 All sleeve styles are based on these three basic types of sleeves.

three-quarters length (just below the elbow), seven-eighths length, wrist-length, or long (just below the wrist bone).

Set-in sleeves are sleeves that are stitched to the garment around the regular armhole. They offer the best fit for most people. They can be tight (fitted), puffy, long, or short. In all cases, there is a seam in the front and back from the underarm curving up to the shoulder. That seam follows the natural body line that connects the arm to the body. Examples of many set-in sleeves are shown in 3-6. Take note of the armhole seam joining the sleeve to the garment bodice in each case.

Kimono sleeves, as shown in 3-7, are continuous extensions out from the armhole area with no seamlines connecting them to the garment bodice. Kimono sleeves can be long or short and either fitted or loose. Fitted kimono sleeves often have gussets. A gusset is a wedge-shaped piece of fabric added at the underarm to give more ease of movement to the sleeve.

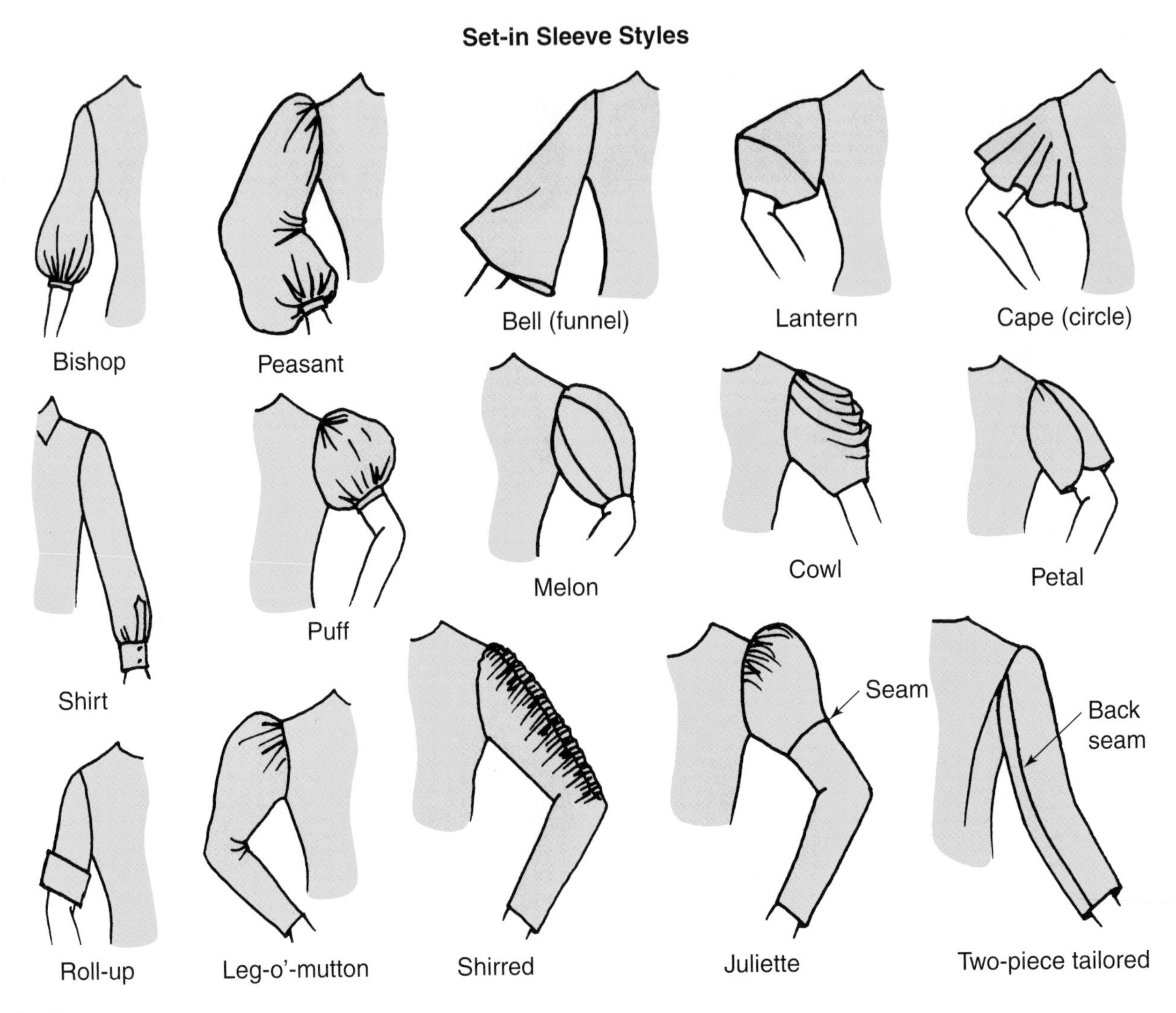

3-6 All set-in sleeves have a seamline around the armhole.

If a kimono type of sleeve is designed lower than usual at the underarm, it is called a *dolman* sleeve. *Batwing* sleeves are very low and loose at the underarm with only a gradual curve, if any, between the side waistline and the sleeve bottom. They are almost capelike.

Stylized kimono sleeves can have many different design lines. In all cases, there is no seam joining the sleeve to the bodice.

Raglan sleeves have a shaped seam in the garment originating from the underarm. It does not continue up to the outer shoulder at the top of the arm. Sometimes it goes directly up to the neckline, as seen in 3-5. Other times it goes across to the center front or jogs down into a bodice princess seam, as in 3-8.

Raglan sleeves can be any length, just like other sleeve types. The raglan seamline is usually a smoothly curving line. However, it can also be shaped in unusual ways. The *saddle* sleeve design is an example of a specific raglan sleeve. It is also called a strap-shoulder or epaulet sleeve.

Some garments are designed to be sleeveless, as shown in 3-9. These are usually for hot weather or for jumpers or vests that are worn over a blouse or shirt. The basic sleeveless armhole hits the top of the shoulder where a set-in sleeve would ordinarily join the bodice. Sometimes, for design interest, it is cut in toward the neckline.

Cap or *French* sleeves are very short. They are illustrated in 3-10. They are like a sleeveless armhole at the underarm and a short kimono sleeve going out from the shoulder. They can also be an extension of the shoulder

Kimono Sleeve Designs

With gusset

Dolman

Batwing

3-7 A tight-fitting kimono sleeve may have a gusset at the underarm. Dolman and batwing styles are kimono sleeve variations.

Raglan Sleeve Design

Raglan to center front

Raglan-princess

Saddle sleeve design

3-8 A raglan sleeve has a seamline coming up from the underarm in front and back. The seam then turns to form one of a variety of design options.

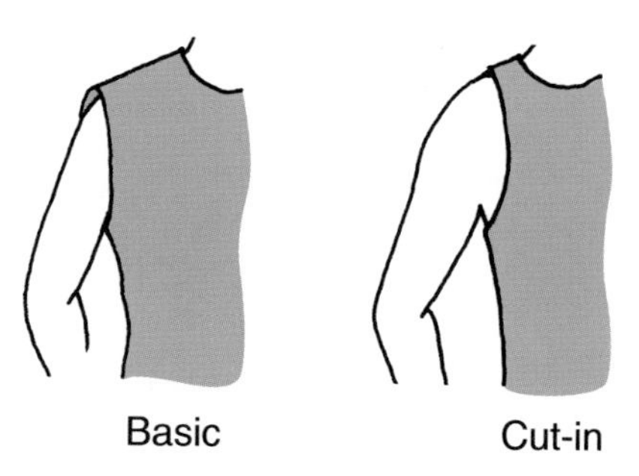

3-9 Sleeveless garments do not have sleeves. The armhole is sometimes designed with varying shapes.

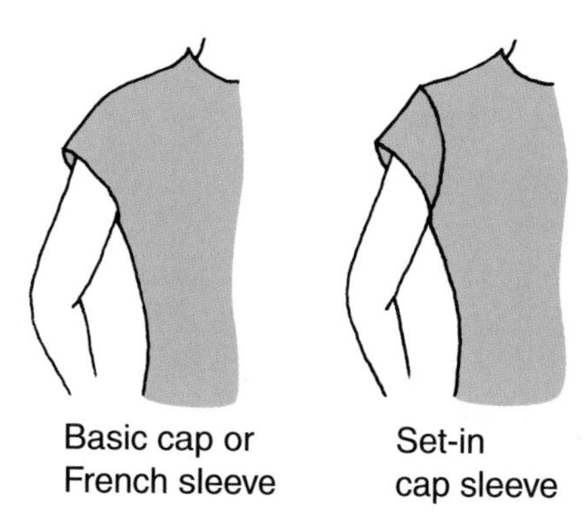

3-10 Cap sleeves have a small amount of length down from the shoulder and no sleeve at all at the underarm.

dropping down over the upper arm. A very short set-in sleeve is also a cap sleeve.

Dropped shoulder designs, as in 3-11, have a horizontal seam around the upper part of the arm. The lower sleeve can be any length. It can also have any amount of fullness.

A ***cuff*** is a band at the bottom of the sleeve. A *vent* is an opening that goes from the open end of the cuff up into the sleeve. It enables the cuff to overlap and button. It is often finished with a *placket* that is a decorative strip of fabric over the vent. Often the placket has a point at its upper end. Study these parts in 3-12.

Skirt Styles

Several skirt styles are illustrated in 3-13. Become familiar with their shapes and their names.

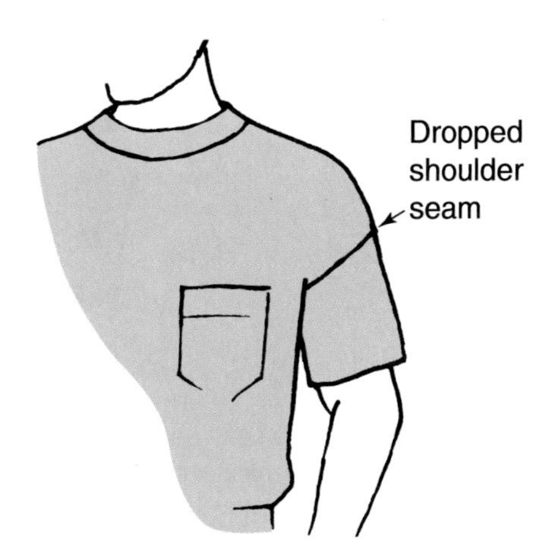

3-11 Dropped shoulder designs have a horizontal seam around the upper part of the arm.

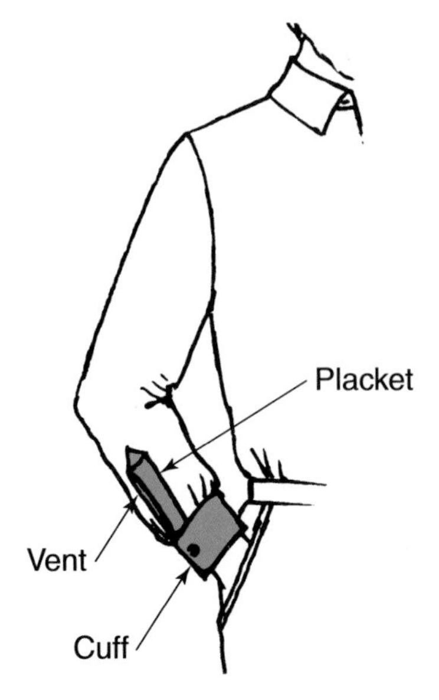

3-12 A long shirt sleeve usually has a cuff, a vent, and a placket.

Straight skirts have no added fullness at the hem. They go down straight from the hipline for a very slim silhouette. They are sometimes called fitted skirts.

A-line skirts have extra width at the hem on each side. When viewed from the front or back, the silhouette resembles the letter "A."

Flared skirts have some fullness at the hem all around. There are soft ripples going upward from the skirt bottom toward the waist.

Circular skirts are very full at the hem. When held out at the sides during wearing, this style of skirt forms a half circle. When opened up and laid flat, it forms a circle.

Full skirts are *pleated* or *gathered.* Pleats are structured folds of cloth. The pleats either hang open from the waist or are stitched down for a snug fit from the waist to the hips. Gathered skirts have the fullness of the fabric pulled together at the waist without structured folds.

A slightly gathered skirt that is not very full is called a *dirndl* skirt. It is quite straight and often has pockets in the side seams.

Gored skirts have vertical seams all the way from the waistline to the hem. They are similar to the princess seams in a dress. The seamlines cause the skirt to have several panels or gores. The gores are a bit wider at the hem than at the hipline. Skirts can have four, six, or many gores.

Umbrella skirts have many narrow gores. The gores are pressed to have a narrow silhouette, but when the wearer walks or moves, the gores spread open and closed like an umbrella.

Wrap skirts wrap around the body and overlap at the side-back or side-front. They are most often fastened with a tie or button. They usually have a straight or slightly flared silhouette.

Some skirts are part of a dress, attached to a bodice with a waistline seam. Other skirts are held at the top by a ***waistband.*** A waistband is a band of fabric that goes around the waist and fastens with a button or hook and eye.

All skirt styles also have the option of varying lengths. Notice the names of the most common hem lengths in 3-14.

Pants Styles

Pants are also called *slacks* or *trousers.* Pants can be many different lengths. Some are short shorts. Others are so long that they drag on the ground! Notice the names of the many lengths indicated in 3-15. The seam on the inside of the leg (from crotch to pants hem) is called the ***inseam.*** The inseam length is an important measurement when buying or hemming pants.

Pants can be many different widths as well as lengths. These are indicated in 3-16. *Straight* pants are the same width at the hem as they are at the knee. *Tapered* pants are narrower at the hem than at the knee. *Flared* pants are wider at the hem. Sometimes the flare is just below the knee. Other times it is from the hips on down, or even from the waist. Pants flared from the waist are often worn as fancy evening slacks. They are called *palazzo* pants. Flared pants that are gathered in at the ankles are called *harem* pants. Notice these in 3-17.

Gauchos, knickers, and *culottes* are shown in 3-18. They have some fullness and usually end

3-13 Skirts can be very narrow, extremely full, or shaped in various ways. Fashion interest can be created with gathers, seams, pleats, or an overlap.

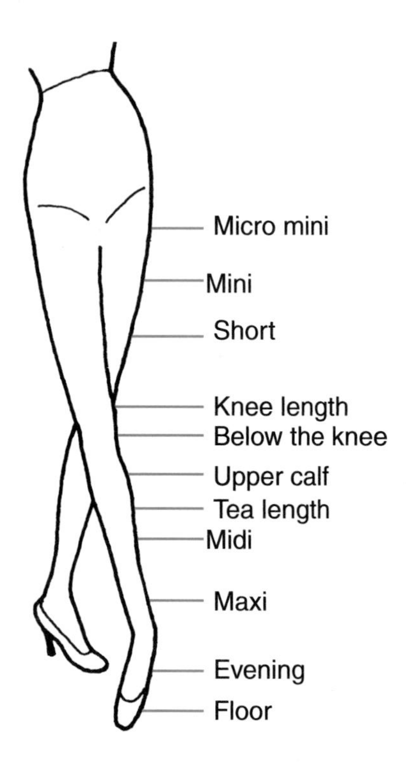

3-14 Skirt designs can be any length that is popular or desired.

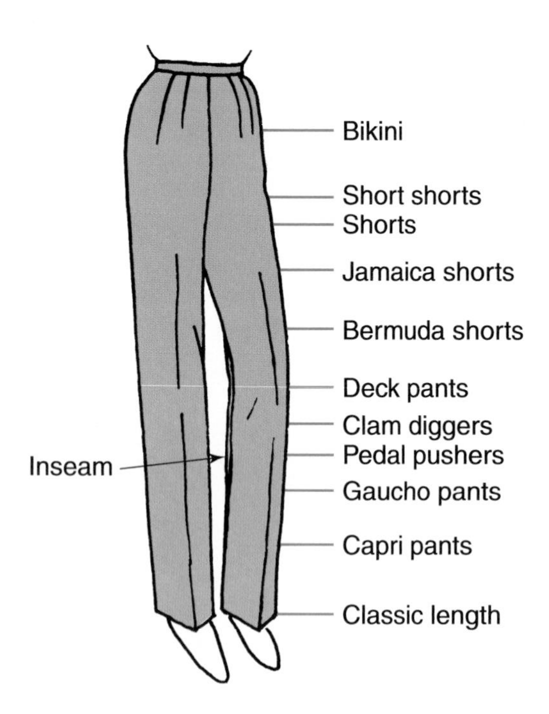

3-15 Pants can be almost any length, depending on the dictates of fashion.

3-16 The legs of tapered pants get tighter as they go down toward the ankle. Flared pants legs get wider.

3-17 Some pants have special design interest. The ones shown here have excess flare.

below the knee. The legs of *gaucho* pants are like wide tubes. *Knickers* are gathered to a band or strap below the knee. *Culottes* are pants that look like a skirt, also called pantskirts. A *skort* combines a skirt with shorts.

One sturdy type of pants is jeans, shown in 3-19. They are comfortable and have become fashionable as everyday wear. They can be casual when worn with a sweatshirt or dressier when worn with a blouse or sport coat. They have heavy double stitching and are usually made of denim. "Cords" are jeans made of corduroy fabric.

The top of *hip-hugger* pants is lower than the regular waistline. The pants "ride" on the upper hips. A garment with a bodice, or top, attached to pants is called a *jumpsuit*. Look at these in 3-20.

Knickers

Gauchos

Culottes

3-18 These pants styles have emerged on the fashion scene at various times through the years.

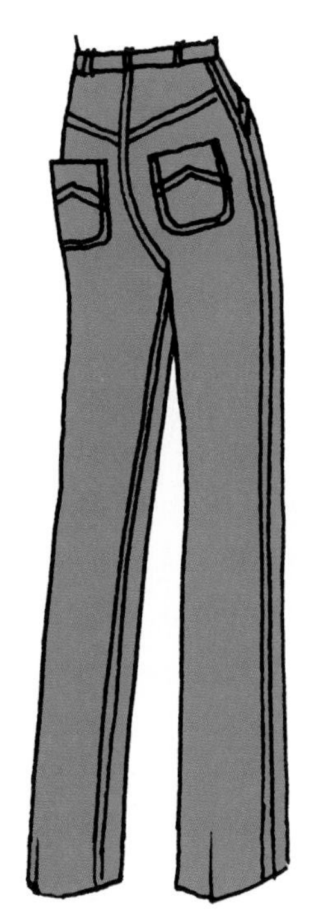

3-19 Jeans have become a classic style of pants worn for their sturdiness and versatility.

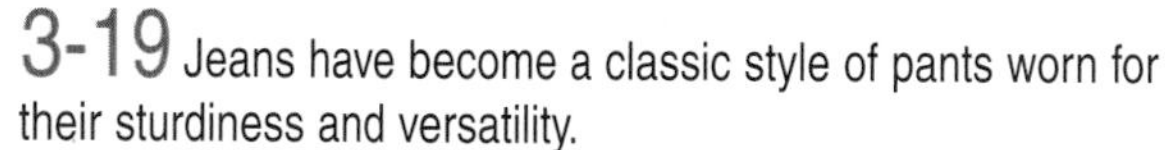

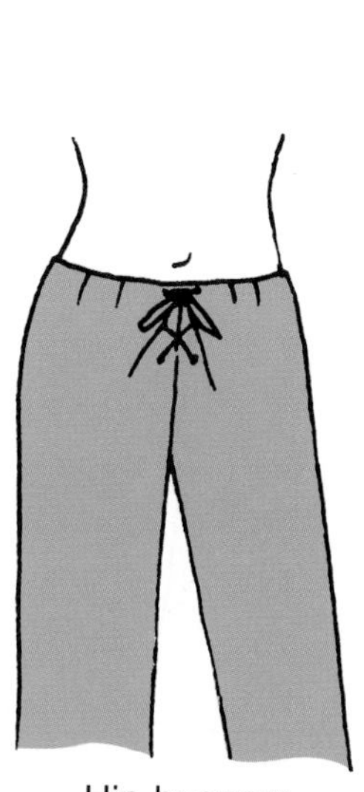

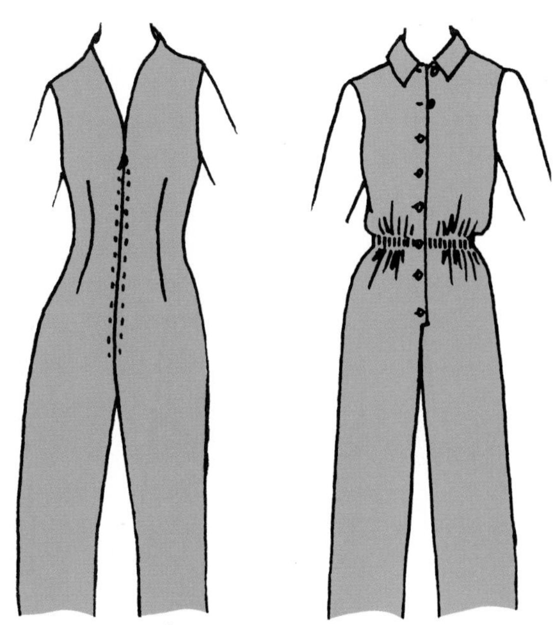

3-20 Like other garments, pants have a wide variety of design options.

Coat and Jacket Styles

Coats are warm or weatherproof garments that are worn over a person's regular clothing. Some popular coat styles are shown in 3-21.

Capes, as in 3-22, are coatlike outer garments that hang from the neck and shoulders over the back, front, and arms. They have no sleeves, thus arm movement is restricted. There are usually slits in each front side of capes for the hands to slide through.

Jackets are short coats. A few common jacket styles are shown in 3-23.

Sport coats or *blazers* are classic jackets that are always in fashion. They, along with suit jackets, are either single-breasted or double-breasted. ***Single-breasted*** garments are held shut with one row of buttons in front, as shown in 3-24. ***Double-breasted*** garments have a wider overlap and two rows of buttons. The suit coat in 3-25 is double breasted. European cut blazers and suit jackets are more

3-21 A coat is needed as an outer garment in cool climates.

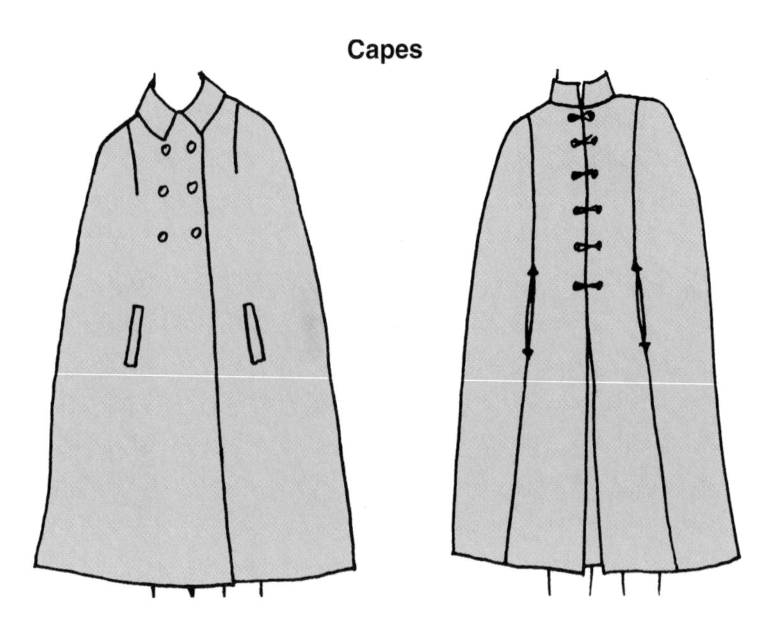

3-22 Capes go in and out of fashion. They are usually not worn as an everyday cover-up.

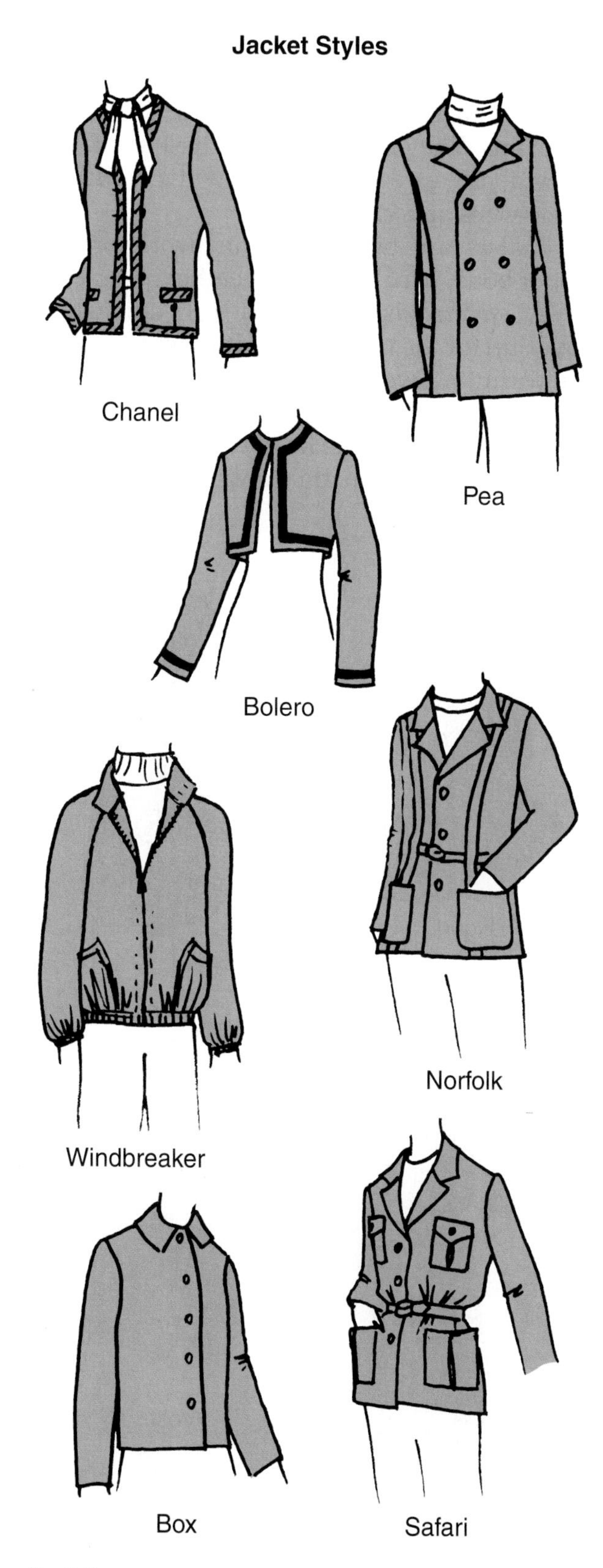

3-23 The variety in jacket styles seems endless.

Hart Schafner & Marx

3-24 This single-breasted suit coat looks stylish and comfortable with one row of three buttons and buttonholes.

Perry Ellis by Hartmarx

3-25 Double-breasted jackets have a wider overlap at center front to accommodate their two rows of buttons.

tightly fitted at the waist than traditional American blazers.

A *poncho* is similar to a blanket with a slit or hole in the middle for the head. It has no sleeves. Sometimes it is made of waterproof cloth and used as a rain jacket. A *parka* is a heavy winter jacket with a hood. It has a warm lining. Sometimes the lining is of quilted fabric or real fur. Look at these in 3-26.

Miscellaneous Styles and Parts

A *hood* is a head covering that is attached at the neckline of a garment. A hood is shown in 3-27 along with garments that are described in the following paragraphs.

A *tunic* is a long blouse or shirt that extends down over pants or a skirt. It is a long upper garment that goes over a lower garment. Tunics are hip-length or longer. Sometimes they are belted.

Caftans are long, flowing, robelike garments. See 3-28.

Vests are sleeveless, close-fitting, jacket-like garments. They cover just the chest and back. They do not extend much below the waistline. Some vests are called *weskits*.

Halters are brief garments worn on the upper body, usually in hot weather.

A *sweater* is a knitted (or crocheted) covering for the upper body. It is usually worn for warmth. Sweaters are either cardigans or pullovers. ***Cardigans*** open down the front. ***Pullovers*** slip over the head when they are put on or taken off. Both are shown in 3-29.

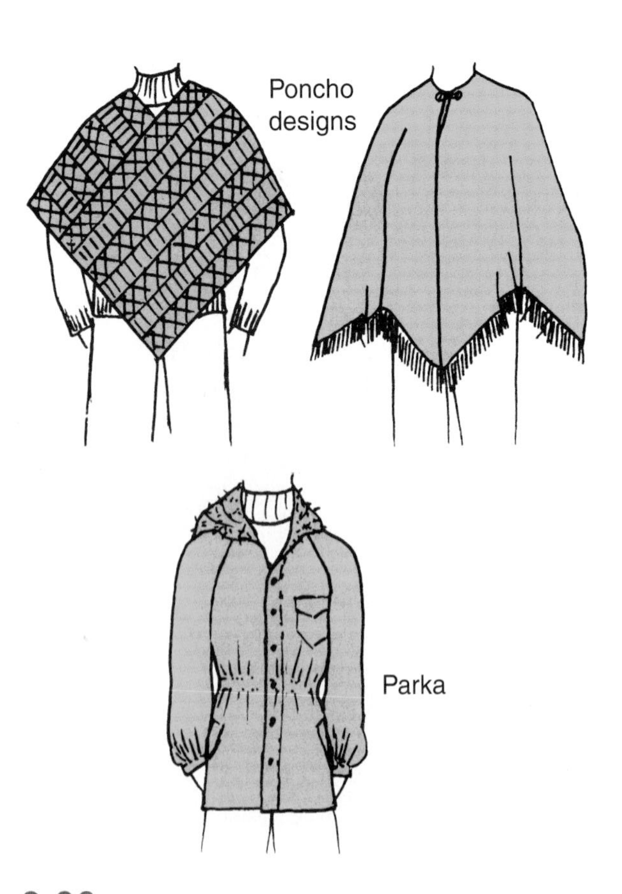

3-26 Ponchos are loose garments that slip over the head. Parkas, on the other hand, can be snug at the waistline and cuffs.

3-27 All these apparel items have specific names and characteristics.

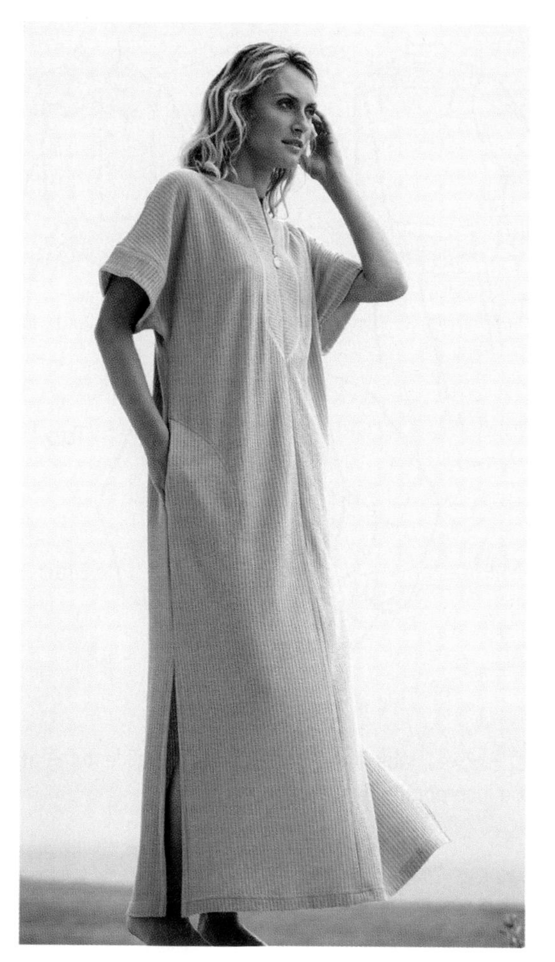

JC Penney

3-28 This robe is a caftan style, which is long and loose-fitting.

Kellwood Company

3-29 This woman is wearing a cardigan sweater. The man is wearing a pullover sweater.

The term *lingerie* refers to undergarments, ladies' slips, and feminine nightwear. *Pajamas, nightgowns,* and *nightshirts* are forms of nightwear. A *robe* is worn over nightwear when the person is not in bed. It is long and loose. Its name originates from the flowing garments of ancient civilizations.

Closures enable the wearer to get into and out of garments. They include zippers, buttons, snaps, hooks and eyes, Velcro®, or any other fasteners that open a correctly sized space and close it again. See 3-30.

A ***yoke*** is a band or shaped piece, usually at the shoulders or hips, that gives shape and support to the garment below it.

Pockets are built-in "envelopes" that hold items. Pockets are usually added onto the outside of garments or are inserted into seams. Pockets sometimes have added tabs or flaps. *Tabs* are decorative fabric pieces that go out from the edge of pockets. *Flaps* are decorative fabric pieces that fall down over the openings of pockets. Illustration 3-31 shows several yokes as well as a pocket, tab, and flap.

Design Options

Fashion designers have almost endless options of garment parts, styles, lengths, and widths to combine in different ways. Thousands—even millions—of new fashions can be created!

Old styles are continuously put together in new ways. Often, new and exciting names are given to the same old garment styles and parts. This makes them sound like they have never been used before! In these ways, the fascinating world of fashion continues.

Men's Fashion Association of America/M. Julian

3-30 Zippers, ties, buttons, and snaps are fasteners that can open or close garments.

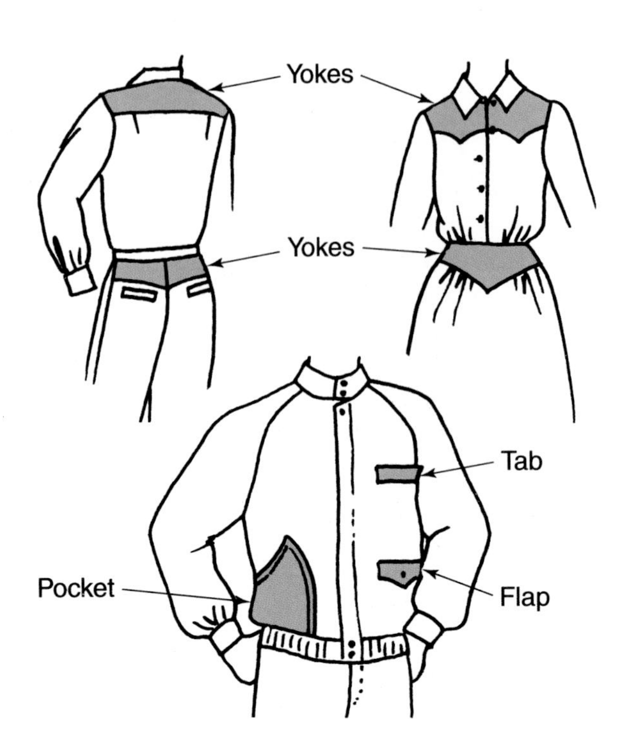

3-31 Yokes, tabs, flaps, and pockets are all features that can be incorporated into apparel designs.

Chapter Review

Summary

Fashions are formed by how garment parts are designed and combined. Fashion designers have almost endless options of garment parts, styles, lengths, and widths to combine in interesting ways.

Basic dress styles include the sheath, shift or chemise, A-line, tent, empire, and lowered waistline styles. Other dress styles include princess, blouson, shirtwaist, coatdress, asymmetrical, jumpers, and sundresses.

Neckline styles include décolleté, jewel or round, boat or bateau, scoop, horseshoe, cowl, and many others. The many different collar styles can have long, short, or rounded points. Collars can be narrow or wide, include lapels, or be a ruffled jabot.

All sleeve designs are based on the set-in, kimono, or raglan types of sleeves. There are also sleeveless designs, cap sleeves, and dropped shoulder designs.

Skirt styles can be straight, A-line, flared, circular, or gored. Fullness can be in pleats or gathers. Some skirts imitate an umbrella or wrap with an overlap.

Pants (slacks or trousers) can also be different widths, such as tapered, straight, and flared. Many lengths offer options for pants styles such as gauchos, knickers, and culottes. Other pant styles are jeans, hip-huggers, and jumpsuits.

Coats, jackets, and capes are worn over other clothing. They can be of many styles and may be single-breasted or double-breasted.

Miscellaneous garment styles include tunics, caftans, vests, halters, lingerie, and nightwear. Additional garment parts include hoods, closures, yokes, pockets, tabs, and flaps.

Fashion in Review

Write your answers on a separate sheet of paper.

Matching: Match the following general categories of styles with the specific terms.

1. Bell.
2. Mandarin.
3. Flared.
4. Leg-o'-mutton.
5. Keyhole.
6. Jewel.
7. Shawl.
8. Horseshoe.
9. A-line.
10. Kimono.
11. Convertible.
12. Gored.

A. neckline style
B. collar style
C. sleeve style
D. skirt style

Completion: Write the word or words that correctly complete the sentences.

13. _____ dresses have vertical seamlines going up and down their entire length.
14. The right sides of _____ garments are different from the left sides.
15. A _____ is a pointed part of the garment that turns back at the front neckline below the collar.
16. _____ sleeves are stitched to the garment around the regular armhole.

True/False: Write true or false for each of the following statements.

17. Raglan sleeves have no seamline connecting the garment bodice to the sleeves.
18. Closures are fasteners that enable the wearer to get into and out of garments.
19. A décolleté neckline is high and usually has a turtleneck collar with it.
20. Gored skirts have vertical seamlines that cause them to have several panels called gores.

Fashion in Action

1. Mount at least six pictures or drawings of current fashions on pieces of paper. Label the particular garment styles and parts (scoop neckline, raglan sleeve, tapered pants, and so on). Catalog descriptions might give you some hints.
2. Look up information on different types of pleats. Examples might be knife, inverted, box, accordion, sunburst, kick, and any others you can find. Illustrate and describe each.
3. Choose three different collar and sleeve styles. Combine them into six totally different shirt designs. (Nine are possible.) Add variety by making the sleeves various lengths and the collar points longer, shorter, pointed, rounded, or whatever your creativity desires. Draw your designs on tracing paper using the form of a drawn (not photographed) figure from a newspaper ad placed under the tracing paper.
4. Investigate the history of jeans. Write an essay on how they originated, the various groups of people that have worn them at different times, and the types of jeans available today.
5. Collect and mount samples of all kinds of fasteners. Label each and describe where it would be used on garments.

PART TWO

Apparel Industries

4 The Development of Fashion

Fashion Terms

rag trade	Coty Awards
trade publications	Tommys
trade associations	designer patterns
collections	house boutique
fashion piracy	franchise
logo	licensing
Chambre Syndicale	prêt à porter

After studying this chapter, you will be able to

- discuss the worldwide importance of the apparel industries.
- list several trade publications and trade associations.
- describe the development of haute couture.
- explain how the couture industry is changing.
- discuss the importance of the designer ready-to-wear industry.

The *apparel industries* center around textiles, garment manufacturing, and retailing. The textile industry includes the business firms that produce the fabrics used for apparel and other end uses. The apparel manufacturing industry includes the firms that do the designing and factory construction of garments. Retailing includes the activities dealing with the direct selling of the items to consumers. Fashion promotion is also an important aspect of the industries. All the business segments concerned with apparel are exciting, yet complicated.

The Scope of the Apparel Industries

The apparel industries have been developing for many centuries. Clothing styles were evident even in ancient Greek and Roman times. However, fashion as we know it today started in Europe during the Renaissance period in the 1500s. Textiles and clothing changed very slowly until about the time of the *Industrial Revolution*, which began in England in the late 1700s and spread to the United States.

In general, the Industrial Revolution caused a switch from handmade garments to mass-produced ready-to-wear clothes. Three major developments contributed to this overall change:

- Mechanized textile mills were able to make fabrics of better quality in less time.
- The sewing machine was invented, and factory manufacturing of clothing was introduced.
- Techniques in mass distribution, advertising, and retail selling were developed and refined.

Today, more than 10 percent of all world business is related to clothing. Textile and apparel production is the largest industry in many nations, especially in developing countries. Retailing is a large industry in all countries, especially in well-developed ones like the U.S.

The influence of apparel businesses goes beyond the design, manufacture, and distribution of textile and clothing items. Many other industries are affected by the fashion field. For instance, advertising firms rely on the apparel industries for a great deal of their work. Publishing companies need fashion news. Also, one part of the apparel industry can greatly affect other sections. Fitted waistline styles increase belt sales. Rising hems boost hosiery sales.

Inside "The Trade"

The garment industry is sometimes called the ***rag trade*** or the needle trades. In the United States, the industry started as small tailoring shops in New York City. Immigrants settled there as skilled tailors and assistants before the turn of the century. The garment industry was located between the textile mills of the North and the cotton fields of the South. Plus, New York was the nation's largest city and the center of fashionable society. Today, New York's fashion industry is centered along Seventh Avenue. Other nearby streets and avenues are home to the textile firms, menswear showrooms, fashion accessories companies, and fur district.

Other cities also have their "garment districts." Apparel businesses provide many jobs in all parts of the U.S. and the world for semiskilled labor. In fact, the industry is a leader in hiring and training unskilled workers. There is increased use of technology as time passes, which requires workers with even higher skills. The industry provides employment for those who like routine jobs, as well as for gifted, creative individuals.

Many fashion magazines and trade publications are important to the apparel industries. ***Trade publications*** are magazines, newspapers, and books that deal specifically with a certain industry or trade. *Women's Wear Daily (WWD)* is one of several trade publications for the apparel industries, 4-1.

The *Women's Wear Daily* newspaper is considered to be the "bible" of the women's fashion trade. It reports all new apparel trends with photos, fashion illustrations, and stories.

4-1 Trade publications give specific news about particular industries.

It also reports on business and financial news of the apparel industry. Its counterpart for the textile and menswear industries is the *Daily News Record (DNR)*. A few examples of the many other specialized trade publications are *Textile World, Apparel Magazine, Footwear News, Fur Age Weekly, Intimate Fashion News,* and *California Apparel News.* To read descriptions of specific trade journals, find their Web sites by searching each newspaper/magazine title through an Internet search engine.

As in other industries, apparel businesses have trade associations to which workers belong. ***Trade associations*** are groups that promote or further the interests of a certain industry or trade. Some examples are the American Apparel and Footwear Association (AAFA), the National Retail Federation (NRF), the Textile Distributors Association, and the National Textile Association. Such associations try to further the interests of their members.

They set standards for their industry and allow constructive communication among their members. They lobby to the government for laws that help their industry and also promote their industry to the public.

Each association has a trade publication that distributes information to its membership. Each also has periodic trade shows. The shows attract thousands of people associated with production of ready-to-wear clothing. Equipment manufacturers, textile firms, and apparel manufacturers participate. They display and give presentations about machines and manufactured goods, in hopes of selling them to others who attend the shows. Meetings and lectures disclose new ideas. With such activities, industry members are kept up-to-date with the latest innovations and products.

Couture

High fashion couture clothes are very fashionable and expensive. Couture designers and their firms serve the small, but influential, high-priced market.

The Development of High Fashion

The high fashion garment industry originated with the superior French dressmaking industry. France has led fashions for hundreds of years. Paris has been an international cultural center and is still thought to be the world's fashion center, as shown in 4-2.

Exquisite haute couture shops originated in Paris centuries ago. At first, Parisian couturiers dressed dolls in scaled-down versions of their latest creations. They sent them to prospective customers throughout the world. Only when fashion magazines and newspapers began to appear did this practice fade away.

Early American dressmakers went to Paris for their designs, rather than doing original work. They brought back many fine fashions. From those clothes, they learned about French cut and workmanship. They copied and adapted what they needed for their purposes.

Several influential French couturiers of the past are listed in chart 4-3. Such designers and their Paris haute couture businesses have always been located in city houses rather than in commercial buildings. Thus, they are called "fashion houses." They had international fashion supremacy until World War II. Then the war cut them off for several years.

As World War II stopped the flow of fashion news and garments from France, the American design industry grew. American fashion ingenuity became recognized in all areas of apparel. Young talents created colorful sportswear. The designers became confident, resourceful, and adventuresome with their

4-2 Many of the best-known haute couture businesses, like Nina Ricci, are in Paris.

French Couturiers of the Past

Charles Frederick Worth—Of English descent; founder of French couture (1860s).

Madame Paquin—Elegant, tasteful, innovative designs; fur-trimmed suits; fine embroidery; greatest influence 1890s - 1920s.

Paul Poiret—Bold colors; looser look; greatest influence 1900s - 1920s.

Madelaine Vionnet—Started bias draping; greatest influence 1920s - 1930s.

Coco Chanel—Famous for Chanel cardigan-style suit look; mixed fabric patterns and textures; greatest influence 1920s -1971.

Molyneaux—Of English descent; classic, elegant Paris designs.

Elsa Schiaparelli—Of Italian descent; many types of apparel; creative with prints; greatest influence 1920s - 1950s.

Nina Ricci—Important feminine collections from 1932; firm now run by son.

Cristobal Balenciaga—Of Spanish descent; outstanding quality of clothing construction; initiated a basic silhouette; greatest influence 1940s - 1950s.

Christian Dior—Wide skirts and small waistlines; greatest influence 1940s - 1950s. John Galliano now designs for this label.

Pierre Balmain—Designed elaborate gowns for famous people until his death in 1982; greatest influence 1950s - 1960s.

4-3 Haute couture, or the creating and selling of high fashion designs, started in the 1800s.

creations. Also, Hollywood movies encouraged original designing and became a vehicle to spread fashion ideas throughout the country. Several American high fashion designers of the past are listed in 4-4.

When Paris couture houses reopened after the war, they were challenged by fashion designers in the United States, Italy, and England. Today many cities join Paris as being influential on the apparel scene. In Italy, couture houses are located in Milan, Florence, and Rome. Other important cities are New York, Tokyo, and London. California has centers known for sportswear. Dallas, Chicago, and Montreal are gaining respect for their fashion industries. The Scandinavian countries are known for their native patterned knitwear. Israel has beachwear and leather coats. Hong Kong is high on the list with silks. Dublin features tweed and knit creations. Madrid excels with leather and beading.

Many designers from other cities have branch salons in Paris. Also, most designers of Paris, Milan, and other fashion centers have offices and salons in New York. Most major cities have several talented, exclusive designers.

Recently, the term *haute* has been eliminated and the term *couture* has come to mean "top of the line designer fashions." Exclusive designers throughout the world now name their highest-priced collections "couture," even though many of each are produced, but in limited quantities. Upscale retailers have "couture departments" where they present expensive ready-to-wear suits, dresses, and evening fashions of top designers.

American High Fashion Designers of the Past

Gilbert Adrian—Hollywood costumes from the 1920s; greatest influence 1940s.

Claire McCardell—Original sportswear; greatest influence late 1930s-1950s.

James Mainbocher—Theater and other clothes; greatest influence 1940s-1960s.

Norman Norell—Won the first Coty Award in 1943; was later the first Hall of Fame winner; fashion influence until the early 1970s.

4-4 American high fashion designers became recognized in the mid-1900s.

The Business of High Fashion

Most successful couture designers have had help getting started from wealthy financial backers. Their talents have been recognized, and they have been able to succeed because of the money and promotion of their benefactors. It is also important for designers to have the support of the press and other media that can spread fashion news quickly and make or break a couturier.

French couturiers used to design all their garments specifically for the clients who would wear them. Sometimes a client even contributed creative suggestions. Each fashion had true originality and presented the self-expression of its wearer.

Today fashion designers do ***collections,*** which include all their designs for a specific season. The designs are created for whomever will buy or order them. The designs are based on the fashion influences that each designer projects as the coming trends.

The couture fashion houses present two major showings of their collections a year. They are the fall-winter showings (held the preceding summer) and the spring-summer showings (held the preceding winter). The showings are held in one common location, often under large tents, with each firm having a specific time slot. This allows for buyers and the press to attend each one.

The showings are glitzy runway extravaganzas. They are like theater productions that feature slender, high-fashion models parading to music. From 50 to 100 outfits might be shown by each designer. The cost of preparing for and producing the showing is exorbitant. The fine fabrics and trimmings, as in 4-5, as well as the quality construction of each of the many designs, are expensive. Additionally, there are modeling fees and costs for hair and makeup experts, lighting, and music. The designer makes a traditional walk down the runway at the end of the show.

The exclusive audiences at the collection showings are made up of wealthy private customers, press representatives, and commercial buyers. (The commercial buyers represent various retailers, apparel manufacturers,

International Linen Promotion Commission

4-5 For collection showings, models wear original fashions that are often unusual and show the flare of the designer.

and pattern companies.) If a private customer buys a design, it is constructed to that person's measurements. Showings are now sometimes videotaped and sent to the couturier's best clients.

Fashion piracy, or the stealing of design ideas, is always a threat to designers. Therefore, clients who are allowed into the showings are carefully screened. Commercial buyers have to pay large cash deposits. The purpose of the deposits is to assure designers that the apparel manufacturers and pattern companies that do attend will buy some designs. Once the designs are ordered, the deposits are deducted from the purchase prices. However, these companies must pay more for each garment they buy than a private customer pays. This is because they are essentially paying for copying rights as well as the garment. Fewer and fewer commercial buyers attend each year because of the high costs.

Even though no sketching is allowed, the clothes are shown on the Internet and copied in cheaper versions by manufacturers soon after the showings. The couturiers do allow photographs for press releases, which gives them free publicity. The publicity keeps the designers' names famous and their designs in demand.

Couturiers must constantly be original and artistic, 4-6. They strive to create well-designed fashions of impeccable quality, distinctive styling, and lasting good looks. Most designers want to produce timeless clothing rather than "trendy" outfits. Some of their creations that are selected to be shown by the press have a touch of the outrageous!

To succeed, couturiers should have business sense as well as artistic creativity. The job is strenuous and competitive. Many designers have business partners who watch "the numbers" of the firms so the designers can strictly design. Many fashion houses, especially in Italy, are family-run with the skills passed on from generation to generation.

Top fashion designers have personal trademarks for which they are known. Each designer strives to have his or her signature, initials, logo, or type of garment instantly recognized. A ***logo*** is a symbol that represents a person, firm, or organization. In the past, logos and labels were put only on the inside of garments. Now they are often "status symbols" placed at visible locations on the outside of apparel. Some couturiers even create their own lining or garment fabrics, with their logos or names woven into the designs.

Fashion Associations and Awards

The trade association for top designers of the Paris couture is the ***Chambre Syndicale*** (pronounced shom'br sin-dee-kall'). It was formed to determine qualifications for couture houses and to deal with their common problems and interests. A couturier must be recognized as talented and successful to become a member. He or she must agree to abide by rules that include a code against copying. The rules also govern minimum numbers of staff models and production workers, dates of showings, and shipping dates.

The Chambre Syndicale coordinates the dates of the showings and issues admission cards to press people for the openings. It sponsors a school for the education of apprentices for the couture industry. It also represents its members in relations with the French government, arbitrates disputes, and regulates working hours and uniform wage arrangements.

The Fashion Institute of Design & Merchandising Costume Museum, California

4-6 Examples of haute couture fashions of the past, from both France and the United States, show tasteful designs of the times and lasting quality of construction.

Most cities or countries that are international fashion centers have similar organizations. For instance, the Tokyo Collection was founded to introduce Japanese designers to buyers, the public, and the press. It operates through a central information office. It has given the designers a structure in which they can compete with each other, such as through scheduled showings. Through such an organization, designers can join together to help make their city or country a leader in the fashion world.

Many female fashion professionals belong to The Fashion Group. It was organized in 1930 by women who were influential in American women's fashions. They were designers, magazine editors, retail executives, and others. They were committed to promoting American fashions. The Fashion Group helped build the entire American fashion industry to the place of prominence it has today.

The prestigious American Fashion Critics' Awards, referred to as the ***Coty Awards***, were presented each year from 1943 to 1978, and then became the *Cutty Sark Awards* through 1988. Such awards were given to the fashion designers recognized as the most creative and outstanding. Awards were given for excellence in women's wear, accessories, and menswear. Voting was done by a national jury of fashion editors of newspapers and magazines. Being named to the *Coty Hall of Fame* was the highest fashion honor in the United States. A designer could achieve that honor only by winning a Coty Award three different years.

Recently, major fashion awards have been presented by the Council of Fashion Designers of America (CFDA), such as for designers of the year in several categories as well as special awards for other fashion achievements. More than 250 designers are members of the CFDA. See 4-7. Various awards are also given by well-known retail stores, businesses, and other

Council of Fashion Designers of America; Patrick McMullan, Photographer

4-7 CFDA President and designer Stan Herman is shown here with Lauren Hutton at an industry event.

American Printed Fabrics Council: Jerry Kean, Photographer

4-8 The printed fabric receiving a Tommy is worn by a model. She is posing between the fabric print designers who are holding their Tommy awards formed from three golden tommy keys.

professional groups. The American awards for the best printed fabrics are the ***Tommys***. The trophy is made of three tommy keys. Tommy keys are tools used to adjust rollers for printed fabrics, as shown in 4-8.

Most countries have their own ways of recognizing talent in fashion design. Japan has The Best annual fashion awards presentation. Taiwan has a ceremony for their Golden Fashion Awards. European countries have their special awards. A worldwide event that distributes the International Fashion Awards is sponsored by the London *Sunday Times*.

Recent Changes for Growth and Income

Today consumers have a great abundance of mass-produced fashions of excellent quality available to them at many different price points. Thus, the expensive couture industry has only a small number of rich and famous customers. People's lifestyles and attitudes toward clothes have changed. Most people want more clothes and don't feel the need to buy exclusive items.

Besides the financial crunch of decreasing sales, haute couture houses have had increasing problems in hiring skilled tailors needed to make the clothes. The old tailors are retiring, and young people are not choosing to go into this field. To survive, many top fashion designers are diversifying. They create ready-to-wear collections and fashion accessories for the moderately priced market.

Several couturiers add to their incomes by selling the patterns of some of their creations to commercial pattern companies. Illustrations of ***designer patterns*** can be seen in pattern catalogs at fabric stores where the patterns are sold. Skilled home sewers can use the patterns to make their own designer clothes.

Couturiers also supplement their businesses with boutiques and franchises in cities around the world. Many of them also have licensing agreements with manufacturers of other types of products, such as perfumes and home linens.

House boutiques are small retail shops that are "branch stores" owned by the couturier, as in 4-9. They are usually in or near the haute couture premises, but may also be located in fashionable areas of other cities.

4-9 This house boutique in Paris sells shoes that are designed by the staff of Christian Dior.

They feature high-priced, high-quality accessories, such as handbags, shoes, and scarves. The merchandise is usually manufactured for the house by outside producers, but it is designed by the couturier or a member of the staff. It bears the designer's label. The couture house pays the rent and hires the salespeople.

Franchises are arrangements in which a firm, such as a couturier, provides retailers with a famous name and merchandise. In return, the firm (or couturier) receives a certain amount of money. The franchise owners do not work for the designer directly. However, they have been granted the right to use the designer's name and trademark to market goods. The designer does not help run the franchises, but the use of his or her exclusive name helps the businesses prosper. Franchises of designer fashions are often located in exclusive shopping areas of major cities. Sometimes they exist as boutique areas within large department stores.

Licensing is an arrangement whereby manufacturers are given exclusive rights to produce and market goods that bear a famous name as a stamp of approval. In return, the person or firm whose name is used receives a percentage of wholesale sales. These are royalty fees. The royalties received by couturiers in licensing agreements are very lucrative. Major couture houses license their names to different manufacturers on a large variety of products. The merchandise includes eyeglasses, sheets, perfume, cosmetics, luggage, candy, shoes, china, and automobile interiors. Ralph Lauren, shown in 4-10, is a highly successful apparel designer who licenses his name for many other types of products.

Licensing arrangements began to multiply in America in the 1970s, though they were common in Europe much earlier. It has been called the "name game." Some designers find it hard to keep tabs on the design or quality of all the products that bear their names. Some have established design studios. Employees of the studios arrange the licensing deals and check to be sure that the products are worthy of their labels. Inferior products would lower the value of designers' names.

The primary purpose of today's couture operations is to maintain the prestige of the designers. The houses must be successful with their couture creations to stay in demand for

Council of Fashion Designers of America; Mary Hilliard, Photographer

4-10 Ralph Lauren received the CFDA "Womenswear Designer of the Year" at the same CFDA Awards Gala that Lauren Bacall received an industry "Special Award."

licensing. Only a slim profit, if any, is made on the couture clothes. The extravagant collection showings often lose money. However, the publicity from the showings keeps the designers' names in the fashion spotlight, so other big business ventures can be pursued.

Charts 4-11 through 4-14 list many of the currently popular American and foreign fashion designers, and two are shown in 4-15. Become familiar with the names of the designers and their apparel specialties. In some cases, the well-known name of a designer is maintained for the fashion house after that person has died. A new designer is hired to continue the tradition of high fashion with new creations. This is true for the houses of Chanel, Dior, Nina Ricci, Anne Klein, and Perry Ellis.

Designer Ready-to-Wear

The designer ready-to-wear (RTW) industry has become more important than couture. Showings of these "bridge lines" are held as giant trade exhibitions. They take place in New York City, Paris, and other cities. In New York City, designer fashion week shows are held twice a year in tents at Bryant Park in Manhattan. The event is called "7th On Sixth." The fall/winter collections are presented in early spring, and the spring lines are shown in the fall. For each fashion season, over 50 shows are staged in six days. Each show is 20-25 minutes long. Retail buyers, celebrities, and press attend. Other RTW lines are presented in corporate showrooms on or near Seventh Avenue. Fashion (market) week activities are glamorous, exciting, and exhausting for everyone involved! A newer, lower-key trend is to have a breakfast or brunch showing with models walking around the tables wearing collection designs. A great deal of publicity is generated by the designer RTW showings, and new fashion ideas are noted. Depending on orders from retailers, only about half of the designs will be put into production.

There are many RTW designers listed in Charts 4-11 through 4-14. These include familiar names such as Coty Hall of Fame members Donna Karan, Calvin Klein, and Ralph Lauren.

Some Top French Fashion Houses

Balenciaga —Simple, elegant women's fashions.

Chloe —Youthful, edgy women's fashions now designed by Phoebe Philo.

Louis Feraud—Excels in colorful, artistic, luxurious women's attire.

Gianfranco Ferre—Italian who now has the design responsibility at the House of Dior; elegant, romantic, chic designs.

Jean-Paul Gaultier—Designs for the young market and does "single-sex" dressing that both men and women can wear.

Hubert de Givenchy—Especially big in the 1950s; glamorous evening wear, ladies' suits, and dresses; perfection of cut and lasting quality.

Kenzo—Of Japanese descent; colorful, casual, multilayer silhouettes.

Karl Lagerfeld—Of German descent; creates many inventive, wearable designs in France, Italy, and the United States for his own collection, the Chanel label, and others.

Guy Laroche—A master of understated, quality fashions for mature, elegant women; also menswear, accessories, and perfumes.

Claude Montana—Strong on silhouette, color, and texture.

Hanae Mori—Of Japanese descent; fashions with poetic colors and patterns; also accessories.

Thierry Mugler—Often creates sexy, clingy dresses.

Sonia Rykiel—The red-haired "queen" of knits and sweater dressing.

Yves Saint Laurent—Known for stylish, elegant, trendsetting designs; has been called "the king of fashion;" Creative Director is Tom Ford.

Emanuel Ungaro—Known for sensuously draped dresses; feminine prints and textures.

Valentino—Italian who moved business from Rome to Paris; ladylike, sophisticated fashions for famous clientele.

Louis Vuitton — Luxurious women's and menswear, leather goods, accessories, and shoes.

4-11 France is still considered to have a great deal of design talent for high fashion apparel.

Some Top American Fashion Designers and Firms

Adolfo—Cuban born; started with dramatic hats; now tasteful, classic, feminine dresses and suits; some licensing but no RTW collection.

Geoffrey Beene—A Coty Award winner; sophisticated elegance with unusual details since the 1940s.

Dana Buchman—Beautiful women's fashions for busy lifestyles.

Liz Claiborne—Retired, but firm continues with comfortable, fashion-right clothes, especially for working women.

Oscar De La Renta—Dominican of Spanish descent; creates romantic, colorful, glamorous fashions with ruffles and flourishes for own label and others; Coty Hall of Fame.

Louis Dell-Olio—Designs several lines with different labels; does furs, shoes, suits, and sportswear.

Carolina Herrera—Elegant women's creations.

Tommy Hilfiger—Classic American styles and jeans updated with unique twists.

Betsey Johnson—Originally based on dance costumes, now many stores worldwide with fresh, unique designs.

Norma Kamali—Designs avant-garde fashions with unusual materials and ease of movement.

Donna Karan—Designs simple, sensual women's clothes that stretch and move well; also DKNY label at a lower price point; Coty Hall of Fame.

Calvin Klein—"All-American" sportswear designs with sophisticated simplicity; status jeans; has many famous clients; Coty Hall of Fame.

Michael Kors—Provocative, comfortable women's clothes.

Ralph Lauren—Coty Hall of Fame; uses blazers with looks from "prairie rugged" to "English gentry" to romantic; home fashions; retail stores; Polo logo.

Michael Leva—Young designer known for easy "experimental" classics.

Bob Mackie—Designs for Hollywood stars; lots of beaded work.

Mary McFadden—Unusual, artistic, eccentric designs in decorative fabrics; luxurious evening clothes; Coty Hall of Fame.

Nicole Miller—Beautifully cut, contemporary, whimsical clothes; generous charitable giving.

Josie Natori—Lingerie and accessories with fine detailing.

Tom & Linda Platt—Irreverent fabric and color combinations; elegant but uncomplicated garments.

Arnold Scaasi—Glamorous gowns; Coty Award winner.

Adrienne Vittadini—Patterned knits; several lines; licenses; boutiques.

Vera Wang—Elegant, sexy wedding gowns and other fine apparel.

4-12 Many American designers have built up their design businesses into large, successful firms.

In Paris, the designer RTW industry is called ***prêt à porter*** (pronounced pret ah por-tay'). Prêt à porter means "ready-to-wear" in French. (The literal translation is "ready to carry.") Prêt à porter collections are mass-produced designer fashions. They are often merely called the "prêt collections" (to differentiate them from the couture collections).

The womenswear prêt showings are held twice a year in Paris. The showings are always exciting and busy. They last for a week to 10 days. Over 100,000 fashion professionals come to these shows. Orders are placed by retail buyers. The buyers and media reporters also assess new directions that the influential French fashions are taking. Menswear, children's wear, and knitwear showings are at other separate times.

The Italians promote their fashions often, with different specialties at different times. In Milan, designers show their lines at prestigious fashion fairs just prior to the Paris prêt showings. The exhibits are sponsored by two large industrial groups. One group is for ready-to-wear, and the other is for knitwear. There are also individual fabric fairs for silk, wool, leather, and home furnishing fabrics. In Florence, there are major exhibitions for children's wear and menswear. Italy even has fashion fairs devoted to uniforms and work clothes! Other countries do some of this, but to a lesser degree.

Some Italian Fashion Names

Giorgio Armani—Milan; did menswear first, creates masculine, unstructured, layered designs for women; witty; relaxed refinement.

Hugo Boss—From Spain; distinctive menswear.

Benetton—Knitwear; many worldwide outlets; corporate advocacy advertising.

Nino Cerruit—RTW knitwear.

Dolce and Gabbana—Design team with sometimes outrageous womenswear and menswear, often in knits.

Fendi—Rome; sisters started with leather accessories; now also furs and RTW.

Ferragamo—Florence; a family business started with shoes.

Gucci—Florence; known mainly for leather goods; uses "GG" logo.

Christian Lacroix—Daring use of color.

Andre Laug—Rome; very original, wearable women's ensembles.

Mariuccia Mandeffi—Milan; designs the "Krizia" collection with husband, Aldo Pinto; started "hot pants" in the 1960s; now does dresses, suits, animal knits, and other sweater dressing.

Rosita and Ottavio Missoni—Children and grandchildren join them in creating multicolored knitwear and other products.

Emilio Pucci—Simple structural garment shapes for women; strong artistic prints and signed fabrics.

Gianni Versace—Milan; known for slinky, chain mail evening dress; also very original daytime outfits; since his death, designing has been done by his sister, Donatella.

4-13 Italy has become well known for fashion, including leather products, knitwear, and garments with international flair.

Other Fashion Names Around the World

Efrin Africa—Philippines; specializes in women's suits with classic lines.

Ika Bouton—Hong Kong; decoratively sequined and other artistic dresses.

Robert Burton—Creative Australian designer.

Scott Crolla—British; uses upholstery fabrics and tapestries in apparel.

Victor Edelstein—British; a favorite of the late Princess Diana.

David Emanuel—Feminine British fashions and gowns.

Katia Filippova—Moscow; designer of avant-garde, eclectic combinations of colors, textures, and shapes.

Diane Freis—American-born designer in Hong Kong; romantic, feminine dresses with frills, feathers, beads, and tassels; also one-size shirred waistband creations.

Rei Kawakubo—Japanese; unconventional, somber designs in neutrals or earthtones; firm is Comme Des Garcons.

Jurgen Lehl—German who designs a full line in Japan.

Pat McDonagh—Canadian fashion designer.

Antonia Miro—Leader of Spain's men's elegant, sporty lines.

Mitsuhiro Matsuda —Japanese; mixes many colors and layers of apparel.

Issey Miyake—Combines classic Japanese design with unusual knits, draped skirts, and eccentric "sculptural" layering; uses fabrics with interesting textures.

Pitoy Moreno—Philippines; creates conservative, elegant gowns.

Jean Muir—British; best known for ladylike, jersey knit dresses; also does eccentric classics in silk, suede, and leather.

Bruce Oldfield—Original British designs for women.

Zandra Rhodes—British; trendy, off-beat women's clothes; innovative surface decorations; unusual hemlines.

Jill Sander—German; produces and shows designs in Italy.

Sybilla—New York City-born designer who lives and works in Madrid, Spain.

Alfred Sung—Leading couturier in Canada.

Tamotsu—Japanese; soft, curvaceous designs reveal the body shape.

Zang Toi—Malaysian-born designer creatively uses bright color combinations.

Roberto Verinno—Spanish; simple, elegant, professional women's clothes.

Yohji Yamamoto—Japanese; colorful, avant-garde, jovial creations.

4-14 Original and innovative design talent exists throughout the world. Fashion is truly international big business.

Council of Fashion Designers of America/Roxanne Lowit, Photographer

4-15 Fashion designers Pauline Trigere and Bob Mackie enjoy a social event with other top industry colleagues.

Carnaby Street in London has boutiques where colorful, uninhibited fashions for young men were first launched. However, England's ready-to-wear strength today lies in their high quality men's tailored apparel, fine rainwear, and far-out women's clothes. British designers show their lines at the London Designer Collections Fair.

Other countries are eager to export their fashions, too. Most countries have their own semiannual trade fairs and exhibits. They also have showings and promotional trade offices in cities in the United States and around the world.

California Mart

4-16 Many fashion designers, including Norma Kamali, design ready-to-wear collections.

Unknown designers are hired to work on the ready-to-wear lines under the big name designers in the fashion houses. Eventually, some of the best ones branch off into businesses of their own. Some designers specialize in jewelry, handbags, belts, or shoes. The designer labels in ready-to-wear clothes and accessories indicate to consumers that the items should be of fine quality and in good taste. The showroom in 4-16 contains the ready-to-wear collection of Norma Kamali for retail buyers to see.

Chapter Review

Summary

The apparel industries of textiles, garment manufacturing, and retailing developed slowly throughout the world until the Industrial Revolution. Then there was a switch from handmade garments to mass-produced ready-to-wear clothes. In the U.S., the "rag trade" started in the garment districts of New York and other cities. There are many fashion magazines, trade publications, and trade associations tied to the segments of the apparel industries.

Expensive, high fashion clothes originated in Paris haute couture houses. American dressmakers copied French designs until World War II. Then designers in all major cities of the world became known independently for various types of fashions.

Most designers have wealthy financial backers or business partners. They present glitzy showings of their collections and promote their exclusive logos. Fashion piracy is common. The designers belong to fashion associations. Excellence is rewarded with prestigious awards. Recent changes for design firms to achieve more growth and income include designers sewing patterns, house boutiques, franchise agreements, and licensing arrangements.

The showings of designer ready-to-wear (RTW) lines are now more important than couture. Collections of these mass-produced designer fashions are presented on runways and in showrooms in various fashion cities during market weeks. Designer labels in RTW clothes and accessories indicate to consumers that the items should be of fine quality and in good taste.

Fashion in Review

Write your answers on a separate sheet of paper.

Short Answer: Write the correct answer to each of the following questions.

1. What three major developments during the Industrial Revolution caused a switch from handmade garments to mass-produced ready-to-wear clothes?
2. What are two common names for the garment industry?
3. Name two trade publications of the apparel industries.
4. Name two trade associations of the apparel industries.
5. The high fashion garment industry originated with the superior dressmaking industry of what country?
6. How did World War II affect the American design industry?
7. How many major showings of collections do the couture fashion houses present each year?
8. Why was the Chambre Syndicale formed in Paris?
9. Name two designers in the Coty Hall of Fame.
10. Name two reasons for the recent changes in the haute couture industry.
11. A retail shop owned by a couturier that carries accessories designed by the couturier or a member of his or her staff is a ___________.
12. In a ___________ arrangement, a couturier provides a retailer with a famous name and merchandise in return for a certain amount of money.
13. In a ___________ arrangement, a manufacturer pays for the exclusive rights to produce and market goods that bear a designer's name as a stamp of approval.
14. In New York City, the fashion industry centers around ___________ Avenue.
15. Which do more retail buyers attend—showings of the couture collections or showings of the prêt collections?

Fashion in Action

1. Use additional resources to learn more about a famous fashion designer of the past. Prepare an oral report (with visual aids) on the life, work, and major contributions of the designer.
2. Make a booklet showing the fashions of at least three current well-known designers. If available, show each designer's unique signature or logo.
3. Look for articles and pictures of fashion showings in the United States or abroad in recent newspapers and magazines or on the Internet. Where were the showings held? What designers were featured? What types of apparel were shown? What distinguishing characteristics and fashion trends were evident?
4. Ask local apparel businesses and libraries if they subscribe to trade publications for the apparel industries. Bring copies of as many different publications as possible to class. Notice what kinds of information are presented in each. Compare the content of these publications to the fashion news in newspapers and consumer magazines.
5. Prepare a visual presentation on the influence of Hollywood movies on the fashion industry in the United States.

5 The Textile Industry and Home Sewing Patterns

Fashion Terms

greige goods
textile converters
microelectronics
robotics
innovation
forecasting services
marketing
imports
exports
trade deficit
guide sheets
croquis
sloper
prototype
grading

After studying this chapter, you will be able to

- explain the workings of textile businesses.
- describe how new color and fashion trends begin and are marketed.
- discuss the worldwide textile industry of today and the future.
- write a report about the home sewing pattern industry.

The textile industry is an extremely vital part of the overall apparel industries. Textile companies produce fibers, yarns, and fabrics for fashion and other products.

Home sewing patterns are a small part of the fashion industry, but deserve to be studied. Pattern companies design and make patterns for home sewers.

The Textile Industry

The textile industry produces fashion fabrics for garments. It also develops and manufactures fabrics and other textile products for home decorating and many industrial uses. The industry is very large and vital to the economy of the United States.

Fabric Production and Distribution

There are four main steps in the production of finished fabrics. They are

1. *fiber production* from natural or synthetic materials
2. *yarn production* at spinning mills
3. *fabric (cloth) manufacturing* by weaving, knitting, or other methods
4. *fabric finishing* to satisfy the look, feel, and performance demands of the market

These processes involve highly specialized machinery and great skill. All firms strive for peak production and maximum quality at the lowest cost. They are also concerned about not damaging the environment. After production, the finished fabrics must be sold and distributed.

Fiber Production

Different raw materials are processed into various hair-like fibers.

Agricultural industries supply natural fibers such as cotton, wool, flax, and silk. They are grown in fields or on animals.

Chemical companies produce manufactured fibers such as rayon, nylon, spandex, acetate, and polyester. Most are liquid chemical mixtures that form into thin "threads" as they solidify, as in 5-1. They are made to have characteristics that suit their end uses. They were first invented in the early 1900s and have grown in popularity and use. Millions of pounds of fibers are sent to mills by fiber producers.

Yarn Production

Mills spin fibers into yarns, as shown in 5-2. Several fibers are twisted together to form the long strands of yarns used to make fabrics. Most U.S. textile mills are in the Southeast. The states of North Carolina, South Carolina, and Georgia have especially large numbers of textile mills. A great deal of yarn production for U.S. consumption is done in other countries.

Fabric Manufacturing

Textile manufacturing plants (also called mills) weave or knit yarns into fabrics. Huge, mechanized looms, as shown in 5-3, and knitting machines produce great amounts of yard goods very fast. However, the cloth is still unfinished. The unfinished fabrics are called ***greige*** (pronounced "gray") ***goods.***

Fabric Finishing

Finishing is done by bleaching, dyeing, printing, or applying special coatings to the greige goods. This imparts color, texture, pattern, ease of care, and other characteristics

Trevira Polyester- Hoechst Fibers Industries

5-1 Manufactured fibers are made with special equipment under controlled conditions by many chemical companies.

International Linen Promotion Commission

5-2 Spinning mills have many rows of specialized machines that twist fibers into yarns quickly and precisely.

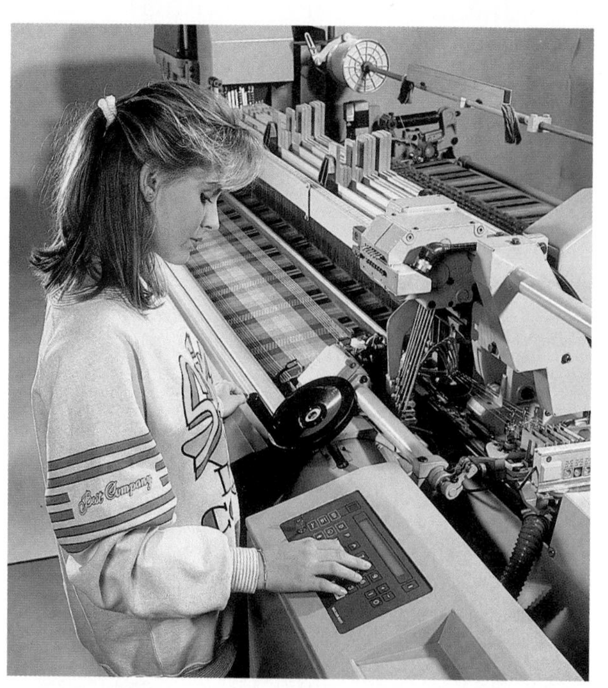

Sulzer Ruti, Inc.

5-3 This automated weaving machine can insert yarn colors in desired sequences to produce fashion fabrics.

to the fabrics. The fabric in 5-4 has been finished to have a soft, luxurious feel.

Textile converters are firms, or individual merchants, who buy or handle greige goods for finishing. They keep close tabs on fashion trends to anticipate demand. They contract with others to dye, print, and finish the goods to their specifications and then sell the finished products to apparel manufacturers or fabric retailers. Usually they do not own fabric plants or finishing facilities. Instead, they serve as middle agents between these stages and the final-use markets.

Distribution

Finally, sales offices are necessary to market the finished fabrics. They sell the fabrics to apparel and accessory producers, fabric retailers, and specialists for home and industrial uses. Many textile firms have showrooms in New York City near the garment district.

The Development of Textile Corporations

Historically, each textile company specialized in a single stage of production. Spinning mills, weaving mills, knitting mills, finishing plants, and selling agents all worked independently. However, after World War II, problems of scarcity and price made this fragmented operation outdated. The industry began to integrate itself. Textile firms started to do several, or all, processes, including direct selling. In some cases, selling agents bought manufacturing and finishing plants. Mergers and acquisitions created large textile corporations.

Today, some companies specialize in one phase of textile production, such as producing manufactured fibers. Other companies specialize in one type of fabric, such as brocades or knits. However, the textile industry is dominated by corporations that handle all processes of production and distribution. Steps from fiber production to selling are done in different plants and locations belonging to, or as divisions of, one corporation. This enables a firm to control its goods through as many processes as are potentially profitable. Also, the many products of a firm are distributed by its different, specialized marketing divisions.

Burlington Industries, Inc.

5-4 The modern dry-finishing equipment shown here can create fabric surfaces that are soft to the touch.

Textile Technology, Fashion, and Marketing

Textile companies must have technology, fashion, and marketing skills to succeed with their textile lines. The people who work in these individual areas must coordinate their efforts closely with the other areas. All are interdependent.

Technology

The textile industry's technology is kept up-to-date by efforts in research and development. "R & D" has been a vital part of the textile industry for many years. New manufacturing machinery and procedures have been invented, as in 5-5. Fibers, finishes, and other textile products have been continuously developed and improved.

The development of computerized, electronic machines has made textile manufacturing faster and better. There are computerized spinning, knitting, weaving, dyeing, and finishing machines. Waste is reduced since only the needed amounts of materials are used.

Some universities and schools of textile science do research in textile ***microelectronics*** (computer-related procedures and technology).

Sulzer Ruti, Inc.

5-5 Research and development of new weaving machines is now done at interactive graphic computer workstations.

5-6 Advances in technology have led to new textile products for end uses varying from apparel to space-age electronics.

They also conduct research in the field of ***robotics*** (mechanically accomplished tasks done by automated equipment). At manufacturing technology centers, skills are taught on the latest equipment.

New manufactured fibers and chemical finishes have revolutionized the textile and garment industries. Extra-fine fibers and stretch fabrics have enabled new types of garments to be designed. Home decorating fabrics are stronger and more resistant to fading and abrasion. Many advances have also been made for specific industrial and medical end uses of textiles. Examples are inflatable buildings, human artery replacements, fiber optic procedures, and space suits, as shown in 5-6. Quality standards rate textiles according to levels of defects. Performance standards rate the suitability of textiles for specific end uses.

Often the apparel industry requests the production of fabrics with certain characteristics to satisfy market demand. At other times, developments in the textile industry inspire new apparel designs. For instance, water-repellent fabrics have enabled raincoats to be styled for all-purpose wear.

Textile producers want to do more than introduce new products into the market. They try to see that apparel and industrial manufacturers use those products in innovative, marketable ways. ***Innovation*** is the creative, forward-thinking introduction of new ideas. Sometimes whole new realms of applications are opened up for textiles.

Textile companies and their researchers subscribe to such magazines as the *Textile Research Journal* and *Textile World*. Most belong to the American Fiber Manufacturers Association, Inc. and/or the American Textile Manufacturers Institute, Inc. Through these channels, they learn about the latest textile developments and communicate them to others.

Fashion

Before the textile companies can begin to develop their fiber characteristics or fashion fabrics, they must come up with some early projections of colors, textures, and weaves. Their textiles must meet public fashion acceptance at the time they reach the market. Textile designers decide on colors and patterns about a year and a half before the public sees them. For instance, during the fall of one year, textile people are designing their lines for apparel that will sell to consumers the spring after next.

Color must be decided first, as in 5-7. In the final garment, color will be the most important factor in drawing the consumer to the item. A color line is determined by logic, research, and "gut feeling." The best and worst color sellers of the last season are reviewed with apparel manufacturers and retailers. Colors are often altered by evolution rather than by sharp changes. The good ones might be kept and altered to go through several seasons. Burgundy may become purple, then magenta, then pink. Colors that have not been used for quite a while may be brought back if they now seem new.

The fabric designer of a textile firm must be aware of upcoming fashion shapes. The weight and construction of the fabrics must complement the styles of garments that will be manufactured. To determine these "fabrications," market research is done. Market trends are analyzed formally (through statistics) and informally for an awareness of what is coming.

There is an information network of fashion, fabric, and color development in the industry. ***Forecasting services*** look about two years ahead to predict coming trends. They are highly skilled consultants. They foresee the colors, textures, silhouettes, and accessories of the future, as in 5-8.

Forecasting services sell their information to companies that subscribe to their publications and services. They have branch offices in major cities around the world. They use color cards, fabric swatches, videos, slides, and sketches to present their ideas.

Forecasting services are invaluable to textile firms, as well as to apparel manufacturers and retailers. They provide general design or merchandise advice, or specific information that targets the needs of the firms buying their services. Companies often subscribe to more than one service. Through these services, all segments of the industry get the same information. Each company uses the information in its own way.

Textile firms design "collections" of distinctly different groupings of fabrics. Textile designs are created by using different weaves, textures, colors, and applied patterns. The finished fabrics must have the right combinations of colors, patterns, and fabric weights for changing fashion trends.

Ideas for prints and textures may come from sources other than forecasting services. They may be inspired by nature, modern trends in art and entertainment, or historical research.

Textile designs used to be shown in technical drawings on graph paper. Now many are designed on computers. The designer points or writes on an electronic "tablet" with a handheld wand, as in 5-9. With proper computer commands, the computer can increase color brightness or fade it. The computer can reduce or enlarge motifs. It can combine many multiples of

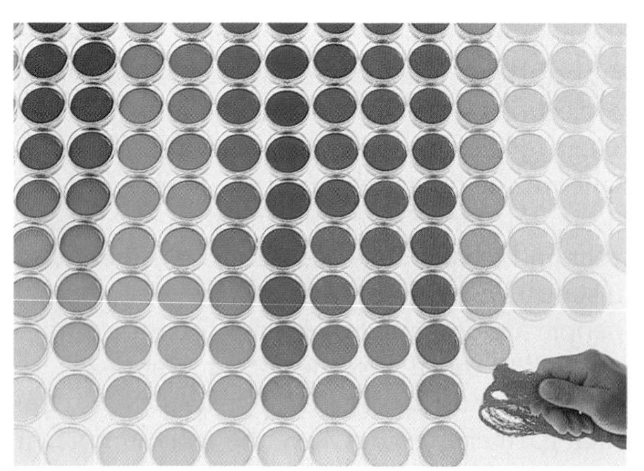

BASF Corporation, Fibers Division

5-7 The colors for textile products are planned well ahead of when the fabrics will be made into apparel and sold to consumers through retail stores.

The Doneger Group

5-8 Forecasting services sell expert advice about future color trends, fabric textures and patterns, and apparel shapes and designs.

Fashion Institute of Technology

5-9 Trained textile stylists use computer-aided design systems to create new patterns and give related manufacturing specifications quickly.

the same motif together and give related manufacturing specifications quickly. The design can be given a name, saved in the computer, and brought back later. It can also be photographed by a special camera to make a color slide that enables permanent viewing of the design.

Fabric collections are created for certain end uses such as children's wear, men's sport shirts, or bedsheets. The collections are given distinctive names. This enables a strong marketing message to be conveyed to potential clients. Some textiles are developed exclusively for a designer's line of apparel or household linens through licensing agreements.

Marketing

Marketing is finding or creating a market for specific goods or services. It identifies the customer and determines what products will satisfy the customer's needs while, at the same time, providing a profit for the company. To create and sell successful fabric collections, the marketing strategies of textile firms ensure that the right fabrics are available at the right time and at the right price. Marketing goes hand in hand with the technology and fashion functions previously described.

Most textile producers maintain "fabric libraries" in their showrooms for use by the companies that buy their textiles. The textile manufacturers produce and display small amounts of the fabrics in their lines, so they can promote them. Fashion boards with sketches and fabric samples are set up as selling aids. They show the fabrics that are currently available or are scheduled for mass production for an upcoming season. Also, sample garments are often made that feature a company's fibers, yarns, or fabrics.

Apparel manufacturers and retailers are invited to see and discuss the new fabrics with the textile producers. Apparel designers, retailers, and fashion reporters use this as a source of information for upcoming fashion products.

Fiber and fabric producers assist apparel manufacturers and retailers in every way they can. They help to create final products that please consumers. They help manufacturers with needed market information and other planning data, as shown in 5-10. They assist the manufacturers in selling their new apparel lines to retailers. They help retailers sell to consumers with publicity and advertising about their fibers and fabrics. They

Burlington Industries, Inc.

5-10 Through computers, textile firms can allow apparel manufacturers access to their data concerning planning information.

help train store sales personnel. They provide consumer education programs and educational materials. They also provide textile trademark tags and labels to be attached to the apparel.

Textiles Worldwide

The textile industry is of worldwide scope. The United States is efficient at textile production. Nevertheless, the U.S. imports great amounts of textile fibers, yarns, and fabrics from other countries. (***Imports*** are goods that come into the country from foreign sources.) Some of the imports are superior fibers and fabrics from countries known for producing them well. Most other textiles are imported simply because the prices are low. They are from countries whose low labor and factory operation costs enable them to sell at lower prices, even after shipping costs and tariffs are added.

Manufactured fibers and knitted goods come into the United States from Japan. Wool fibers and fabrics come from the United Kingdom and Belgium. Cotton fibers are imported from Egypt, and flax from Ireland. Silk fabrics come from China and Italy. Laces come from France.

Countries promote their textile industries to encourage trade, as shown in 5-11. International fabric fairs and trade shows are held in many countries. They promote textile machinery as well as fibers, fabrics, and apparel.

Firms in the United States also export some textile goods to other countries. (***Exports*** are commercial products sent out of a country to other countries.) However, imports far outweigh exports. This creates a ***trade deficit***—a condition in which the amount of imports is greater than the amount of exports.

Growing competitive pressures, largely from imports, have severely cut into the textile industry of the United States. Large amounts of profits and many jobs have been lost. Textile firms have been forced to make drastic changes. Plants have been closed. Low-profit and diminishing product lines have been discontinued. Improvements in technology and production efficiencies have been implemented. Textile firms are striving to create goods that will please consumers, sell well, and make money.

Seattle International Trade Center

5-11 This fashion show is promoting the silk products of China as a whole rather than the specific collections of individual manufacturers.

The Future of Textiles

Due to increasing global competition, the U.S. textile industry is innovative and progressive. This competitive spirit is helping it adapt to rapid changes, resulting in a high degree of computer technology, 5-12. Textile firms are closing outdated plants, automating newer ones completely, and establishing American-owned plants in foreign countries.

In the future, small, weak firms will fail or be purchased by larger companies. Such acquisitions and mergers will create fewer, stronger companies. Those with top-quality products and customer service will flourish.

Demand for textiles will continually increase as worldwide standards of living go up. Experts predict polyester to be the predominant fiber. Cotton will remain popular. Modifications to existing fibers and newer mixtures of fibers are more likely to emerge than totally new fibers. Nonwovens and "disposable" textiles will become more common, especially in medical and food service uses. Also, textile imports will

Info Design, Inc.

5-12 Future computer automation, being developed today, will simulate and oversee color accuracy, technical design, and all production machinery operations.

continue to increase as more production is done in low-wage countries.

Predictions imply that the production of knits will grow, with fewer wovens produced. Knitting machines use less energy than weaving looms. Also, knitting mills have lower capital requirements than weaving mills. (*Capital* is the financial worth, or accumulated investment cash, needed to start, expand, or run a business.) Yarn production will create greater popularity of novelty yarns.

Successful textile firms of the future will use industry change to their advantage. They will learn to work with imports. More sophisticated marketing techniques will give more attention to pleasing retailers and consumers with the right products.

Flexibility and versatility for shorter production runs of different fabrics will be needed to quickly satisfy changing market needs. New computerized methods will enable fast manufacturing changes to provide smaller, more customized orders of unique products. These orders will be filled faster and more accurately, with direct computer communications between textile firms, apparel manufacturers, and retailers. Production will go up while waste decreases and environmentally friendly processes are used.

Restructured firms will have fewer levels of workers and management. Automated equipment will require continued retraining of lower-skilled employees. Ambitious young workers can become leaders.

Home Sewing Patterns

Pattern making companies design and make patterns for the home sewing market in the United States and around the world. Every year, new styles appear in the major pattern catalogs. Thousands of fashion and crafts patterns are available to sewing enthusiasts.

The Business of Patterns

Many people have sewing skills. Few have pattern making skills. Therefore, most people buy patterns to use as guides for their sewing projects.

Pattern making companies offer a large choice of designs for home sewing consumers. The thick counter catalogs at fabric stores clearly show all designs and their variations. The catalogs have sections by category of figure and garment types. Examples are children's patterns, formal wear, misses' dresses, and menswear.

To help home sewers, patterns have printed ***guide sheets***, with illustrated directions for all cutting and sewing steps. Fabric layouts are shown. Needed notions, such as zippers, trimmings, and buttons, are listed on the envelope for each design. Appropriate fabrics are suggested, too.

The pattern companies tabulate sales of the patterns monthly to see how well each is selling. This tracking system tells them what designs are in demand and in what sizes. Pattern preferences generally follow the trends of ready-to-wear fashions. Most patterns stay in the catalogs for a year or two. Those patterns that continue to sell well are kept in the catalogs longer. Those that do not sell are dropped to make room for new patterns.

The Breadth of Pattern Companies

As mentioned in Chapter 4, *designer patterns* are offered to give home sewers a touch of couture, as shown in 5-13. Sophisticated designer lines are replicas of actual couturier fashions. They are bought from designers through licensing agreements.

Signature lines have the endorsement of celebrities. Usually the celebrity is an actor or

5-13 As shown on the envelope of this Vogue Pattern for home sewers, the dress is by fashion designer Oscar de la Renta.

5-14 Easy-to-sew patterns from different companies have various names, such as "Easy McCalls," "Jiffy" by Simplicity, "Fast and Easy" by Butterick, and "Very Easy, Very Vogue."

model. The styles in such lines are typical of what that person likes and wears. The celebrity helps develop the line through a licensing agreement. The celebrity is shown wearing the fashions in photographs in counter catalogs and promotional materials.

Easy-to-sew patterns are of designs that are simple to cut out and make. They are developed for beginning or busy sewers. They are also specially marked, as shown in 5-14.

The major pattern companies act as information centers. They research fashion trends and spread fashion news to the retail fabric stores that sell their patterns. Education departments of pattern companies provide services to help teach sewing. Pattern companies have mail, telephone, and e-mail answering services, too. They solve sewing, fitting, and wardrobe problems of consumers.

Several pattern companies publish fashion magazines that provide fashion news and sewing advice. Each magazine also promotes its company's patterns.

Designing Home Sewing Patterns

Home sewing patterns are designed in much the same way as ready-to-wear fashions. The staff designers are inspired by American and foreign fashions. They follow the advice of trade publications and fashion magazines, and they use information they gather at fabric fairs. They consider requests from home sewers and imitate the latest in ready-to-wear. They also evaluate what has or hasn't sold well in previous lines.

Both easy-to-sew and challenging designs are created. The company also tries to come up with the right mix of dressy, casual, athletic, and specialized patterns. Design ideas are coordinated with marketing plans to create something for almost everyone.

The rough first sketch, or ***croquis*** (crow-key), of each design shows both front and back views. It also shows any special details the designer wants included in the pattern. It is presented to management and pattern makers in a "construction meeting." Details about fullness and length of garment parts, structural needs (linings, interlinings, etc.), and sewing methods for the design are discussed. Then notes are written on the croquis concerning when the design will appear in the counter catalog. The figure type category and special construction techniques are also noted.

Suitable fabric samples are chosen for each design from the pattern company's fabric library, as shown in 5-15. The library has swatches of fabrics that are available to home sewers. Then the design is put into a final artist's sketch showing it in a fashionable fabric.

Perfecting the Patterns

Making patterns requires a combination of design talent and technical expertise. A member of the pattern making staff makes the first pattern, usually with flat paper. To keep sizing consistent, the work is done from a basic pattern, called a sloper. The ***sloper*** is in the company's basic size from which all designs in a category start. Sometimes the first pattern is made by draping muslin onto a body form of the basic size, rather than from the sloper.

Next, a partial muslin "proof" of the pattern is sewn and fitted on a body form. Only one side of the design is made, as in 5-16. Then the designer is asked to check the work. The original sketch has been transformed into three-dimensional form. Fit and design corrections are made as needed.

Next, a full muslin prototype is sewn together. (A ***prototype*** is the first full-scale trial garment of a new design.) The prototype is worn by a model who has the same proportions as the pattern company's standard client. The model walks, bends, and sits in the garment to test its wearing comfort. The designer makes final fit, drape, and design changes at that time.

The next step is to gain the approval of a committee of company management. Then all planned views (showing, for example, sleeve and collar variations) are penciled onto the muslin sample. Notches and other construction markings are added to the muslin garment. Notes are made about notions and fabrics.

The muslin pattern pieces are used to make a hard paper master pattern for the design. The pattern pieces are marked with notches and construction symbols. A descriptive name is put on each piece, such as pants front, back pocket, or waistband.

The garment is then cut out and constructed in fashion fabrics for photographs

The McCall Pattern Company

5-15 Design staff members of major pattern companies consult with their in-house fabric libraries as they create new designs for the sewing public.

The McCall Pattern Company

5-16 The muslin proof of each pattern is checked to be sure it accurately represents the designer's ideas.

and fashion shows. If the design is planned for striped or plaid fabric, it is made up with that type of fabric. If it is "for stretchable knits only," a sample is made in a stretchable knit. Every effort is made to show consumers how to achieve good results as they use the pattern.

Finishing the Patterns

Next, the pattern is graded. ***Grading*** is making each pattern piece in all the sizes that will be sold. It is a scientific process that is now done mostly by computer. Each pattern piece in each size is made in heavy paper.

Fabric layout charts are made using outline drawings of the individual pattern pieces. They show how the pieces should be placed on fabric when being cut out, as in 5-17. The layout charts are devised for various standard widths of fabric. Also, yardage requirements are recorded.

A guide sheet is prepared with sewing instructions and drawings. Most pattern instructions are stored in computers. Each company has a "standard procedure book" on how to construct different kinds of garments and details. It is constantly being updated. The appropriate codes are entered into the computer, and generic sketches with "how-to" copy appear on the guide sheet.

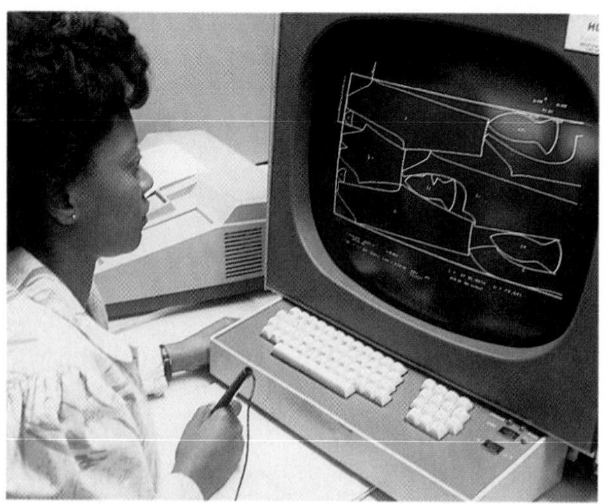

The McCall Pattern Company

5-17 Fabric layout charts are made by computer and printed in the pattern's instructions. They show home sewers how to cut out the garment pieces with the least amount of fabric waste.

The pattern envelope must be developed, too. The design is drawn in watercolor as it would look if it were made in a specific fashion fabric. The artist's painting or a fashion photograph is used on the envelope as in 5-13 and 5-14 and in the counter catalog.

The tissue pattern pieces, guide sheets, and envelopes are printed on printing presses. Machines fold the patterns and guide sheets and insert them into the envelopes.

Innovations for the Future

The patterns of the major companies will probably become more generalized in the future. Pattern markings and instructions will be bilingual. This eliminates the printing costs of separate sets of pattern pieces and guide sheets in different languages for international use. European patterns will be readily available in the United States. There will also be more multisize patterns. These are patterns with several sizes printed together on the same pieces.

More patterns in the future will probably be for use with knits, since textile companies will be producing more knit fabrics. Also, guide sheet directions may be written for *serger* (or overlock) sewing machines. These sewing machines duplicate ready-to-wear manufacturing techniques. They are very effective in sewing knits and newer types of fabrics.

Custom pattern companies have already come into existence. They make patterns to fit individual measurements. Some are designs that are hand drafted. Others are basic patterns that are done by computer for use as slopers that fit an individual's measurements. The customer fills out information in a measuring kit from the company. From that, "electronic patterns" are made for basic pants, skirt, and bodice versions.

Someday, basic and design patterns may be made by home computers. Home sewers will give commands of what they want. They will be able to see on the screen how all angles of a garment will look. They will then be able to make changes and have the pattern printed out with their proper measurements—all from a computer program!

Chapter Review

Summary

The textile industry produces fibers, spins yarns, manufactures fabrics, finishes fabrics, and distributes the finished fabrics. Some textile corporations do only one of those steps, while others handle all processes of production and distribution. To succeed with textile lines, companies combine technology, fashion, and marketing skills.

Today, the textile industry is worldwide. The U.S. imports great amounts of textiles, creating a trade deficit. In the future, the textile industry will be more innovative and flexible to meet increasing demands and changing market needs.

Companies that make home sewing patterns are an important segment of the home sewing industry. Pattern companies sell designer patterns, signature lines, and easy-to-sew patterns. The patterns are designed by sketching ideas and calculating construction details. After samples of the designs are sewn and approved, the patterns are graded into sizes and printed, along with guide sheets and envelopes.

In the future, there will be more multisize patterns and duplication of RTW manufacturing techniques. Someday, sewing patterns may be made by home computers.

Fashion in Review

Write your answers on a separate sheet of paper.

Short Answer: Write the correct answer to each of the following questions.

1. List and briefly explain the four main steps in the production of finished fabrics.
2. Give three examples of how new manufactured fibers and chemical finishes have revolutionized the textile and garment industries.
3. When textile companies develop their lines, what factor do they decide first? Why?
4. List three sources of ideas for prints and textures in textile designs.
5. Explain how textile patterns can be designed on a computer.
6. To develop successful textile lines, companies must have technology, fashion, and _____ skills.
7. Usually, are more textile products exported to other countries or imported into the United States?
8. In the future, which is likely to grow more, the production of knits or the production of wovens?
9. Explain how the major pattern companies act as information centers.
10. What does a croquis show?
11. True or false. The grading area of a pattern company sews prototypes.
12. In the future, are the patterns of the major companies likely to become more generalized or more specific?

Fashion in Action

1. Be creative and pretend to be a textile designer. Make colored drawings or paintings on paper of a collection of at least five different textile designs. Make sure the motifs are clear and easy to reproduce. Describe in words the texture and fabric weight of each. Give the collection a distinctive or catchy name. Write a paragraph describing the intended end uses of the fabrics.
2. Try to stretch your brainpower! Pretend to scientifically develop a new textile for a fashion or industrial end use (or both). Describe the textile's technological specifics in a short report. Give hypothetical marketing information describing how it might be promoted.
3. Select one aspect of the worldwide trade of textile products (low cost fabrics or garments from developing nations, a textile specialty from a specific nation, impacts of having a trade deficit, etc.). Do further research and present your findings in an oral report.

4. Visit a local fabric store. Take notes on what types of fabrics are available. What colors, textures, and fabric weights are most common? Study the pattern catalogs. What pattern companies are represented? Note the categories of patterns available. Are designer, signature, and easy-to-sew patterns offered? What categories of patterns interest you the most? Are magazines and pamphlets from pattern companies available? Write a report on your findings and personal viewpoints.

5. Research the history of sewing patterns. Prepare a written or oral report on the influential people and important developments in the industry. Include details about the first patterns. Explain when patterns started to be used by the general public. Describe the standardization of size measurements by the industry.

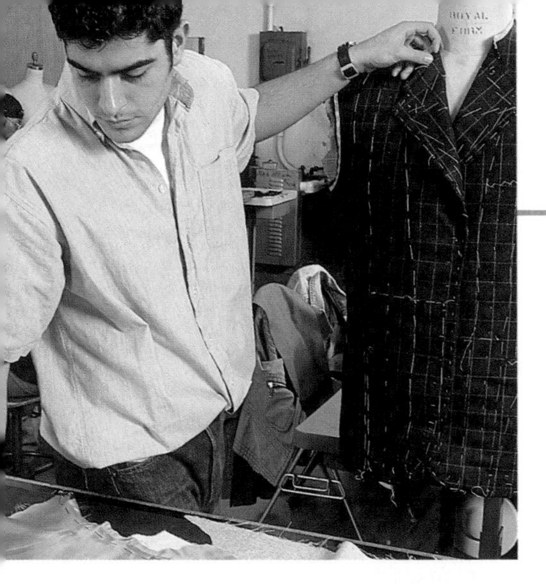

6 Apparel Production

Fashion Terms

profit
contractor
merchandising
sample
samplings
stylist
ford
costing
computer-aided design (CAD)
marker
laser cutter
tailor system
piecework system
unit production system (UPS)
computer-aided manufacturing (CAM)
modular manufacturing
quotas
offshore production
domestic production
joint venture
computer imaging
computer-integrated manufacturing (CIM)
Quick Response (QR)
business-to-business ("B2B") e-commerce
agile manufacturing

After studying this chapter, you will be able to

- discuss the business aspects of apparel production.
- distinguish between the production processes of ready-to-wear clothing.
- describe selling methods used by apparel manufacturers.
- compare various viewpoints about overseas manufacturing.
- discuss possible future directions of apparel production.

The factory production of garments includes all the steps needed to turn fabrics into finished wearing apparel. These steps are the buying of fabric and the designing, pattern making, cutting, sewing, assembling, and distribution of garments. Ready-to-wear garment manufacturers produce apparel in large quantities according to projected amounts of consumer demand within given price ranges. Consumer whims present high risks. Most companies have some very good years and also some lean years.

The Business of Clothing Production

Clothing manufacturers must struggle to keep their costs low, quality high, and selling prices competitive. Manufacturers must buy fabrics and pay their employees' wages. They must buy new equipment and pay for the upkeep of old equipment. They have overhead expenses, such as gas and electricity, for their factories and offices. They must pay rent or a mortgage, as well as taxes.

Manufacturers try to sell products at or lower than their competitors' prices, while maintaining their standards of quality and design. However, the companies must be able to make a profit to stay in business. ***Profit*** is the incoming money that is left over after all the outgoing costs have been deducted. Firms use their profits for expansion and improvements. The money is used to buy new equipment, hire more people, conduct research, and pay dividends if the firms have stockholders.

Apparel producers vary widely as to size, variety of products, and type of business operation. The largest firms are publicly owned by stockholders. They produce many lines under different labels. Smaller garment manufacturing companies are often privately owned. They usually specialize in one or two apparel categories in a particular size range. They also stay within a certain price level.

Over the years, apparel manufacturing firms have expanded and diversified by adding or acquiring subsidiary divisions. Some small manufacturers were bought to become divisions of large conglomerates, or parent firms. Many are global companies with production plants located in low-wage countries around the world and sales offices in major cities. However, there are still many prosperous small businesses run by enterprising, creative individuals.

Inside and Outside Shops

Firms that do all stages of production, from fabric purchasing to the distribution of finished garments, are called *inside shops.* Firms that handle everything but the sewing, and sometimes the cutting, are *outside shops.* They contract those parts of production to other, independently-owned sewing factories called ***contractors***. Contractors produce the garments according to the manufacturing firm's designs and specifications.

The contracting process has been done since women were hired to do sewing in their homes over a century ago. It places the burden of production peaks and valleys on the hired factory. The hired factory must also deal with labor costs and maintenance of the sewing machines. A contractor may do work for several different companies, or it may rely on just one company to keep it busy.

The contracting system enables a variety of products to be made by the original manufacturing firm. An outside shop may contract many different sewing factories at the height of its production season. The firm can meet changing consumer demands without making large financial investments in new equipment. Also, newcomers to the industry who have design ideas can go into production quickly without much capital.

Inside shops even contract-out some of their sewing for extra capacity during busy times. The photo in 6-1 shows monogramming being done.

Establishing Merchandising Plans

Merchandising is the process through which products are designed, developed, and promoted to the point of sale. It involves varying degrees of planning, buying, advertising, and selling. It is aimed at the final consumer of the goods, even though retailers are the next step in the chain, or "pipeline," after apparel manufacturing. Consumer needs and demands must be determined in order to provide retailers with the right goods. The proper amounts of supplies must be acquired to do the manufacturing. Production must be planned, so the finished garments can be delivered at the right price and at the proper time.

Clothing producers develop merchandising plans to help them estimate and meet these needs. The planning involves many steps. In small companies, this is done by the president or the designer, or both. In large

Lands' End Direct Merchants

6-1 Monogramming initials onto apparel is a specialized job done by trained operators who use machines made specifically for the task.

firms, a merchandising staff carries out these duties.

First, companies look at their primary consumer markets to see what kinds of clothes their clients prefer. They study past sales. They analyze the "winners" and "losers" of previous seasons. This has become easier with the use of computers. Sales by style number, color, fabric, size, and so on can be calculated quickly and accurately. Sudden or gradual changes in customer preferences can be noted. Sometimes changes show up in one part of the country and not in other parts. Production is planned accordingly.

The merchandising staffs of manufacturing firms do research to decide what directions fashions will take. To do this, they note the trends of couture fashions. They study the current arts, including movies, television, and museum showings. They subscribe to fashion forecasting services. They even buy "hot" items from retail stores—to copy!

Successful clothing manufacturers seem to have a "sixth sense" for judging fashion trends. Although the element of "gut feel" is unscientific, it is very important in fashion. Occupational guessing is a game that puts millions of dollars at stake.

Calculations and Decisions

Merchandising people must also figure out manufacturing details, as in 6-2. They must decide when and where to order fabrics, and at what price. Production capabilities are calculated for inside shop factories. Outside shop firms find out where their manufacturing can be done by contractors. Some of the work might be done in Kentucky, Alabama, Hong Kong, Korea, Mexico, or other places.

With this information and the ideas from the firm's designers, the preliminary lines are developed. The *lines* are collections of styles and designs that will be produced and sold as the firm's new selections for a given season. Each line contains a group of garments that are somewhat similar. The garments are usually all for the same time of day or type of occasion. Sometimes they are color coordinated. Large manufacturers might have several related lines. For instance, they could produce ski, tennis, sweat suit, and swimwear lines.

Textile/Clothing Technology Corporation

6-2 Training is being given at an apparel industry education facility, so industry employees can learn how to do effective manufacturing planning.

After the lines are established, samples are made of the chosen designs. ***Samples*** are trial garments made up exactly as they are intended to look when sold.

The next step in merchandising is communicating with the firm's key retail accounts. All samples are shown. The retail buyers respond with specific remarks about strong and weak points in the line. After these meetings, designs may be altered and manufacturing details may be changed. The amounts of each style to be manufactured are decided, as well as what items should be dropped from the line.

Finally, merchandising people must decide when, where, and how to show the lines. Should they be shown in several cities? Should the showings be big, splashy events, or quiet and small? How many trade showings should there be? Should road salespeople take the lines around to show retail buyers at stores? What advertising should be done?

Merchandising people keep tabs on garment acceptance all the way to the final consumers who buy from retail stores. They try to reduce risks with market surveys, computer analysis, and samplings. ***Samplings*** (not sample garments) are small quantities of garments that are made up and placed in retail stores. At the first indication of consumer reaction, acceptable styles are reordered by the stores and produced in large quantities by the manufacturer. Unpopular styles are discontinued. Consumer reaction is the most important input.

The Designing Process

One or many apparel designers may work for a garment manufacturing firm, depending on its size. They work one year ahead of when the apparel will be bought by consumers. They coordinate their ideas with the merchandising plans of the company. They use computers to experiment with design ideas, as shown in 6-3. They can print out their designs as color sketches. Expensive dress designs might be draped over a mannequin (dressmaker's form).

When one line is ready for production, designing plans are started for the next season. The styles that are selling well in one season are modified and repeated for the next season. Sometimes the same design is done in cotton with short sleeves for summer and in a wool blend with long sleeves for winter. Some new garment designs are added to each new line.

Sources of Inspiration

Even though ready-to-wear designers borrow ideas from others, the upscale lines also include new ideas of their own to move fashion forward. Designers use many different sources for inspiration. Forecasting services are very influential. They even offer private design work and consultations, for a fee, to manufacturing firms. Historic revivals and art movements influence fashion designs. Nature contributes to some designs. A popular movie or television series can cause great consumer demand for certain fashions. The design of the shorts in 6-4 is based on military wear from the past.

News events create interest in clothing. For example, when Alaska and Hawaii became states in the mid 1900s, Eskimo jackets and muumuus became fashion trends. When the United States resumed relations with China, Mandarin collars became popular. Wars often create a military influence in fashions. New living patterns also influence fashion. For instance, the fitness craze has brought about the popularity of exercise attire.

Royalty and prestigious people sometimes set fashion trends. A rock group might wear something that starts a design idea. "Fashion forward" young people on the streets sometimes invent their own extreme looks that

Gerber Garment Technology, Inc.

6-3 Computer systems allow for enhanced creativity of design ideas, including the ability to change silhouettes, garment parts, colors, fabric patterns, and textures of fashion designs.

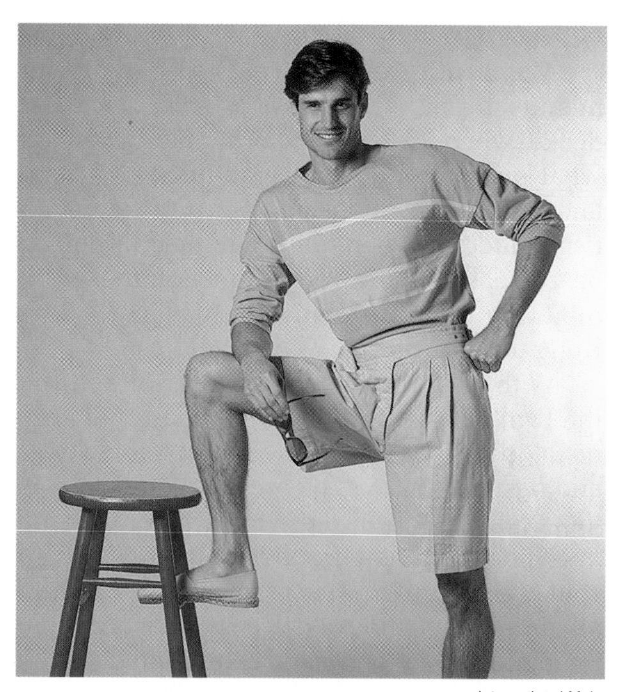

International Male

6-4 A rugby jersey (shirt) design is combined here with "Nepal gurkha" shorts, inspired by those worn by the British military in the past.

designers notice. Designers are keen observers of people and the way they put outfits together.

Designers also get inspiration from foreign and American fashion magazines and fashion shows. Fabrics from textile firms have a large influence. Trade shows enable designers to show their lines and to see the lines of other manufacturers. The designers also talk to store owners and buyers about what kinds of garments their retail customers would like.

Designers must have a feel for changing economic, social, and political conditions which affect fashion. Today's fashion designs must be functional. For sports, they must allow freedom of movement. A garment designed for jogging on the road at night must reflect light for safety. A jacket designed for sailing must repel water. These garments must also have a pleasing look for social acceptance. The fabric, design, and construction must all be in harmony.

Adapting Designs

The designing for the ready-to-wear garment industry includes much adapting of higher-priced fashions. Specific designs are simplified for less expensive production, or toned down so they will have widespread appeal. Clothing that reflects the latest trends is manufactured in all the price and size ranges to suit the current consumer market.

Manufacturers of low-priced garments employ fashion ***stylists*** instead of designers. The stylists redesign existing garments into knockoffs, rather than create new designs. They adapt current fashions to meet the mass production abilities and price ranges of their companies. This reduces the risk of producing unsuccessful designs, and it saves time and money.

Fashion piracy is considered to be a way of life to ready-to-wear designers. Copying dominates the industry. In the United States, there is no legal copyright protection against it. Apparel designs are not registered with the government.

There are practical reasons why copying has become so prevalent. Producers must get fast-selling lines out immediately, or the time span of popularity will be lost. Another reason is that firms are highly specialized in terms of apparel type, size, and price range. Many different firms must produce versions of a design if that design concept is to reach all potential customers.

Fashion pirating is so common that the industry has a name for such garments. A style or design that is produced at the same time by many different manufacturers at many different prices is called a ***ford***.

To finalize a design, a stylist may make several different sketches of a garment until it looks just right and is suitable for production. The sketch is then shown to managers who suggest changes or approve it. They consider if it will sell in their market. They also do preliminary ***costing*** to figure the expenses of producing the design. They tabulate the price per yard of fabric and how much yardage will be needed, as in 6-5. They figure labor costs and production capabilities of the firm.

Contractors' costs are calculated for outside shops. If there will be foreign production, shipping costs must be included. Design and management personnel consider the costs of trimmings, such as embroidery and buttons. They calculate production costs for design details, such as pleats, tucks, yokes, pockets, and topstitching. Sometimes design details must be eliminated. Inexpensive copies of

Burlington Industries, Inc.

6-5 Many calculations can be done by computer to help with costing and figuring specifics for apparel production.

higher-priced garments have simplified style lines, fewer seams, no linings, and cheaper fabrics and trimmings.

Preparation for Production

The fashion designer or a pattern maker makes the first pattern, as in 6-6. Also, high-tech companies have computers that can print out the pattern from a sketch on the monitor screen. ***Computer-aided design (CAD)*** uses electronic equipment to create new designs for textiles, apparel, or other products. Hundreds of options are available on a computer. When they are combined just right, the computer produces the pattern.

CAD has significantly reduced the time and number of people needed to make a pattern and prepare it for cutting. Data is entered into the computer. A knowledgeable pattern maker develops the pattern on a monitor screen. The computer can be instructed by the designer or operator to move darts or to include into gathers, pleats, or flares. It can make design changes, add seam allowances, and make a hard copy of the final pattern pieces. Many previously time-consuming hand manipulations are done quickly and accurately with the CAD system.

People who are trained to do CAD are increasing. There is a growing demand for designers schooled in the technique, such as the work being done in 6-7. Universities and trade schools have put CAD instruction into their apparel curriculums. Companies realize it is a must if they are to stay competitive.

CAD is best for sportswear and children's clothes that are designed with flat patterns. It is also good for knitwear, such as sweaters. CAD equipment can be used to design the knitwear and then, in a matter of hours, program the knitting machines to produce them. The programming used to take a few weeks. The procedure is not as good for garments that are designed by draping, such as evening gowns made of soft, flowing fabrics.

Making and Using Samples

Sample garments are made by sample makers, or *sample hands*. These people are skilled in all sewing techniques. They make up all design samples (prototypes) in their firm's one standard size. The sample garments are made in fashion fabrics. The sample hands work closely with the designer to make any needed revisions.

Styling and fit are checked by putting the samples on a mannequin or a live model. Samples are revised until they are just right. Specifications are finalized, and tricky construction details are made clear. Management must approve the new styles. They recheck the costs, availability of materials, and profit potential. The approved samples are later used

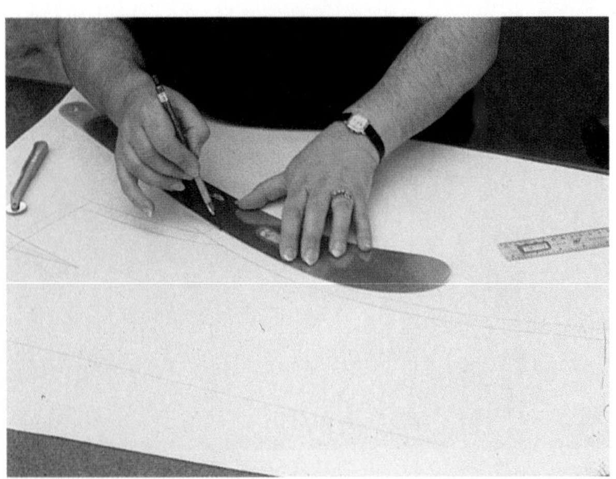

The McCall Pattern Company

6-6 This designer is making the pattern for a specific garment.

Gerber Garment Technology, Inc.

6-7 Computer-aided design (CAD), which enables patterns to be designed and graded quickly, is being used more and more in apparel production today.

as examples for production and for sales promotion.

Apparel firms have four main production seasons. They are usually called Spring, Summer, Fall I, and Fall II (Winter). The Winter line is often directed toward the holiday season including Christmas, Hanukkah, and New Year's celebrations. Sometimes a beach or resort wear line is considered to be a fifth production season.

Only a fraction of the designs first intended for a line are actually produced. Some are eliminated because the merchandising plan indicates they will not sell well. Other designs cost too much to produce in relation to the retail price range of the company's market.

The final garment selections are given style numbers. A master production pattern is made for each design in the firm's standard size, either by hand or computer. Final production details are arranged.

Preliminary Production Procedures

The master pattern in the firm's standard size is *graded* into smaller and larger sizes. Most grading is done by computer, as in 6-8. Computer grading is especially good for producers who use similar styles over and over again. The computer can then be programmed to grade that type of garment expertly. Grading that used to take three days by hand can be done in about an hour by computer.

Final fabric selections are ordered from textile firms whose quality can be trusted and whose prices are reasonable. The fabric suppliers must be able to deliver the goods on time. They must also be able to supply more fabric at a later time if production is increased.

Fasteners, thread, and trims are ordered in volume for the chosen designs. Special sewing machine attachments are sometimes ordered or made for new design processes.

The layout of pattern pieces is drawn to become a marker. A ***marker*** is a long piece of paper that lays out all of the various pieces and sizes of the pattern for cutting. It is the plan of how every pattern piece will economically fit onto the fabric. It is used as a guide for the cutter. It has to be planned for the most efficient cutting of garment parts. No fabric should be wasted.

High-tech manufacturing firms use computers to determine the best layout, as in 6-9. Miniature pattern pieces and their layout arrangement are displayed on the computer screen. The layout is calculated for the fabric width. The program can determine where the end of one bolt and the beginning of the next falls. It can match plaids and stripes, and it avoids flaws (fabric defects). It minimizes waste.

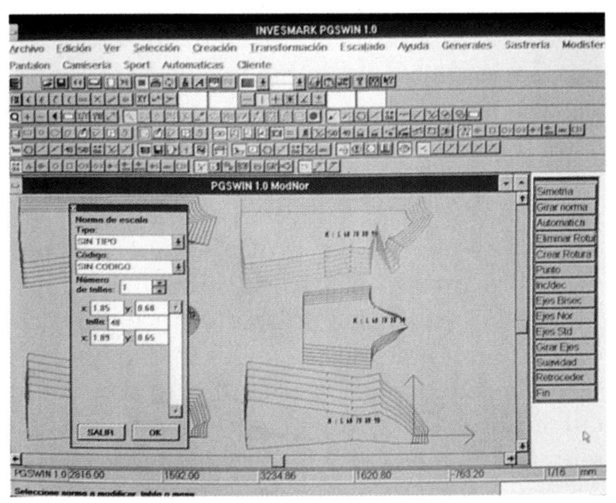

Investronica Systems, Inc.

6-8 The different lines along the edges of these pattern pieces show the edges of the various smaller to larger sizes in which the garment will be produced.

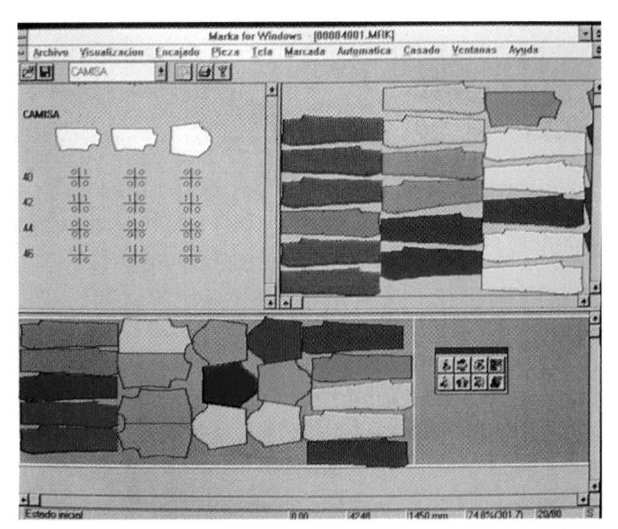

Investronica Systems, Inc.

6-9 Specialized computer systems, in multiple languages to be used globally, allow those trained to use them to judge marker efficiencies and reduce fabric waste.

When the best layout is established, the computer prints out the full-size marker. The pattern number and size is on each drawn pattern piece. Some systems have a special telecommunication feature. It allows a company to do the grading and marking at its main location and instantaneously send markers to a manufacturing operation elsewhere.

Smaller or less modern manufacturers make their markers by hand. This is a slow process. Pattern makers and graders who work for firms that are modernizing with computers must be retrained or replaced. There are also computer pattern firms that can be hired to make the patterns and the markers for a manufacturer's designs.

Factory Production

Apparel manufacturing is done throughout the United States and around the world. The cutting and sewing of wearing apparel begins about six months before the clothes are in retail stores. Usually large numbers of orders are taken by the apparel firm before production of a garment begins. This assures that a market exists for what is made.

The garments are mass-produced. They are manufactured and sold by dozens. They go step-by-step through the assembly line. Production costs are high. Every unnecessary procedure is eliminated in an effort to keep production costs down.

Apparel production labor costs are based on companies' wholesale prices. Companies that produce cheaper, lower-quality garments pay lower wages and require less skill than upscale garment manufacturers. Although most apparel manufacturers run good operations, there is an ongoing campaign against sweatshops in the U.S. and abroad. Sweatshops are manufacturing plants that pay less than the required minimum wage, use child labor, do not recognize overtime worked, have unclean conditions, or violate workers' rights in other ways. It is believed that solving this problem requires cooperation and coordination throughout the fashion industry. Government crackdowns as well as consumer awareness and response are also needed.

The U.S. government raised the conditions of manufacturing workers many decades ago by setting labor standards. Unions have won higher pay and better workers' rights. The public must also understand that these improvements create higher apparel manufacturing expenses that have, in turn, increased consumer prices.

Cutting

To begin production, cutters unroll layers of flat fabric into high stacks. They control large machines, called spreaders, to move bolts of yard goods back and forth on very long tables. The stacks are quite high. They might have 100 layers of fabric. The long paper marker is placed across the top.

Electric straight-knife cutting machines, like large power saws, are manipulated by hand along the outlines of the pattern pieces of the marker. The number of garments cut at one time varies, depending on the thickness of the fabric, the cutter's skill, the price of the garments, and the number of orders. The entire thickness of the stack is cut at once. This results in fabric stacks of garment parts.

Computers are now used to control cutting. Computerized knife cutters, like manual cutters, cut multiple layers of fabric. Some even sharpen themselves automatically. Water-jet cutters cut smaller stacks of fabric layers. Laser cutting is done on a single layer of fabric.

A ***laser cutter*** is a device that generates an intense, powerful beam of light. It vaporizes the fabric almost instantaneously. Laser-beam fabric cutters cut one garment a piece at a time. They are economical because they are very fast and accurate. They can cut as many garments per unit of time as a human operator using a manually controlled multiple-layer cutter. They are precise at cutting intricate shapes. They offer great style flexibility since they can cut 100 or more different styles in one hour.

Computers preset the cutters to accurately cut around the garment parts automatically, eliminating the need for a marker. See 6-10. The shapes and sizes of the garment parts are programmed into the computer.

The Tailor System

The ***tailor system*** is a manufacturing system in which all sewing tasks for a garment are done by one person. Today it is only used for high-priced, custom sewing jobs or for structured, tailored suits and coats, as in 6-11.

The tailor system is hardly used commercially in the industry because it is slow and expensive. The person sewing the garment must be skilled in all sewing and tailoring tasks. Much handwork and extensive pressing are done.

Faster methods that involve less handwork have been developed for suits. They include more built-in shaping from the pattern, rather than from hand sewing or steam pressing. Also, some garment parts are fused together. Unstructured styles are offered in sizes such as small, medium, and large, which are less precise than regular sizing. "Section" construction is sometimes used by manufacturers, in which sections of a garment are made by operators rather than one operator sewing an entire garment.

The Piecework System

The ***piecework system*** is a manufacturing procedure in which each specific task is done by a different person along an assembly line. This divides the total manufacturing process into small, individualized jobs. Every sewing machine operator does only one job on a specialized machine, as shown in 6-12. This procedure attempts to make production time per item as short as possible while the work moves along.

Piecework has been done by the progressive bundle system since mass production began in the industry. In this system, the cut garment parts are packaged into bundles of dozens to go through the sewing operations. Work tickets are put on the bundles. The bundles are put into canvas bins with wheels called "handling trucks."

The handling trucks are rolled to the different production stations in the factory.

Fashion Institute of Technology

6-11 The tailor system, in which a person with expert sewing skills makes the entire garment, is sometimes used for expensive suits and coats.

Investronica Systems, Inc.

6-10 This fabric does not need to have a marker placed on top of it because the cutting instructions are programmed into the computerized system of the cutter.

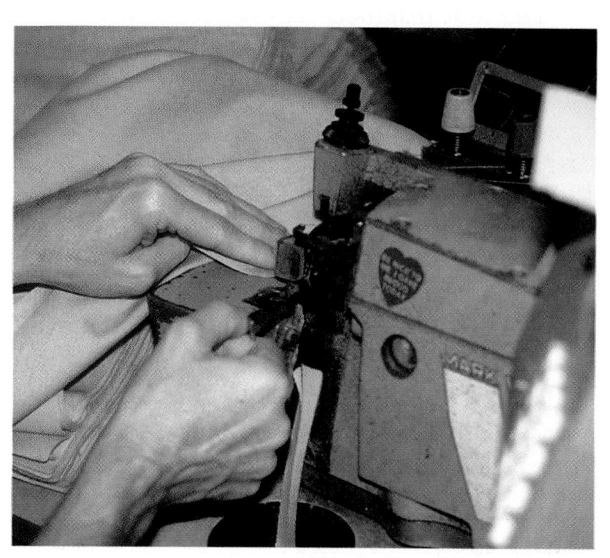

National Cotton Council

6-12 Machine operators use specialized sewing machines to do specific tasks along an assembly line.

Bundles of collars are delivered to one station; sleeves to another; fronts and backs to still other stations. The bundles are moved through the assembly line. Pockets are put on by one person; buttonholes are made by another worker. Further down the line, the parts come together and are joined by other workers.

The most efficient use of time and people on the assembly line is figured out ahead of time by engineers. Workers are paid according to a rate that is determined by the difficulty and time required to do their specific sewing tasks. This rate is multiplied by how many times they accomplish the task, such as how many collars or sleeve cuffs they assemble.

The goods are inspected periodically during the production process. Almost no hand finishing jobs are done. Hems and fasteners are put in by machine during assembly. Any hand sewing that is required falls under the heading of trimming. It is done at the end by workers who specialize in such tasks.

The ***unit production system (UPS)*** is a computerized piecework arrangement. The cut pieces of each garment are hung (loaded) together onto an overhead product carrier, which is moved through the manufacturing steps by a conveyor system. Time is saved since bundles do not need to be untied, retied, ticketed, or manually moved along as work is done. Also, operators can better see how the garments look overall.

The unit production system is a method of ***computer-aided manufacturing (CAM)***, which utilizes electronics for the production of apparel. Each workstation has a computer terminal. When a task has been completed, the operator presses the "send" button on the workstation terminal or scans a bar code on the garment holder. This directs the carrier to automatically take the items to the next station.

With CAM, work is routed in such a way that the sewing line is continuously and automatically balanced. Garment parts are fed to the workstations with full consideration of the skill and speed of each operator. Each terminal feeds information into the main system. The computer tracks all inventory and piece rate data for better production planning. Supervisors can continually monitor the production line with a central computer, as in 6-13. They can make changes when needed. They can also check on operator efficiency, payroll information, operating costs, inventory control, and style data.

The Latest Production Methods

Modular manufacturing is the latest method being used in apparel production. This method is mostly used by large manufacturers who can afford to implement it and train employees to use it. The modular system divides the production employees into independent teams, or module work groups. The module members choose a leader and hold weekly meetings (usually one hour long) on company time to sort out problems and agree on their own work assignments and schedules.

Modular manufacturing has improved the flexibility and productivity of apparel manufacturing. *Productivity* is a measure of how efficiently or effectively resources, such as labor supply, machines, and materials, are used. Improved productivity results in higher amounts of output in relation to the amounts of input.

Modular manufacturing also improves product quality and employee morale, while reducing union grievances and production throughput time. Peer pressure reduces absenteeism and encourages nonperformers to leave

Gerber Garment Technology, Inc.

6-13 This computer allows a supervisor to analyze all work-in-process data and make production decisions as the sewing work moves from station to station.

on their own. Companies share the resulting profits with the operators through higher pay. Individual piece rate pay is replaced with pay based on total module production.

The modular manufacturing system "empowers" the workers. In other words, it gives them the authority and autonomy to make the decisions needed to ensure the highest group productivity. For this to be successful, top managers and production supervisors must be willing to trust their employees will meet the manufacturing challenges.

Emphasis is also now being placed on ergonomics and benchmarking. *Ergonomics* matches human performance to the tasks being done, the equipment used, and the environment. It is also called human engineering. To prevent repetitive motion injuries, occupational health problems, and worker fatigue, equipment is designed and arranged for the most effective and safest interaction with those who use it. For instance, the height of sewing machine stands are adjustable, worktables can tilt, and footrests have been added.

Benchmarking is the continuous process of measuring a company's products, services, and practices against other companies' extremely good products, services, and practices. The best ideas of the apparel industry and related industries are identified and improved upon. For example, one company might have an innovative method for attaching sleeves. Other companies notice that "benchmark" and adopt it. Each company strives to do the best job in each operation, so the total manufacturing process is as efficient as possible. Also, industry training opportunities are available to workers who want to learn the latest techniques, as in 6-14.

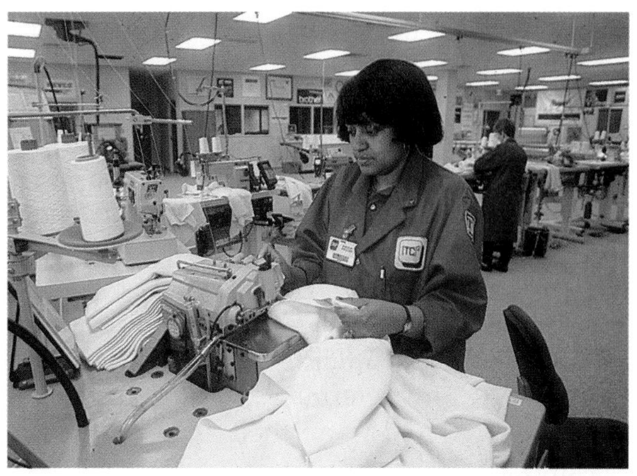

Textile/Clothing Technology Corporation

6-14 Training in the latest apparel manufacturing techniques is offered at universities and industry technology centers such as this one.

Finishing

After manufacturing, finished garments are inspected inside and out. If flaws or mistakes are found, and if they can be fixed, the garment is returned to the operator responsible. In some cases, the inspector may decide it would take too much time or effort to fix the problem. Then the garment is sold cheaply as a second.

To finish production, pressers give a complete final steam pressing to the garments. Labels or hangtags listing fiber content and care instructions are attached to all garments. If a large order is being made for a particular store, the factory may attach the store's price tags. The garments are finally ready for distribution to the stores that ordered them.

To ship garments to other cities and states, folded garments are packed flat in corrugated containers. Hanging garments are hung on bars in stand-up boxes. Some items have special packaging, as in 6-15. Then the cartons of garments or accessories are sealed and addressed to retail stores for shipping.

For apparel manufactured in the New York City garment district, rolling hang-up racks, as in 6-16, are used to move the clothes. The racks are rolled along the streets from factories to shipping rooms. Sometimes the racks go directly to nearby department stores that are desperate to receive their orders.

Selling the Apparel

Sales promotion by apparel producers is far ahead of the wearing season. Apparel companies present their new lines to retail store buyers by showing them sample garments about six months before the merchandise will be bought by consumers.

Because of the speed needed to get fashions into the stores, most apparel goes right from producer to retailer. The wholesale "jobbers" of past eras, who distributed manufacturer's goods to retailers, have almost disappeared. Modern transportation systems move merchandise to stores in fast and dependable ways. This direct-shipment system also helps to keep prices as low as possible since no middle people have to be paid.

Showing the Lines

During the period known to the trade as "market weeks," retail store buyers from all over the country go to fashion market centers. They look at the new collections in the manufacturers' showrooms. Some manufacturers show their collections accessorized and professionally modeled in elaborate fashion shows. Others simply hang their garments on racks in their showrooms and show them to individual retail buyers, as in 6-17. Some firms hold press previews for fashion reporters and editors in hopes of getting publicity. Some produce videos of their lines.

Wemco Fashion Apparel

6-15 Packaging and shipping is an important aspect at the end of apparel production.

6-16 In the garment district of New York City, racks of garments are rolled from place to place.

Seattle International Trade Center

6-17 Retail buyers visit apparel manufacturers' showrooms to look at lines from which they might place orders.

All companies hope that buyers will place orders for their retail stores before they leave the showrooms. As orders are taken, production is scheduled. Many manufacturers start producing some of the goods before the openings. They rely on their merchandising projections to estimate initial market demand.

Some companies show their lines in Dallas before New York City. Dallas seems to be representative of the whole country. It gives feedback about the lines. Styles can then be altered so a stronger statement can be made in New York a week later. Also, regional trade shows are usually held in such cities as Atlanta, Chicago, Los Angeles, Seattle, and Miami.

Apparel marts are buildings or complexes that house permanent showrooms. They offer apparel manufacturers the opportunity to be in one convenient location, so buyers can see the lines at any time. Marts are usually departmentalized to concentrate the same type of merchandise, such as children's wear, in a particular area. Apparel marts are located in several cities. The Chicago Apparel Center and the Dallas Apparel Mart are shown in 6-18 and 6-19.

Chicago Apparel Center

6-18 Apparel marts are located in major fashion areas of the United States.

In addition to their periodic new lines, firms also introduce new styles as a season progresses. Highly-accepted styles may be made in a wider assortment of fabrics or colors than originally shown. New fads may come into demand and be produced. Unpopular styles are dropped from the line completely.

Besides having selling staffs in their showrooms, some firms have sales representatives ("reps") showing and selling their lines around the country. The manufacturers show the new styles to their sales representatives at fashion shows and sales meetings. The reps then travel within their assigned territories with sample lines to show to retailers. Sometimes a certain label of goods from a manufacturer is "confined." That means it is sold to only one retailer in a certain trading area on an exclusive basis.

For small manufacturers without their own sales reps, there are independent sales reps. They maintain permanent showrooms in different parts of the country. They sell collections of several different manufacturers' lines that are not in competition with one another.

Promoting the Lines

Manufacturers advertise in trade publications to bring their names and products to the attention of retailers. These ads are less expensive than ads in consumer publications. Some apparel producers do *cooperative advertising*

International Apparel Mart, Dallas

6-19 This six-story apparel mart covers four city blocks on a 20 acre site with about 2,000 showrooms under one roof.

with retail stores. They might advertise their merchandise in national fashion magazines along with a list of the regional stores that carry their lines. Other times, retailers advertise in local newspapers and promote a particular manufacturer's products. Either way, the businesses share the advertising expenses and both benefit.

Manufacturers sometimes provide retailers with selling aids. Large photographs for store displays, newspaper advertising mats, and customer mailing pieces are examples. They also have in-store programs for the retailers who sell their lines. They provide training talks to salespeople, personal appearances by the firm's designer, fashion shows, as in 6-20, and trunk shows. During a *trunk show*, a complete collection of samples is brought into a store for a limited amount of time. There is heavy local advertising. Orders are taken directly from customers by a key company salesperson.

Manufacturers' promotions in stores provide consumer feedback. It is important for companies to listen and respond to constructive criticism. If orders are not coming in as well as expected, or fall off, there must be a cutback in production. Manufacturing has to be stopped as early as possible to keep costs down. On the other hand, production must be increased for popular items. Reorders provide the best chances for making big profits since all the designing and preliminary details have already been done.

6-20 Apparel producers sometimes present fashion shows in stores that carry their lines in order to increase customer interest and retail sales.

Overseas Manufacturing

Much manufacturing competition is coming from outside the United States. Taiwan, Hong Kong, and South Korea are major apparel producing locations. The quality of these goods has steadily improved. Other developing countries will become clothing manufacturing centers in time. Mainland China has great potential. Caribbean and Latin American countries are gaining in production, as shown in 6-21. Also, developing countries in the Middle East, Africa, and other parts of Asia have increased apparel manufacturing.

Imports of goods that come into the country from foreign sources have risen because of the intense competition in the apparel industry. In some cases, imports fill certain voids in our domestic industry for small, unusual, or low-priced items.

6-21 These marketing materials are promoting the apparel manufacturing facilities and expertise of a Latin American country to fashion companies in other parts of the world.

Differences in Cost, Fashion, and Construction

The first void in American-made apparel is created by the high costs of production in the U.S. Manufacturers worldwide compete for price advantages. Low-wage countries can mass-produce garments at a low cost. This allows them to sell their garments at lower prices than comparable items produced in the U.S. Thus, they have a market in the U.S.

The lower labor rates overseas include no overtime pay or *fringe benefits*, such as vacation time or health insurance. Foreign government export incentives, to encourage sending products out of the country, also reduce companies' production costs in those countries. Examples of export incentives are tax exemptions, rebates, and preferential financing plans offered by governments to producers who export goods. Conversely, U.S. manufacturers must comply with government regulations regarding such matters as safety and benefits for workers.

The savings from foreign production more than compensates for the shipping costs of bringing the goods to the United States. The savings are great even after tariffs, or import duties, are paid to our government when the merchandise enters our country to be sold. Most goods travel across the oceans via container shipping freighters.

The second void in our domestic apparel industry is a lack of fashion innovation. Firms in the United States tend to mass-produce "safe" styles that are sure to sell. Foreign sources, on the other hand, often produce small amounts of more adventurous garments. Producers worldwide compete for new and different fashions. European fashions, for instance, are often innovative and of fine quality. They have appeal just because of their European origin. The name of the designer or firm on such imported apparel is a drawing card for American demand, even though the price may be higher.

The third void in our industry is in the area of handwork. Hand embroidery, beading, and knitting are examples. Our highly paid labor force cannot spend time on such tasks. Our industry is most profitable with large scale, fast, machine production.

Competition of International Trade

Clothing imports and exports have become major economic issues in international trade. In the United States, imports have been steadily and rapidly increasing for many years.

Apparel production is a labor-intensive industry. Since many people are needed, wage rates make a big difference in production costs. There is little difference in the amount of time it takes to make apparel here or abroad. It takes about the same number of minutes to make a shirt in Tennessee as it does in Haiti.

Apparel manufacturers in the United States are at a disadvantage when competing for international trade against low-wage countries. Textile and apparel industries are among the first to be set up in developing countries. Garment production requires what they have: plenty of cheap labor and low capital expenses. Clothing factories can be started in places where technology and money are limited.

In recent years, the importance of the apparel industry, in relation to overall manufacturing in the United States, has declined. Its share of the nation's total industrial activity has fallen. The number of people working in the industry has been reduced. Experts cite imports as the major cause for these economic facts.

The textile and apparel industries have fought for varying degrees of government protection to control the level of imports. The United States has some agreements with countries to limit imports by a quota system. ***Quotas*** are limitations established by the government on certain goods that can enter the country during a particular time span. They have been established on categories of products, such as specific fibers or types of finished garments.

The government, however, has to be careful about restricting trade. Curbing imports of fashion merchandise may bring retaliation with other countries refusing to accept goods we want to export to them. The government must look at the total picture of trade. International diplomacy, which involves the balancing of political and economic issues, is important in trade negotiations. Borders of most countries are becoming more open, encouraging trade.

Using International Trade

The apparel industry in the United States sometimes *uses* cheap foreign labor instead of competing against it. Some American firms now have their own production facilities in developing countries. However, most contract with locally owned companies in those countries to manufacture their garments. Large firms may have an American supervisor stationed in a foreign country. This person keeps tabs on all of the contract production in that country or in that particular area of the world. The production for an American apparel firm may be spread among several different countries. All garments cannot be produced in the same country because of production capabilities and U.S. import quotas or other regulations.

Sometimes garments that are cut in America are shipped overseas for sewing and finishing. Then they are shipped back. In this procedure, only a minimum duty is paid on the value that has been added while the merchandise was out of the United States.

Manufacturers who have their goods produced overseas (called ***offshore production***) are constantly looking for countries with lower wage rates, better innovations, and unrestricted quotas. Today, the fact that a garment has the name of an American designer or firm is no assurance that the item is American made (called ***domestic production***).

American apparel producers do export some goods. However, they have not been able to match their exports to the rising tide of imports. Thus, the deficit in the balance of trade has continued to grow.

Instead of exporting goods from the United States, some of our major apparel producers have set up licensing agreements with foreign producers. Apparel is produced overseas with an American firm's label and specifications. It is sold overseas without coming to the United States. In return, the firm receives a percentage of sales.

Another method is with a ***joint venture*** arrangement. That is when two firms form a partnership for combined advantages. For international trade, an American firm and a foreign producer enter into a partnership for production and sales overseas. The American company provides the designs, patterns, technical expertise, and use of its brand name. The foreign producer employs the labor to produce and market the goods in that country.

Major American apparel producers have licensing agreements or joint ventures in many countries. Their goods are made and sold worldwide. Some even have wholly owned manufacturing plants in foreign countries.

No Clear-Cut Answers

There are no clear-cut or easy answers in dealing with apparel imports. *Economic* views differ. Some people feel there should be stronger government controls to try to help the apparel industries. Others feel that the industries must make themselves competitive in world markets or else dissolve. They say it is a fact that developing countries can produce textiles and apparel more cheaply than the United States. They also say that many of the developing countries critically need apparel industries. Therefore, they believe that the United States should be moving into industries where it would not compete with developing countries. We could be more competitive in world markets, they predict, if we would focus on industries that require more capital and technology.

Political views question whether import protection of the apparel segment would alienate other countries that trade with us. Textile and apparel manufacturing companies and unions have campaigned against imports to save their jobs. They want government help.

Humanistic viewpoints compare the jobs of apparel workers in this country with those in developing countries. Without apparel jobs, many people in developing countries would have no way of obtaining the basic necessities of life. Unemployment in our apparel industry has sad consequences as well. Hard times spread to nearby stores and businesses and whole communities. Government expenses increase for the relief and welfare of many people. Also, the pride and self-respect of the unemployed workers are lowered.

Country of Origin Labeling

Legislation in the mid-1980s required textile and apparel products to have labels indicating where they were made. "Made in the USA" labels are placed in domestically produced apparel.

At the same time, the apparel industries unified themselves by forming the Crafted with Pride in U.S.A. Council. They conducted nationwide efforts to bring the industries' message to the American people. The council, made up of every segment of the textile and apparel industries, sponsored public awareness campaigns. It spread the "Buy American" message through television and other media. It also encouraged the use of hangtags and labels such as those in 6-22.

Ongoing Innovation for Apparel Production

In the future, the trend of apparel imports is expected to continue, but the pace of increase should slow down. To compete, American production facilities will become even more automated. They will use more computerized pattern designing, laser-beam cutting, sonic joining, and robotic sewing.

High-tech equipment and the competition of imports will reduce the number of workers in the industry. However, the economic hardships of the resulting unemployment should eventually subside. Some present workers will retire, and others will be retrained to operate computerized equipment. Fewer new workers will seek jobs in apparel production.

Firms from other countries, such as Japan, may set up apparel production operations in the United States. On the other hand, American firms will also continue to invest in overseas production facilities. The apparel industry's scope is becoming totally worldwide.

Crafted with Pride In U.S.A. Council, Inc.

6-22 American textile and apparel industries hope that labels such as these will encourage consumers to notice the country of origin and buy domestically produced apparel.

Computer Automation

High-tech computer automation is revolutionizing the entire apparel production industry. Computer-aided design enables ***computer imaging*** to be done on the screen in three dimensions. The graphics allow the sketch to turn to any angle so all sides of the fashion design can be seen. Fabric draping grids, as in 6-23, help with "fluid" designing. Then the computer automatically converts the three-dimensional work into a two-dimensional flat pattern.

Unlike machines that must be retooled to change, computer-controlled automation is flexible. The programming is changed on software rather than altering the machinery. This reduces setup costs and time. It allows smaller quantities of apparel to be produced economically. Forecasting mistakes can be fixed faster. Fewer risks are taken. Automated sewing, which can rapidly produce small quantities of garments in various styles, also allows for more innovative designing.

Robotic machines are being developed to do whole piecework operations with little or no human intervention. Computerized "workstation" machines first pick up a garment part. They align it, sew it, trim it, and take it to the next station. One specialized machine might place a pocket onto a garment, stitch around it, and pass it to the next machine.

Gerber Garment Technology, Inc.

6-23 Fabric draping can be shown on photos that are scanned into the computer, using a flexible three-dimensional grid.

The best present applications for robotics are in repetitive sewing procedures. A pillowcase can be totally made and packaged by a machine. A cowboy boot machine just has to be loaded with leather. Finished, decoratively stitched boots, as in 6-24, are unloaded.

Computer-integrated manufacturing (CIM) is revolutionizing the industry. It combines CAD, CAM, robotics, and company information systems. It is being called "hands-off production." When it is perfected in the future, planners foresee bolts of fabric entering one end of a robot assembly line and emerging at the other end of the line as finished garments. Fashion designers would create their designs on computers, and the machines would do the rest!

Reducing Lead and Response Times

Industry-wide cooperation has reduced *lead times* (for customers ordering ahead) and *response times* (for manufacturers' production and delivery). All segments of the apparel industries have joined forces to make the whole supply chain work better. The program is called ***Quick Response (QR)***. It ties together the entire textile-apparel-retail pipeline as one unified industry rather than as individual segments.

In the Quick Response system, *electronic data interchange (EDI)* linkages are set up between textile, apparel, and retail concerns, as shown in 6-25. A textile firm sets up a computer system that measures, color codes, and inspects fabric as it comes off the loom. The information is transmitted to computers at the apparel factories before the bolts of cloth arrive. Thus, the sewing factory does not need to inspect, color code, or measure the fabric. The fabric moves directly from the delivery truck to the cutting room for swift processing without any repetition of tasks. The sewing plant then uses computers to inform retail stores about production and delivery of their orders. Invoicing, or billing for the materials sent, is also done automatically by computer.

Information also flows backward through the chain. As point-of-sale data are recorded through check-out computers at retail stores,

6-24 After looking at these boots, imagine all the specific procedures done robotically from the beginning to the end of manufacturing.

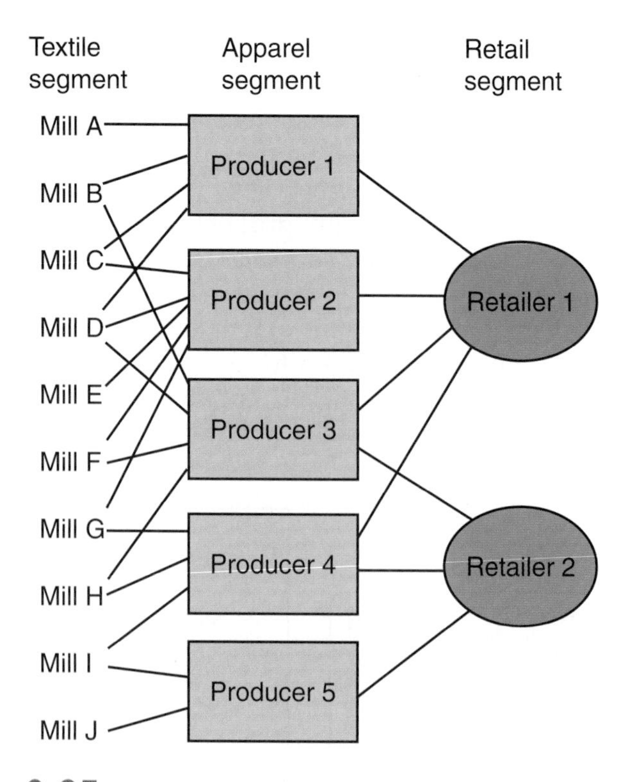

6-25 This simplified illustration shows long-term Quick Response partnerships that link businesses from each segment into chains of suppliers and buyers.

they trigger manufacturers to produce more of the items that are selling. This also prompts textile firms to automatically send more of the required fabric to the apparel manufacturers. Because products are automatically made to replenish those that have been sold, inventory levels can be kept lower.

As a result of Quick Response, communications are improved and processing costs are reduced. There is less redundancy of functions, which means a faster response to market trends than before. Therefore, the right merchandise is available for demand and costs are lower. Retailers have fewer missed sales and forced markdowns. Also, with long-term partnerships, suppliers maintain the promised quality and expected on-time delivery. This makes domestically produced apparel more competitive against imports. The domestic apparel manufacturing process that has normally taken one year (longer for offshore production) is now much shorter, as well as more accurate.

Quick Response is being used more each year. It is expensive initially, but pays for itself with savings in just a few years. The fear of sharing information electronically with other companies is diminishing. Also, standardized software programs have been developed so all companies' computers can communicate with one another.

Other Industry-Wide Cooperation

The Textile/Clothing Technology Corporation, called (TC)^2, researches high-tech innovations in apparel production processes and helps the industry implement them. This results in better quality, service, and cost from American apparel sources than from their overseas competitors. By pooling industry resources, financial responsibility and leadership are shared. Research and development duplication is eliminated, since companies are not working independently from one another. Industry unity and strength are improved.

(TC)^2 has a "Tech Center" in Cary, North Carolina that offers education courses, demonstrations, and books and manuals. Its interactive video training media for mechanics and operators is shown in 6-26.

The Future of the Industry

Business-to-business ("B2B") e-commerce is the transacting of business between companies on the Internet. It is called e-commerce because it is done electronically (with computers). It will have a large influence on the future of the apparel industry.

B2B e-commerce has many advantages. It operates globally, with worldwide scope. It does not require set-up time because the Internet is already in use and available. Direct communications can be easily established between companies, rather than having to link computer systems into partnerships. Thus, e-commerce is replacing some EDI linkages. Also, time zone differences are insignificant since people can receive and send messages at their own convenience. Language differences are less of a barrier, since nations are teaching and using more English, which has become the universally accepted language of the Internet. Cultural differences are being minimized. Additionally, news or updates can be sent out as they occur so there is no "lost time" in business dealings. However, some security of information privacy may be compromised.

In the future, the industry will export more products. American products with the "Made in the U.S.A." label will be in greater demand. Companies will concentrate on individual market approaches for each country.

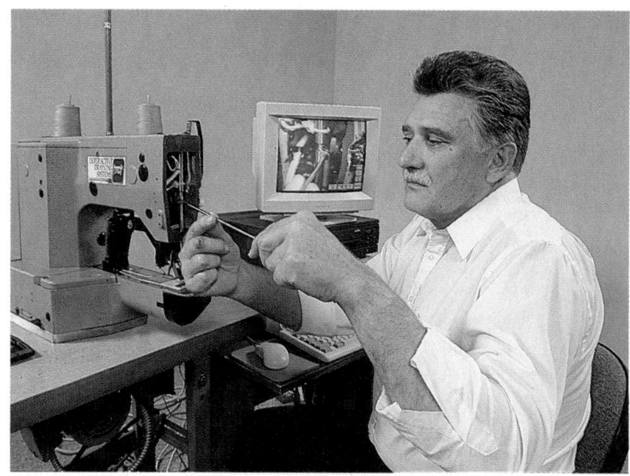

Textile/Clothing Technology Corporation

6-26 Interactive videos combine videos with computer text and graphics into portable "tutors" used to train industry mechanics and operators.

Apparel production will become a marketing rather than a manufacturing industry. The focus will be pleasing customers, not making and selling products. Company-owned facilities will be cost effective and use the latest technology. CAD work will interface with sourcing and quality specifiers that are on-line via company worldwide computer networks. Textile providers, product developers, manufacturing sites, and retailers will be networked. Unified tracking systems will tell at what stage and where products are at all times.

U.S. manufacturers will move into ***agile manufacturing***, a "seamless" data capture system of information, production, and delivery. The system will combine QR partnership alliances, EDI or e-commerce linkages, and wireless (radio frequency) transmission of consumer purchase data and trends. It will provide custom garments with design features selected by the consumer and to fit that consumer's body.

Agile manufacturing will use precise three-dimensional body scanning and interactive design stations. Customized production will remove forecasting risks while meeting specific needs of individual consumers. Live "fit sessions" will take place via computers. A garment will be cut and produced exactly for each person's dimensions. Production of the custom garments will use single-ply laser cutting as well as textile printing onto the garment parts. More seams will be fused together rather than sewn. This will be done by sonic (radio-wave, supersonic sound) joining. Another method, electronic injection sewing, is done with liquid chemical thread that solidifies. Also, technology may fabricate apparel automatically in the form of seamless garments. These garments might be formed, or molded, rather than sewn.

Another possibility for the future is apparel that might contain heat-sensitive body coils to heat or cool the wearer. Climate-controlled suits could be powered by solar cells. At any rate, all these theories provide directions for research to investigate!

Chapter Review

Summary

Clothing manufacturers try to keep their costs low, quality high, and selling prices competitive to make a profit. Companies may be inside shops that do all stages of production or outside shops that hire contractors.

Merchandising plans try to satisfy consumer needs and demands so retailers have the right goods at the right price at the right time. Manufacturing details and production capabilities are calculated to develop the apparel lines. Merchandising people keep tabs on consumer acceptance of each garment.

Garment designers do computer sketches or drape ideas onto mannequins. Their ideas are from forecasting services, history, art, nature, popular entertainment, news events, and other influences. Much adapting of higher priced fashions is done, simplifying designs for lower production costs.

Garment patterns are made by CAD equipment. Samples are sewn to check styling and fit. Production procedures, such as grading and marker-making, are done for the final garment selections.

Factory production in world locations depends on costs and quality. Cutting garment parts might be with hand saws or by computer-driven cutters. The piecework system is preferred by most manufacturers over the tailor system of manufacturing. Bundles or dozens of garment parts move among operators of specialized sewing machines. CAM systems are being used, and modular manufacturing incorporates ergonomics and benchmarking. Finally, finished garments are inspected and pressed.

To sell the apparel, manufacturers show their lines to retail store buyers during market weeks. This is done in company showrooms or apparel marts, or continually by sales representatives. Manufacturers promote their lines to retailers in trade publications, and may offer benefits of cooperative advertising, selling aids, and trunk shows.

Much labor-intensive apparel manufacturing is done overseas, where labor costs and fringe benefits are lower. There may be better fashion innovation overseas, since the U.S. mass-produces "safe" styles. Many American apparel producers are now using offshore production through licensing agreements and joint ventures. There are no clear answers about apparel imports because of varying economic, political, and humanistic viewpoints. Country of origin labeling is required on apparel.

In the future, apparel production will be even more global. High-tech computer automation will utilize robotics and CIM. Lead and response times will be reduced as all parts of the industry cooperate through "B2B" e-commerce for agile manufacturing.

Fashion in Review

Write your answers on a separate sheet of paper.

Matching: Match the following aspects of apparel production with the appropriate terms.

1. Piecework system.
2. Stylists.
3. Trunk shows.
4. Fords.
5. Grading.
6. Joint venture.
7. Market weeks.
8. Marker.
9. Low labor rates.
10. Tailor system.

A. designing
B. preparation for production
C. factory production
D. selling and promoting
E. overseas manufacturing

Short Answer: Write the correct answer to each of the following questions.

11. Name three reasons why a manufacturer might choose to function as an outside shop.
12. What are samplings, and what is their function?
13. Name five factors that are considered during the costing process.
14. True or false. CAD is especially useful for designing evening gowns and other garments of soft, flowing fabrics.

15. True or false. CAM is being used in more and more tailor system sewing shops.
16. True or false. Apparel marts are huge factories for apparel production.
17. Which provide better chances for making big profits, first-time orders or reorders? Why?
18. List three voids in our domestic apparel industries that are filled by imports.
19. Why are textile and apparel industries among the first to be set up in developing countries?
20. What is the goal of $(TC)^2$?

Fashion in Action

1. Collect three pictures of current garment designs you think were influenced by particular people or events. For each, describe what you think the influence was and how it is shown in the garment. Also, write a paragraph about some current events that could influence upcoming apparel designs.
2. Write a report on the production of a pair of jeans. List all the different steps you think took place to produce the jeans. What different skill levels were needed for the different jobs? What types of specialized machines or attachments do you think were needed?
3. Target a group of customers and develop an imaginary line of men's or women's sportswear. Trace fashion illustrations from newspaper ads or pattern catalogs to show your line of garments. Gather color and fabric swatches for your line from fabric stores, or make drawings of them. Mount them to make an attractive display. Describe why you think the styles, fabrics, and colors are good for your line.
4. Team up with other students to hold a class debate about apparel imports. Decide who will be pro-American and who will be pro-imports. Debate the economic, political, humanistic, and other areas of concern about the subject. Ask the class to vote on whether they favor more or fewer government regulations on trade.
5. Do further research related to recent technological advances that are affecting apparel production. Interview people who work in the industry, if possible. Present an oral report to the class.

7 Fashion Promotion and Retailing

Fashion Terms

indirect selling
direct selling
promotion
advertising
publicity
visual merchandising
video merchandising
channel of distribution
markup
markdown
open-to-buy
resident buying offices
private label
telecommunication retailing
offshore sourcing

After studying this chapter, you will be able to

- describe fashion promotion in terms of advertising, publicity, visual merchandising, and video merchandising.
- define many retail terms.
- distinguish between different types of apparel outlets.
- discuss the pros and cons of retail imports.
- describe possibilities for the future in retail sales of apparel.

Once textile and clothing items are designed and produced, they must be sold. This is done with a combination of indirect and direct selling. ***Indirect selling*** is promotion to the general public. ***Direct selling*** is the exchange of merchandise to individual consumers in return for money or credit.

Promotion

Promotion is indirect, or nonpersonal, selling. It is aimed at a large general audience. It tries to catch the public's eye by appealing to the needs and desires of people.

The purpose of fashion promotion is to make people interested in particular apparel products, so they will want to buy them. Promoters want people to feel as if they "have to have" the latest items. Fashion promotion includes advertising, publicity, visual merchandising, and video merchandising.

Advertising

Advertising is a paid promotional message by an identified sponsor. It appears in media such as newspapers, magazines, television, and radio. Local and regional retail stores use mostly local newspaper and radio ads. The ads are usually prepared by each store's advertising department. Nationwide retailers and textile and apparel manufacturers often use television and national magazine ads, which are more expensive than local newspaper and radio ads. The TV and magazine ads are generally prepared by advertising agencies.

Sometimes textile and apparel manufacturers share the costs of cooperative advertising with retailers. This is usually done on a local level. The retail store benefits by paying only part of the advertising costs. The manufacturer benefits by having its name or product linked with a known and respected store.

Publicity

Publicity is free promotion. It includes nonpaid messages to the public about a company's merchandise, activities, or services. An example might be a newspaper story about a store or its branches. Another example might be a fashion magazine's layout of photographs taken at a manufacturer's collection showing, such as in 7-1.

Public relations departments of businesses try to get publicity through various media. Sometimes individuals promote their firms during speeches to live audiences. Sometimes they hold press conferences to promote their products or services.

Press kits are distributed by some manufacturers. The kits include photos and written copy about the companies' lines in hopes of being included in fashion news articles. A press kit is shown in 7-2.

Visual Merchandising

Visual merchandising is presenting goods in an attractive and understandable manner. The way goods are placed on view can be a key to achieving high sales. Displays and exhibits are ways that clothing items are visually promoted. See 7-3. Mannequins are commonly used in both window displays and interior displays. Some displays have special effects such as moving parts, lights, or mirrors, as in 7-4. Fashion shows, contests, and other special events add visual excitement. Malls often have special displays, such as in 7-5.

Some department stores include theatrics in their promotional activities. Models dance in the store to rock music. Colored spotlights accent displays. Celebrity appearances are scheduled. Purchases are placed in high-profile

International Linen Promotion Commission

7-1 Including this photograph as part of a fashion story in a magazine would give the manufacturer publicity, or free promotion.

7-2 Press kits include pictures and words that can easily be placed into newspapers or magazines, sent out to try to obtain publicity for the company.

7-3 Visual merchandising is done through eye-catching presentations of the merchandise in stores.

JW Robinson's

7-4 Visual merchandising can be especially effective when props, such as mirrors, are used.

shopping bags. Departments have catchy names. Floors are arranged as arcades or streets of shops. Some stores hold art shows. Stores want to be regarded as exciting or entertaining places to be. Then there is a better chance that they will sell more goods.

Retailers are becoming more talented and aggressive with visual merchandising activities. They often have their own special-events departments. They are becoming entertainment, service, and news-making businesses. They are developing "images" for their stores. Having an image helps a store attract customers. Since many stores may carry the same merchandise, the presentation of the goods is often what makes the difference.

Video Merchandising

Video merchandising uses videos in retail stores to show new fashion trends, promote merchandise, and build customer traffic. Videos

Great Mall of the Bay Area

7-5 Malls create visual excitement with themes and displays that draw shoppers to their locations.

are sometimes set up in retail store departments near the merchandise they are showing. They attract the attention of passing customers with sound and movement. Videotaped fashion shows and interviews with designers encourage shoppers to stay in the department or store. However, they have not had a strong sales effect.

Large retailers have their own videotape centers. Small retailers rent or buy fashion videotapes. At least one video sales service offers monthly fashion tapes by subscription, as shown in 7-6.

Some fashion tapes are educational to train salespeople. They describe the philosophy, facilities, and lines of manufacturers. Other videos are educational for consumers. They may explain how to select jewelry or how to determine good fit in garments. Some show people how to build wardrobes that will help them look successful. Videos may play silently, or they may be accompanied by spoken words or music. Short videos hold customer attention best.

Fashion "rock videos" tell a story while presenting promotional information. They are very creative and usually only three to five minutes long. They do not contain obvious advertising, such as a sales pitch. Instead, they have lots of action and music with bright, fun clothing. The clothes might have a subtle outside logo or label. A manufacturer's name might be flashed casually in the background. This is "soft sell."

How Retail Works

Retailing is direct selling. It is the exchange of merchandise in return for money or credit. Retailers sell goods directly to the final consumers who will use them. This completes the channel of distribution of apparel items shown in 7-7. A ***channel of distribution*** is the route that goods and services take from the original source, through all middle people, to the ultimate user.

Retailers buy large amounts of apparel and other goods, usually directly from manufacturers. They, in turn, sell small quantities of goods or individual apparel items to many different people.

Videofashion

7-6 Fashion videos are available for television programming or for use by retailers for video merchandising.

Retail Terms

The difference between the store's cost of goods bought (purchase price) and the retail price of goods sold is called the ***markup***. Some of that money is profit. Most of the markup, however, must be used toward operating expenses and overhead. It must also cover losses from damaged goods, unsold merchandise, and shoplifting. Shortages from customer and employee theft can be shockingly high.

Markdowns are price reductions made in hopes of getting rid of certain goods. Markdowns are most often made at the end of a season. They enable stores to move otherwise unsalable, excess merchandise. They are necessary, but they cause retailers to lose a great deal of potential revenue. Sometimes retailers receive "markdown money" from manufacturers to compensate for losses when the selling prices of the goods must be reduced.

Odd-figure pricing is the retail pricing of merchandise a few cents less than the next higher dollar. Examples are $2.99 and $19.98. The purpose is to make the merchandise psychologically seem less expensive.

Many stores offer *loss leaders*. These are low-priced articles on which stores make little or no profit. They are popular items to attract shoppers into the stores. Retailers hope that shoppers will also buy other goods at regular prices from which profits will be made.

Stock control or *inventory control* is the receiving, storing, and distributing of merchandise. In modern stores, computerized

Channel of Distribution

Textile Producers
↓
Apparel Manufacturers
↓
Retailers
↓
Consumers

7-7 The retailer is near the end of apparel's channel of distribution.

conveyors move the received merchandise into the proper sections of warehouses or storage areas. Inventory is marked with bar codes, so it can be recorded electronically. The apparel is prepared for the selling floor and is taken to the proper area when requested.

Basic stock is merchandise that is constantly in demand. It is mostly staples, such as underwear or men's business shirts. It has a predictable and constant customer demand.

Retailers place *purchase orders* with manufacturers. These are written documents authorizing manufacturers to deliver certain goods at specific prices. Each purchase order has a particular number for identification. When a manufacturer finishes producing the items specified for a purchase order, the goods are shipped to the retailer.

Retailers can also buy *odd lots* (sometimes called job lots). These are incomplete assortments of goods. They might be items that are being discontinued or overruns that a manufacturer has left at the end of a season. Not all sizes or styles are available. They are bought at reduced prices to clear out the manufacturer's inventory. Retailers use these goods as sale items.

The *completion date* is a date designated on a purchase order by a retailer. Any merchandise that has not left the manufacturer by the completion date is subject to cancellation. If the completion date cannot be met, the store might grant the manufacturer an extension.

As ready is an expression used by manufacturers regarding the estimated delivery time of merchandise ordered by retailers. This is a promise to fill (ship) orders when they are completed. It takes the place of the commitment for an exact shipping date. Decisions about how firm or open the agreement should be depend a great deal on the past relationship between the manufacturer and retailer.

Retail Buying

Thousands of buyers go to New York City or to regional apparel marts to place orders for retail stores. Marts have permanent manufacturers' showrooms, as shown in 7-8. They may also rent additional space to manufacturers for use during market weeks. Fashion shows, as in 7-9, feature manufacturers' lines during market weeks.

Regional marts save retailers the time and expense of going to New York City for buying trips. This especially matters to small, independent retailers who may be located far from New York. In addition, manufacturers' salespeople periodically call on retail buyers at their stores.

Retail companies limit buyers to a certain open-to-buy. ***Open-to-buy*** is the amount of merchandise (in dollars or units) that buyers are permitted to order for their store, department,

California Mart: In-Wear/Matinique, Inc.

7-8 Each showroom in a regional mart contains the line of a specific apparel manufacturer.

In Cooperation with the Fashion Association

7-9 Numerous promotional activities, like fashion shows, are scheduled for the weeks when retail buyers "go to market."

or apparel category. The open-to-buy is for a specified time period, such as for fall or spring goods.

Retailers order apparel about six months ahead of the wearing season. It is promoted to consumers about three months ahead. Each buyer specializes in a certain type of merchandise with a targeted customer audience in mind.

Resident buying offices are service businesses located in major market centers. Their client groups consist of noncompeting member stores from many parts of the country, as in 7-10.

Resident buying offices provide their member retailers with market information. They cover specific wholesale markets and present surveys about trends to their member stores. They send out news bulletins about new items, best sellers, price changes, supply conditions, and other developments. They sometimes set up market appointments for buyers with certain apparel firms. They may recommend suppliers and accompany buyers on calls. They also provide office space, clerical help, and mail and telephone service for their visiting buyers.

If requested, resident buying offices will do actual buying for stores. They will follow up on shipments of orders. They will place reorders and fill-in orders for fast delivery. They will prepare sample advertising, direct mailing pieces, and promotional ideas for their member stores.

Frederick Atkins, Inc.

7-10 The retail store names and logos that appear here are all associated with the same resident buying firm.

Fees charged by resident buying offices vary in relation to the types and amounts of services rendered. The information exchange they offer is most important to small merchants. Group buying with other stores through such offices can offer lower prices through quantity discounts. Group buying also helps stores receive more reliable deliveries.

Timing and Pricing for Demand

Merchandise is put onto the floor of retail stores well ahead of the wearing season. Thus, consumers are urged to buy their summer garments in the spring. Stores try to sell all of their shorts, swimsuits, and other summer attire by July 4th. Fall clothes are in the stores in midsummer. Stores like to be sold out early as "insurance" against consumers changing their minds when the season arrives. However, consumers are now objecting to this type of timing. They prefer items to be available during the season of wear rather than only before the season.

Retailers must constantly be aware of consumer desires. They must notice changes in consumer living patterns and tastes. Consumers make the final decisions about what will or will not sell. If the public is not ready for certain styles, they will not move. To encourage sales, the presentation of items in the store must educate consumers about how to wear or combine articles of clothing. Retailers must show how to put together and accessorize outfits. In this way, they can create consumer demand through understanding.

Each store's buyers must try not to order fewer *stock keeping units (SKUs)* than needed. The items that sell quickly may not be available for reorder because manufacturers may already be working on the next season's line. That causes the store to lose the opportunity of additional sales dollars. Buyers must also try not to order more than consumers will purchase. Goods left over will be sold at reduced prices, as in 7-11. This causes financial losses. Good buyers seem to have instincts about what and how much their stores will need. Also, point-of-sale data from check-out computers tells them what items are increasing or decreasing in popularity.

Some markdowns are planned for sales promotions, in hopes of attracting more shoppers. However, some retailers have been criticized for overdoing price promotions. In the effort to be able to mark down merchandise to please consumers, stores often first mark up the goods to higher-than-normal prices. This allows their *margin* (money made) to be acceptable even when goods are marked down to what the stores call "sale" prices. In the long run, if such promotions are done constantly, stores lose their credibility with customers. The pricing policies and ethics of the stores are questioned.

Private Label Brands

Private label merchandise is produced specifically by or for a retailer. The trademark or brand name on the label is owned by the store, or group of stores that sell the goods, rather than by the manufacturer. The merchandise is produced exclusively in the colors, patterns, styles, and fabrics the retailer specifies.

Retailers provide specifications ("specs"), and usually samples, of the designs they want produced by the contracted sewing shop. Sometimes they supply the fabrics. Often the goods are very similar to national brands and made by the same factory. Sometimes retailers purchase a manufacturer's line and have it labeled for them when made. Some retail firms even have their own manufacturing plants.

Through private labels, retailers are entering the apparel production business. *Product development directors* are assisted by buyers in establishing the basic items and fashion copies to be made for their stores, as in 7-12. Product developers need a thorough knowledge of fabrics and manufacturing methods.

There are various *advantages* to private label programs for retailers:

- Retailers have exclusive design control. They can specify the quality, fit, price, and packaging of their items.
- Retailers pay lower than regular prices for the goods. Expenses for the merchandising, designing, and selling functions of a manufacturing firm have been eliminated. The stores can then have higher markups and better profits.
- Many lower-priced knock-offs are made of fast-selling, higher-priced styles. Slight changes can be made to update them for each future season if they stay popular.
- Retailers have an opportunity to build an image for themselves through distinct private label goods.

7-11 Retail stores "move" excess garments by putting them on sale at lower prices.

Gerber Garment Technology, Inc.

7-12 Retail product developers plan private label merchandise and arrange for production of the items.

There are also several *disadvantages*:

- Private label manufacturing requires large financial investments. It ties up money for production that could be used to buy goods from other manufacturers.
- A great deal of private label manufacturing is done overseas. Thus, there may be a long time from making design and production arrangements until the goods actually arrive at the stores. This can cause problems with fashion timing as well as price changes if money exchange rates fluctuate.
- There may be slow customer acceptance of products until the labels are recognized.
- No markdown money is available from manufacturers. The retailers must absorb the losses if there are quality, fit, or other problems.

Retail Business Considerations

Location is an important aspect for retail businesses. The most successful stores are in locations that are convenient for customers. That results in a high volume of sales. With a continuous fast stream of sales, a store can afford to use smaller markups and offer lower prices, while still making good profits.

Mass retailing has encouraged customers to shop without help from salespeople. Most stores no longer wait on clients one at a time. Instead, vast amounts of merchandise are put out on racks, shelves, or in bins. Customers are free to look at and try on garments. Clothes are *package priced*. That means that each garment, or package of items, has its individual price marked on it. Customers can do their shopping without asking for help. This impersonal selling is done by retailers to save money, since fewer salaries have to be paid. The savings are passed on to customers in the form of lower prices.

In most retail stores, computers record what goods are purchased as inventory and sold to customers. Through cash register computers, selling trends are detected hourly and daily. Central computers record sales information immediately from all chain or branch stores in the system. Weekly, monthly, and yearly sales reports are also calculated.

Retailers measure a store's business in different ways with such facts and figures. They calculate the productivity of each square foot of selling space. They check stock turnover, financial assets, promotional efforts, and salespeople's activities. They also try to determine merchandising success in terms of correct predictions on goods ordered, pricing, and timing for their customers.

Besides attending trade shows and subscribing to trade publications and resident buying services, retailers belong to other organizations. Examples of these are the National Retail Federation (NRF), local Chambers of Commerce, and shopping mall merchant groups. *STORES Magazine* is the trade publication of the NRF.

Types of Apparel Retail Outlets

There are many sizes and types of apparel retail stores or outlets. Through them, manufactured goods are sold to consumers. Specific retail stores may fall under more than one of the following categories.

Department Stores

Department stores are retail establishments that offer large varieties of many types of merchandise placed in appropriate departments. Almost all clothing and household needs are sold in a wide range of colors, sizes, and styles. The goods are categorized into such areas as menswear, juniors, infants, jewelry, and shoes. See 7-13. The departments

7-13 This section of a department store handles all types of women's apparel items.

may also define various price ranges. For instance, there may be a "ladies' better dresses" department and a "budget sportswear" area. Each department has salespeople and a place to pay for the goods being purchased. Each department is a separate profit center for record-keeping purposes.

Department stores may offer credit, return or exchange privileges, and home delivery. Other conveniences may include liberal payment plans, gift wrapping, pleasant rest rooms, and customer service desks. They may even have bridal registries and personal shopping assistants.

Department stores stress customer satisfaction. They try to maintain a reputation for quality and integrity. They want customers to regard them as being dependable. Often they serve as *anchor stores* in shopping malls. Anchor stores provide the attraction needed to draw customers to malls.

Department stores sell to many income levels because of their wide range of products. However, most department stores target the majority of their merchandise to people in the middle to upper-middle income brackets.

Department stores have somewhat higher operating expenses than most other stores. That is because they have more personnel, bookkeeping, and extra services. They advertise and have promotional activities. Their prices must cover the costs of the extra services they offer. However, they also have large buying and sales volumes that help to keep their prices low.

Branch Stores

When a well-established department store opens a store in another location, the new one is called a *branch*. It operates from the original *flagship store* from which it receives its merchandise and direction. The buying, advertising, and control of all branches can be done centrally to save on operating expenses. If an item is not available at a branch store, it may be obtained on short notice from the flagship store.

Retailers first established suburban branches in widening circles around a main downtown store. Now branches are located far away from the home base. Many large stores have expanded nationally.

Chain Stores

A *chain* is a group of stores owned, managed, and controlled by a central office. See 7-14. All the stores in the chain handle the same goods at similar prices. All of a company's chain stores look very much alike. No store is considered to be the main store.

Some chains are centered in various regions of the country. Others are nationwide. Merchandising and operating decisions are made in a central office. Some chains are big enough to have regional offices to which their stores report, as well as the central office to which all regions report.

Prices in chain stores are usually lower than those in department stores. Since large chains have enormous buying power, manufacturers can afford to produce merchandise to the chain's specifications. Thus, private label merchandise may be styled and made exclusively for them at lower prices.

7-14 Chains may have stores throughout the country, all with similar appearance and the same merchandise.

Discount Stores

Discount stores, as in 7-15, sell clothing and other merchandise in large, simple buildings with low overhead. Large amounts of garments are crowded onto racks and shelves. Some items are well-known brands, and some are unknown or private label brands of the discount chain. The merchandise is sold at low prices.

Discount stores often stock up on manufacturers' overruns and end-of-season goods. They buy odd lots of specialty apparel. They sometimes buy irregulars. They get special deals from manufacturers for buying in large quantities. Discount stores sell huge amounts of lower-priced apparel and accessories to customers with modest clothing budgets.

Discount stores earn profits with only small markups because they sell large quantities of merchandise and have lower expenses. They use mass retailing methods. They expect customers to shop without assistance. Merchandise from all departments is paid for at checkout counters just inside the store exits. No extra customer services, such as telephone orders, gift wrapping, and home deliveries, are offered.

In the past few years, discount stores have been growing rapidly. Most are chains. They are open in the evenings and on Sundays and most holidays. They undersell other kinds of stores, thus creating stiff competition.

Off-price discounters have well-known brand name merchandise at lower than normal prices. These specialized apparel discounters are growing in popularity. They focus on moderate to higher-priced merchandise for middle to upper income families. They carry quality goods, often with the labels cut out to protect the manufacturers' regular goods sold elsewhere.

7-15 Discount stores sell large volumes of items at low prices in simple buildings.

Off-price discounters do not place orders with manufacturers for specific merchandise to be produced. They buy whatever is available, such as overruns. They pay for the goods rather than asking for credit. They buy this season's extra items when most other stores are thinking about the next season. They put the goods on the floors of their stores to be sold and worn immediately.

Factory outlets are discount stores owned by manufacturers. Each store sells only the merchandise the manufacturer makes. The goods include seconds, overruns, and items from the previous season's line. Some manufacturers now produce some lines just for their factory outlet stores.

A factory outlet was originally a single, small store near a factory. The store sold the manufacturer's surplus production. Recently, however, many new factory outlet malls have opened that offer manufacturers' lines nationwide. They are usually located in outlying areas and must be careful not to compete against the full-price retailers to whom they sell their lines.

Other types of discount stores are wholesale warehouse clubs and hypermarkets. *Warehouse clubs* specialize in bulk sales of nationally branded merchandise. *Hypermarkets* are large stores that sell almost every type of merchandise, including apparel and groceries.

Specialty Stores

Specialty stores might handle only apparel, or they might specialize even further into a specific kind of apparel. Examples are maternity shops, shoe stores, bridal boutiques, children's apparel stores, and T-shirt shops, as in 7-16.

Many small specialty stores have lower volumes of sales and must charge higher prices than larger stores. Small quantities of goods are stocked. The employees must perform all of the stock control, promotion, and direct selling functions that big stores split among their specialized personnel.

Specialty stores provide many personal services that appeal to some shoppers. They stress customer service and personal contact. They give advice about the selection of goods within their specialties. The personal preferences of regular clients are acknowledged. Returns are usually accepted, but customers may be given credit toward other purchases rather than cash.

Many specialty stores are located in shopping centers or malls. Many of them are independently (privately) owned. Sometimes they are small, with only one or a few employees.

Boutiques are a type of specialty store. They are small shops that sell few-of-a-kind apparel and accessories. Their merchandise is often fashion-forward and presented to their very specific customer base in a creative way. They have a distinct image and give individualized attention to their clients.

7-16 This specialty store concentrates on selling only T-shirts.

Small specialty stores that are not well established sometimes go out of business. Others are starting all the time. Successful specialty stores become loyal customers of particular manufacturers. The buyer for the store is the same year after year, quite different from large firms in which employee turnover is common.

Store ownership groups are corporations formed by individual stores joining together. Each store retains its own identity. However, all share the information and expertise of the corporation. Such corporate ownership groups continue to acquire more stores, consolidate, or merge with other retailers. They can then use centralized volume buying and effective merchandising techniques.

Franchises

Franchise stores are individually-owned businesses that use the name and merchandise of an established firm. The franchisor (parent company) provides a franchisee (owner-operator) with exclusive use of the name and goods in a specified city or area. Sometimes the parent company offers assistance in organizing, training, and management. After meeting the financial obligations to the parent firm, the franchisee keeps the remaining profits.

The franchisee can expect quick acceptance because of the well-known name attached to the merchandise. The parent company, on the other hand, can spread the locations of its goods without large capital expenditures. Designers and manufacturers have franchise shops all over the world.

Mail-Order Houses

Mail-order houses sell to consumers through catalogs. They offer shopping at home for customers who cannot, or prefer not to, go out. Some specialize in a single line of merchandise. Others offer a full line of items similar to department store choices.

Mail-order customers select merchandise by looking at catalog pictures and reading the descriptions. Orders are placed either by mail or telephone. Catalog firms usually have toll-free numbers and extended order-taking hours, sometimes around the clock. Most accept credit card payment over the phone.

In many mail-order distribution centers, computers are used to track inventory and control warehousing. Careful packaging techniques are used to distribute the merchandise. Sometimes gift-wrapping is offered, as in 7-17.

Mail-order houses often sell goods at low prices. They purchase goods in large quantities, have low overhead, and sell to a mass market. They have a large volume of business and offer few services. They usually offer money-back guarantees if customers are not satisfied.

Mail-order retailing (also known as *direct-mail marketing*) is growing faster than conventional retail sales. It especially suits two-career families and affluent singles whose busy schedules leave them little time to shop. These people can study the catalogs during their spare moments. Elderly consumers can also shop without leaving home. The appeal of mail-order shopping is an answer to the crowding, lack of service, and inadequate parking at some conventional retail outlets. Most mail-order retailers also have Internet Web sites.

Lands' End Direct Merchants

7-17 Apparel items must be carefully packaged to prevent wrinkling, and are sometimes gift wrapped, before being sent to the recipients.

Mail-order houses offer every type of merchandise—from practical, low-budget items to unique, luxurious ones. The trend in mail-order is toward specialty catalogs aimed at narrow segments of the consumer market. Computerized mailing lists provide companies with names of people in particular market niches, such as various income levels or hobbies. Catalogs are prepared with target audiences in mind. Catalogs showing inexpensive basic items for a general audience may be printed on newsprint. Expensive merchandise is shown in elegant, colored catalogs of magazine quality.

Some retail department stores and chains offer catalog shopping in addition to their regular store selling. Such stores distribute their catalogs through the mail or as newspaper inserts. Also, large mail-order houses sometimes have one or more retail stores.

Telecommunication Retailing

Telecommunication retailing offers shopping via communication devices. This includes television and computer retailing. Both are still in their marketing and technological infancy. Telecommunication retailing gives instant computer calculation of ongoing sales to fashion companies as the sales occur—in units and dollars—about style, size, and color preferences of the market.

Television retailing involves showing and describing merchandise on certain television channels. See 7-18. Viewers order by telephone, pay by credit card or other means, and have the merchandise sent to them.

Web-based retailing, done electronically via the Internet and referred to as *"e-tailing,"* is gaining rapid acceptance. This online retailing allows consumers to view merchandise on their computer monitors. Consumers click onto Web sites of specific retailers or general marketplace portals that include many retailers. Three-dimensional digitized images of apparel and other items, with detailed written descriptions, are viewed on the monitor screen. Users are able to comparison shop for features and prices.

Retail Web sites offer a "shopping cart" into which consumers electronically place their selected items. To make the final purchase, after reviewing the selections in the

QVC Network, Inc.

7-18 This retailing show is describing cosmetics prior to a make-up demonstration on a stage set in front of television cameras. The live image is transmitted to home viewing audiences.

shopping cart, a checkout system requires input of credit card information for payment. An electronic verification is given of the order immediately, and also later via the customer's e-mail address. The merchandise is delivered to them according to instructions they provide.

Online customer service has been a problem for e-tailers. Consumers cannot go to an actual salesperson to ask questions about the merchandise or to clarify computer buying procedures. To combat this, companies are placing customer service representatives at computers to answer e-mail questions from consumers.

Some retail companies offer consumers the choice of purchasing from either their Web sites or physical stores. Since merchants who only have stores are called "bricks-and-mortar" retailers, multi-channel merchants who sell both through the Internet and through stores are said to be "clicks-and-mortar" retailers!

Other Types of Retailing

Some less common types of retailing are variety stores, personal selling, catalog showrooms, and leased departments.

Variety stores started as "five-and-ten-cent" stores. They now offer wide assortments of lower-priced merchandise. They have inexpensive clothing as well as stationery, plants, and toys. The checkout counters are near the exits.

Personal selling is done without a store. It moves cosmetics, jewelry, clothing lines, and other merchandise directly to customers through parties or showings in homes. See 7-19. Orders are taken and the items are delivered to the customer later. The merchandise is usually of high quality, but is also quite highly priced.

Catalog showrooms display items in the stores that are also shown in their catalogs. Merchandise must then be ordered by catalog number on a special telephone or computer. In a short while, the order arrives on a conveyor belt (from the warehouse area) at the pickup location. Low overhead at such stores enables prices to be low.

A *leased department* is a department within a store operated by an outside firm. Usually the store supplies the space and essential services in return for a fee or percentage of the sales. Name designers sometimes lease a department within a retail store. Shoes, fine jewelry, and restaurants are other examples of leased departments.

Retail Imports

Offshore sourcing is the term used when goods are bought from overseas producers. Most large retailers do some direct importing

Doncaster

7-19 Some clothing lines are shown privately to the clients of "consultants" who sell the items from their homes.

from abroad. Some retailers produce their own private label goods in overseas contract factories.

There has been a fast rate of growth of apparel imports. A larger percentage of offshore sourcing is done each year.

Many retailers are in favor of imports. Imports are cheaper for retailers to buy, and they can be sold to consumers at about the same prices as domestic goods. Thus, imports allow retailers to have higher markups and bigger profits.

There are other reasons why retailers like imports. Sometimes a retailer can import new and different merchandise that no one else has. Also, garments that require much handwork may not be available from domestic sources. If retailers want these garments, they may have to be imported.

There are risks for companies doing offshore sourcing. Long lead times require early buying decisions since orders are placed much farther ahead of when they will be received and sold. Retailers are forced to order next year's seasonal merchandise before they know what is selling well this year. It is hard to order the right merchandise or the right amount of merchandise. With early ordering, imports must also be financed for a longer time than domestic orders. This leaves less money available for retailers to buy other goods to sell in their stores.

Imports limit retailer's abilities to respond quickly to market developments. Often stores cannot reorder. There are shipping delays, magnified by customs inspections, which result in late merchandise arrivals. There are quota uncertainties. Often surprises occur with the quality of goods delivered. Communications are difficult between the retailer and offshore producers. Staff travel costs are high.

Commissionaires are hired around the world by retailers. They are usually natives of the countries in which they reside. They know the customs, laws, and production capabilities of their countries. They also know the tariff quotas of the United States. They function like an overseas buying office, often working for many firms.

As discussed in Chapter 6, retailers have been encouraged by American apparel industries to seek domestic sourcing. Manufacturers are striving for more efficient production to give fast response and delivery times. For retailers, that would mean lower inventory costs and much more accurate predictions about what items will sell and how many of each to order.

Retailers' Future

In the future, retailers will reposition themselves by recognizing market changes. Retailers will probably attract shoppers in three main ways:

- use technology and efficiency to satisfy busy consumers' necessary shopping tasks faster, easier, and less expensively
- provide entertainment along with shopping for consumers' leisure time
- offer unique products to satisfy out-of-the-ordinary tastes of specific consumers

Technology will greatly enhance retail efficiency. Shoppers will scan items themselves in retail stores, either to purchase them or to record them for friends and relatives to give them as gifts. Web-based shopping will offer more products and be easier for consumers to do from home. Technology will reduce long checkout lines and facilitate check and credit card processing. Small, easy-to-shop strip malls will gain favor over huge, crowded courtyard malls that have parking hassles and long walking distances. One-stop shopping will offer goods that are instantly available. Store organization will make it easier to find products quickly.

Consumers become bored with stores that look alike and sell similar merchandise. They tend to make more purchases if they are in a good mood. Shoppers want socialization, recreation, and an enjoyable "escape." This *shoppertainment* draws people for the entertainment value, such as high-tech interactive computer simulations and amusement rides.

Retailers will offer more unique products of better quality instead of overusing the term "sale." They will market to tiny niches of specific customers identified by credit card purchases and other database information. *Niche retailing* is dividing the total consumer market into narrow target markets with specific tastes or lifestyles. There is less competition in niche markets, but also fewer customers. By clearly understanding and focusing on their target customers, niche retailers can operate with less stock, make more sales, and gain higher profits. In this way, small retailers will complement mass merchandisers rather than compete against them. They will have to stay flexible and creative to fill observed gaps with innovative merchandising methods.

The concentration of large retailers will continue, with buyouts of smaller retail chains resulting in fewer, bigger corporations. They will have more efficient central buying and distribution procedures. More private label brands will be carried. Merchandise and shipping containers will be marked with Universal Product Code (UPC) bars to be scanned for instant computerized inventory and sales tracking.

Nonstore formats such as direct (mail-order) marketing, TV retailing, and on-line computer commerce will take market share from traditional retailers in the future. Consumers will find the desired factors of convenience, entertainment, and unique merchandise in these consumer-direct formats. By 2015, as much as 50 percent of retail merchandise could be sold this way, compared to about 25 percent now. However, consumers will lose the experience of actually seeing, touching, and trying on merchandise. Although these experiences will be computer simulated, customers may still return large amounts of merchandise.

Through the Internet, apparel manufacturers as well as retailers will sell directly to consumers around the world. Increased numbers of consumers will have and use computers, and technology will become more user-friendly. Also, on-line security systems and encrypted credit card numbers will provide confidentiality for shoppers. More and more screen-based catalogs and electronic shopping malls will evolve.

Eventually, televisions and computers may combine into one piece of electronic retailing equipment. Body scanning, as shown in 7-20, will collect individual sizing information electronically. The information will be stored on "smart cards" that can be updated as often as needed and inserted like credit cards to be read.

www.landsend.com

7-20 Body scanning will electronically collect individual sizing information in retail centers and through some home equipment.

Consumers will be able to redesign garments electronically. See 7-21. Lengths, colors, and patterns can be changed until the perfect fashion selection is created. When garments are just right, shoppers can see how they look in them, as in 7-22, with virtual reality images of themselves on the screen wearing the creations! When approved, a radio-frequency transmission of data will begin the fast computer manufacturing process. The finished items will be shipped directly to the consumers who ordered them.

www.landsend.com

7-21 Consumers will be able to change the sleeves, collars, fabric types, prints, and all other aspects of a garment by computer.

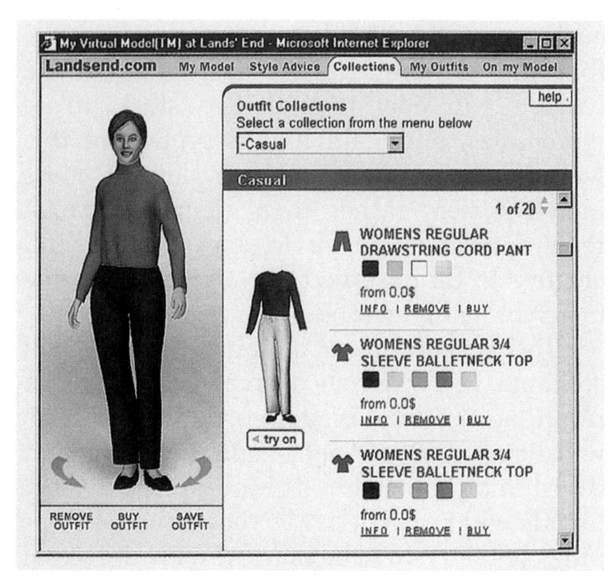

www.landsend.com

7-22 With the scanned measurements of the body and a realistic model shown wearing the chosen garments, consumers will be able to order exactly what they want through the technology of the future.

Chapter Review

Summary

Fashion promotion, or indirect selling, is done through advertising, publicity, visual merchandising, and video merchandising. Advertising is done locally, nationally, or cooperatively. Public relations departments seek publicity by distributing press kits. Visual merchandising presents goods with displays, exhibits, and special events. Videos attract and educate shoppers, train salespeople, and try to sell merchandise.

Retailing, or direct selling, completes the apparel channel of distribution to consumers. Retail vocabulary includes pricing, stock, and purchasing terms. Retail buying is done in New York City or regional apparel marts, often assisted by resident buying offices. The merchandise must be timed and priced for demand. Product developers plan private label goods and arrange for production of the items. Many business details must be considered for retail success.

There are many types of apparel retail outlets. Chain stores or branches of department stores often serve as anchor stores in malls. Discount stores sell large quantities of goods at low prices. Specialty stores cater to specific markets. Franchise stores have well-known names and merchandise. Mail-order and telecommunication retailing enable consumers to shop from home and are increasing in popularity. Other types of retail outlets include variety stores, personal selling, catalog showrooms, and leased departments.

Retail merchandise will continue to be sourced offshore. In the future, retailers will provide more shopper efficiency, entertainment with shopping, and unique products for specific consumer tastes. Both apparel manufacturers and retailers will increase selling worldwide through the Internet. Eventually, consumers will redesign garments electronically for themselves.

Fashion in Review

Write your answers on a separate sheet of paper.

Matching: Match the following terms with their definitions.

1. Low-priced articles on which stores make little or no profit; used to attract shoppers into stores.
2. Store merchandise that is constantly in demand.
3. Price reductions made in hopes of getting rid of certain goods.
4. The retail pricing of merchandise a few cents less than a dollar denomination.
5. The receiving, storing, and distributing of merchandise.
6. An expression used by manufacturers indicating a promise to fill (ship) orders when they are completed.
7. Incomplete assortments of goods that retailers buy from manufacturers.
8. A designated time after which any merchandise that has not left the manufacturer is subject to cancellation.
9. The difference between the store's cost of goods bought and the retail price of goods sold.
10. A written document authorizing manufacturers to deliver certain goods at specific prices.

A. odd-figure pricing
B. markup
C. markdowns
D. loss leaders
E. stock control
F. basic stock
G. purchase order
H. odd lots
I. completion date
J. as ready

Short Answer: Write the correct answer to each of the following questions.

11. What is the purpose of fashion promotion?
12. Describe two ways an apparel firm might get publicity.
13. How does developing an image with visual merchandising help stores attract customers?
14. Name three reasons why retailers use video merchandising.
15. List the four steps in the channel of distribution of apparel items.
16. List five services offered by resident buying offices.
17. Name two advantages and two disadvantages of private label programs for retailers.
18. Name and describe five types of apparel retail outlets.
19. What do commissionaires do?
20. In the future, in what three ways will retailers probably attract shoppers?

Fashion in Action

1. Find examples of apparel promotion by walking through stores and looking through newspapers, magazines, and fashion publications. In an oral report, describe examples of advertising, publicity, visual merchandising, and video merchandising.
2. Visit a shopping center or mall. What types of stores are included? What other attractions are offered? Prepare a written report about your findings.
3. Look at the apparel price ranges and labels in various retail outlets. Do you see markdowns, imports, and private label merchandise? Can you identify some loss leaders? What goods are package priced? Share your findings in a class discussion.
4. Working in small groups, make appointments to talk to local store managers. Ask them if their stores belong to chains or store ownership groups. How is their buying done? Do they work with resident buying offices? Do they go to New York City often, or do they deal mostly with regional marts? How are the stock control functions done? Present the groups' findings to the class.
5. Collect five mail-order catalogs that offer apparel. Describe size, type of paper, presentation of merchandise, and price range of each catalog. How does a person place orders and make payments? How are goods delivered? What are the return policies? Present your findings in a written report.

PART THREE

Textiles: The "Science" of Apparel

Chapter 8
Textile Fibers and Yarns

Chapter 9
Fabric Construction and Finishes

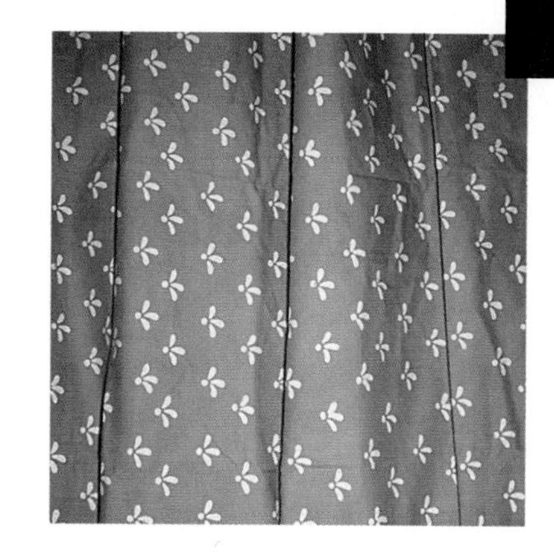

8 Textile Fibers and Yarns

Fashion Terms

- fibers
- woolen yarn
- worsted yarn
- filament
- sericulture
- polymer
- generic name
- denier
- spinneret
- staple fibers
- wicking
- yarns
- ply yarns
- combination yarn
- blend
- texturing

After studying this chapter, you will be able to

- describe the sources and processing of natural fibers.
- explain how manufactured fibers are made.
- identify the characteristics and uses of individual fibers.
- discuss how different types of yarns are made.

Fabric starts with fibers. ***Fibers*** are long, thin, and hairlike. They are very small in diameter. They are the basic units in making textile products.

Fibers are spun into continuous strands called yarns. Yarns vary in size. Some are coarse and fluffy, while others are finer than sewing thread. Yarns are woven, knitted, or pressed into fabrics. Then fabrics are finished with dyes, prints, and coatings before being made into apparel or other items.

Some fibers are obtained from natural sources. They are called *natural fibers*. Others are made in factories from chemical and other sources. They are called *manufactured fibers*.

No fiber is perfect. Each has some good, fair, and poor characteristics, or properties that make it suitable for certain uses. The basic characteristics of a fiber can be slightly altered but never totally changed. Fiber properties determine the appearance, performance, and comfort of fabrics. Likewise, fiber and fabric properties influence the appearance, performance, and comfort of finished garments. Thus, an understanding of fibers, yarns, and fabrics is basic to the study of apparel.

Natural Fibers

Natural fibers are made from natural sources, mainly plants and animals. Natural fibers are listed in 8-1.

Cellulosic natural fibers, such as cotton and linen (flax), are from plants. Less common cellulosic natural fibers are jute, kapok, sisal, straw, hemp, and ramie.

Protein natural fibers, such as wool, are from animals. Specialty hair fibers, such as mohair, vicuña, alpaca, cashmere, camel's hair, and angora are protein fibers. Silk is also a natural protein fiber.

All cellulosic and protein natural fibers are absorbent, or able to take up moisture. This provides wearing comfort, since perspiration is absorbed, and body heat is controlled. Absorbency also provides good dyeability.

Natural fibers vary in quality due to weather conditions, soil fertility, and animal disease. Also, their quality depends on their varieties or breeds.

Cotton

Cotton fibers come from plants. The United States leads the world in cotton production. This crop is important to the nation's economy. It is grown in the South, sometimes called the "cotton belt." Some cotton is exported from the United States to other countries. Cotton is also grown in India, China, the Commonwealth of Independent States, Brazil, Egypt, Mexico, Pakistan, and Turkey.

Cotton Production and Processing

Cotton fibers come from the seed pods of the plant as it ripens. First, flower blossoms ripen, wither, and fall off. Then they leave green seed pods, called cotton *bolls*. Inside the bolls, moist fibers grow and push out from newly formed seeds. Either chemicals or natural frost causes the leaves to fall off the plant. Once exposed to more sunlight and air, the fibers continue to expand until they split the boll apart, as in 8-2. The cotton boll is then plucked off the plant, usually by a mechanical stripper or picker. See 8-3.

After harvesting, the cotton bolls go to the gin where the fibers are separated from the seeds. Cleaning machines remove burrs, dirt, and leaf trash. The ginned fiber is called *lint*. It is pressed together and made into large *bales*. Before going to a spinning mill, samples are taken from each bale to determine the cotton's value, as shown in 8-4. Each bale is classed according to length and fineness of fibers, as well as strength, color, luster, and cleanness.

At the mill, cotton is processed by a continuous, automated system. Cotton from several

Natural Fibers

Cellulosic (from plants)

- Cotton—most common
- Linen (flax)—quite common
- Others:

Hemp	Ramie
Jute	Sisal
Kapok	Straw

Protein (from animals)

- Wool—most common
- Silk—quite common
- Specialty hair fibers:

Alpaca	Guanaco
Angora	Llama
Cashmere	Mohair
Camel's hair	Vicuña

8-1 The most popular natural fibers are cotton, linen, wool, and silk.

National Cotton Council of America

8-2 The cotton boll bursts open to expose fluffy cotton when it is ready for harvest.

National Cotton Council of America

8-3 When cotton bolls are ready to be picked, they can be mechanically harvested.

bales is blended together. This provides uniformity and maintains constant yarn quality. Also, any fiber lumps are reduced.

The loosened fibers then receive more cleaning and fluffing. The cotton is formed into a continuous wide sheet called a *lap*. The lap is rolled around and around a cylinder. From there it is fed into the *carding* machine where it receives final cleaning and is separated into individual fibers. It is eventually drawn through a funnel-shaped device that molds the fibers into a sliver. A *sliver* is a round, rope-like strand of fibers about the diameter of a person's finger.

Combing is done for high-quality fibers of exceptional smoothness, fineness, and strength. This combs out short fibers. Long fibers remain. This high-quality fiber is again formed into a sliver.

After carding, and possibly combing, cotton goes through *drawing*. Several slivers are combined into a strand that is drawn out, without twisting, and reduced to about the same diameter as the original sliver. This further blends the fibers, arranges them in parallel order, and increases uniformity.

Slivers are then fed into the *roving* frame where the cotton is twisted slightly and drawn into a smaller strand, as shown in 8-5. The roving, or twisted sliver, is fed to the *spinning* frame where it is drawn out to its final size and twisted into yarn. The twist holds the fibers together and gives strength to the yarn that is then wound onto bobbins. Refer to 8-6 to review the steps of cotton processing.

Pima cotton fibers are of very high quality with naturally long fibers. Cotton

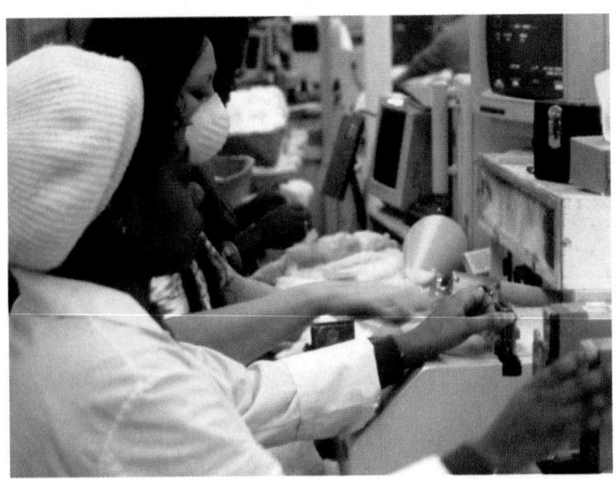

National Cotton Council of America

8-4 Cotton fiber samples are classed according to their quality and value before being processed into yarns.

National Cotton Council of America

8-5 When slivers of cotton fibers continue to be twisted and pulled, a smaller roving strand results. Then it is spun into even finer yarn.

Cotton Processing

Harvesting of cotton crop
↓
Ginning of cotton bolls to separate fibers from seeds
↓
Blending of fibers from several bales
↓
Cleaning and forming into lap
↓
Carding into rope-like sliver → (optional) Combing of high quality fibers into combed sliver
↓
Drawing to combine and pull several slivers into a drawn sliver
↓
Slight twisting and pulling makes a smaller roving strand
↓
Spinning draws out and tightly twists the fibers into yarn

8-6 After cotton is harvested from the field, it goes through several processes before becoming yarn.

fibers termed *Egyptian* (originally grown in Egypt) have a smooth, silk-like texture.

Cotton Incorporated is the marketing and research organization for American cotton growers. It acts as a product research, development, and promotional center for the cotton producers. It prepares and distributes fashion forecast information to designers, manufacturers, the fashion press, and retailers. It advertises cotton products in consumer and trade media. It also encourages manufacturers and retailers of cotton products to use the *Seal of Cotton*, shown in 8-7, on their identification and promotional materials.

A different organization, the National Cotton Council, distributes educational materials and lobbies for trade legislation on behalf of the cotton industry.

Cotton Characteristics and Uses

Cotton has many good characteristics. It is relatively inexpensive. It is soft and absorbent. Cotton is comfortable to wear in hot weather. It takes dyes and prints very well. It does not cling. Cotton launders well and is strong even when wet.

Cotton can be treated to become fire-resistant, mildew-resistant, and water-repellent. It is strong and durable to wear for a long time. It does not *pill*, or form an accumulation of little balls on the surface. It also blends well with other fibers.

A disadvantage of cotton is that it wrinkles easily. To counteract this, it is often treated or blended with manufactured fibers to reduce its wrinkling characteristics. Another disadvantage is that it shrinks. Because of its absorbency, it tends to pick up spots and stains. It is not elastic (stretchable) or resilient (able to spring back to its original condition).

Cotton is very versatile. In other words, it can be used in many different ways. It is used for clothes ranging from baby diapers to high fashion garments to heavy denim, as in 8-8. Household uses range from towels to upholstery to awnings. Apparel and household items are the largest users of cotton, but thousands of bales are also used by industry.

Linen

The fibrous materials of the stalk of the *flax* plant produce *linen*. Today, Belgium and Ireland are the countries most often associated with the growing and processing of linen. Flax is also grown in France and the Commonwealth of Independent States. In the United States, some is grown in Oregon.

Linen Production and Processing

Flax is a grass that looks like tall, slender reeds. The linen fibers are taken from the long, wiry stem. Long ago, the stem was beaten by

Cotton Incorporated

8-7 The Seal of Cotton is the registered trademark/service mark of Cotton Incorporated for products made of 100 percent cotton.

Draper's & Damon's

8-8 Cotton's comfort, durability, and versatility make it suitable for many types of apparel including denim garments, sweaters, and slacks.

hand to produce the fiber that could be spun into yarns.

Harvesting of flax plants takes place three to four months after planting. When the plants are mature, they are pulled out of the ground by machine, as shown in 8-9, rather than being cut.

Threshing machines remove the seeds and leaves. The flax stalks are allowed to dry in the sun. Then they are tied into bundles and soaked in *retting tanks*. This lasts for one or two days. Bacterial action loosens the outside flax fibers from the rotting woody center stem. Then the flax is untied and dried again out in the fields.

Later, the flax is *scutched*, as in 8-10. In this step, rollers crush the stalks to complete the separation of the soft flax fibers from the harsh, woody parts. Then they are *hackled* or combed, to straighten and clean the fibers. A drawing machine then combines the fibers into a continuous wide ribbon. The drawing process is repeated until all the fibers are parallel in small, rope-like slivers ready for spinning. Illustration 8-11 shows flax spinning.

Linen fibers are drawn and twisted together in a way similar to cotton. Finally, a continuous, strong, uniform, and workable strand of yarn wound on bobbins results. Refer to 8-12 to review the steps of linen processing.

International Linen Promotion Commission

8-9 At harvest time, machines pull the flax plants from the ground, remove seeds and leaves, and lay them flat to dry.

Linen Characteristics and Uses

Linen fibers look like bamboo sticks when they are magnified. They have a great variety of thicknesses. Fiber "bundles" tend to cling together, giving linen its characteristic thick-and-thin appearance.

Compared to cotton, linen is expensive. Its higher price is due to its limited production and the hand labor still involved in its processing. Also, it is imported from high-wage countries, which adds to its price.

The quality of linen is determined by the length and fineness of the fibers. The finer and longer fibers (from the taller plants) are used in better quality linen fabrics. They are thinner, smoother fabrics. Also, the degree of bleach, if white, or the fastness of the dyes, if colored, affect the quality. The colors of pure linen are neutral.

Linen is very strong. It is also known for its durability and luster. Linen wears evenly;

International Linen Promotion Commission

8-10 This man is working at a scutching machine, which crushes stalks to separate the fibers from the unwanted parts.

International Linen Promotion Commission

8-11 In flax spinning, the fibers are drawn out and then tightly twisted into yarns.

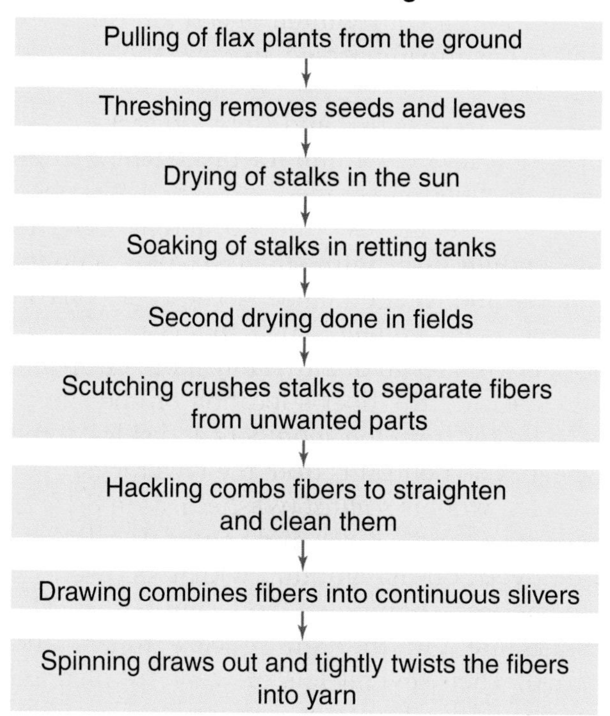

8-12 The fibrous materials from the stalk of the flax plant are processed into linen yarns.

it is more likely to get thin than to develop holes. The luster of linen often increases with continued use. It gets softer the more it is washed and worn. It is also cool and absorbent.

Linen wrinkles and creases even more easily than cotton. Thus, it is popular for apparel designed to have a "crushed look." Since flax fibers are longer than cotton ones, there are fewer fiber ends in a yarn. Thus, less lint, or fuzz, from fiber ends is at the fabric surface. Therefore, linen does not attract or hold soil the way cotton does. It sheds stains and dirt easily.

Fabrics made of linen usually *ravel* (come unwoven at the edges) easily since linen yarns are coarse and must be fairly loosely woven. Seam allowance edges in garments should be finished so no raw edges are exposed. Linen garments should be dry-cleaned, or washed if they have been preshrunk. For a smooth surface, they should be pressed while damp, with a hot iron, under a pressing cloth.

Linen fibers are made in lightweight to heavyweight fabrics. Some are thin and airy; others are crisp and smooth, as in 8-13. Still others are heavier or thick and rustic. There are sheer linens, linen suit fabrics, as in 8-14, and heavy linen canvas. Besides clothing, linen is used for handkerchiefs, tablecloths, kitchen towels, upholstery, and draperies.

Linen is often blended with other fibers, especially cotton, polyester, rayon, and silk. These combinations can give better wrinkle-resistance and softer, richer textures. Using certain finishes on linen can also improve its wrinkle-resistance.

Other Plant Fibers

Other plant fibers are related to flax. Most are *bast* fibers that lie in bundles just under the bark in the stems of various plants. They are strong, woody fibers. Many are used for twine, rope, braided rugs, and burlap. For apparel, they are used for hats, bags, belts, and shoes.

International Linen Promotion Commission

8-13 Fancy bedsheets and pillowcases are sometimes made of high-quality linen fabrics.

Talbots

8-14 Linen suiting fabric has the right weight to be made into jackets and pants.

Ramie is a fiber that is sometimes used as a linen or cotton substitute. It has been nicknamed "China grass." Examples of other bast fibers are *hemp* and *jute*. Fibers from leaves or seed pods are *sisal* and *kapok*.

Wool

Wool is the most commonly used animal fiber. It is from the hair of sheep or lambs. Merino sheep are said to produce the finest wool fibers. Sheep are raised for wool throughout the world. The leading producer is Australia. The United States ranks sixth in wool production, but first in the consumption (use) of wool. The U.S. raises sheep in every state, but the industry is concentrated in the West.

Wool Production and Processing

To be processed, first the fleece is *sheared* off the sheep by experts who use large power clippers. It is removed close to the skin and in one piece. The fleece is usually one year of wool growth for the sheep. Most shearing is done during the spring months, just before warm weather arrives and after the fleece is thick from a cold winter. Sheep are put back out to pasture unharmed, to grow another coat of wool.

The fleece is tied and stored in sacks. Then it is sent to a wool mill for processing. At the mill, the fibers are graded and sorted. A trained worker separates the different grades depending on length, strength, color, amount of curl, and feel of the fibers. The fibers vary in length from one-half inch to over one foot. Quality is a result of the health of the sheep, the climate, and the fiber's location on the sheep. The best is from the shoulders and sides of the sheep. The poorest is from the lower legs.

The wool is *scoured* (washed), as in 8-15, to remove sand, dirt, and natural oil. The removed oil is lanolin, which is used in cosmetics, shampoos, and ointments. The fibers are put through squeeze rollers and dried. Then several lots of wool are blended mechanically to attain uniformity. The wool may be dyed at this point or later during processing.

Carding further removes impurities while straightening the fibers. The carding machine has a series of various-sized rollers with fine

The Wool Bureau, Inc.

8-15 This New Zealand scoured wool is clean and soft, ready for processing into yarns and fabrics.

wire teeth that revolve at different speeds in opposite directions. The fibers are straightened and interlaced. They then leave the machine in the form of a thin, wide, continuous web.

The short fibers (less than two inches) are put aside for less expensive ***woolen yarn***. The long staple fibers will become ***worsted yarns***.

For woolens, the web is divided into narrow strips that form roving for spinning. The yarns contain short and long fibers that lie in random directions. They have a twisted, loose effect. Fabrics from these yarns are relatively dense and have soft, fuzzy surfaces.

For worsteds, combing straightens the long fibers and places them parallel to each other. The resulting sliver is drawn through a series of machines that gradually reduce it to a thin, slightly twisted roving. It is placed on large spools, ready for spinning.

Yarns for both woolens and worsteds go through spinning processes. In each case, the roving is pulled (drawn) and twisted into the final yarn. The finished yarn is wound onto revolving bobbins or spools after twisting. Refer to 8-16 to review the steps of woolen and worsted processing.

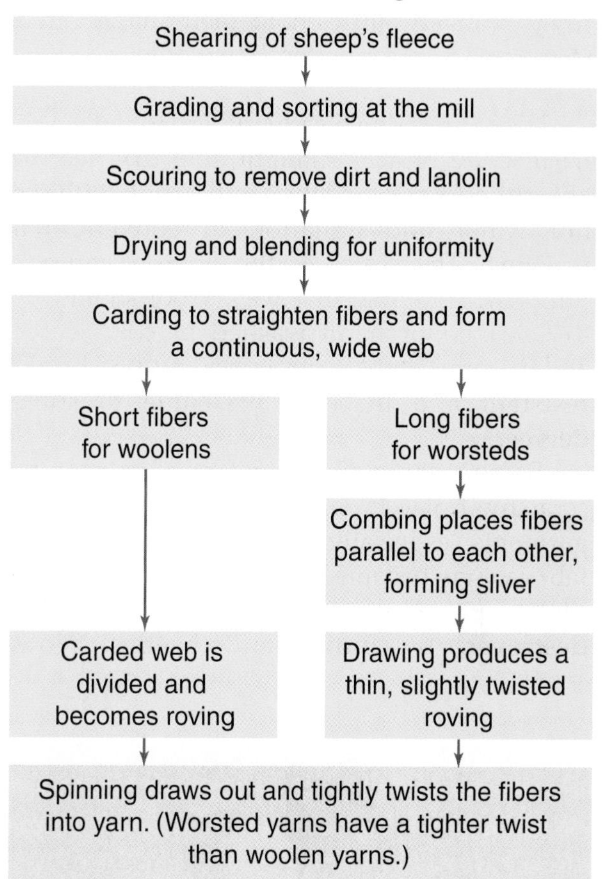

8-16 Wool processing turns the sheared fleece of sheep into either woolen yarns or worsted yarns.

Wool Characteristics and Uses

The wool fiber has great versatility. It is made into featherweight fabrics, bulky tweeds, and plush carpets. Wool fabrics have endless varieties of weights, textures, finishes, and colorations.

Wool fibers have natural *crimp*, or curl that looks like wavy lines or coil springs under a microscope. Better quality wool fibers have more crimp per inch. It gives the fiber elasticity to stretch, compress, and recover to its original shape without damage. This resiliency enables wool garments to return to their original size and shape after a "rest." Crimp is what causes wool to be wrinkle-resistant.

Wool is a warm fiber. It has good insulating qualities since it traps air among the fibers. This inhibits the rapid transfer of temperature. It is lightweight, too. Wool garments have always been popular for military use because of their light weight and warmth.

Wool is somewhat water-resistant as well as absorbent. Wool fibers have a thin film over scalelike cells that overlap like roof shingles. This covering helps to shed water, as well as spots and stains. However, with prolonged exposure to wetness, wool is absorbent. Wool fabric can absorb moisture, or perspiration, to quite a degree without making the wearer feel wet. Like all natural fibers, wool fibers allow the fabric to breathe. This makes wool comfortable to wear. Wool takes readily to dyes. It can be dyed in a wide range of colors and intensities. Wool also resists fading. Wool is long-wearing because it is strong. However, when soaked in water, wool fibers weaken.

Wool is static-resistant because of its moisture content. It does not "spark" or cling from static electricity. Wool also has flame-retardant characteristics. It will extinguish itself if it is not kept in direct contact with a flame. Dense weaves are especially flame-retardant, since they have less air in them.

Wool is easy to tailor. During garment construction, it eases, stretches, and shapes well. With steam, creases can be removed and reset. Worsted fabrics hold their shape and a sharp crease much better than woolens do. See 8-17.

There are some disadvantages to wool. It is harmed by bleach, sunlight, and dry heat. It absorbs odors. Wool fabric will shrink with the hot water and agitation of conventional washing. The scales of the fibers compress, interlock, and stay that way. Wool should be dry-cleaned or hand washed in cold water. Today some wool is processed to be shrink-resistant and machine washable to some degree.

Some people find wool garments to be scratchy if placed next to the skin. Some wool garments, especially those made of worsted fabrics, may become shiny with wear. Another disadvantage of wool is that it can be destroyed by moths and beetles. Now, however, wool can be chemically treated so moths and other insects shun it. Also, wool can be reused rather than discarded.

Hart Schaffner & Marx/Hartmarx Corporation

8-17 Top brand sportcoats, such as this one, and other high-quality wool garments are made of 100 percent worsted wool.

Categories of Wool

There are different categories of wool. Also, wool is often blended with less expensive fibers. To inform and protect consumers, the Wool Products Labeling Act requires the label on wool products to give the percent of fiber content and also to give the source. The Act does not say anything about quality. It is enforced by the Federal Trade Commission (FTC).

The terms *virgin wool, pure wool,* and *100% wool* are used interchangeably. They indicate fabrics that are made from all-wool fibers or yarns that have never before been used. These wools are softer, stronger, and more resilient than recycled wool.

Recycled wool refers to wool fibers from previously-made wool fabrics. They might be from cutting scraps, mill ends, or garments. These are shredded back into fibers. They are not as soft, strong, or resilient as those from new wool. They are used mostly in thick, stiff utility fabrics. They often go into winter gloves, interlinings for coats, and picnic blankets. Wool products must be labeled "recycled" when not made of virgin wool.

The Wool Bureau, Inc. is an association of U.S. wool growers. Located in New York City, its purpose is to assist mills, manufacturers, and retailers in promoting wool and wool products. It develops special advertising and marketing programs. It researches new products and new processes for economical production. It also functions as a fashion forecasting service and consultant to its members. Its *Woolmark* and *Woolblend* certification marks are shown in 8-18 and 8-19. To be labeled with these symbols, samples of the products must have been tested by Wool Bureau inspectors for strength, color fastness, fiber content, and quality.

Specialty wools are from such animals as goats, camels, and llamas. Their supplies are limited and from the outer regions of foreign lands. They are more expensive than sheep's wool and used less. Some specialty wools are from coarse, long, outer animal hairs. They are used in interlinings, rough coatings, and

The Wool Bureau, Inc.

8-18 As stated by the trade organization, "The Woolmark label is your assurance of quality-tested fabrics made of the world's best...Pure Wool."

The Wool Bureau, Inc.

8-19 As stated by the trade organization, "The Woolblend Mark label is your assurance of quality-tested fabrics made predominantly of wool."

Mohair Council of America

8-20 Mohair is a specialty wool from the long silky fleece of Angora goats. Fashionable apparel made from mohair fibers is warm and soft.

upholstery fabrics. Others are fine, soft, undercoat hairs. They are used in luxury coatings, sweaters, shawls, suits, and dresses.

Some examples of specialty wools are luxurious cashmere from the cashmere goat and *camel's hair* from the two-humped Bactrian camel. Silky *mohair* is from the Angora goat. Find out about mohair in 8-20. *Angora* (from the Angora rabbit), *vicuña*, *alpaca*, and *guanaco* are other specialty wools.

Silk

Silk is obtained by unwinding the cocoons of silkworms. The silk fiber is the only natural filament fiber. A ***filament*** is a very long, fine, continuous thread, rather than being many short lengths. Silk filaments may be a thousand yards or more in length.

Silk Production and Processing

Silk originated in China. Its revered silkworm cocoon and method of production were kept a secret for over 2,000 years. Eventually, silkworm eggs and mulberry seeds were smuggled to other countries. Today, Japan is the leading producer of silk fiber. Korea, Taiwan, Thailand, Italy, and France also produce it. The United States consumes more silk than any other country of the world. Many fine silk products are made in the United States from the imported fibers.

The raising, or cultivation, of silkworms is the science of ***sericulture***. It carefully controls the cycle from moth to silk fiber.

A silk moth lays several hundred eggs that eventually hatch. The resulting tiny worms feed on the leaves of mulberry trees, as in 8-21, until they are fat and about two inches long. Then they stop eating and spin a protective cocoon around themselves, shown in the upper part of 8-22. A silk filament is released from each silkworm to spin the cocoon. The filament hardens in the air. A gum, called *sericin*, is released at the same time and holds the cocoon together.

In silk processing, the worm inside the cocoon is intentionally killed by heat before it turns into a moth and forces its way out of the cocoon. That would ruin the continuous silk filament. The cocoon is soaked in hot water to allow for the *reeling process*, as in 8-23. During this process, the cocoon becomes soft and easy to unwind. The unbroken silk threads are unwound from cocoons and put onto large reels.

The sericin that originally bound the fiber to the cocoon is boiled off either before or after weaving. Silk does not need to be spun into a yarn like cotton or wool because it is a continuous filament. However, most silk, except coarse sizes, must have many filaments tightly twisted together for a heavier weight before it is woven into fabric. This process is called *throwing*. It is solely to provide the weaver with a yarn of large enough width and proper weight for the particular fabric being made. Refer to 8-24 to review the processing of silk.

Used alone, the term "silk" refers to the cultivated silk from carefully tended silkworms. *Wild silk* fibers come from uncultivated silkworms. The wild silkworms eat the leaves of trees other than mulberry. They spin

China External Trade Development Council, Taiwan

8-21 Cultivated silkworms eat 150 to 200 pounds of mulberry leaves during the 40 days until they are large and mature.

China External Trade Development Council, Taiwan

8-22 A mature silkworm takes about three days to spin a cocoon around itself, held together with sericin. Silk fibers for apparel are obtained from this.

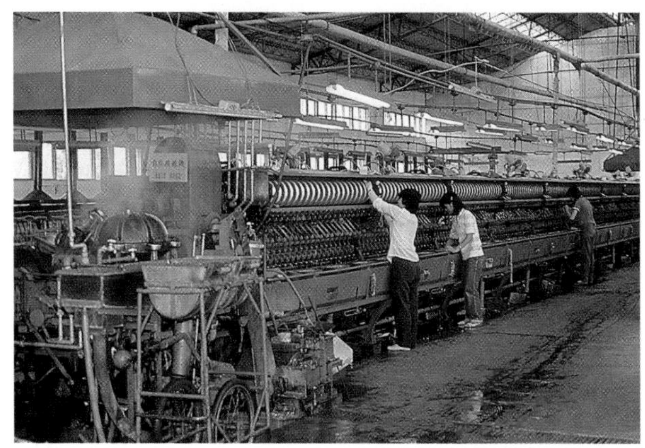

China Trade Development Council, Taiwan

8-23 Hot water is used to loosen the silk filaments, so they can be unwound by automatic reeling machines.

Silk Processing

Silkworm spins cocoon around itself

↓

Cocoon is soaked in hot water; silkworm dies; silk filament loosens

↓

Reeling of several filaments off cocoons while slightly twisting them together onto reels

↓

Sericin is boiled off now or after weaving

↓

Throwing tightly twists more filaments together into yarn

8-24 Silk, known for its luxurious beauty, is the only natural fiber in filament form.

cocoons under natural conditions. The fiber is strong with a distinctive rugged appearance when woven. It is uneven in texture and is somewhat stiff and coarse. It is usually used in its natural tan color. It is difficult to bleach or dye. *Tussah* fabric is woven from wild silk fibers.

Doupion silk comes from two silkworms that have spun their cocoon together. It has irregular, thick-thin filaments producing a slubbed effect. It is used in making *shantung* fabric.

Spun silk is made from pierced cocoons (cut filaments) or waste silk. The tangled fibers on the outer layers of leftover cocoons, along with imperfect ones, are used to make spun silk. The fibers are several inches long. They are cleaned, combed, and spun into yarn the same as cotton and wool. Spun silk is not as strong or lustrous as reeled silk.

Silk Characteristics and Uses

Silk has always been a status symbol. In past eras, it was known as the "cloth of kings." Elegant, expensive silk fabrics are synonymous with luxury.

Silk is made into fabrics of various degrees of crispness and softness, thickness or transparency, as shown in 8-25. It has a natural shine or luster. Silk fabrics drape nicely. They are absorbent and can resist permanent wrinkles. Silk fibers are strong and will last a long time.

Silk is a natural insulator. It is warm in the winter and cool in the summer. It is light and comfortable. Also, silk takes dyes very well, since it is highly absorbent.

Silk requires some special care. Silk garments should be dry-cleaned unless the care label says they can be hand washed. Silks can shrink and some dyes that are used may bleed during laundering. Sunlight and alkaline soaps can damage silk. High iron temperatures weaken silk and cause the color to fade or to yellow. Also, silk has a tendency to waterspot. Perspiration harms silk, and deodorants may cause it to deteriorate.

When the sericin is washed out, some silk becomes very thin and lightweight. Salts from tin, lead, or iron can be added to make the resulting fabrics heavier. These *weighted silks* often rustle when shaken. Labels must indicate the percentage of weighting. Weighted silk is usually cheaper, weaker, and less serviceable. These days silk is most often blended with other fibers for weight or performance improvements rather than being weighted with metal salts.

Talbots

8-25 This silk pin-tucked shirt is worn with a silk organza jacket and silk shantung pants.

Pure silk must contain no metallic weightings. It can have no more than 10 percent by weight of dyes or finishing materials. However, black silk may have 15 percent.

A chart of the four most popular natural fibers is shown in 8-26. The characteristics of each fiber can be reviewed and compared to the others.

Other Natural Fibers and Materials

Asbestos is a natural mineral fiber found in veins of rocks. Asbestos does not burn. Thus, it was used for firefighters' clothing, fireproof curtains in theaters and school auditoriums, insulation in buildings, and industrial clothing. However, constant exposure to asbestos has been found to cause health problems. Today, nonflammable manufactured fibers have replaced asbestos.

Miscellaneous materials such as fur and leather, as shown in 8-27, are from the hides of animals. They are not fibers; however, they are used for apparel. They are quite expensive since their supply is limited. They have been used since before recorded history. They are durable and weatherproof. Today they are highly fashionable and controversial. Artificial substitutes have been developed to copy their looks.

Popular Natural Fibers

Fiber Name and Source	Fiber Advantages	Fiber Disadvantages	Typical Apparel Uses
Cotton Boll of cotton plant	strong, durable absorbent, cool quite inexpensive versatile uses soft, comfortable no static buildup stands high temperatures dyes and prints well	affected by mildew wrinkles unless treated or blended shrinks in hot water weakened by finishes, perspiration, sun burns readily no elasticity	underwear socks shirts jeans sportswear dresses blouses outerwear
Linen Stem of flax plant	very strong resists dirt, stains absorbent, cool comfortable durable over time stands high temperatures smooth, lustrous lint-free: for dish towels and medical cloths	affected by mildew, perspiration wrinkles easily hard to remove creases expensive if good shines if ironed on right side burns readily ravels, shrinks	handkerchiefs suits dresses skirts shirts
Wool Fleece of sheep	very warm lightweight durable very absorbent comfortable resilient resists wrinkles creases well dyes well resists static easy to tailor can be reused	weakened when wet affected by moths shrinks and mats with heat, moisture, and agitation needs special care, dry cleaning absorbs odors harmed by bleach, perspiration scratchy on skin pills	sweaters suits coats skirts socks slacks outerwear
Silk Cocoon of silkworm	lustrous smooth, luxurious very strong absorbent lightweight resists wrinkles, soil, mildew, and moths comfortable dyes well drapes well	expensive needs special care, dry cleaning water spots yellows with age weakens with perspiration, sun, soaps attacked by insects, silverfish	evening gowns wedding gowns lingerie blouses scarves dresses neckties suits

8-26 All natural fibers have characteristics that make them suitable for various types of apparel.

Down is a light, fluffy feather undercoating that protects geese and ducks from extremely cold temperatures. The best quality is on the breast and underbody of large birds during the winter months. It is lightweight and extremely effective as an insulator. Down's microscopically thin fibers have millions of tiny insulating air pockets in each ounce. Down-filled comforters, sleeping bags, ski jackets, winter coats, and other products offer maximum warmth. The down is placed between the outer fabric and the lining of these items, as shown in 8-28.

Manufactured Fibers

Manufactured fibers are made from substances such as wood cellulose (the fibrous substance in plants), oil products (petroleum), and chemicals. They try to duplicate and improve upon the characteristics of natural fibers. They are often cheaper, stronger, and more durable than the natural products they replace. They are more uniform in size and quality than natural fibers, since their processing is controlled. They can be carefully engineered and modified to have characteristics that fit specific end-use needs. As with all fibers, each manufactured fiber has both advantages and disadvantages.

In general, manufactured fibers tend to offer easy-care *resilience*, retaining their shape or bouncing back when crushed. They are wrinkle-resistant and often require no ironing. In fact, high temperatures will soften those that are heat-sensitive. Fabrics of heat-sensitive fibers have a low melting point and can be heat-treated to set pleats, mold shape, or emboss fabric designs.

Most manufactured fibers are versatile. Most are nonallergenic, strong, and resistant to abrasion (surface wear and rubbing). Most do not absorb moisture or "breathe" for wearing comfort. Thus, sometimes they feel clammy when worn, especially in hot, humid weather. However, this same characteristic enables them to dry quickly.

Saxony Sportswear Company

8-27 Leather and fur have been used for warmth and protection since before recorded history. They are exquisite but controversial materials.

JC Penney

8-28 The puffiness and warmth of these ski parkas is provided by down feathers inserted between the inside and outside fabric layers of the jackets.

Manufactured fibers that are nonabsorbent build up static electricity that causes them to "spark" and to cling to the wearer. They can be surface cleaned with a damp sponge since their smooth, nonporous surfaces do not allow dirt and grime to become imbedded. However, oil and ground-in stains are hard to remove. Moths and mildew do not usually affect these fibers.

The trade organization for producers of manufactured fibers is the American Fiber Manufacturer's Association, Inc. (AFMA)

The Development of Manufactured Fibers

After thousands of years with only natural fibers, experimental production of manufactured fibers started in Europe around 1850. By copying nature, scientists first changed natural fiber materials into liquids. Then they copied the silkworm by forcing the liquid out through tiny holes. Cellulosic material was taken from plants and "formed" into fibers.

The first commercial manufactured fiber was produced in the United States in 1910. This fiber was sold as "artificial silk" for several years. It was given the name *rayon* in 1924. That same year, commercial production of an *acetate* fiber was started.

Later, textile chemists made their own fiberlike materials by duplicating the molecules of plant and animal fibers. In 1939, *nylon* was introduced. It was the first noncellulose, "test-tube" fiber. It was made totally from chemicals. Women's hosiery was one of the first products made with it. It was used for military purposes during World War II because of its strength.

In the 1940s, '50s, and '60s, more manufactured fibers were developed. Fiber blending became widespread, bringing together the best properties of two or more fibers into fabrics that offered easy care. Also, sophistication of manufactured fibers took place for specific end-use products. The fibers were modified to give greater comfort, more flame-resistance, better soil release, easier dyeability, and so on. Qualities were built into fibers to satisfy consumer textile needs.

As time passed, there were also more production facilities for manufactured fibers. By the start of the 1970s, fibers produced at chemical plants accounted for well over half of America's textile usage.

Manufactured Fibers Present and Future

Today, manufactured fibers account for about three-quarters of all fibers used by American textile mills. They are made into a wide range of textile products. They are used in medicine, transportation, space travel, environmental control, and many other industries.

Fiber innovations and new fabrics are being developed by technology to fill our apparel needs. For instance, swimsuit fabrics now resist fading from sunlight, saltwater, and swimming pool chlorine. They are fast drying and resistant to mildew, mold, and silverfish.

Other fabrics, used in exercise and active outdoor wear, do not absorb moisture. Instead, they pull it away from the skin to the surface of the fabric where it can bead up and evaporate. The wearer stays fairly dry.

Gore-Tex® membrane is a fluorocarbon material. It contains billions of tiny pores per square inch as a result of stretching it as thin as possible during production. The membrane is too thin to be used alone, so it is *laminated* (joined with an adhesive) to a sturdy outer fabric. Garments made from Gore-Tex® laminate are waterproof, yet breathable. This means they block wind and rain, but perspiration can pass through the garment, keeping the wearer dry and comfortable.

New trademarks appear as modified manufactured fibers are developed. Old ones are discontinued and replaced.

The future of textile technology may be even more amazing. Apparel may be made of synthetics that can automatically adjust to different environmental conditions, such as heat, cold, or rain. The possibilities are endless!

Categories of Manufactured Fibers

There are two basic types of manufactured fibers: cellulosic and noncellulosic.

Cellulosic fibers are produced from cellulose, the fibrous substances found in plants. These fibers are made with a minimum of chemical steps. Most of the cellulose used in them comes from softwood trees. The four cellulosic fibers are rayon, acetate, lyocell, and triacetate.

Noncellulosic fibers are made from molecules containing various combinations of carbon, hydrogen, nitrogen, and oxygen. These are obtained from petroleum, natural gas, air, and water. Textile chemists link these molecules into chemical compounds called ***polymers***.

Manufactured fibers have been grouped for identification. A ***generic name*** is given to a group of fibers of similar chemical composition. Manufactured fibers are now classified into about two dozen generic groups by the Federal Trade Commission. A new generic name is established only when a fiber is developed that is different in chemical composition from other fibers. The present generic categories are listed in 8-29. Some are more familiar than others. Also, not all of these are commonly used in apparel. Some are used mainly for home furnishings, industrial, or medical products.

Within each generic group, there can be many individual fibers. They are modified versions of the generic composition and are called *variants*. Each is slightly different from the others. They are engineered to have specific properties for certain purposes. They may offer greater comfort, flame resistance, soil release, or blending qualities. Each brings specific desirable qualities to a finished product.

When a fiber producer develops a new variant, it is given its own brand name, or *trademark*. Such names are always capitalized. The trademark, and the process for making the particular fiber, are registered with the U.S. Patent Office. The individual fiber producers own, advertise, and should stand behind their trademarked fibers. Sometimes fiber producers have several trademarks within a generic category because they have several fibers with slight differences. No other producers can make that exact variant unless they are licensed to do so by the company holding the patent. Some generic fiber categories, such as nylon and polyester, have many different manufacturers and trademarks.

Manufactured Fibers	
Cellulosic	**Noncellulosic**
acetate	acrylic
lyocell	anidex*
rayon	aramid
triacetate*	azlon*
	glass
	lastrile*
	metallic
	modacrylic*
	novoloid *
	nylon
	nytril*
	olefin
	polybenzimidazole
	polyester
	rubber
	saran
	spandex
	sulfar
	vinal*
	vinyon*

* Not currently produced in the U.S.

8-29 Manufactured fibers can be grouped as cellulosic or noncellulosic, depending on their composition.

Producing Manufactured Fibers

All manufactured fibers are made by the same process. It is done by forcing a syrupy substance through tiny holes. Solid fiber-forming substances (the raw materials) must first be converted into a liquid state. To do this, they are dissolved in a chemical solvent or melted with heat.

Extruding

The syrupy liquid is *extruded*, or forced out, in the desired thickness. The term used to describe fiber thickness or diameter is ***denier***. The smaller the denier of the fiber, the softer and more pliable it is. An 840-denier is used in truck tires, while a 15-denier filament is used in sheer panty hose.

Fine, *microdenier* fibers that are half the denier of silk are now being produced. They are soft, luxurious, and drapable. Fabrics made from them are wrinkle-resistant and water-repellent, yet breathable.

A device called a ***spinneret*** is used to extrude the liquid. It is similar in principle to a bathroom showerhead. It has from one to hundreds of tiny holes. It is usually made from corrosion-resistant metals. Each tiny hole forms one fiber.

Manufactured fibers can be extruded from the spinnerets in shapes such as round, octagonal, trilobal (three-sided), and so on. The holes in the spinneret are shaped the way the fiber will be formed. The shape of a fiber is changed to make it more or less lustrous, or to help it hide dirt. (Variations of shape are not possible for natural fibers.)

Certain additives put into the extruding solution can give the finished fiber such characteristics as resistance to static or flame-retardency. *Solution dyeing* is the process of adding color to the solution before it is extruded. This gives a high degree of colorfastness, since the pigment is an integral part of the fiber.

Chemical Spinning

Chemical spinning causes the extruded solutions to become fibers. The filaments of continuous strands coming from the spinneret are hardened or solidified. There are three methods of spinning manufactured fibers: wet, dry, and melt spinning. Some fibers may be produced by more than one method.

Wet spinning hardens the filaments in a chemical bath. As the filaments emerge from the spinneret, they pass directly into the bath where they are solidified or regenerated. Acrylic and rayon fibers are produced by this method, shown in 8-30.

Dry spinning solidifies the extruded filaments by drying them in warm air. The chemical that was used to change it to a liquid is evaporated away. This process, shown in 8-31, is used for acetate, acrylic, modacrylic, spandex, triacetate, and vinyon.

Melt spinning extrudes melted substances. They are hardened by cooling, as indicated in 8-32. Nylon, olefin, polyester, aramid, and glass fibers are produced by melt spinning.

Stretching

During or after hardening, manufactured fibers are stretched. This makes the denier smaller. It also causes the fiber molecules to

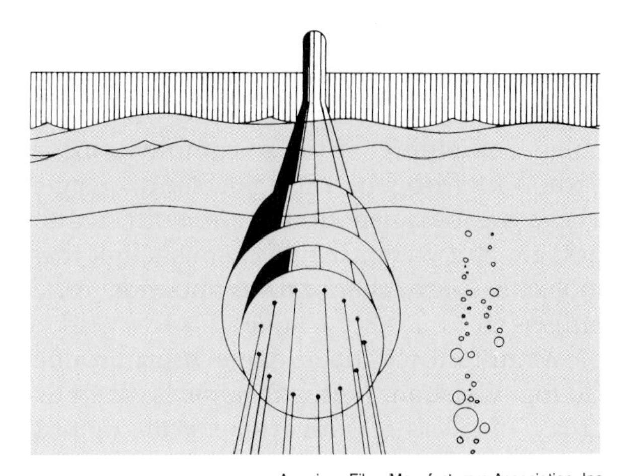

American Fiber Manufacturers Association, Inc.

8-30 In wet spinning, the filaments emerge from the spinneret directly into a chemical bath for solidification.

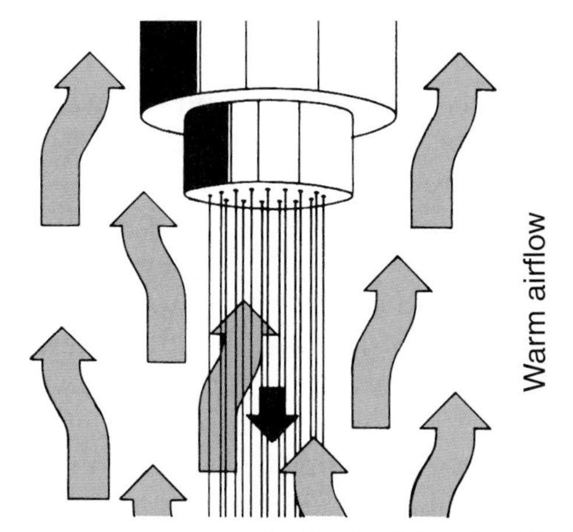

American Fiber Manufacturers Association, Inc.

8-31 In dry spinning, the filaments coming from the spinneret solidify by drying in warm air. The original liquefying chemical evaporates away.

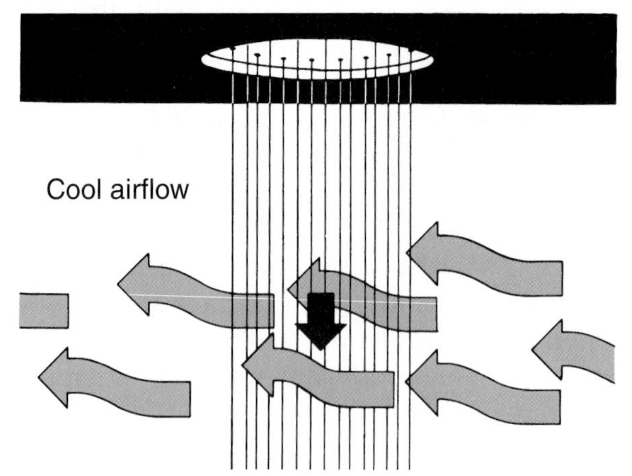

American Fiber Manufacturers Association, Inc.

8-32 Melt spinning solidifies the melted, extruded filaments by cooling them.

become arranged, or *oriented*, into a more orderly pattern. This gives added strength. Fibers can be stretched to a consistent diameter. They can also be made thick and thin by varying the stretching pressure during spinning.

Cutting

Sometimes manufactured fiber filaments are cut into staple. ***Staple fibers*** are short pieces, usually between one and four inches. Cotton, wool, and linen fibers occur naturally as staple. All staple fibers (natural or manufactured) are classed into length categories. Longer staple fibers are better for weaving into yarn than short staple lengths.

Manufactured fibers are extruded from the spinneret in large continuous filament bundles called *tow*. Filament tow may contain thousands of parallel filaments in rope form, as shown in 8-33. The tow is usually mechanically *crimped* (curled or waved) and cut into the desired staple lengths. Some crimped staple is used as filling without being made into yarns. It is called *fiberfill*. Study 8-34 to review the steps in manufactured fiber processing.

Trevira Polyester Hoechst Fibers Industries

8-33 Each of the lines going into these barrels contains many continuous filaments of a manufactured fiber.

Generic Characteristics

Each generic classification of manufactured fiber has its own chemical composition and set of characteristics. The most widely used manufactured fibers in apparel are polyester, nylon, rayon, and acrylic.

Polyester

Polyester has become the most widely used of all manufactured textile fibers. It has outstanding wrinkle-resistance and easy care qualities. It does not shrink or stretch. It is

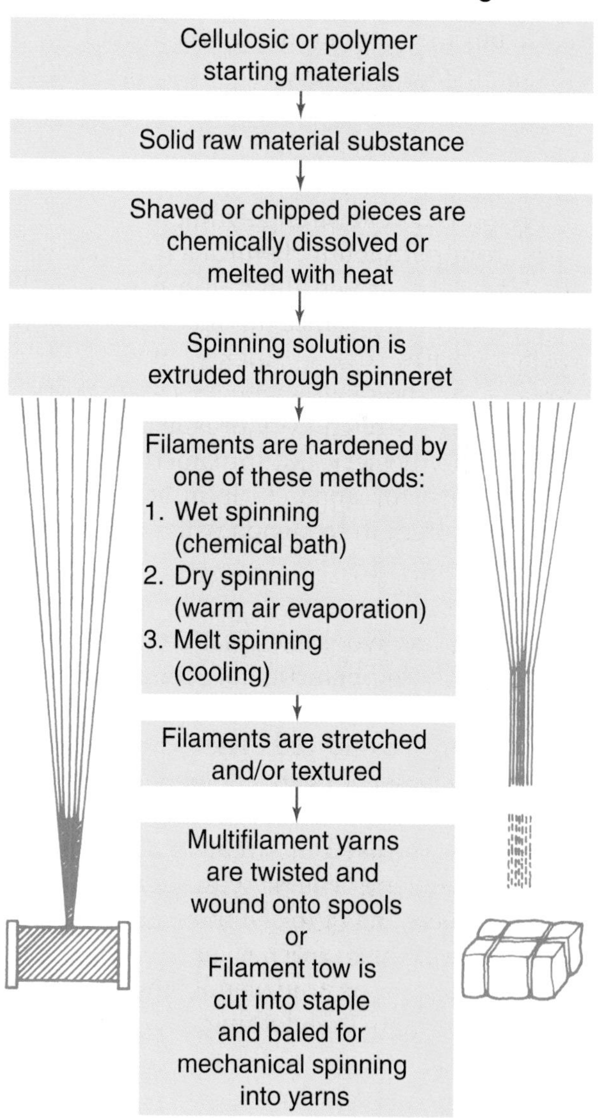

8-34 Manufactured fiber processing uses specific raw materials that are melted or dissolved into a liquid. The liquid is extruded through a spinneret, solidified, stretched, and twisted or cut.

strong and quick-drying. Heat-set pleats and creases stand up extremely well under everyday wear. Also, water-based stains are easy to remove from fabrics made of polyester fibers. Polyester gives these properties to "permanent press" blends, especially when combined with cotton, rayon, or wool.

Polyester is extremely versatile. It can be made to have the look and feel of any natural fiber while adding easy care, durability, and low price. It is used in woven, knitted, and nonwoven fabrics. It is used for all kinds of apparel, home furnishings, and industrial products. Polyester insulating fiberfill fibers have hollow cores that act as tiny air-trapping shafts. They are soft and compressable like down, but they remain fluffy when wet. Some fiber companies produce many different variations of polyester.

Nylon

Nylon is a very strong synthetic fiber. It is made from petroleum chemicals (petrochemicals). It has great versatility. It gives strength and abrasion resistance to many blends. It washes easily, dries quickly, needs little pressing, and holds its shape well since it does not shrink or stretch. Monofilament nylon is knitted into hosiery. Multifilament nylon is used as yarn for apparel. Spun staple nylon yarns add strength to blends with other fibers.

Acrylic

Acrylic has wool-like qualities as well as easy care. These fibers are not scratchy and do not shrink or mat. However, they do build up static electricity and pill. They can be made into crisp fabrics, but they are more commonly used to make soft, high-bulk, textured yarns. Such yarns are used in sweaters and fur-like fabrics. They are warm with light weight. Acrylic fibers have resistance to common chemical solvents, bleaches, and weathering. They are used alone or in blends, and from fine fabrics to heavy work clothing.

Modacrylic

Modacrylic is a modified acrylic fiber. Its heat sensitivity permits it to be manipulated for different results. It can be stretched, embossed, and molded into special shapes. It is often made into dense, high pile, fur-like fabrics. It also resists flames and most chemicals. It is often blended with other fibers to reduce their degree of flammability. It is used extensively for flame-retardant sleepwear.

Rayon

Rayon is composed of regenerated cellulose, mainly from wood pulp. Due to its high absorbency, it dyes very well. Rayon fibers are soft and comfortable, as shown in 8-35. They are versatile, creating fabrics with different looks, weights, and performance. They can have beautiful luster. When given a high twist, they are used in crepe fabrics. Rayon fibers also blend well with almost all other fibers.

Acetate and Triacetate

Acetate and triacetate are made from a derivative of cellulose, cellulose acetate. Fabrics from these fibers drape beautifully, and look and feel luxurious. They often have crisp body. Their natural luster can be adjusted from dull to bright for a wide variety of rich textures. They take special dyes well and are colorfast. Also, they blend well with other fibers.

Acetate and triacetate have low wet strength and low resistance to abrasion. They must be dry-cleaned or carefully hand washed. They may be heat-treated to help them resist

Draper's & Damon's

8-35 Rayon is terrific when used by itself, such as for this reversible skirt, or when blended with other fibers, such as with nylon in this sweater.

wrinkles and maintain their stability. In fact, triacetate has resistance to damage or glazing by heat, thus it may be safely ironed at higher temperatures. Heat-setting treatments can give triacetate pleat retention by permanently changing the crystalline structure of the fiber. Triacetate is stronger than acetate and more shrink-resistant.

Olefin

Olefin is very nonabsorbent. Thus, it is shrink resistant and difficult to dye. It is available only in limited colors and resists fading. Its great advantage is its ***wicking*** power. In other words, it can pull moisture away from the body to the surface of a fabric where it can evaporate quickly. Because of this, it is used next to the skin for exercise and outdoor wear, ski underwear, and knitted sportswear. It is also very effective in disposable diapers.

Olefin is very lightweight, yet it is bulky enough to trap air for insulating purposes. It is less expensive than other manufactured fibers. It does not deteriorate from chemicals, mildew, perspiration, rot, or weather. Consequently, it is used to make protective clothing for many industries. Its disadvantage is that it is sensitive to light and to heat, which causes it to melt. However, modifications to the olefin fiber have improved these characteristics.

QVC; Will Rutledge, photographer

8-36 With only a small amount of spandex, this pants fabric has good stretch and recovery, providing comfortable fit and freedom of movement.

Spandex

Spandex has great elasticity. It is stronger, lighter, and more durable than rubber in apparel. Garments containing spandex retain their holding power through constant stretching and recovery. Spandex was first used in swimsuits and foundation garments since it provides softness and comfortable fit with freedom of movement. However, now it is popular in all types of apparel, including high-fashion designs. Spandex fibers are always used with other fibers since only a small amount is needed to give the desired holding power and recovery characteristics. See 8-36.

Lyocell

Lyocell is a cellulosic manufactured fiber, the most recent generic category to be developed, was introduced in the mid-1990s. It is biodegradable and heralded as being environmentally friendly. It is made with wood pulp from trees grown in managed, constantly replanted forests. The chemical agents used to manufacture it are recycled. Lyocell fibers are strong, highly absorbent, and blend well with other fibers. They are used in many types of fashion fabrics.

Metallic

Metallic fibers are made of metal, plastic-coated metal, metal-coated plastic, or a fiber core completely covered by metal. They resist weather, mildew, and moths. They are not absorbent, and they do not shrink or stretch. These fibers are limited to decorative uses.

Aramid

Aramid fibers are lightweight, tough, and resistant to flames and chemicals. They can stop bullets. These fibers are used in protective clothing for firefighters, police officers, military personnel, and others in hazardous occupations.

Anidex

Anidex is an elastic fiber. It is very resistant to chemicals, sunlight, and heat. It has excellent feel and easy-care properties. It offers stretch and recovery in blends with other fibers without changing the natural look and feel of the basic fabric.

Polybenzimidazole

Polybenzimidazole does not burn, melt, or drip. It provides garment comfort, since it retains high amounts of moisture. It has excellent resistance to chemicals, solvents, fuels, and steam. It resists oils and stains and has good insulating properties. It is used for firefighters' protective apparel and industrial work clothing. Flight suits and racecar driver uniforms are its other uses.

Vinyon

Vinyon has low strength and poor spinning qualities. It is very sensitive to heat. It is resistant to chemicals and has good elasticity. It is used mostly in bonded fabrics and nonwovens. It is used to make industrial clothing, but not for fashion apparel.

Rubber

Rubber fibers are made from the sap (latex) of certain plants. They are usually used as a core around which other fibers are wrapped to protect the rubber from abrasion. They have good elasticity to stretch and recover. They have low resistance to bleach, heat, and sunlight. They are used in such items as rubber boots, raincoats, and elastic.

Glass

Glass fibers have a mineral source. They are not used in apparel because of their heavy weight, low abrasion resistance, and poor bending strength. They also may cause skin irritation or prickly cutting. For industrial purposes, *fiberglass* provides excellent insulation qualities, reinforcement for molded plastics, and fire resistance.

Saran

Saran has low absorbency, high resiliency, and little reaction to chemicals or weather. It has low resistance to heat and poor stability. It has household and industrial uses rather than being used in apparel.

A chart of the more popular manufactured fibers used in apparel is shown in 8-37. Use it to review and compare the characteristics of the different fibers. Basic knowledge of these fibers is important for smart clothing decisions.

Yarns

Natural and manufactured fibers are made into yarns. ***Yarns*** are continuous strands of textile fibers in a form suitable for knitting, weaving, or otherwise processing into yard goods. Raw fibers are converted into several different types of yarns. The yarns, in turn, are made into fabrics.

Yarn Types

Monofilament yarns are simply single filaments, usually of a high denier. One example of monofilaments is the single strand yarns in women's hosiery. Another is the clear, plastic-like sewing thread used to hem some ready-to-wear garments.

Most often, filaments are combined and used as *multifilament yarns* in apparel. The yarns are formed by twisting the many continuous strands of fiber being extruded through the spinneret at the same time. As the degree of twist is increased in any yarn, the yarn becomes harder, more compact, and less lustrous. A low twist is used for most multifilament yarns, so they are soft and lustrous.

Spun yarns are made with staple fibers. The fibers are usually bound together by *mechanical spinning*. This process pulls (draws) and twists the fibers together to obtain a continuous length sufficient for weaving or knitting. Generally, for spun yarns, a tighter twist produces a stronger yarn. However, too tight a twist can weaken the final yarn.

Yarns spun from staple fibers are more irregular than filament yarns. The short ends of the fibers produce a fuzzy effect on the yarn surface. Spun yarns are more bulky than filament yarns of the same weight. They are also more porous and warmer. The resulting fabrics often have a more natural feeling and snag less than fabrics made from filament yarns. However, spun yarns have a tendency to *pill* (form little balls of fiber on the surface) from rubbing and wearing. Look at 8-38 to see a monofilament yarn, a multifilament yarn, and a spun yarn.

Ply yarns are formed by twisting together two or more single yarns. Each yarn strand is called a *ply*, as in "three-ply" yarn. Ply refers to the number of yarns twisted together. This is

Popular Manufactured Fibers

Generic Name: Some Trademarks	Fiber Advantages	Fiber Disadvantages	Typical Apparel Uses
polyester Coolmax Dacron Diolen Fillwell Fortrel Loftguard Microlux Microtherm Serelle Tarilin	resilient colorfast strong, durable easy care resists wrinkles, abrasion, bleach, perspiration, mildew, moths can be heat set easy to dye does not stretch or shrink very versatile	low absorbency spun yarns pill takes oily stains static buildup	permanent press fabrics fiberfill insulation shirts blouses dresses slacks suits underwear sportswear children's wear
nylon Anso Antron Cordura Micro Supplex Silky Touch Stainmaster Supplex Tactel Ultron Wear Dated	very strong lightweight dries quickly durable resilient resists mildew, moths, chemicals, wrinkles lustrous colorfast elastic	low absorbency surface pills damaged by sun picks up oils and dyes in wash static buildup heat sensitive	sweaters hosiery lingerie skiwear slacks windbreakers dresses raincoats swimwear blouses
acrylic Acrilan Biofresh Creslan Duraspun Evolutia MicroSupreme Pil-Trol Wear-Dated WeatherBloc	resembles wool lightweight soft, fluffy warm, bulky resilient resists weather, moths, chemicals, mildew, wrinkles shape retention easy care colorfast	low absorbency static buildup surface pills heat sensitive	sportswear sweaters infant wear socks knitted garments pile fabrics jackets skirts slacks bathrobes
rayon Fibro Galaxy	very absorbent dyes/prints well soft, pliable, drapable comfortable/versatile inexpensive colorfast no static or pilling	wrinkles/shrinks unless treated low resiliency heat sensitive will mildew stretches weak when wet	blouses dresses sport shirts lingerie sportswear neckties jackets

(continued)

8-37 The many manufactured fibers available today allow great variety in the production of apparel and other items.

usually done when extra strength, added bulk, or unusual effects are desired. When a number of ply yarns are twisted together, they form a *cord,* as seen in 8-39.

Combination Yarns and Blends

Combination yarns are ply yarns composed of two or more different yarns. They can be formed by putting spun staple

Popular Manufactured Fibers *(continued)*

Generic Name: Some Trademarks	Fiber Advantages	Fiber Disadvantages	Typical Apparel Uses
acetate Celanese Celstar Chromspun Estron MicroSafe	silk-like look/feel drapes well does not shrink resists moths, mildew, pilling inexpensive easy to dye versatile	poor abrasion resistance weak heat sensitive needs special cleaning care dissolved by nail polish remover	dresses blouses linings lingerie shirts scarves neckties
olefin Angel Hair Crowelon Duron Herculon Impressa Marquesa Lana Nouvelle Trace	resists abrasion, chemicals, stains, mildew, pilling, wrinkles, static not affected by aging, weather, perspiration excellent wicking thermal warmth strong, durable very lightweight	heat sensitive poor dyeability nonabsorbent	knitted sweaters, socks, sportswear nonwoven fabrics for industrial apparel filler in quilted goods disposable diapers
spandex Dorlastan Glospan Lycra	very elastic resistant to lotions, oils, perspiration, sun, flexing lightweight strong, durable soft, smooth	yellows with age heat sensitive harmed by chlorine bleach nonabsorbent	swimwear/skiwear foundation garments support hose slacks exercise and dance wear fashion apparel
lyocell Tencel	biodegradable absorbent strong resists sunlight, aging, abrasion	attacked by mildew	soft denims shirts reusable nonwovens many fashion fabrics
aramid Kevlar Nomex	strong no melting point resists abrasion, most chemicals resilient, supple flame-resistant	no stretch nonabsorbent	protective clothing
polybenzimidazole PBI	flame-resistant comfortable chemically stable		firefighters' coats astronauts' space suits

8-37 *Continued*

and filament yarns together in different ways. This gives softness from the spun staple and added strength from the filament fibers.

Another form of combination yarn is made with two or more yarns that vary in fiber composition, content, or twist level. They are put side by side to twist into a yarn.

A ***blend***, on the other hand, is made when two or more fibers (usually in staple form) are put together before they are spun into yarns. This uniformly mixes fibers with different physical characteristics. The best performance feature of each fiber is sought. When blended properly, the positive qualities of one fiber can decrease the negative properties of another fiber. The resulting fabrics have better performance, nicer appearance, or lower prices.

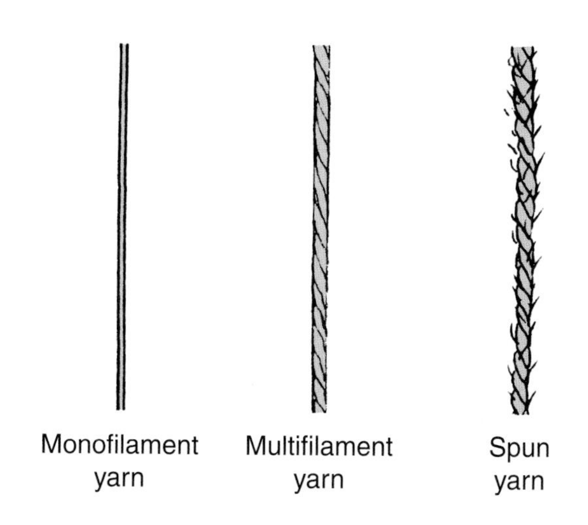

8-38 Yarns may be monofilament, multifilament, or spun.

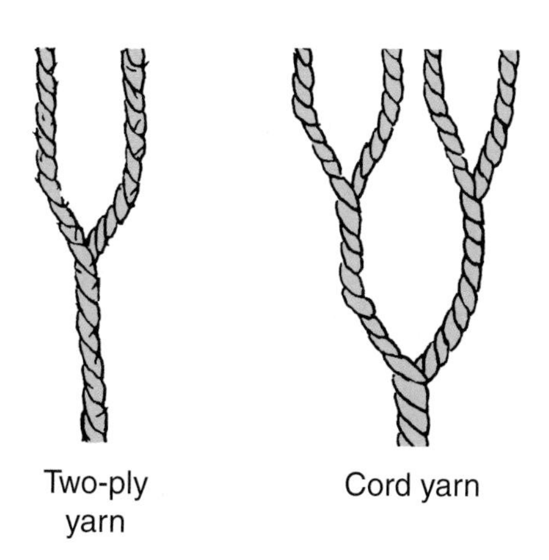

8-39 When single yarns are twisted together, a ply yarn is made. If two or more ply yarns are twisted together, a cord yarn is produced.

Most natural and manufactured fibers blend well. This can increase strength and durability, prevent shrinkage, or provide other advantages. The most desired performance characteristics sought in apparel fabrics are wrinkle-resistance, shrinkage control, and colorfastness. The cost of the fabric is lowered when natural fibers are blended with less expensive manufactured fibers.

By knowing the advantages and disadvantages of individual fibers, the performance of fabrics containing them can be judged. For instance, when blended, nylon adds strength and stability to fabric. Rayon and cotton offer absorbency and comfort. Acrylics improve softness and warmth without adding weight. Spandex gives elasticity. Silk gives beautiful luster. Acetate adds drapability and texture. Polyester contributes easy-care, permanent-press qualities, as well as abrasion-resistance. Polyester/cotton blends are popular for easy-care, comfortable apparel.

Yarn Textures

Textured yarns are illustrated in 8-40. ***Texturing*** of manufactured filaments is done by processing with chemicals, heat, or special machinery. Texturing turns the straight, rodlike filaments into crimped, coiled, or looped forms. It creates a different surface texture, since the filaments no longer lie parallel to each other in the yarn. It is a permanent treatment that gives the yarns bulk, softness, stretch, and wrinkle resistance. When made into fabrics, textured filament yarns have more resemblance to spun yarns but give better durability.

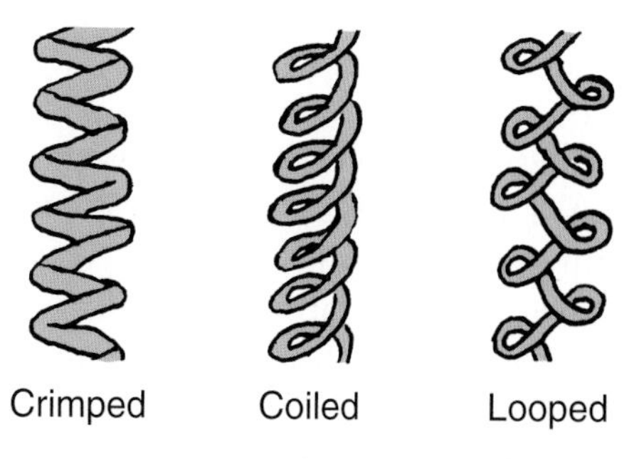

8-40 Textured yarns have more bulk, stretchability, and wrinkle resistance than yarns of straight filaments.

Chapter Review

Summary

Cotton and linen/bast fibers are cellulosic natural fibers from plants. Cotton is harvested from seed pods (bolls), cleaned, and drawn into strands for spinning. It is quite inexpensive, soft, absorbent, and versatile, but also wrinkles. Linen is from the stalk of flax plants. It is more expensive than cotton, and is strong, durable, and lustrous. Linen wrinkles and ravels.

Wool and silk are protein natural fibers from animals. Wool is from the fleece of sheep, which is sheared and scoured before being made into woolen or worsted yarn. Wool is versatile, warm, easy to tailor, and has resiliency from its crimp, but shrinks and can be scratchy. Wool can be used, pure, blended, or recycled. Specialty wools, from other animals are also available. Silk is a cultivated filament fiber unwound from silkworm cocoons. Other silks are wild, doupion, and spun silks. Silk is luxurious and strong, but requires some special care.

Other natural materials are fur, leather, and down. They are not fibers, but have specific uses in apparel. Artificial substitutes have been developed to copy them.

Manufactured fibers are engineered for improved characteristics. Textile chemists developed them by duplicating the molecules of plant and animal fibers. Today, innovative manufactured fibers have many apparel, medical, and industrial uses.

The two basic manufactured fiber types are cellulosic, from plant substances, and noncellulosic, from chemical polymers. Most generic groups contain many trademarks. Generic groups include polyester, nylon, acrylic, modacrylic, rayon, acetate, olefin, spandex, lyocell, and others. All manufactured fibers are made by extruding liquid materials through a spinneret. The material is then dried by chemical spinning (wet, dry, or melt spinning), stretched, and cut if staple is desired.

Yarns are continuous strands of fibers suitable for making fabrics. They might be monofilament, multifilament, spun, or ply yarns. Combination yarns and blends can change performance characteristics. Textured yarns have added bulk, softness, stretch, and wrinkle resistance to resemble spun yarns.

Fashion in Review

Write your answers on a separate sheet of paper.

True/False: Write *true* or *false* for each of the following statements.

1. Cotton, linen, jute, and ramie are protein natural fibers.
2. Combing is an extra processing step done for high-quality cotton fibers.
3. The United States leads the world in the production of linen.
4. Linen is known for its durability and luster.
5. Short wool fibers of less than two inches are used to make worsted yarns.
6. The terms virgin wool, pure wool, and 100% wool can be used interchangeably.
7. Silk is the only natural filament fiber.
8. Highly absorbent manufactured fibers tend to build up static electricity.
9. Denier is the term used to describe fiber length.
10. A blend is made when two or more fibers are put together before they are spun into yarns.

Multiple Choice: Write the letter of the best response to each of the following questions.

11. A fiber's resiliency refers to its _____.
    - A. tendency to form little balls of fiber on the surface
    - B. ability to spring back to its original condition, size, and shape
    - C. ability to pull moisture away from the body to the surface of a fabric where it can evaporate quickly
    - D. tendency to resist soil and stains
12. The most widely used manufactured textile fiber is _____.
    - A. cotton
    - B. wool
    - C. polyester
    - D. nylon

13. Silk that comes from two silkworms who have spun the cocoon together is _____.
    A. doupion silk
    B. weighted silk
    C. spun silk
    D. wild silk
14. The type of yarn that can be made from manufactured fibers is _____.
    A. monofilament yarn
    B. multifilament yarn
    C. spun yarn
    D. All of the above.
15. Which statement is NOT true of a blend?
    A. Fibers with different physical characteristics are uniformly mixed.
    B. Fabric cost is generally increased, since natural fibers are blended with manufactured fibers that are more expensive.
    C. The positive qualities of one fiber can decrease the negative properties of another fiber.
    D. Polyester/cotton blends are popular for easy-care, comfortable apparel.

Short Answer: Write the correct answer to each of the following.

16. Name two advantages, two disadvantages, and two end uses of cotton.
17. Name two advantages, two disadvantages, and two end uses of linen.
18. Name two advantages, two disadvantages, and two end uses of wool.
19. Name two advantages, two disadvantages, and two end uses of silk.
20. Name a manufactured fiber. Then name two advantages, two disadvantages, and two end uses of that fiber.

Fashion in Action

1. Find five advertisements for textile products in magazines and newspapers. In an oral report, identify the generic names and trademarks of the fibers used in the textile products. Describe any special properties or advantages of the fibers that are noted in the ads. Discuss the end uses of the products.
2. Read the fiber content labels of five different garments. In a written report, explain how the fibers used affect the appearance, performance, and comfort of each garment.
3. Examine three pieces of scrap fabric with various textures. Note the appearance and feel of each. Then unravel the yarns to find textured filaments, staple fibers, or whatever. Analyze the twist and fiber lengths of each type of yarn. Mount the fabric pieces, yarns, and fibers on paper along with your written descriptions of them.
4. Research the history of a manufactured fiber. Write a report on its development and its current uses.
5. Research the source of a specialty wool fiber. In an oral report, describe the animals that produce the fiber. Tell where the animals are raised. Try to find out how many animals are needed to give enough fibers for one sweater or one coat. Explain why you think these fibers are fashionably desirable and why they are so expensive.

9 Fabric Construction and Finishes

Fashion Terms

weaving
loom
warp yarns
shuttle
filling yarns
selvage
plain weave
twill weave
satin weave
knitting
wales
courses
gauge
weft knits
warp knits
dyeing
roller printing
screen printing
finishes
hand

After studying this chapter, you will be able to

- describe how fabrics are made from fibers and yarns.
- identify what fabric characteristics result from the different types of fabric construction.
- explain how fabrics are given color and surface designs.
- discuss fabric finishes, how they are applied, and the effects they give.

The appearance and performance of textiles depends on their fiber content, type of yarn, fabric construction, and added color and finishes. These factors can be varied to make millions of individual fabrics that have different characteristics.

Fabric Construction

Fibers and yarns can be held together in various ways to make fabric. The two most common methods of making fabric are weaving and knitting. Other methods, such as felting, bonding, braiding, netting, and lace-making, are used less often. Most apparel is made of either woven or knitted fabrics. Many modern textile mills start with raw fibers at one end and go through all processes that result in finished yard goods at the other end.

Weaving

Weaving is a procedure of interlacing two sets of yarns placed at right angles to each other. The yarns are sometimes also called threads. The machine for weaving is called a ***loom***. A basic hand loom is shown in 9-1.

Two sets of yarns are used for weaving. The lengthwise yarns are threaded onto the loom side-by-side and pulled tight. These are called ***warp yarns***. They must be strong and durable to withstand the strain of the weaving process.

The crosswise yarns weave back and forth from edge to edge with a ***shuttle*** that pulls the threads through the warp yarns. This forms the fabric. These threads are called the ***filling yarns*** (or "weft" or "woof" yarns). They pass over and under the warp yarns. Where they turn at the fabric's edge to go back the other direction, is called the ***selvage***, as shown in 9-2. It runs along both edges of the fabric. It is strong and will not ravel.

Automatic looms, as shown in 9-3, have power shuttles that move at tremendous speeds. When the appropriate warp yarns are

Fashion Institute of Technology: John Senzer, Photographer

9-1 This woman is weaving fabric using a simple loom. Huge, mechanized looms work on the same principle—weaving a set of yarns over and under a perpendicular set of yarns.

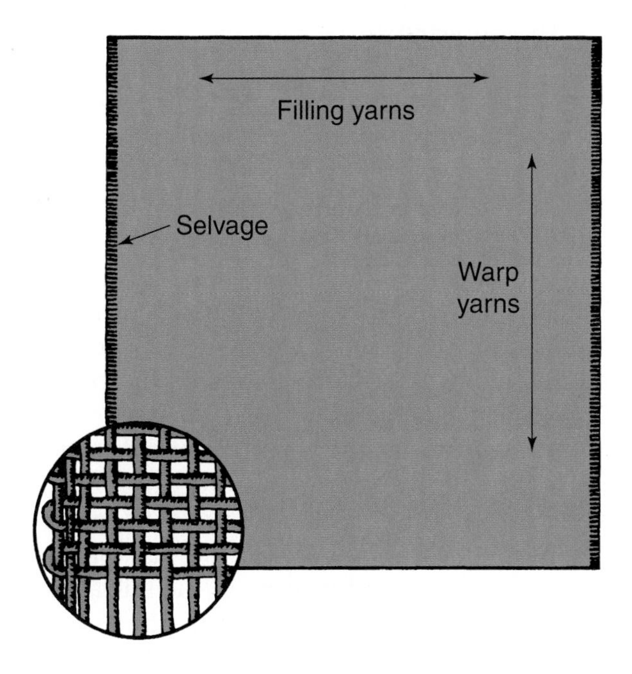

9-2 Filling yarns pass over and under warp yarns to weave fabrics. They turn and go back the other way at the selvage.

raised, the filling yarn shoots from one side to the other. Different warp yarns are then raised by the loom, and the shuttle propels the filling yarn back to the original side. This continues as the fabric is woven. It all happens so quickly that it appears to be a smooth, continuous operation.

A *shuttleless loom* carries the filling yarns by steel bands attached to wheels on each side of the loom. A *dart loom* has many darts that

Pendleton Woolen Mills

9-3 Automatic looms are guided by people and computers to produce plain or fancy woven fabrics. Half the warp yarns have been lifted for the shuttle to take a filling yarn to the other side.

shoot across at the right time with individual lengths of yarn from a package beside the loom. In still other looms, the filling yarn is carried by a tiny jet of water or air.

Weaving plants have rows of machines (looms) producing fabrics at high speeds, as shown in 9-4. Sophisticated, computerized weaving techniques are becoming more widely used all the time.

Fabric Grain

The direction that yarns run in a fabric is called the *grain*. Warp yarns form the *lengthwise grain*. They run parallel to the selvages. Filling yarns run along the *crosswise grain*. In woven fabrics, the filling (crosswise grain) stretches more than the warp (lengthwise grain). The grain is important in apparel, since garments need to stretch more around the body than up and down. Warp yarns should go up and down garments for strength and stability.

Bias grain goes diagonally across the fabric. *True bias* runs at a 45 degree angle, or halfway between the lengthwise and crosswise grains. The greatest amount of stretch in a woven fabric is along the true bias. Note the grainlines indicated in 9-5.

Weaving the yarns closely together results in a strong, dense fabric. When the yarns are woven loosely, the fabric is lighter and more porous. If the fabric has low-twist or "special effect" yarns, they are in the filling direction. A textured effect is achieved by weaving some low-twist and some high-twist yarns together. "Stretch" woven fabrics are woven of stretchable yarns.

Sulzer Ruti

9-4 Large looms produce woven fabrics quickly and efficiently for use in garments and other finished textile products.

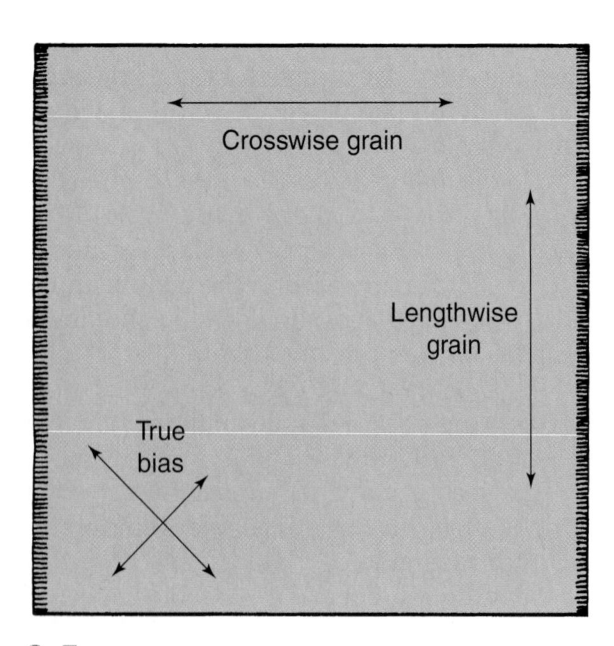

9-5 Various grains can be determined by the direction of the yarns in woven fabrics.

Types of Weaves

There are three basic weaves for making fabric. They are the plain, twill, and satin weaves. They differ by passing the filling yarns over and under different numbers of warp yarns. They also differ in appearance and durability. Every other weave is a variation of one or more of the three basic weaves.

The simplest and most common weave is the ***plain weave***. It is shown in 9-6. In most cases, the lengthwise and crosswise threads are the same denier. The filling yarns pass alternately over and under the warp yarns as in a tennis racquet, only much closer together. This creates a fabric that is strong, reversible, and durable. Examples of plain weave fabrics include muslin, percale, gingham, chiffon, poplin, and broadcloth.

The appearance of a plain weave can be changed by using large yarns with small ones, textured yarns, or a special finish. The *rib weave*, shown in 9-7, uses coarser yarns along with regular ones. A corded effect results. The *basket weave* is a common variation of the plain weave. It is formed by using two or more yarns as one. Illustration 9-8 shows a common 2 x 2 basket weave. Two filling yarns pass over and under two warp yarns. Hopsacking, monk's cloth, and oxford cloth are examples of basket weave fabrics.

In the ***twill weave***, a yarn in one direction *floats* (passes) over two or more yarns in the other direction at regular intervals. Each float begins one yarn over from the last one. A dominant yarn can be seen on the surface of the cloth. It creates a diagonal rib or cord pattern, called a *wale*, as shown in 9-9.

9-6 The plain weave is the simplest weave. Each filling yarn passes alternately over and under each warp yarn at right angles to it.

Twill weave fabrics are usually very firm and tightly woven. The weave is often used to produce strong, durable fabrics. Twill weave fabrics resist wrinkles and hide soil. Examples of twill weave fabrics are denim, gabardine, and surah.

There are many ways to vary twill weaves. The floats can be long or short. The angle of a wale may vary from a reclining slope to a very steep slope. Large, high-twist, or textured yarns also create different looks.

A common variation of the twill weave is the *herringbone* pattern. In this weave, the wale changes direction at regular intervals to produce a zigzag effect.

The ***satin weave***, as shown in 9-10, has long yarn floats on the surface. They go over four or more opposite yarns and under one. Each float begins two yarns over from where the last float began. The face of a satin weave fabric is composed almost entirely of yarns running in only one direction. This creates a surface that reflects light and has a lustrous sheen.

The satin weave is smooth, slippery, and drapable. It is good for linings and formal wear. If the yarns are woven together closely, the fabric is stiffer and stronger. If they are woven more loosely, they resist wrinkles and are more pliable. However, in general, this type of weave is the least durable of the three weaves. When worn, the floating threads are likely to catch other surfaces, which may cause snagging, pulling, and friction. Friction causes the fabric to pill and thus dulls the sheen.

The satin weave is used to make both satin and sateen fabrics. In *satin*, the floats run in the warp direction. For the highest luster, the long floats are made of filament fibers with low twist. In *sateen*, the floats run in the filling direction. Spun yarns are used. Sateen usually has less luster than satin.

Variations

Pile fabrics have loops or yarn ends projecting from the surface. An example of a fabric with loops is terry cloth. Some fabrics

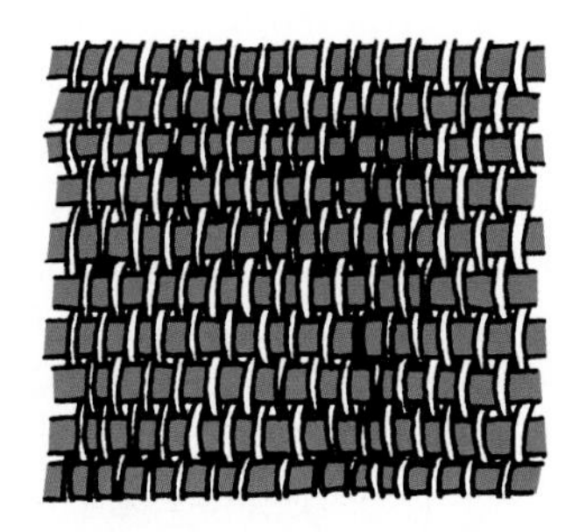

9-7 The rib weave uses filling yarns that are quite different in size from the warp yarns. Thus, a ribbed look results.

9-9 A twill weave causes the fabric surface to have a diagonal wale.

9-8 The basket weave is a plain weave with multiple yarns.

9-10 The smooth surface of the satin weave is created by floating yarns.

with clipped yarn ends are corduroy, velvet, and fake furs. Notice the difference between the looped pile illustrated in 9-11 and the cut pile in 9-12.

Woven pile fabrics are made by varying either a plain weave or a twill weave. Extra filling yarns are woven into the basic weave. In some cases, such as for velvet, fabrics are woven double, face-to-face. They are then cut apart between the layers to make two separate pieces of pile fabric.

Pile and brushed-surface fabrics have nap. *Nap* is a layer of fiber ends raised from a fabric surface. Nap appears different when viewed from different directions. This is because fiber ends lying at various angles reflect varying amounts of light. In apparel, the nap must run the same direction throughout an entire garment. If one garment part is cut from fabric going the opposite direction, a difference in color, shading, or pattern will result.

Other variations of the three basic weaves create patterns in woven fabrics. Stripes, checks, and plaids, as in 9-13, are the oldest woven patterns. They are made on a *box loom.* Shuttles carrying different colors of yarn are housed in boxes at the sides of the loom. Specific shuttles are used at certain times to produce the desired pattern in the fabric. Other symmetrical or geometric patterns can also be produced in this way.

Large and intricate designs are woven on a *Jacquard loom*. The fabric pattern is programmed on punch cards. Each card has a unique pattern of holes for a specific warp yarn. Together, the cards control which of the warp yarns are raised and which are lowered for each passage of the shuttle. This procedure is now also computerized. Damask, tapestry, and brocade fabrics are woven in this way. See 9-14. Fabrics with a woven pattern that contains a name or logo are also done on a Jacquard loom.

9-11 A looped pile fabric, such as terry cloth, has loops of yarn extending out from the fabric surface.

9-12 Cut pile fabrics, such as corduroy or velvet, have yarn ends projecting from the fabric surface. To achieve this, extra yarns are woven into the fabric and then cut.

Eagle Shirtmakers/Men's Fashion Association of America

9-13 A plaid pattern can be woven into fabrics by inserting yarns of different colors where the lines are wanted.

Smaller designs, such as geometric forms in a sequence, can be woven into a fabric with a *dobby attachment*. It is placed on a simple loom and is less expensive than Jacquard weaving. A plastic tape is punched with holes. These holes control the raising and lowering of the warp yarns to produce the design. An example of a dobby weave pattern is illustrated in 9-15.

In the *leno weave*, the warp yarns do not lie parallel to each other. Instead, they are used in pairs, and one crosses over the other before the filling yarn is inserted. The leno weave is illustrated in 9-16.

A special loom attachment is needed to produce leno weave fabrics. The fabrics have an open effect. The crossing of the warp yarns adds strength and prevents slippage of the yarns. Leno weave fabrics are used to make shirts, curtains, thermal blankets, mosquito netting, and bags for laundry, fruits, and vegetables.

Escada by Margaretha Ley

9-14 Large, intricate fabric patterns, such as brocades, are woven on Jacquard looms.

Knitting

Knitting is a fabric construction method of looping yarns together. In knitting, one yarn can form the entire fabric. A series of loops of the same, continuous yarn are interlocked. This can be done a row at a time or in a continuous circular pattern. Loops of yarn are pulled through other loops of yarn. The knitted fabric, or garment, gets longer as rows of loops are interlocked into the previous loops. The loops, or stitches, can be varied to create numerous patterns and textures.

Lines of loops that run the length of knitted fabrics are called ***wales***. Lines of loops that run crosswise are called ***courses***. Both are illustrated in 9-17. ***Gauge*** is the number of stitches, or loops, per inch in a knitted fabric. A higher gauge number indicates a closer and finer knit.

Knit fabrics are popular for apparel, 9-18.

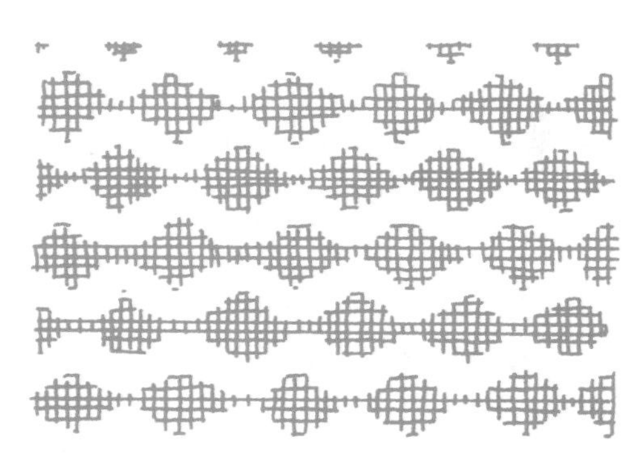

9-15 This weave is one of several patterns produced on a loom with a dobby attachment.

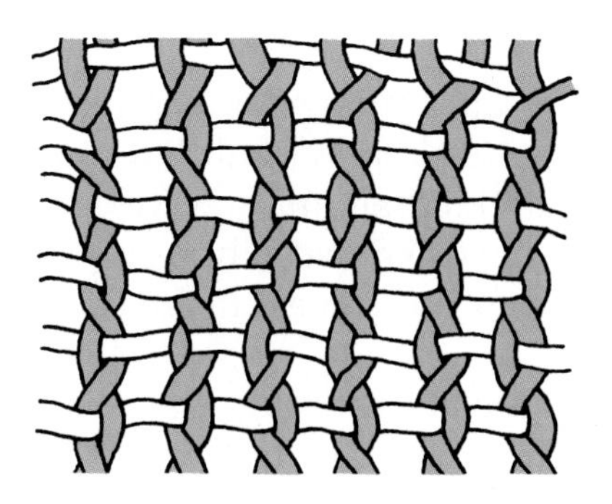

9-16 Because leno weave fabrics have open spaces, they may look fragile. However, the crossings of the warp yarns give them strength and stability.

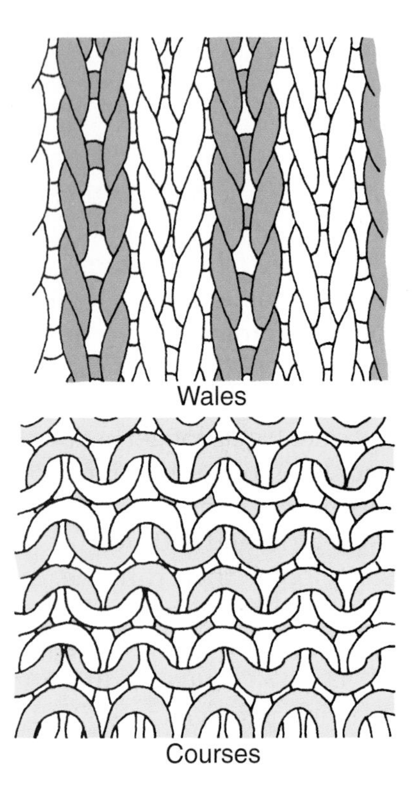

9-17 In knits, wales run in the lengthwise direction, and courses run across the fabric.

Monsanto Fibers

9-18 Knitwear can be casual or high fashion. It is popular because of its comfort, versatility, and ease of care.

They tend to move with, and fit, the body shape comfortably. They are versatile and wrinkle resistant, so they are good choices for travel clothes. Most natural and manufactured fibers are suited to knitting. Also, woven fabric designs can be copied by knits.

Knits do not ravel the way woven fabrics do. However, they can get lengthwise "runs" from broken threads, as seen in nylon stockings. They can also snag or come unknitted if the yarn is pulled. They are usually flexible and stretchy. Their stretch is built in by the knitting process. (This differs from stretch wovens that are made with elastic fibers or yarns.)

Almost all knitting for garments sold in retail stores is done on knitting machines in commercial knitting mills. Circular knitting machines make knitted fabrics in tube form. Flat knitting machines make flat knitted fabrics.

Commercial knitting is faster than weaving on looms. In fact, it is done faster than the eye can follow. Technicians develop knit patterns on computers. The computers control the knitting machines. The result might be knit yard goods, or finished socks, sweaters, or other knitted garments.

Types of Knits

The two methods of knitting fabrics are weft knitting and warp knitting. The difference is in the way the loops are formed. Their structures and characteristics are described in chart 9-19.

Weft knits (or *filling knits*) have one strand of yarn that runs across, forming a horizontal row of interlocked loops. The commercial process is essentially mechanized hand knitting.

Weft knits can be made either on a circular machine or on a flat machine. Specially designed circular knitting machines are used to make the right-sized tubes for such items as hosiery, underwear, and T-shirts. Large widths of circular knits are cut open and used as flat fabric.

With flat machines, the number of stitches can be automatically increased or decreased to shape the finished fabric. Shaping during the knitting operation reduces garment seam construction and is considered to make a finer garment. The resulting shaped product is

Basic Knits and Their Characteristics

Type of Knit	Structural Drawing	Fabric Names	General Characteristics
Weft Yarns run across		Single knit Jersey Purl knit Ribbed knit Double knit Interlock knit Sweater knit	Made on circular or flat machine Can be full-fashioned Stretchy May run if snagged Cut edges may curl Great versatility
Warp Yarns run lengthwise		Tricot Raschel knit	Made only on flat knitting machine Stable and durable Does not ravel; has selvage edges Quite run-proof Limited stretch Versatile

9-19 In weft knitting, the yarn forms interlocking loops in horizontal rows across the fabric. In warp knitting, interlocking stitches carry the yarn lengthwise in a zigzag pattern.

described as being *full-fashioned*. Where stitches have been dropped or added is slightly noticeable, as shown in 9-20. Sweaters and dresses are often full-fashioned, as well as having separate shaped collars, cuffs, and trim pieces.

Single knit fabrics, such as jersey, are usually made of cotton, rayon, nylon, or blends. They are made on a single needle, weft knitting machine. They stretch in both directions and may tend to stretch out of shape. They have lengthwise wales on the right side and crosswise courses on the wrong side.

To determine whether a fabric is a single knit, pull it crosswise. If the fabric rolls toward the right side, it is a single knit. This type of knit is lightweight, soft, and drapable. Its cut edges may curl or roll, and it will often run if snagged. It is most often found in formal wear, lingerie, and T-shirts.

Double knit fabrics are made on a weft machine with two needles and two yarns. Loops are drawn through from both directions to knit two fabrics as one. Plain double knits look the same on both sides, with rows of fine wales. The crosswise direction stretches more than the lengthwise one. They are more firm and stable than single knits. They have "give," but they will not sag or stretch out of shape. They are usually made from polyester, triacetate, or wool. They are used for garments such as suits, coats, slacks, and dresses.

9-20 Full-fashion marks occur where shaping is done through the knitting process rather than by the stitching of seams.

To identify a double knit, look at the edge and pull it crosswise. It will not roll, and you will see two separate layers of loops. However, it cannot be separated into two individual layers.

There are many varieties of double knits, including *interlock knits*. They are lightweight and stretchy. Interlock knits have a smooth surface on both sides and a very fine lengthwise rib. Most will run along an unfinished edge. They are used to make underwear, T-shirts, golf shirts, and dresses.

Purl knits have pronounced horizontal (crosswise) ridges. They have superior stretch and recovery in both directions. Also, they are usually reversible since the back is identical to the face of the knit.

Rib knits have pronounced vertical (lengthwise) ridges. They have great crosswise stretch. They are usually used for waistbands, neckbands, and cuffs rather than for whole garments.

Sweater knits are very stretchy and are usually loosely knitted. They have large denier yarns to resemble hand-knitting. *Textured knits* are made from filament yarns that have been permanently crimped, coiled, curled, or looped.

Other patterns are created by changing the placement of smooth and bumpy stitches. Sometimes stitches are dropped, added, alternated, or crossed. A Jacquard knitting machine makes complex designs and textures. Tremendous pattern variety can be achieved with combinations of these stitches.

Warp knits are made only on flat knitting machines. Many yarns and needles are used to produce them. Interlocking loops are formed in the lengthwise direction, a whole row at a time. Each yarn is controlled by its own needle and follows a zigzag course.

Warp knits are stable, durable, and relatively run-proof. They can be produced fast, inexpensively, and in great quantity. Most warp knits are lightweight and tightly knit. They stretch only in the crosswise direction. In yard goods form, warp knits can be recognized by their straight selvage edges.

Tricot is the most familiar warp knit. It is used for drapable or clingy dresses, shirts, and lingerie. It does not run or ravel, but raw, cut edges have a tendency to curl.

Raschel knits are made on the raschel knitting machine. The machine can make all kinds of fabrics from heavy crochet-like knits to sheer net or lace effects. It knits stripes, checks, and diagonal patterns. Raschel knits can be made of every possible fiber, and they usually have a lot of texture. They often have limited stretch.

Fabrics from Other Construction Methods

Although weaving and knitting are the most popular ways of constructing fabrics, yard goods can also be produced by several other methods.

Nonwovens

Nonwoven fabrics are made directly from fibers rather than from yarns. The fabrics are made from a compact web of parallel, cross-laid, or randomly dispersed fibers. The fibers are held together through a combination of moisture, heat, chemicals, rubbing, and/or pressure. Sometimes a chemical "binder" is added as a gluelike substance. Manufactured fibers can be melted together. Also, a few nonwoven fabrics are produced by machines that mechanically tangle the fibers into a mat. They are used for padding and as *batting* in quilting.

Nonwovens have no grainline and limited stretch. Their edges do not ravel. They are relatively inexpensive. Many are used for industrial and medical purposes. They are often disposable. Others are used in garment interfacings. Interfacings are placed inside certain parts of garments, such as collars and cuffs, to give strength and support.

Felt is a type of nonwoven fabric. Wool fibers are used to make some felt fabric. Wool is the only fiber that "felts" naturally. When heat, moisture, and pressure are applied, the scales of the wool fibers interlock and hold themselves matted together. Cotton, rayon, and other fibers are also used in felts.

Felt fabrics are thick and somewhat stiff. They are not as strong as woven or knitted fabrics. They can be molded into shapes. They are often used for hats and craft projects.

Vinyl and urethane *films* are thin, nonwoven "sheets" that are not made from fibers. However, they can be finished to look like leather or woven fabrics. Often they are used as coatings on other fabrics. The coated fabrics are used for raincoats, umbrellas, purses, shoes, and household textiles, such as tablecloths.

Artificial suedes, such as Ultrasuede® and Facile®, are nonwoven fabrics composed of polyurethane with fine fibers of polyester. When they are made into garments, raw cut edges can be left exposed like real suede or leather, since they do not ravel or fray. Also, they can be stamped, embossed, painted, or perforated for unique textures.

Some nonwovens are *needle punched* to hold their fibers tangled together. Needle punching creates a mechanical interlocking of fibers with a needle loom, as shown in 9-21. Small holes, regularly placed, are evident in the fabric.

Bonded Fabrics

Bonding is a method of permanently fastening (laminating) together two layers of fabric in some way. A chemical adhesive may be used, or the layers may be fused with a web of fibers that melts when heat is applied. Bonding provides stability, strength, body, or opacity. One layer of fabric may serve as a lining to eliminate the need for a separate lining in the garment.

The layers of bonded fabrics may be similar or different. For instance, two knitted fabrics or two woven ones may be fastened together, or a knitted fabric may be bonded to a woven or lace fabric. Sometimes a layer of foam is attached to a layer of fabric to add warmth. Sometimes a stable fabric is bonded to a knit to limit stretch. Tricot knit is often used as a backing fabric. It is inexpensive and allows some stretch. A bonded fabric is shown in 9-22.

BASF Corporation, Fibers Division

9-22 This bonded fabric, made for skiwear, is warm, stretchable, and colorfast. The different outer and inner layers of the fabric can be seen in the upper right corner.

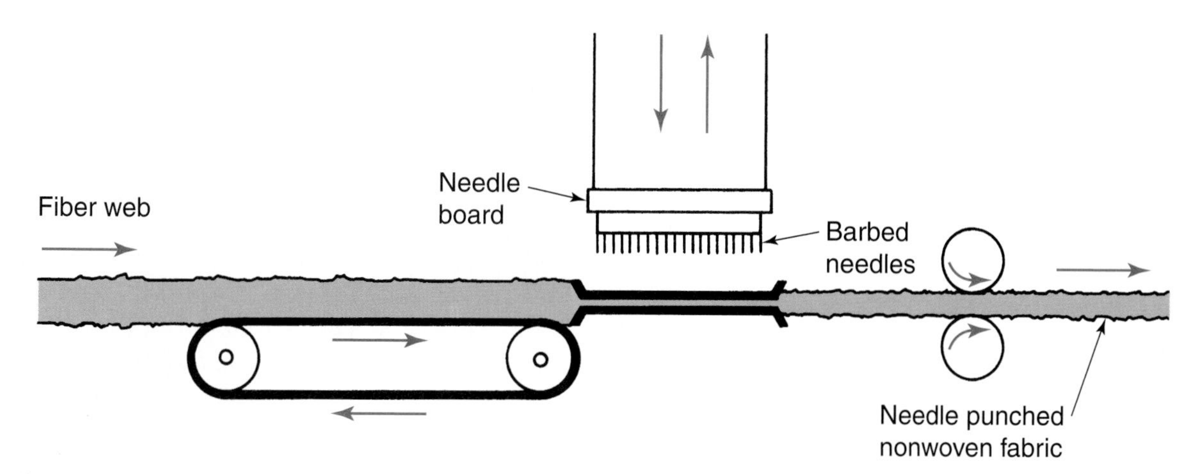

9-21 A needle punching loom punches needles into the fiber web and then withdraws, leaving the fibers entangled as a nonwoven fabric.

Bonded fabrics wrinkle less and have more body than single fabrics. They do not fray or ravel. They are easy to sew and do not need to be lined. They are warm without being heavy. They are often used for skiwear and winter coats.

Fusible web is sold in fabric stores for home use. It is a sheet of binder fibers that can act as an adhesive because its softening point is relatively low. It can fuse two layers of fabric together when it is placed between them and pressed with a heated iron. It is sometimes used for hemming or other garment construction processes. Iron-on patches and interfacings have a fusing substance only on the wrong side. This enables them to stick to other fabrics when applied with a hot iron.

Quilted Fabrics

The construction of quilted fabrics consists of a layer of padding (or batting) sandwiched between two layers of yard goods. The three layers are usually held in place by machine stitching around decorative areas or shaped spaces. Sometimes fabrics are quilted by *pinsonic thermal joining*. That is done by a machine that uses ultrasonic energy to join layers of thermoplastic materials. The ultrasonic vibrations generate localized heat by causing one piece of material to vibrate against the other at an extremely high speed. This results in a series of "welds" that fuse the materials together in a chosen "quilted" design.

Braided Fabrics

Braided fabrics are made by *braiding*, also called *plaiting*. It is the process of interlacing three or more yarns to form a regular diagonal pattern down the length of the resulting cord. Braided fabrics are usually narrow. They are used for decorative trims and shoelaces. Braids are often joined together to make rugs.

Laces and Nets

Laces and nets are openwork fabrics made by knotting, twisting, or looping yarns. Knots may hold threads together when they cross each other. In other cases, continuous coils of thread loop through each other to form a mesh. Laces and nets can be constructed by hand or machine. They can be fine or coarse.

Fabric Coloring and Printing

Coloring and printing processes include bleaching, dyeing, and printing. Sometimes fibers or yarns are dyed before being made into fabrics. However, the largest volumes of fabrics, especially wovens, are bleached and/or dyed after the fabric is produced.

Bleaching

Bleaching is a chemical process that removes any natural color from fibers or fabrics. This is done if white yard goods are wanted. Fabrics may be bleached before being dyed or printed to make sure the applied colors are "true." This procedure, or a boiling rinse, is needed for some cottons and linens to remove impurities. Such impurities may include sizings, oils, and waxes that may have accumulated during the manufacturing process.

Dyeing

Dyeing is a method of giving color to a fiber, yarn, fabric, or garment. See 9-23. It is done with coloring agents called *dyes*. Some dyes are from natural sources, such as bark, roots, leaves, flowers, insects, and animals. Others are produced chemically. No one dye is best for every textile fiber. Some dyes that

BASF Corporation, Fibers Division

9-23 These nylon yarns were dyed in bright, clear colors while they were in fiber form.

color one type of fiber are useless on others. Dyes are applied according to what kind of fiber is being used and what kind of result is desired.

Dyes can create millions of different colors. Computers are programmed to mix specific colors. This technology enables uniformity and saves time.

The term *colorfast* implies that the color in a fabric will not fade or change. Conditions that test colorfastness include laundering, dry cleaning, sunlight, perspiration, and rubbing. The care label on a garment should state if the product is not colorfast under certain conditions. Colorfastness depends on the chemical makeup of the dye, the fiber content of the textile product, and the method of dyeing. Textiles can be dyed at the fiber, yarn, yard goods (piece), or garment stages.

BASF Corporation, Fibers Division

9-24 These freshly extruded acrylic fibers have been solution dyed. The pink dye was part of the solution that came through the spinneret when the fibers were made.

Fiber Dyeing

Fiber dyeing imparts color to fibers before they are spun into yarns. This allows the mixing of colored yarns for almost unlimited fabric patterns and effects. Natural fibers are said to be *stock dyed*. Color is added to the loose fibers before spinning. It penetrates the fibers to produce uniform color and good colorfastness. Manufactured fibers are *solution dyed*, as in 9-24. Dye is added to the syrupy fiber solution before it is extruded through the spinneret. This results in clear, rich colors with excellent colorfastness, since the dye becomes a structural part of the fibers.

Burlington Industries, Inc.

9-25 Large amounts of yarn are dyed together in baths containing the desired color, which is usually mixed by computer.

Yarn Dyeing

Yarn dyeing is done by placing spools of yarns into a dye bath, as shown in 9-25. Yarn dyeing gives good color absorption and an opportunity for fabric designing. The threads of woven stripes, checks, plaids, and Jacquard patterns are yarn dyed before the fabric is made.

Piece Dyeing

Piece dyeing is the most common and least expensive method of coloring textiles. Greige goods are placed in a dye bath before being cut and made into garments. Piece dyeing is a very fast process. The fabric travels at a high speed through the steps of dyeing, setting, rinsing, and drying. Piece dyeing achieves less complete dye penetration than fiber or yarn dyeing does. However, it does allow fabric manufacturers to respond quickly to fashion trends. Large volumes of fabric can be stored undyed.

Then, just before the fabric is cut into garment pieces, it can be dyed in the latest fashion colors. Most piece-dyed fabrics result in solid colors.

Cross dyeing is a form of piece dyeing. Two or more fibers with different dyeing properties are combined in a fabric. The dye bath may contain one dye that each fiber takes in a different way or a separate dye for each fiber. Fibers react to the dyes in different ways to create different looks. A single dyeing operation can dye the background of a woven or knitted fabric one color and the pattern a different color. Simple plaids and checks or two-tone heathery looks can be produced from cross dyeing.

Vat dyeing is used mainly with cellulosic fibers. It is done by the piece dyeing method. The term vat refers to the dyes that are used. The color develops (oxidizes) after the vat dyes are inside or have combined with the fiber. Vat-dyed fabrics are very colorfast.

Garment Dyeing

Garment dyeing is becoming more common, enabling garment producers to make fast deliveries of orders to retailers. Apparel is manufactured in various styles, but of undyed yarns. Garments manufactured as "P.F.D." (prepared for dye) are made larger to allow for shrinkage. They are made with fabric, thread, and trimmings that will have identical appearance from dyes to be used. When specific orders are received, the garments are dyed in the requested colors, labeled, and shipped. Fabrics are sometimes also printed after garment construction.

Garments can be dyed at home with powdered or liquid dyes that are sold in stores. This can be tricky and may or may not produce the desired results.

Also done on a very small scale is *resist dyeing*. In this procedure, areas that are to be colored are left to be exposed to the dye. Other areas that are not to be colored are restricted from the dye. *Batik* is a form of resist dyeing in which wax is used to cover the area where dye is not wanted. *Tie dyeing* is also resist dyeing. Garments to be dyed are twisted, tied, or knotted, so the tight folds of the fabric form barriers to the dye. This creates patterns on the fabric.

Printing

Printing is a process for adding color, pattern, or design to the surface of fabrics, as shown in 9-26 and 9-27. This is different from woven or knitted designs that are part of the fabric structure. On printed fabrics, the design can be seen distinctly only on the right side. The wrong side is much lighter. If printed poorly, the design may not be straight with the grain of the fabric.

Printing is most successful on fabrics made of absorbent fibers. The dyes in the print can penetrate deeply into the fibers of the fabric, as for the shirt in 9-28.

Info Design, Inc.

9-26 Patterns to be printed onto fabrics can now be designed by computer, automatically programmed into the printing equipment, and coordinated with yarns for matching sweaters or skirts.

International Linen Promotion Commission

9-27 This electronically controlled printing bed has been programmed by computer to do its work.

Overall prints are the same across all of the fabric. *Border prints,* as in 9-29, have a design that forms a distinct border, usually along one or both sides of the fabric. *Directional prints* have an up and down direction to them. They must be used going the proper direction. Even or *balanced plaids* are the same in both the lengthwise and crosswise directions. The design will match if a corner is folded back across the center of any repeat. *Uneven plaids* are different in one or both directions. For instance, they may be different to the right and up than they are to the left and down.

Roller printing and screen printing are the two basic methods of printing. Screen printing is considered to give better quality designs than roller printing.

Roller Printing

Roller printing is sometimes called *direct, calender,* or *cylinder printing.* When roller printing is done, color for the design is applied directly to the fabric as it passes between a series of metal rollers. The design to be printed is engraved onto the rollers. A different roller is used for each color. The engraved part is full of dye paste and transfers the color directly onto the fabric. A sharp blade scrapes the paste from the unengraved portion. Machine settings are critical to ensure exact matching in multicolor patterns. A diagram in 9-30 shows how roller printing is done.

Roller printing is a simple, high-speed printing method. It is used to produce large

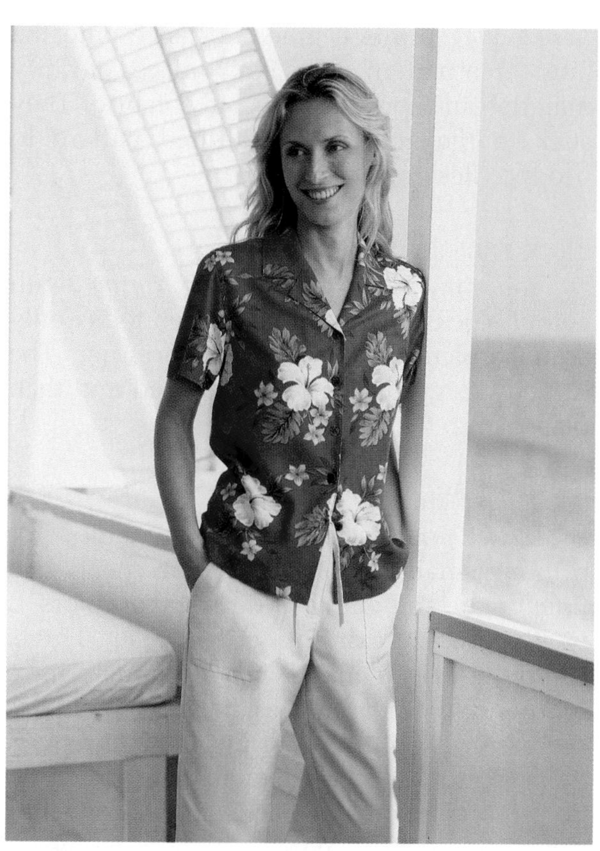

JC Penney

9-28 This rayon camp shirt takes print dyes very well because the fiber is absorbent.

China External Trade Development Council

9-29 Border prints show a definite edge to all four sides of most silk neck scarves. In yard goods, the border might be along one or both of the selvage edges.

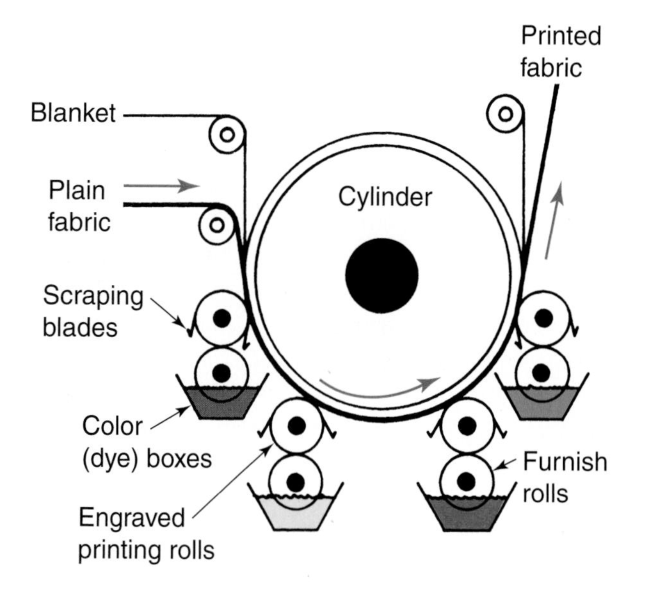

9-30 In roller printing, the fabric moves around a large cylinder. Carefully calculated small printing rolls put dye onto it in the right amounts and at the right places.

quantities of designs inexpensively. The majority of printed fabrics on the market today are roller printed.

The quality of roller printing depends on the skill of the engraver who etches the design on the roller. It also depends on the care taken in feeding the fabric through the rollers. Some roller-printed fabrics are made with more than a dozen separate rollers and colors. When the design is printed off-grain, or the color looks blurred, the print is "out of register."

A variation of roller printing is *discharge printing,* (sometimes called *extract printing*). First the fabric is piece dyed a solid color. Then a bleaching paste is applied to the rollers. It removes dye from certain areas that usually results in a white design on a colored background, as in 9-31.

Screen Printing

For many years, ***screen printing*** has been an art done by hand on flat screening frames. It is similar to stenciling. Each color requires a separate stretched screen made of a sheer fabric. The background of the screen is painted with a paste or insoluble film to resist the dye. The untreated area of the screen allows the dye to pass through onto the fabric. Today the screening process is used for printing logos and symbols onto garments. It is also used for very large designs and high quality apparel prints.

9-31 This fabric was first piece dyed to a blue color. Later, a discharge printing process formed the design by bleaching the dye away from the motif areas.

Rotary Screen Printing

Rotary screen printing is a combination of the other two methods. It has become a fast, modern printing method. The dye is transferred to the fabric through porous, cylinder-shaped nylon screens. A separate screen is still needed for each color of the design. Each screen has some design areas open and others closed, and it rolls over the fabric separately. Dye is worked through the open sections, printing the design.

Heat Transfer Printing

The *heat transfer printing* method is not used in great amounts. In this process, the desired colors and patterns are placed in special ink on special paper. The paper is placed on the fabric and, through the use of heat and pressure, the colors and patterns are transferred from the paper to the fabric. Complicated designs with a good deal of depth and color can be produced. T-shirts are often printed with heat transfer designs.

Flocking

Flocking prints a glue substance onto the fabric in a pattern. Then small pieces of fluffy material are sprinkled over the fabric. They stick to the glue in the desired pattern to produce design with texture.

Block Printing

You may have done *block printing* with cutout blocks in art class. It is used today only as a handcraft type of fabric printing. The carved design on the block is inked and then placed onto the fabric.

Ink-Jet Printing

Ink-jet printing is the newest method for printing small lots of fabric more cheaply, in finer detail, and with a wide range of colors. It is like ink-jet printing onto paper. Huge, computer-driven ink-jet textile printers have thousands of micronozzles that spit droplets of the basic colors onto fabrics moving through them. This method eliminates the preparation of printing plates, gives flexibility for short runs, and greatly reduces response time to fashion trends. Experimentation is also being done with photocopying machine technology.

Fabric Finishes

Finishes improve the appearance, feel, and/or performance of textiles. In general, finishing refers to any of the processes through which fibers, yarns, and fabrics are passed in preparation for their end uses. Finishes have a great influence on how the final textile products, or garments, will perform.

Finishes are applied mechanically or chemically to textiles to make them more acceptable. Less desirable characteristics of fibers can often be controlled by special finishes. However, finishes cannot improve the basic quality of fabrics. That depends on the quality of the fibers, yarns, and method of construction.

Unfinished cloth may be finished at the mill. It may also be sent to different types of finishing plants to undergo a variety of finishing processes. These processes impart texture, ease of care, and other characteristics. Only after finishing is the fabric ready for sale to apparel manufacturers, fabric retailers, or other end-use producers.

Most textile finishes are *permanent*. They last the life of the garment. Some are *durable*. They last through several launderings or dry cleanings, but they lose their effectiveness over a period of time. A *temporary* finish lasts until the fabric is washed or dry-cleaned. Starch is usually temporary. A *renewable* finish is temporary but can be replaced or reapplied. Some stain- or water-resistant finishes on such items as trench coats and upholstery fabrics are renewable.

A term that ends in *proof*, such as waterproof, means complete protection. A term with *resistant* or *repellent*, such as water-repellent, means that the finish provides partial protection.

A combination of fabric construction and finishing creates the hand of a fabric. ***Hand*** is the way fabric feels to the touch. Characteristics such as drapability, thickness, softness, firmness, crispness, elasticity, and resiliency can be judged by feeling a fabric carefully. The hand of a fabric can determine how appropriate it is for a certain type of garment or end use.

The great many finishes used on fabrics today fall into two main categories: mechanical finishes and chemical finishes. Fabrics can receive finishes from one or both categories, in addition to being colored and printed.

Mechanical Finishes

Mechanical finishes affect the size and appearance of fabrics. They alter such factors as the amount of surface fiber and fabric thickness. They can give the fabric surface a smooth and flat look, or a napped or flocked texture. They are done by mechanical, rather than chemical, methods.

Drying and Stretching

After being made, cloth is usually given its correct width and length by drying and stretching, as in 9-32. This is called *heat setting* when applied to manufactured fibers, *crabbing* when applied to wools, and *tentering* for other fabrics. It gives the fabric its final shape by passing it through heat and sometimes moisture while it is in a stretched position.

Compressive Shrinking

Trademarked as Sanforizing®, *compressive shrinking* is a process of shrinkage control of fabrics made of cellulosic fibers. It assures consumers that the product will shrink less than one percent in final use. The fabric is compressed in a lengthwise direction while it is in a damp and softened state.

Preshrinking is done with heat and moisture. Garments labeled preshrunk will not shrink more than three percent in either direction.

Pendleton Woolen Mills

9-32 Large fabric drying equipment is used to straighten fabrics after they have been produced.

Fulling shrinks some wool fabrics lengthwise and widthwise under carefully controlled conditions. It makes the fabrics stronger and gives them body by tightening the weave. *Sponging* of wool fabrics also shrinks them, so they will not get smaller with dry cleaning. The fabric is dampened to allow the fibers to relax. The fabric is then dried. These procedures add to the price of quality woolens and worsteds.

Calendering

Calendering presses the surface of fabrics flat with heat and pressure. The fabric is passed between heated cylinders or rollers. This gives the fabric a lustrous, smooth, polished surface.

Calendering can also *emboss* the fabric, which gives a permanent raised and indented design. For embossing, the heated rollers have engraved sections that form the design. Calendering can also produce a watermark, or moiré, pattern on fabrics.

Rubbing is a friction caused by passing fabric between calender rolls that are revolving at different speeds. The side of the fabric against the faster roll results in a soft surface, such as in polished cottons.

Singeing

Singeing is passing the cloth over a series of gas jets, a flame, or heated copper plates to singe off any protruding fibers. This gives a smooth, uniform surface. When done to worsted fabrics, it gives them their characteristic hard finish.

Cutting

Cutting is done for some napped fabrics, such as corduroy, to create a cut pile. The ribs of floating filling yarns are cut down the center with razor-sharp cutting discs. This cutting exposes the yarn ends and makes them stand up.

Brushing

The *brushing* process, sometimes called *napping,* raises the fabric surface by pulling up fiber ends. It tends to hide the weave. The fabric is swept mechanically with stiff metal-bristled brushes. This removes loose fibers, threads, and lint from the surface of the fabric. More importantly, it pulls up low-twist, spun yarn fibers from the fabric surface. This produces a soft, fuzzy, napped finish. Flannel fabrics have brushed finishes.

Shearing

Shearing, as shown in 9-33, trims any fiber or yarn ends that are sticking out from the fabric. If done close to the fabric surface, it causes the weave to become more distinct. If done after surface nap has been raised on a cloth, it gives the pile a uniform length much like mowing a lawn!

Pressing

The *pressing* procedure places the fabric between heavy, electrically heated plates or cylinders. The hot plates press the cloth, with steam if necessary.

Beetling

In *beetling,* linen or cotton fabric is pounded to give a flat effect. It gives a harder surface with increased sheen.

Chemical Finishes

Chemical finishes generally affect fabric performance. They enable fabrics to better serve their intended purposes. The finishing agents become part of the fabrics through

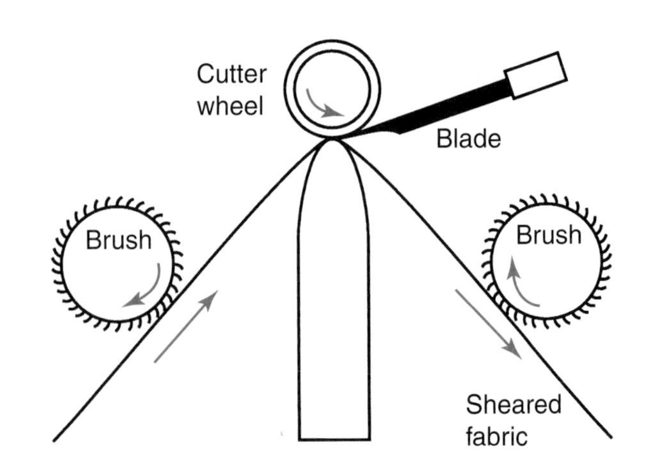

9-33 For shearing, revolving brushes raise fiber or yarn ends, so the sharp blade can trim the surface for the desired appearance.

chemical reactions with the fibers. Some finishes for manufactured fibers are designed to lessen their plastic feeling by making them more absorbent and softer to the touch.

Antistatic

An *antistatic* finish prevents the buildup of static electricity, so garments will not cling to the body of the wearer.

Crease-Resistance

With *crease-resistance*, some fabrics, especially cottons, rayons, and linens, are baked with a *resin* that helps them resist and recover from wrinkles. However, the fabrics become weaker and less absorbent with the finish. Stains are also harder to remove.

Flame-Resistance

A *flame-resistance* finish prevents fabric from supporting or spreading a flame. The fabric is self-extinguishing when removed from the source of the flame, but it is not fireproof. As the fabric burns, the finish cuts off the oxygen supply or changes the chemical makeup of the fibers. The Flammable Fabrics Act sets standards for the flammability of household items and clothing, especially for children's sleepwear. The finish must be durable to laundering. Flame-retardant fibers, such as wool, aramid, modacrylic, and polybenzimidazole fibers, do not need this finish.

Mercerization

Mercerization is a caustic soda treatment used on cellulosic textiles, such as cotton, linen, and rayon. It is also used on cotton thread. It increases the luster, strength, absorbency, and dyeability of the fibers.

Mildew-Resistance

For *mildew-resistance*, a metallic chemical is applied to fabrics to prevent mildew from forming.

Moth-Resistance

A *moth-resistance* finish discourages moths and carpet beetles from attacking wool fibers. The moth-resistance chemicals can be added to the dye bath.

Permanent Press

Permanent press is sometimes called *durable press*. A resin is applied to polyester/ cellulosic blends and other fabrics. It is heat-set onto the fabric. This finish helps the fabric retain its original shape and resist wrinkling. It also retains creases or pleats that have been heat-set in manufacturing. Ironing should not be necessary if care instructions are strictly followed, such as removing from the dryer when still tumbling, and hanging on a hanger. Sometimes stains are difficult to remove from fabrics with this finish. New generations of wrinkle-free fabrics wrap cotton fibers around a polyester core rather than using a baked-on resin.

Shrinkage Control

Dimensional stability, comparable to that obtained by compressive shrinkage, can be gained through the use of special *shrinkage control* chemical finishes.

Sizing

Sizing is a solution of starch or resin used to fill up spaces between yarns. It is applied to fabrics to increase weight, body, and luster. It is sometimes added to cheaper fabrics to improve their appearance. This is becoming less popular. It is usually not a permanent finish.

Soil Release

Soil release finishes make it possible to remove stains. It is often used on polyester or permanent-press fabrics. It helps water-resistant fibers be more absorbent, so detergents can release soil and oily stains from them during laundering.

Stain-Resistance

Stain-resistance makes fibers less absorbent, so it is easier to lift off or sponge away spills of food, water, and other substances. Protection is given against water-based and oily stains. Scotchguard® and Zepel® are trademarks for stain-resistant finishes.

Waterproof

Waterproof finishes fill the pores of a fabric, so water cannot pass through it. A rubber or plastic coating is usually used. Waterproofed fabrics are often uncomfortable to wear since air cannot pass through the fabric for circulation or the evaporation of body moisture.

Water-Repellent

A *water-repellent* finish coats fabrics with wax, metals, or resins. It causes fabrics to shed water in normal wear, but it does not make them completely waterproof. It repels water-based stains while remaining porous. It may have to be renewed after laundering.

Glossary of Popular Apparel Fabrics

Use this section for reference in your study of textiles:

Argyle (arg-ile): fabric of diamond design with contrasting diagonal overstripes. Originally knitted, now also woven or printed.

Barathea (bar-a-thee-uh): fine cloth, originally of silk or wool, with a broken rib pattern.

Barre (bah-ray): knit or woven fabric with crosswise stripes of texture or color.

Batiste (bah-teest): sheer, lightweight fabric that has high texture, softness, and fine plain weave construction.

Bedford cloth (bed-ford kloth): strong woven fabric with lengthwise ribs.

Benares (beh-narh-eez): lightweight fabric from India, usually woven with metallic threads.

Bengaline (beng-uh-leen): strong fabric with defined crosswise ribs.

Birdseye (berdz-eye): has a diamond-shaped weave with a center dot that resembles a bird's eye.

Bouclé (boo-klay): fairly thick fabric woven or knitted with a rough looped or nubby surface, usually with a spongy effect.

Broadcloth (brawd-kloth): closely woven fabric with fine imbedded ribs. It is made in many weights, fibers, and blends.

Brocade (bro-kayd): heavy, luxurious Jacquard weave fabric that has an elaborate raised design, usually of contrasting colored threads.

Brocatel (bro-kuh-tel): brocade-type fabric with blistered or puffed appearance.

Buckram (buck-rum): coarse, stiff, open-weave fabric used for stiffening.

Burlap (burr-lap): a coarse, plain weave fabric made from jute.

Butcher linen (butch-ur lin-en): coarse home-spun linen or an imitation made with manufactured fibers.

Calico (kal-ih-ko): smooth, plain, closely woven, lightweight cotton cloth with small print design.

Cambric (kaym-brik): plain weave fabric with slightly glossy surface.

Canvas (kan-vus): strong, heavyweight, hard-wearing, plain weave fabric.

Cavalry twill (kav-ul-ree twill): strong twilled fabric for uniforms and riding breeches.

Challis (shal-ee): soft, lightweight, plain weave fabric, often printed with delicate floral, Persian, or necktie effects on a dark background.

Chambray (sham-bray): plain weave cotton fabric combining colored warp and white filling yarns, usually creating a pastel color.

Cheviot (shev-ee-ut): rough-surfaced coating fabric, usually with heavy nap.

Chiffon (shiff-ahn): dressy, plain weave fabric that is very light, transparent, and drapable.

China silk (chy-nuh silk): lightweight silk or manufactured fiber lining fabric.

Chino (chee-no): twilled cotton fabric made of combed, two-ply yarn, with slight sheen, used for uniforms, sportswear, and hobby clothing.

Chintz (chintz): glazed, closely woven, plain weave cotton fabric often printed with birds or florals.

Ciré (seer-ay): fabric with a smooth, flat, glossy, and slippery surface produced by heat calendering or a wax process.

Cloque (kloh-kay): fabric with a raised design.

Cord (kord): a strong fabric of heavy wool or cotton warp yarns and lengthwise ribs, often woven in stripes.

Corduroy (kor-duh-roy): durable cut-pile fabric with wide or narrow wales formed with extra filling threads that are cut and napped.

Covert cloth (koh-vert kloth): durable, medium weight fabric of tightly twisted two-ply yarns in a twill weave with a finely flecked look.
Crash (krash): coarse fabric with a rough, irregular surface made from thick, uneven yarns.
Crepe (krayp): fabric with a dull, crinkled surface obtained by chemical treatment, tightly twisted yarns, or novelty weaves.
Crepe de chine (krayp duh sheen): fine, lustrous, lightweight fabric of silk or manufactured fibers that is made like crepe.
Crinoline (krin-uh-lin): stiff, open fabric with heavy sizing applied.
Damask (dam-usk): firm, heavy, reversible fabric woven in Jacquard designs.
Denim (den-um): strong, washable cotton or blend "jeans" fabric in twill weave. The warp yarns are colored, and the filling yarns are white.
Dimity (dim-uh-tee): quite sheer, lightweight cotton fabric with fine woven stripes, checks, or designs.
Dobby (dah-bee): fabric with small figures, such as dots or geometric designs, in the weave. Made on a dobby loom.
Doeskin (dough-skin): fabric of wool or manufactured fibers with a soft, often napped, finish.
Donegal (dohn-eh-gahl): woolen, homespun tweed with colorful woven slubs from uneven yarns.
Dotted Swiss (daht-ted swiss): fine, sheer fabric with small dots applied by weaving, flocking, or using a clipped spot loom.
Double knit (duh-bul nit): weft knit fabric of double construction.
Drill (drill): firm, sturdy, durable, tightly woven cotton fabric used for work clothes, pockets, and khaki or olive-drab uniforms.
Duck (duck): heavy, plain weave fabric that is very durable.
Duffel cloth (duf-ful kloth): thick, heavy, napped coating fabric, usually tan or green.
Eyelet (eye-lit): fabric embroidered with openwork patterns.
Faille (file): dressy fabric of silk or manufactured fibers with flat crosswise ribs.
Fake fur (fayk fur): slang term for pile fabrics that imitate animal fur.
Felt (fehlt): compact sheet of matted fibers that are interlocked by a combination of heat, moisture, and pressure.
Flannel (flann-il): soft, plain or twill weave cloth with a brushed surface made in many fibers and weights.
Fleece (fleese): a bulky knitted or woven fabric used for sweat suits and other soft, warm apparel.
Foulard (fool-ard): soft, lightweight, silky necktie or scarf fabric of plain or twill weave, printed with a small all-over pattern.
Gabardine (gab-ur-deen): hard and smooth or soft and dull surfaced twill weave fabric of medium to heavy weight.
Gauze (gawz): thin, sheer, open-weave fabric.
Georgette (jor-jet): Sheer, creped, dressy fabric with good stiffness and body for its weight.
Gingham (ging-um): cotton or blended fabric, usually with woven check design, but sometimes with woven stripe or plaid.
Glen plaid (glen plad): squares of small woven checks alternating with larger checks to simulate a plaid in one or two muted colors and white. Sometimes called a glen check.
Granite cloth (gran-ut kloth): fabric that has pebbled effects like the grainy surface of unpolished granite.
Harris tweed (hair-iss tweed): soft, flexible, all wool tweeds handwoven on islands off the coast of Scotland. They are expensive and of high quality.
Herringbone (hair-ing bone): fabric with broken twill weave giving a zigzag effect.
Homespun (home-spun): heavy, nubby plain weave fabric from uneven yarns.
Honan (hoh-nan): silk or manufactured fiber pongee with occasional thick-thin effect.
Hopsacking (hop-sak-king): rough, open fabric with ply-yarn basket weave.
Houndstooth (howndz-tooth): fabric with medium-sized broken check that resembles a four-pointed star.
Illusion (il-oo-shun): very fine mesh of silk or nylon used in bridal veils.
Insertion (inn-sur-shun): narrow fabric, often lace, finished on both edges to use as a decorative strip.

Intarsia (inn-tar-see-uh): reversible flat knit with a geometrical colored design.
Interlock (inn-tur-lok): closely knit, smooth, stretchy, fluid, double knit fabric identical on both sides.
Jersey (jur-zee): knitted fabric, with smooth, dull finish, that is elastic and drapable.
Kersey (kur-zee): woven fabric with a fine nap, used for overcoats and uniforms.
Khaki (ka-kee): fabric of various fibers in an earthy tan or green.
Lace (layss): decorated openwork fabric, often of floral design.
Lamé (lah-may): dressy fabric, woven or knitted with metallic threads.
Lawn (lawn): fine, sheer, crisp-finish fabric in a plain weave.
Leno (leen-oh): open, lacy, woven fabric with a mesh effect.
Loden cloth (loh-den kloth): heavy, napped coating fabric woven from rough, oily wool with natural water repellency and a forest green color.
Mackinaw (mack-eh-naw): thick, coarse fabric with some natural water repellency, usually plaid or checked, used for hunting jackets.
Madras (mad-rehss): plain weave cotton fabric, in colored plaids, stripes, or checks with dyes that run together during washing.
Marseilles (mahr-say): firmly-woven, reversible, cotton or blend fabric with raised geometric designs.
Matelassé (mat-la-say): fabric with raised patterns that look quilted or puffed.
Matte jersey (mat jur-zee): dull tricot made of fine crepe yarns.
Melton (mell-tun): thick, heavy, nonlustrous coating fabric with a nap that is raised straight and then sheared.
Mesh (mesh): open fabrics of various construction and fibers.
Middy twill (mid-ee twill): sturdy, often navy blue cotton, twill weave fabric.
Moiré (mwa-ray): fabric, usually taffeta, produced by engraved rollers with a wavy, rippling watermarked appearance.
Momie cloth (mahm-ee kloth): fabric with a crepey pebbled effect from the weave.
Monk's cloth (munks kloth): heavy, coarse, loosely-woven basket weave fabric.
Moss crepe (mawss krayp): crepe with a moss-like surface.
Mousseline (moo-seh-leen): sheer, crisp evening wear fabric of different fibers.
Muslin (muhz-lin): inexpensive, durable, plain weave, cotton type fabric.
Nacré velvet (na-kray vel-vit): velvet with back of one color and pile of another giving a changeable appearance.
Nainsook (nayn-sook): soft, lightweight, plain weave, cotton fabric.
Net (net): open, knotted, sometimes stiff, fabric of various weights, as shown in 9-34.
Ninon (neen-ohn): sheer, open mesh, plain weave fabric.
Organdy (oar-gan-dee): sheer, lightweight, stiff, plain weave cotton.
Organza (oar-gan-zah): sheer, crisp organdy in silk or manufactured fibers.
Ottoman (ot-teh-man): fabric of various fibers with wide horizontal ribs.
Outing flannel (owt-ing flan-el): soft, lightweight cotton fabric napped on both sides.
Oxford cloth (oks-ford kloth): smooth shirting fabric of basket weave, often with colored warp and white filling threads.
Paisley (payz-lee): any fabric with a multicolored cone or scroll-like design.
Panné (pan-ay): very lustrous velvet with pile pressed flat in one direction.
Paper fabric (pay-per fab-rik): slang for nonwoven fabrics, often disposable.

Fashion Institute of Technology

9-34 Net is sometimes used for multiple layers in bodices and skirts of formal dresses.

Patchwork (patch-wurk): fabric resulting from joining small geometric pieces of other fabrics together, or printed to look that way.

Peau de soie (po deh swah): soft, closely woven silk-like fabric with dull sheen, usually used in fancy gowns.

Percale (per-kayl): tight, plain weave, smooth cotton fabric, finer than muslin.

Piqué (pic-kay): medium-weight fabric with raised lengthwise woven cords or patterns.

Plissé (plih-say): thin cotton fabric with puckered stripes or pattern in an overall blister effect caused by selective shrinkage.

Polished cotton (pol-isht kot-un): cotton fabrics with a shiny surface, from a slight sheen to a high glaze.

Pongee (pon-jee): fairly lightweight, plain weave fabric made from irregular yarns; originally handwoven of silk.

Poodle cloth (poo-dul kloth): loopy or bumpy knotted-yarn fabric.

Poplin (pop-lin): medium to heavy fabric with a fine crosswise rib.

Power net (pow-er net): elastic net fabric that stretches in one or both directions, used mainly for support undergarments.

Raschel knit (ra-shel nit): a type of coarse or open knit, often lace or net.

Rep (rep): woven, slippery fabric with crosswise design or stripe, usually used for neckties.

Sailcloth (sayl-kloth): strong, durable, cotton fabric.

Sateen (sa-teen): a satin weave fabric of cotton with a high luster finish.

Satin (sat-in): slippery, lustrous fabric of silk or manufactured fibers with floating surface yarns, as in 9-35. The wrong side is dull.

Seersucker (seer-suk-er): medium-weight, cotton fabric with woven puckered stripes formed by tight and loose warp yarns.

Serge (surj): smooth, twill weave fabric with a diagonal on both sides.

Shantung (shan-tung): plain weave fabric of silk or manufactured fibers with a rough, nubbed surface.

Sharkskin (shark-skin): medium- to heavyweight fabric with a textured, lustrous surface.

Hong Kong Fashion Week/Issetti

9-35 The lustrous sheen of satin is appropriate for dressy attire.

Suede cloth (swade kloth): woven or knitted fabric with napped surface that looks like suede leather.

Surah (soir-ah): soft, lustrous, silky fabric of fine twill weave.

Taffeta (taf-eh-tah): crisp, smooth, plain weave fabric used for evening wear and linings. It "rustles" and has a slight crosswise rib.

Tapestry (tap-es-tree): ornamental fabric with slight looped piles and a pictorial design.

Tartan (tart-un): cloth, often wool, woven in distinctive colored plaids of Scottish Highland clans.

Tattersall (tat-er-sol): checked fabric with two sets of dark lines on a light background.

Terry cloth (tair-ee kloth): absorbent fabric with raised, uncut loops on one or both sides.

Thai silk (tye silk): fairly heavy, slubbed, brightly colored silk fabric made in Thailand.
Ticking (tik-ing): strong fabrics, usually having narrow woven stripes.
Toile (twahl): fabric printed with one-color, fine-line pictorial designs.
Tricot (tree-koh): drapable warp knit fabric with fine vertical wales on the front and crosswise ribs on the back.
Tropical suiting (trahp-ih-kahl soo-ting): crisp, lightweight suiting fabric, often of linen.
Tulle (tool): a sheer, fine net fabric.
Tussah (tuss-ah): tan fabric woven from silk of wild, uncultivated silkworms.
Tweed (tweed): rough-surfaced fabric, often wool, with mixed color slubs of yarn forming a speckled effect on a somewhat hairy surface.
Uncut velvet (un-kut vel-vit): velvet with pile left in looped form.
Velour (vel-ouhr): thick-bodied, close-napped fabric with a velvety look, as in 9-36
Velvet (vel-vit): rich, soft fabric of silk or manufactured fibers, with short, thick pile. Usually woven double, face-to-face, and cut apart on the loom.
Velveteen (vel-veh-teen): fabric with dull, velvety pile woven singly with full, close-cut filling yarns, of cotton or manufactured fibers.
Voile (voyl): lightweight, sheer, crisp, woven fabric of high twist yarn.
Whipcord (whip-kord): strong, often twill weave fabric of various fibers used for uniforms, riding clothes, and hard-working apparel.
Woolen (wuhl-en): fabric made of wool fibers, usually somewhat fuzzy, such as wool flannel, fleece, melton, serge, or tweed.
Worsted (wur-stid): smooth, closely constructed fabric of firm, strong, combed, long-staple wool yarns. Examples include gabardine, challis, crepe, serge, and sharkskin.
Zibeline (zib-eh-leen): heavy coating fabric with nap pressed in one direction.

Draper's & Damon's

9-36 Velour is very comfortable. It is usually used in sports warm-ups or at-home lounging attire.

Chapter Review

Summary

There are various methods of making fabric. Weaving interlaces warp and filling yarns at right angles to each other on a loom. Diagonally between the lengthwise and crosswise grains is the bias grain. The plain, twill, and satin weave types have variations for different surface appearances. Variations also include pile fabrics with nap, Jacquard and dobby designs, and leno weaves.

Knitting loops yarns together as wales or courses. Weft knits can be made on circular or flat machines and as single or double knits. Warp knits, made only on flat knitting machines, might be tricot, raschel, or other knits.

Other fabric construction methods include nonwovens, bonded, quilted, and braided fabrics, as well as laces and nets.

Bleaching removes color or impurities from fabrics. Dyeing gives color at the fiber, yarn, piece (yard goods), or garment stages. Printing adds color, pattern, or design to the surface of fabrics by roller printing, screen printing, rotary screen printing, heat transfer printing, flocking, block printing, or ink-jet printing.

Finally, finishes are applied to fabrics. Mechanical finishes affect the size and appearance of fabrics. Chemical finishes affect fabric performance. A glossary of many popular apparel fabrics is provided to aid you in your study of textiles.

Fashion in Review

Write your answers on a separate sheet of paper.

Short Answer: Write the correct answer to each of the following questions.

1. What are the two most common methods of making fabrics?
2. What is the grain of a fabric?
3. Name the three basic weaves.
4. Large and intricate designs are woven on a _________ loom by programming punch cards.
5. Name the two methods of knitting fabrics.
6. What procedure permanently fastens together two layers of fabric with an adhesive or with heat?
7. What does the term colorfast mean? What conditions test colorfastness?

Multiple Choice: Write the letter of the best response to each of the following questions.

8. The most common, and least expensive, method of coloring textiles is _____.
   A. fiber dyeing
   B. yarn dyeing
   C. piece dyeing
   D. garment dyeing
9. The majority of printed fabrics on the market today are _____.
   A. roller printed
   B. screen printed
   C. rotary screen printed
   D. heat transfer printed
10. A finishing process that can press the fabric surface flat, or emboss it, with cylinders that apply heat and pressure is _____.
    A. drying and stretching
    B. compressive shrinking
    C. calendering
    D. singeing

True/False: Write true or false for each of the following statements.

11. Weaving yarns closely together results in a strong, dense fabric.
12. Denim is a plain weave fabric.
13. Satin usually has more luster than sateen.
14. Knit fabrics are usually stretchy, flexible, and wrinkle-resistant.
15. Warp knits are made on circular knitting machines,
16. Stock dyeing and solution dyeing are forms of fiber dyeing.
17. Finishes improve the appearance, feel, and/or performance of textiles.
18. Finishes improve the basic quality of fibers, yarns, and fabrics.

19. Brushing is a finishing process that creates nap by pulling up fiber ends on the fabric surface.
20. Mercerization is a finish that increases the luster, strength, absorbency, and dyeability of cellulosic fibers.

Fashion in Action

1. Using thin strips of paper in two different colors, make examples of the plain weave, twill weave, and satin weave. First lay down the warp yarns in one color. Then weave in the filling yarns of the other color. Glue your examples to paper. Label each weave and list fabrics made with it.
2. Find a piece of woven fabric, at least 4 inches by 6 inches. Label the lengthwise, crosswise, and bias grains. If it has a selvage, compare that to an edge that ravels. Identify the type of weave. Mount the fabric on paper and write a description of it.
3. Find a piece of knitted fabric. Identify the wales and courses. Determine the gauge. Identify the kind of knit. Mount the fabric on paper and write a description of it.
4. Research one of the finishes described in this chapter. How and where was it invented? On what kinds of textile products is it used? What characteristics does it give to textile products? How is it applied? How durable is it? Give a comprehensive oral report to the class.
5. Collect samples of five of the fabrics listed in this chapter's glossary. They might be from fabric stores, unsewn textiles at home, or old garments. Mount them along with descriptions of their fabric constructions, color applications, and finishes.

PART FOUR

Design: The "Art" of Apparel

Chapter 10
The Element of Color

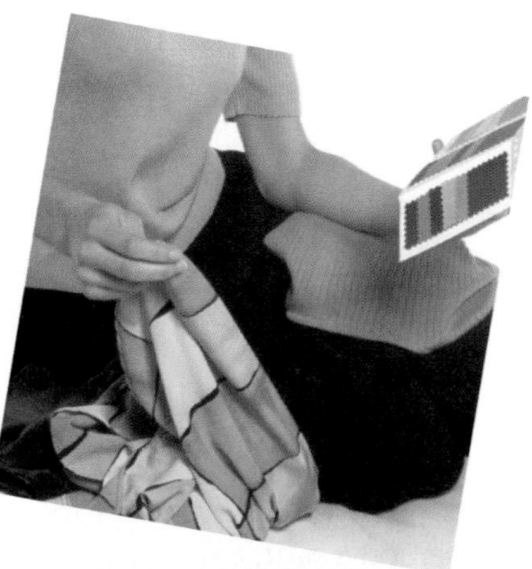

Chapter 11
More Elements of Design

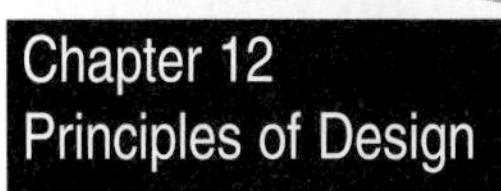

Chapter 12
Principles of Design

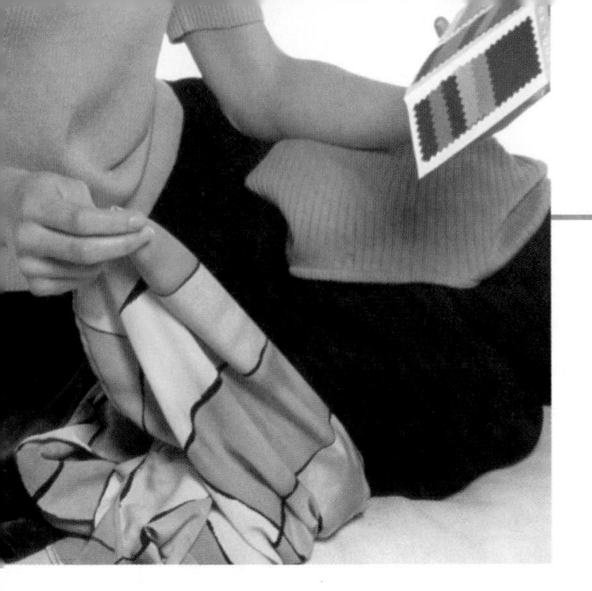

10 The Element of Color

Fashion Terms

hue
value
tint
shade
intensity
neutrals
primary hues
secondary hues
intermediate hues
monochromatic color scheme
analogous color scheme
complementary color scheme
split-complementary color scheme
triad color scheme
accented neutral color scheme
undertone

After studying this chapter, you will be able to

- identify color as a design element.
- explain the symbolism and terms related to color.
- discuss how to use color schemes and illusions effectively in apparel.
- choose clothes in colors that flatter a person's hair, eyes, and skin tone.

The elements of design are color, shape, line, and texture. They are the building blocks of design. They are put together in different ways to form the design of a building, a car, a painting, or a pair of shoes. You can use the elements of design to help you look your best.

Color as a Design Element

Color is the most exciting design element. It is used and noticed everywhere. Just think how drab the world would be without it!

Color enables you to express yourself. It affects how you feel. It can make clothes seem warm or cool, perky or drab. It can create illusions by making you look taller or shorter, larger or smaller. It can emphasize or play down your features. The best colors for you are the ones that make the most of your natural looks.

Knowing how to use color is important in achieving a well-dressed appearance. Color is probably the most personal and important aspect of fashion. It is what first catches the eye of someone who sees you. It sets the stage. Clothing is usually selected because of its color.

The apparel industries like to stress new colors and color combinations each season, as shown in 10-1. When colors become popular, designers and promoters give them fashionable names. Examples include royal blue, salmon pink, sunburst yellow, ruby red, lime green, lilac, peach, coral, and rose. The colors, however, are not new. The new names are used to grab your attention and convince you that you "need" to wear these colors to be in fashion. You should realize that this is a play on your emotions. Having a few items in the new fashionable colors might lift your spirits. However, having a whole new wardrobe in the latest fashionable colors is unnecessary.

Philadelphia University

10-1 New, fashionable colors and color combinations appear in apparel each season.

Symbolism of Color

Some people say the colors you wear reveal your personality. This is true in many cases.

Colors do have definite symbolism. This gives them communication value. Certain colors have come to mean certain things. If you "feel blue," you are sad and lonely. If you "see red," you are very angry! Sometimes we are "green with envy."

Certain religious sects have meanings for various colors. Countries have meaningful colors in their flags. Schools have colors that their athletes and band members wear to represent them.

Some colors have been found to calm our spirits. Some stimulate us. Most people are cheered by bright, clear colors, but not by dark, dull ones. Colors can have a quieting effect, give a happy feeling, or call attention to themselves.

Various colors are specially selected for certain building interiors because of their effects on feelings and emotions. When red is used, it can make you feel good and full of energy. Red can indicate power, spark emotion, attract attention, and activate the appetite. Store promotion areas might be bright to cause people to be cheerful and to buy. Classrooms are usually in soft or neutral colors to encourage serious study.

Chart 10-2 lists some of the common psychological associations of colors. Knowing this information can help you pick the properly colored clothes to suit the feeling of the occasion for which they are worn. In our culture, people wear black for funerals and white for weddings, baptisms, and confirmations. They might wear orange to a party and blue to visit a friend in the hospital. Color has a great deal of influence in our lives and in our clothes!

Common Psychological Associations of Colors

Red—Hot, dangerous, angry, passionate, sentimental, exciting, vibrant, aggressive

Orange—Lively, cheerful, joyous, warm, energetic, hopeful, hospitable

Yellow—Bright, sunny, cheerful, warm, prosperous, cowardly, deceitful

Green—Calm, cool, fresh, friendly, pleasant, balanced, restful, lucky, envious, immature

Blue—Peaceful, calm, restful, highly esteemed, serene, tranquil, truthful, cool, formal, spacious, sad, depressed

Violet—Royal, dignified, powerful, rich, dominating, dramatic, mysterious, wise, passionate

White—Innocent, youthful, faithful, pure, peaceful

Black—Mysterious, tragic, serious, sad, dignified, silent, old, sophisticated, strong, wise, evil, gloomy

Gray—modest, sad, old

10-2 Studies have shown that humans tend to associate certain traits and emotions to specific colors.

Color Terms

Color has three dimensions or qualities. They are hue, value, and intensity.

Hue is the name given to a color, such as red, yellow, green, or violet. It distinguishes one color from another.

Value is the lightness or darkness of a color. The values of colors range on a gradation scale from almost white to almost black. A ***tint*** is made when white is added to the color so it is lighter than the pure hue. Pink is a tint of red. A ***shade*** is made when black is added to make a hue darker. Burgundy is a shade of red. Value is summarized in chart 10-3. You can often see the value of a color better if you squint your eyes to look at it.

Intensity is the brightness or dullness of a color. Very strong, bright colors are said to have high intensity, as in 10-4. Dull, faded, or dusty colors are low in intensity. Intensity can be lowered by mixing a hue with its complement on the color wheel. (See next section.)

Black and white are ***neutrals,*** as shown in 10-5. They are really not colors. White reflects all light so there is no color. It is the "absence of color." Black, on the other hand, absorbs all light and all colors. When white and black are mixed, they become the neutral gray. Beige is also usually considered to be a neutral in apparel, as in 10-6, since it can be used with almost all colors.

White and black are used to lighten or darken colors as well as being used alone. Since white reflects light, it feels cooler in hot climates. In summer, most clothing is white or very lightly colored. That shows ***that*** we want to stay cool. Black brings warmth to the wearer. In winter, we tend to wear dark colors that help us feel warm.

The Color Wheel

A color wheel, shown in 10-7, can be used to show hues and how they can be mixed or used with each other. Use the wheel as a guide to study how to choose and combine colors.

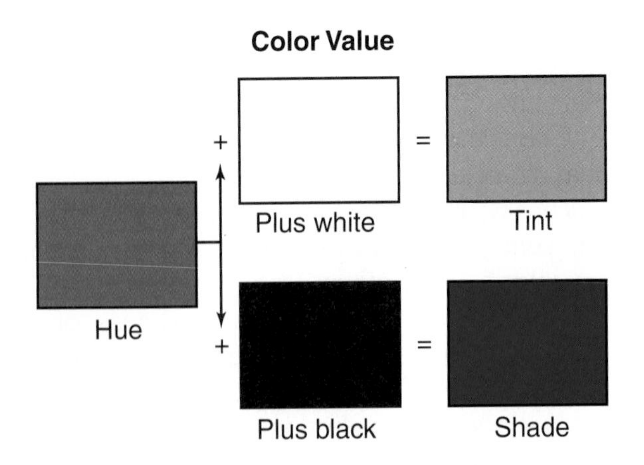

10-3 The value of a color or hue is changed by adding either white to make a tint or black to make a shade.

There are three pure, or basic, hues. They are the ***primary hues.*** They are red, yellow, and blue. They cannot be made from any other colors. They need their own pigments, or coloring matter. All other colors can be made by mixing them. The three primary hues are placed equal distances from each other on the color wheel.

The three ***secondary hues*** are orange, green, and violet (purple). They are made by mixing equal amounts of two primary hues together. They are found halfway between the primary hues on the color wheel. Orange is made by mixing red and yellow. Green is made from equal amounts of blue and yellow. Violet is a combination of red and blue.

Intermediate hues (sometimes called *tertiary hues*) result when equal amounts of adjoining primary and secondary colors are combined. When naming them, it is customary

JC Penney

10-4 The orange in these shirts is bright, or high in intensity.

Escada by Margaretha Ley

10-5 The neutrals black and white are often used in combination with each other in fashionable apparel.

The Fashion Association / Brian McKinney Designer Collections

10-6 Beige, or tan, is really a tint of brown (a low-intensity orange). However, in apparel, beige is usually treated as a neutral.

to state the name of the primary hue first. Intermediate colors are blue-violet, blue-green, yellow-green, yellow-orange, red-orange, and red-violet.

The color wheel can be divided into warm and cool sides. *Warm colors* are red, orange, and yellow. They appear to be hot like the sun, or like fire. *Cool colors* are green, blue, and violet. They remind us of water or the sky. Orange is the warmest color. Blue is the coolest color.

Warm colors give a feeling of activity and cheerfulness. They set an outgoing and lively mood. However, if they are overdone, they can give a nervous impression.

Warm colors appear to advance, or to come toward the observer. They make the body look larger. White and light colors also make objects look larger.

Cool colors give a feeling of quietness and restfulness. They suggest a subdued mood. If overdone, they can be depressing.

Cool colors appear to recede, or to back away from the observer. They make the body look smaller. Designers often use cool colors for garments in large sizes so the people wearing them look smaller.

Color Schemes

Color schemes are the ways that colors are used together. Different results are achieved

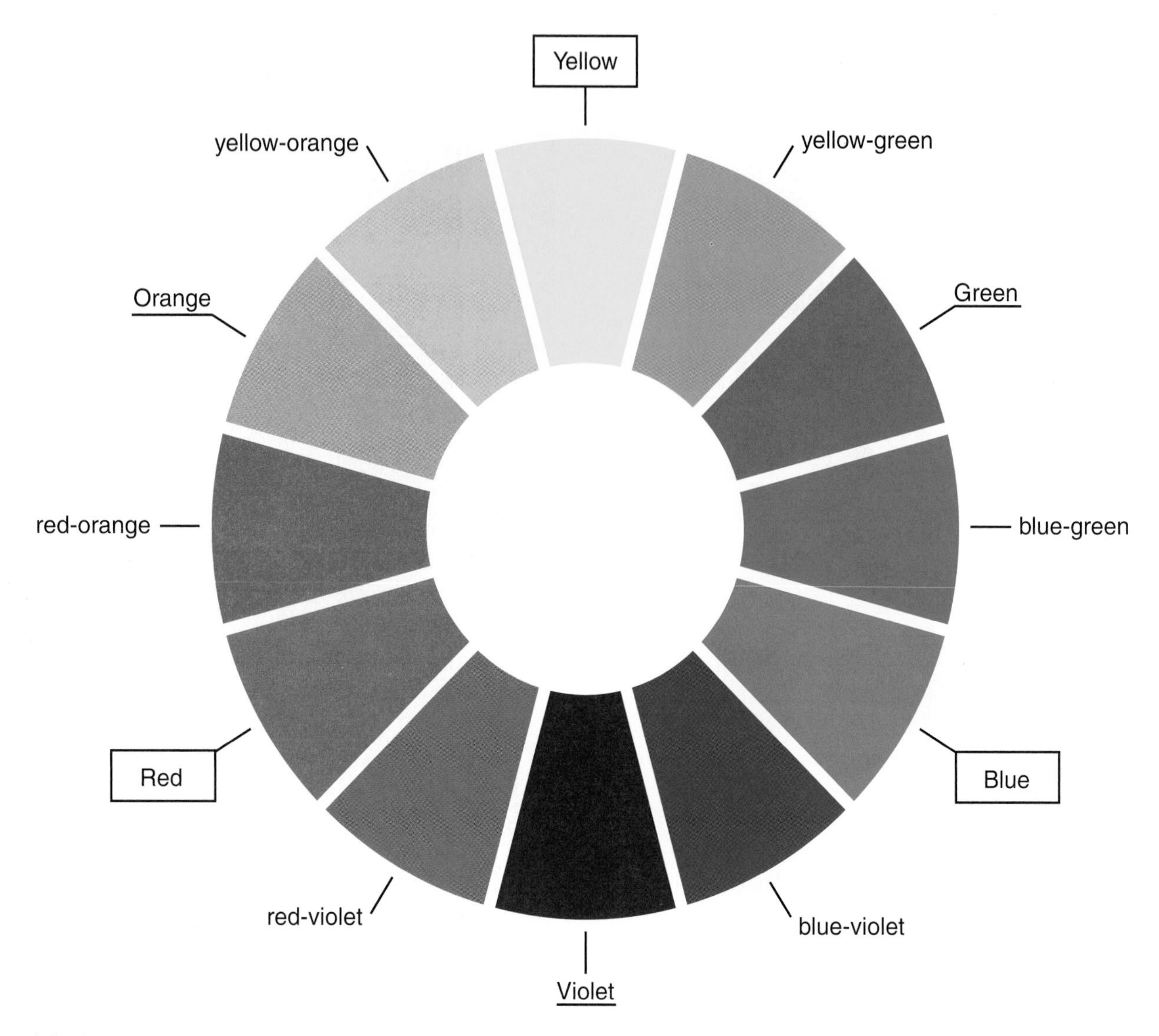

10-7 The color wheel shows how hues are related. The names of the primary hues have boxes around them. The names of the secondary hues are underlined. Intermediate hues have their names written in lower-case (small) letters.

by using different combinations of colors. Successful or harmonious combinations of colors are based on the location of the colors on the color wheel. The following six basic color schemes are diagrammed in 10-8.

Monochromatic Color Scheme

A ***monochromatic color scheme*** is a one-color plan that uses different tints, shades, and intensities of the same hue. It is usually restful to the eyes because of the unity that results from using just one color. A pair of navy slacks with a pale blue shirt is an example of a monochromatic color scheme. Brown slacks with a beige shirt is also monochromatic. See 10-9. Neutrals are sometimes added to a monochromatic scheme for contrast and interest.

Analogous Color Scheme

An ***analogous color scheme*** uses neighboring, or adjacent, colors on the wheel. It is sometimes called a related color scheme since two or three "related" colors are used. To avoid monotony in your clothing, you may want to use different values and intensities for some contrast. This plan is used when blue and green are put together, or orange and yellow, or pink and violet, as in 10-10. The combination of yellow, yellow-green, and green is an analogous scheme with three hues.

Color Schemes

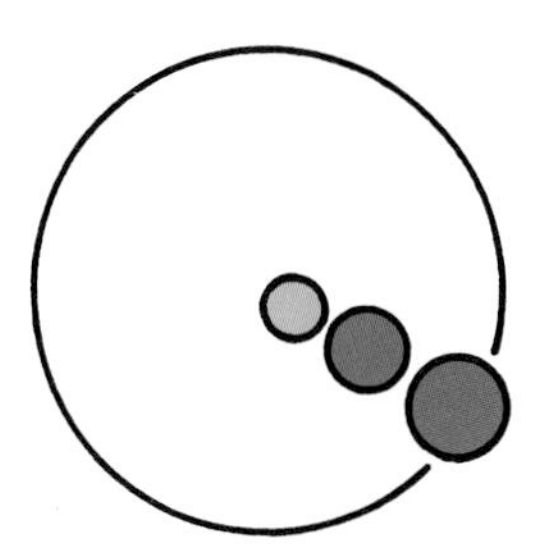
Monochromatic

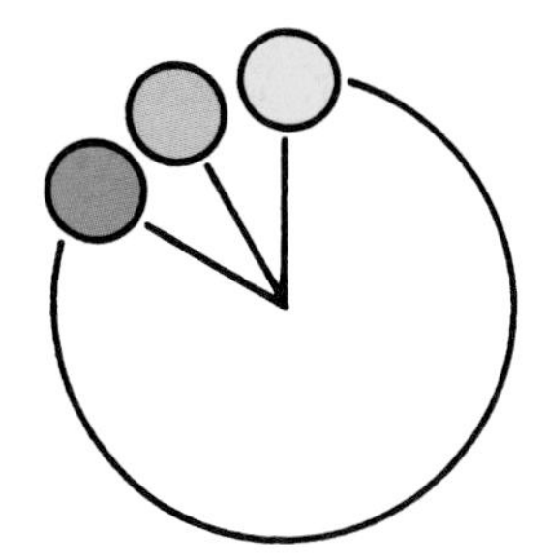
Analogous

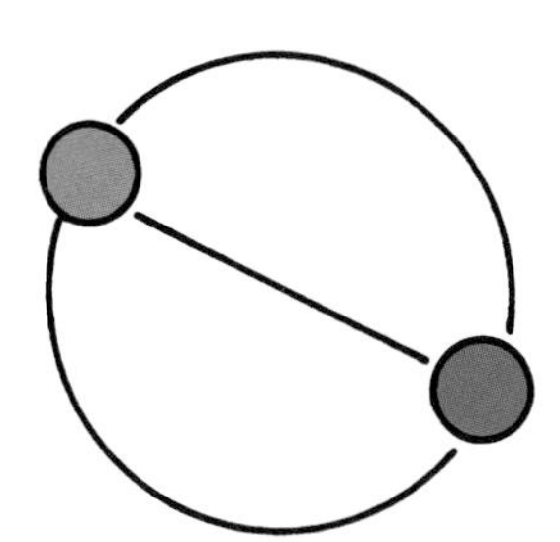
Complementary

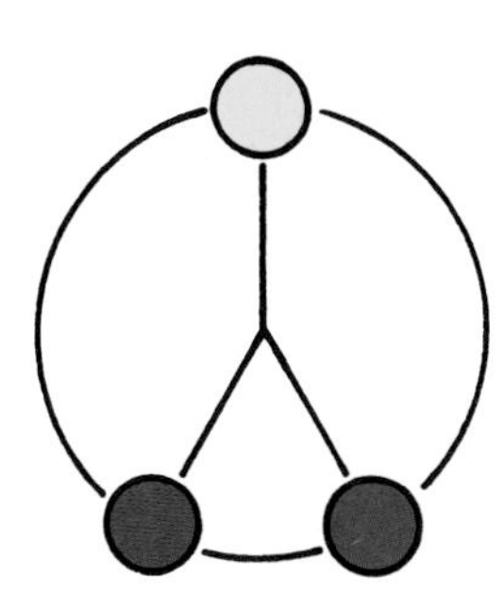
Split-complementary

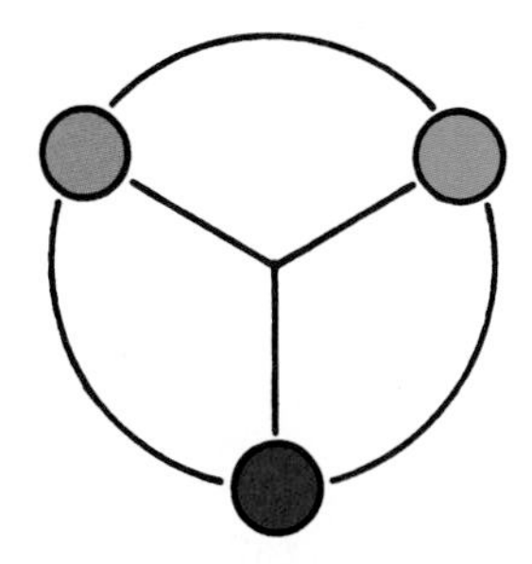
Triad

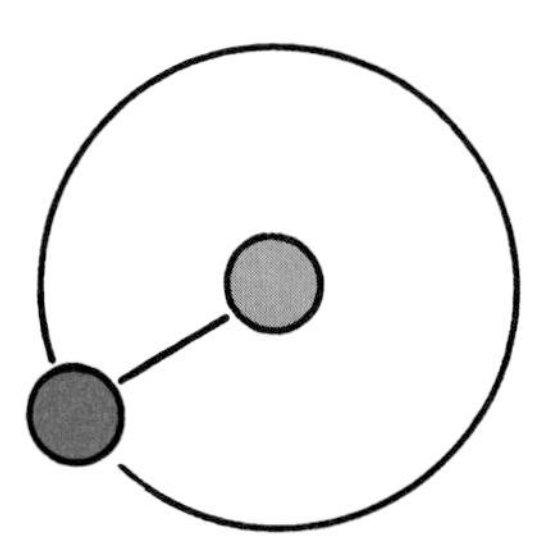
Accented neutral

10-8 Different hues may be used in these six color schemes, but the relative positions of the hues on the wheel do not change.

David's Bridal

10-9 This monochromatic outfit not only has different shades of the same red (burgundy) hue, but also uses different textures to accentuate the color scheme.

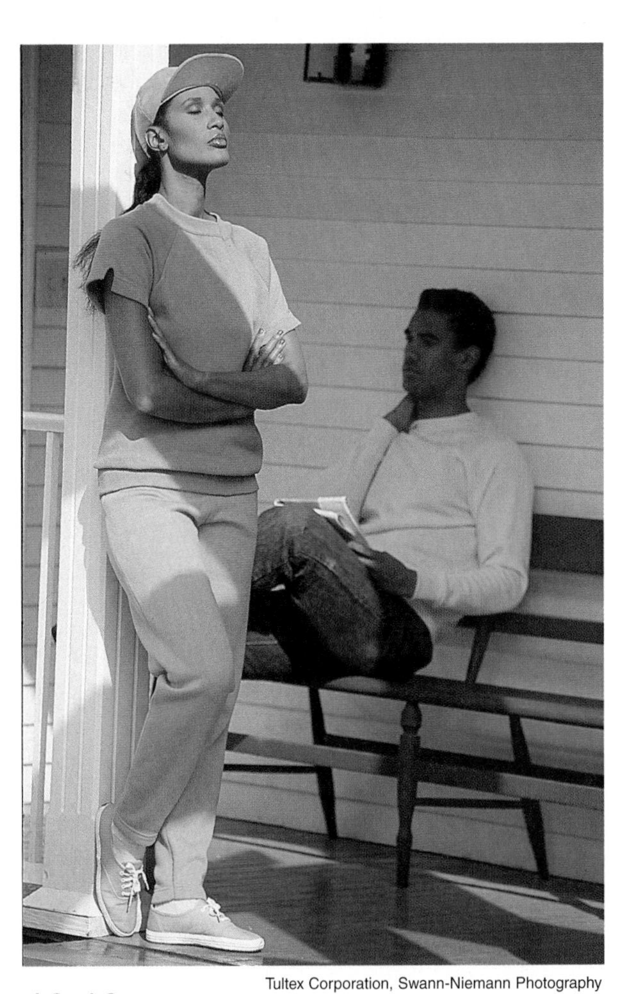
Tultex Corporation, Swann-Niemann Photography

10-10 The tints of adjacent hues, red and violet, are attractively combined here in an analogous color scheme.

In nature, the yellow, orange, and red of autumn is an analogous color scheme. Also, the blue, aqua, and green of sky, water, and grass is analogous.

Complementary Color Scheme

A ***complementary color scheme*** uses opposite hues on the color wheel. Complementary colors are across from each other on the wheel. They have great contrast. In fact, the colors look even brighter when they are used side-by-side. Examples of complementary color combinations are red and green, violet and yellow, and red-orange and blue-green, as in 10-11.

School colors are often of complementary schemes, so uniforms and banners look vivid and exciting. Be careful of this plan for daytime clothing. Worn together in full strength, these colors can give a real jolt! However, when used in tints and shades, they can be sophisticated and pleasing. A soft tint of one is usually attractive with a deep shade of the other. You can also dull the intensities, or use a large area of one hue and a small amount of the other.

Split-Complementary Color Scheme

A ***split-complementary color scheme*** uses three colors. It combines one color with the two colors on the sides of its complement. For instance, blue might be used with yellow-orange and red-orange. This is also a bright color scheme, so use it with care in your apparel.

Triad Color Scheme

A ***triad color scheme*** combines three colors equidistant on the wheel. Examples are red, yellow, and blue, as in 10-12, or violet, green, and orange. It has a great deal of contrast. To

Orvis

10-11 This complementary color scheme combines orange with blue. These hues are opposite each other on the color wheel.

The Fashion Association / Reebok

10-12 The three primary colors used together are an example of a triad color scheme. They are often combined in different values or intensities from each other.

soften the contrast, you may choose to combine pleasing values and intensities. You might wear a large area of one of the colors in a tint or shade and use small amounts of the other two for accents.

Accented Neutral Color Scheme

An ***accented neutral color scheme*** combines white, black, gray, or sometimes beige with a bright color accent. An example would be a gray and black striped suit and white shirt with a red necktie or scarf, or the outfits shown in 10-13. The accented neutral color plan is pleasing to the eye and very versatile in clothing. The color that looks best on you can be used for the accent. Depending on the type of outfit, the accent can be placed at the best place to show off your body—near the face, at the shoulders, or around the waist.

JC Penney

10-13 The three outfits in the foreground are accented neutral color schemes. Two of them use red for the accent, while the middle one uses orange.

Using Colors in Apparel

All colors are beautiful, depending on personal taste. However, if not used wisely or combined well, color can cause apparel to look too gaudy or very drab. Harmony results when hues, values, and intensities are combined in a pleasing way. Otherwise, a clash might occur, which causes discord.

Opinions about color combinations that are considered harmonious change as fashion changes. It is best for each person to strive for a fashionable image with harmony and variety in the color schemes that are best for him or her.

Although fashion often bends the rules, colors in clothing are usually best used according to the following guidelines.

- Black is good for formal wear. It tends to be sophisticated.
- Brown is casual, natural, and informal.
- Navy looks good on almost everyone. It is good for sportswear or classic styles.
- Beige and gray give a professional or tailored image. Both of these neutrals are quiet and unassuming, and they can be accessorized well.
- White looks good with all colors. Off-white is better for most people than pure white.
- Red, green, and blue have many tints, shades, and intensities that make these hues suitable for almost all occasions.
- Yellow is good for casual, fun clothes, but it is not pleasing with many skin tones.
- Bright colors are fun for active sportswear or as accents with neutrals.

The Changeability of Colors

Colors appear to change when viewed under different lights. Additional light makes colors look brighter. Fluorescent and incandescent lights give different effects from those of natural light. Fluorescent light makes colors look bluer. Incandescent light can give a yellow cast and tends to soften or fade some colors. Try to choose colors that look good on you in the environments where you will wear them.

Colors of contrasting value are often exciting when used together. A combination of black and white gives the strongest contrast. Black and yellow is a very distinct, dramatic color combination. Extreme contrast makes colors look brighter.

Using a color with a neutral makes the color appear brighter. Also, white and gray look brighter when placed beside black.

Colors with medium or dark value look even darker when used next to a light area. Light colors look even lighter next to a dark area, as shown in 10-14.

Texture, or the roughness or shininess of a material, affects color. The appearance of a hue in a corduroy fabric would be different from that of the same hue in satin.

From a distance, the colors in narrow stripes and small plaids will appear blended. Small red and white stripes may look pink from across the room. This may or may not be favorable. It depends on the effect you want to give and the colors of the accessories you are wearing.

JC Penney

10-14 When a dark skirt is worn with a white top, the contrast is emphasized. The skirt looks even darker, and the top looks whiter and brighter.

Clothing outfits are generally more attractive if they do not have equal areas of light and dark. They might be mostly dark with an accent of light, or the other way around. In most cases, colors in clothes seem better balanced if light ones are used above dark ones.

Creating Illusions with Color

Colors can set a mood of liveliness or quietness. They can create feelings, brighten eyes, or highlight the hair or complexion. The effects they give depend on how light, dark, or strong the colors are. The effects also depend on how the colors are combined with other colors in a total outfit. Think about these factors when choosing the colors for clothing.

Colors can appear to change the size and shape of the person wearing them. To use color to advantage, there are many pointers you can follow.

Dark, cool, and dull colors make a form seem smaller. Physical shortcomings can be hidden with cool colors, dark shades, and dull intensities. These colors recede and seem to decrease the size of the wearer. Such slimming colors are black, navy blue, dark blue-violet, charcoal gray, chocolate brown, burgundy, and dull dark green, as shown in 10-15.

Light, warm, and bright colors make a form seem larger. Those colors advance and seem to increase the size of the wearer. Such colors are white, yellow, orange, and red.

If a person is heavy in certain sections of the body, dark, dull colors should be worn there. Light, bright colors should be placed on areas that are thin. If a light or bright colored shirt is combined with dark, dull pants, the upper part of the body will look larger. You can use this trick to minimize large hips or to camouflage a small chest.

Bright colors draw attention. Some people look great in them, but it is usually best to use them in small areas. Play up good body features with them since they catch the eye. For instance, people will notice your face if you place a brightly colored neck scarf, tie, or collar near it. See 10-16.

Bright colors are also more easily remembered. Bright articles of clothing cannot be worn as often as more subdued outfits. Less intense colors are easier to combine in outfits and can be worn more often because they are more subtle.

A single color for an entire outfit makes a person look thinner and taller. By wearing all close values, the "eye of the beholder" follows an unbroken line. A business suit of all one color is more slenderizing than slacks and a sport coat of different hues or values. See 10-17.

When combining two colors in an outfit, special precautions are needed. Sharply contrasting colors appear to shorten the body. They seem to break up the body into separate parts. This is because the eye stops at a line of contrast instead of moving up the figure in a vertical direction. Strong contrasts call attention to where the space has been broken. A wide belt or waistband of a sharply contrasting color also makes a person look shorter. If someone is too tall, the outfit can be broken up with different colors on the top and bottom. A brightly colored belt can also be worn.

JC Penney

10-16 A brightly colored neck scarf draws attention up to the face.

Eastern Airlines, Inc.

10-15 This dull shade of green is good for uniforms that are worn by many people with different body shapes. It can make people look slimmer and hide various physical shortcomings.

The Fashion Association / JC Penney

10-17 Although the sport coat and slacks shown here are close in hue and value, the matched suit still gives a more slenderizing appearance.

To emphasize the best physical features, a small amount of a light or bright color can be used in an advantageous location on an otherwise subdued outfit. Accent a nice face or a small waist this way. White collars are often put at the neckline of dark dresses to draw attention. A light or bright stripe is often put on men's sport shirts to emphasize a broad chest. The attention is then automatically pulled away from other, less attractive areas.

In most cases, you should not use more than three major colors in one outfit. It is best to use one color for a large area and another color, or two, for smaller areas. The split-complementary and triad color schemes, described earlier in this chapter, are both three-color schemes. Illustration 10-18 combines variations of red, yellow, and blue, with red being used in the largest amount.

When considering colors for clothing, try to reflect your personality. If you are quiet and shy, you are probably more comfortable wearing "quiet" colors. You may tend to wear pale, cool, and neutral colors. Bright colors would overpower your personality. On the other hand, if you are dramatic and vivacious, bolder colors may suit you. Outgoing people tend to wear warm, bright hues, often with striking contrasts.

Men's Fashion Association of America / Bianculli

10-18 The colors in this outfit look good together because different values and intensities are used and one color (red) is dominant.

Flattering Your Personal Coloring

The proper use of color in a person's clothing will enhance his or her personal coloring. People can wear almost any color and look "okay." However, why not look *super* in your *best* colors? Wearing the right colors makes blemishes seem to fade and enhances the natural skin color. Sometimes we don't pick the right colors for ourselves because peer pressure and fashion advertising influence us. But with awareness, we can choose the most flattering hues, values, and intensities for apparel.

Your Color Tone

Personal coloring includes the color of a person's hair, skin, and eyes. The combination of these results in a total color tone. Clothing colors should be chosen to make your color tone look as pleasing as possible.

The colors in the skin are most important, because they cover the largest area. They are coordinated by nature with a person's eyes and hair. They always harmonize well together. Changing hair color with dyes, or eye color with colored contact lenses, can be tricky. A poorly chosen color may look unnatural. It could take away some of a person's natural good looks.

Skin tones are basically in one color family–orange. However, they may be given labels such as black, red-brown, yellow-brown, yellow, olive, or white. Other labels are fair, medium dark, or very dark. There are many, many variations and different amounts of black and white in people's flesh. The most important consideration is the warm or cool undertones of the skin.

An ***undertone*** is a subdued trace of a color seen through another color or modifying the other color. Everyone's skin color has an undertone of either blue or yellow. The easiest way to test for the undertone is to cut out a two-inch circle from a white sheet of paper. When that is placed over the skin of the inner lower arm, the undertone of blue or yellow should show. Compare your "test picture" arm with others to see the big difference between the cool and warm skin undertones. Warm undertones in the skin have a more yellow cast. Cool undertones are present if the skin looks a bit bluish.

Find Your Best Colors

There are several books and color consulting services that analyze skin tones. They can advise you about what colors are the most flattering for you to wear. This is called *color analysis*. When finished, you are given thirty or more color swatches, in fabric or on printed cards. You carry the swatches, or color cards, with you and use them as guides when you shop for clothes or cosmetics. See 10-19

Color specialists, or consultants, have popped up all around the United States and much of the world. Their slightly different methods approach the subject with different words, but they achieve similar results. The color consulting industry has become big business and has spread into cosmetic lines, personal interior design, and other areas. However, you should be able to do color analysis for yourself and others. Since it may be hard to be objective at first, have your classmates and friends help with this.

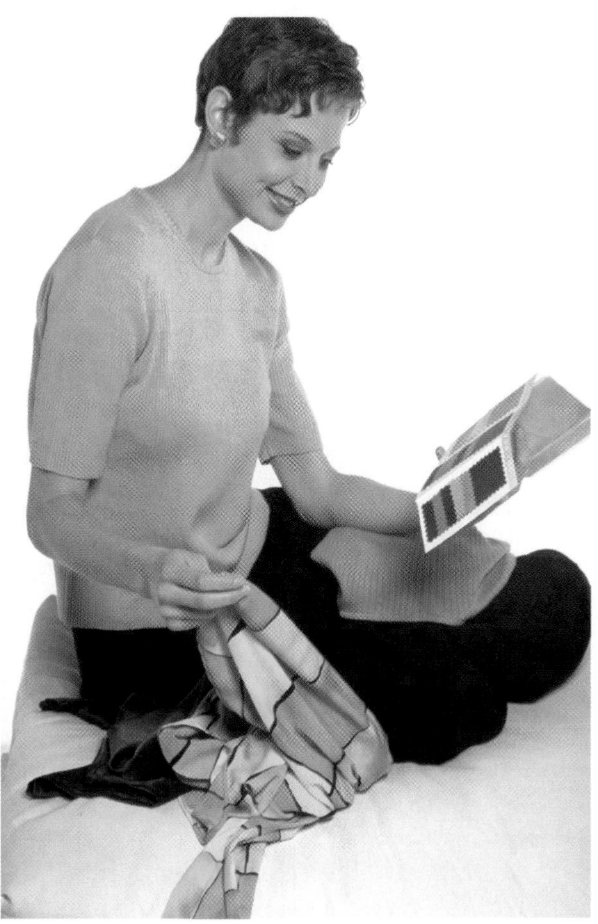

BeautiControl

10-19 Fabric swatches can be used to show what colors look best on you. This can help you to choose the most flattering fashions.

Key your color choices to your personal coloring for visual harmony. Experiment to find the hues, tints, and shades that look best with your coloring. The best colors on you are the ones that get good comments from your family and friends. They make you feel good. They often emphasize your eye color and flatter your skin.

Drape colored pieces of fabric around your neck and shoulders to see which look best. Stand in front of a mirror to analyze the effect of this important experiment. Try to use daylight instead of artificial lighting if possible. Try hues in different values and intensities to find the right ones. Notice the effect of each on your skin tone, hair, and eyes. Have your helpers give you their truthful opinions. What colors make your hair and eyes look bright and lively? Does your complexion look more yellow or brown or red with some colors? The right colors for you will bring out your best natural coloring. They will hide any drab, sallow tones.

Personal Color Categories

For easy reference, personal coloring types have been put into four main categories with the names of the seasons: winter, spring, summer, and autumn. These have nothing to do with a person's birth date! They are just helpful titles. The system was developed in the 1930s and '40s and has grown in acceptance ever since. Two seasons have cool (blue) undertones and two have warm (yellow) undertones. There are varied shades of skin, eyes, and hair within each season.

The largest number of people in the world are of the *winter season* type. Their ancestry is Asian, Indian, Polynesian, South American, African, or Southern European. Their hair is usually dark and may turn gray prematurely. Most have brown eyes. A few have hazel, gray, dark blue, or green eyes. Their skin has a blue (cool) undertone.

Winter people can wear true black or white in their clothing. Clear, true vivid colors from light to dark make them look good. Any colors with blue undertones are recommended. Dull or dusty colors should be avoided. Silver jewelry looks best. See the colors for "winter" in 10-20 for the most flattering hues.

People in the *spring season* type have heritage from Scandinavia, Britain, and Northern Europe. Their hair is flaxen or strawberry blond to medium or reddish-brown. Most have blue eyes. A few have aqua, golden

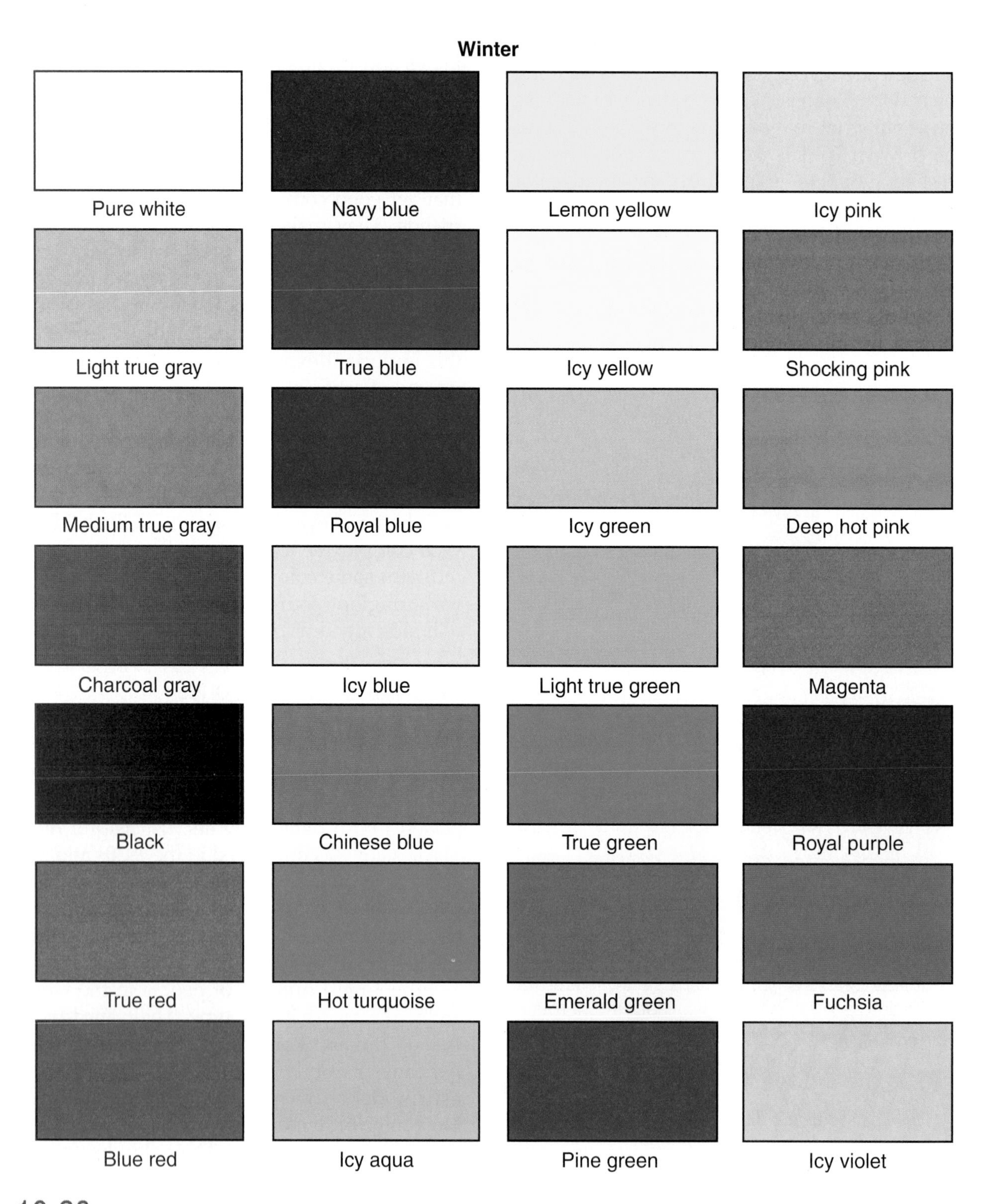

10-20 Colors for people in the winter season category are most flattering if they are true and vivid with blue undertones.

brown, or green eyes. Their skin has a yellow (warm) undertone. They should wear hues with yellow undertones. Medium to light colors are better than dark shades. Gold jewelry looks best. See colors for "spring" in 10-21.

People in the *summer season* type also have backgrounds from Scandinavia and Northern Europe. They have rosy, delicate coloring with a blue undertone. They may blush and sunburn easily. Their hair is ash blond, which often

10-21 The colors in this chart are best for "spring" people with yellow (warm) undertones.

darkens with age. Blue eyes are the most common. A few have green, grayish, hazel, or brown eyes. Their skin has cool coloring. They should wear dusty, muted shades with blue or rose undertones. Cool, soft colors are the best. Silver jewelry is most flattering. Look at the colors for "summer" in 10-22.

People in the *autumn* season type are from many diverse racial backgrounds. Redheaded Irish are typical of this category. People with

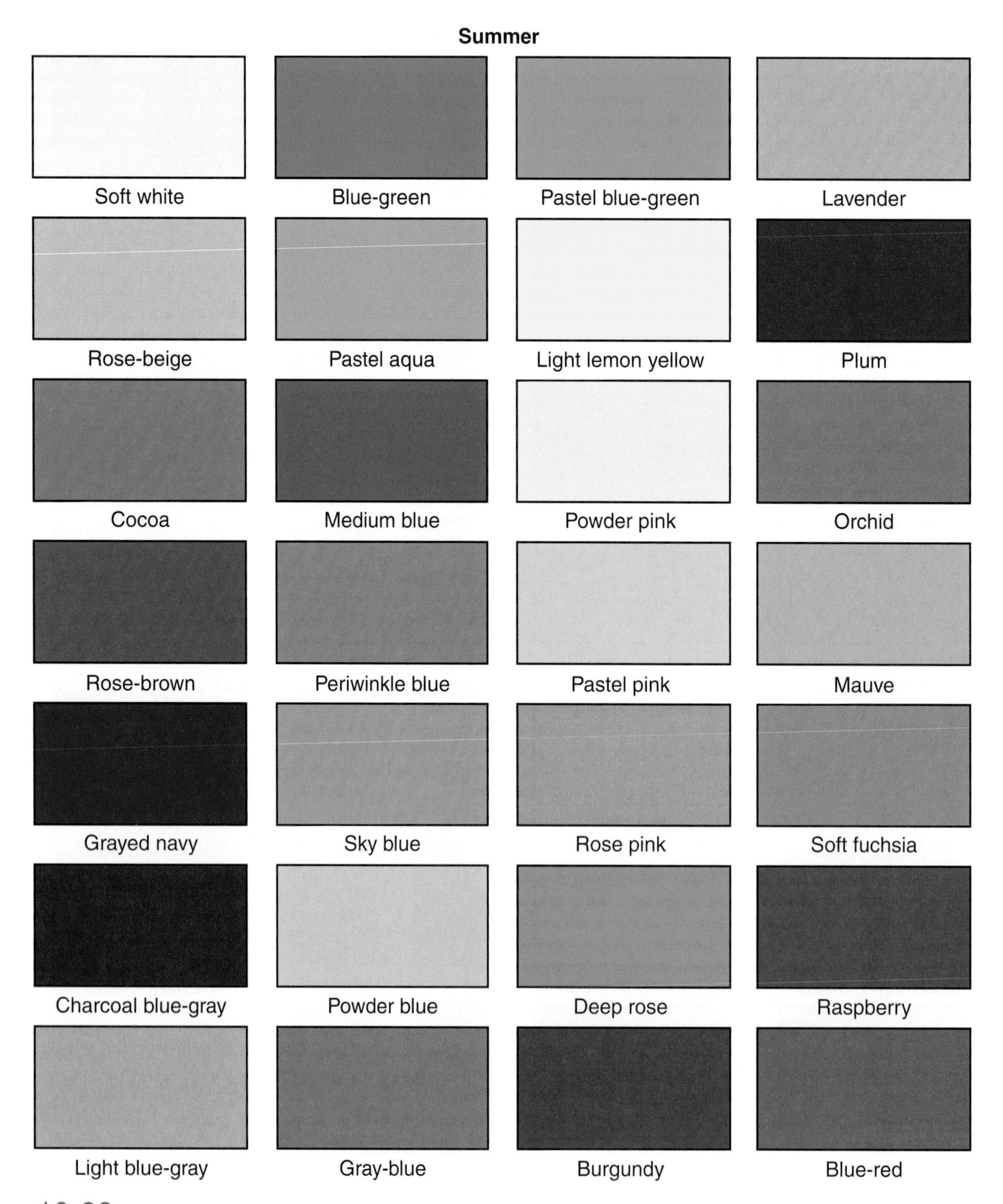

10-22 Apparel in the cool, soft colors shown in this chart make a "summer" person look his or her best.

dark skin tones are autumns if they have a truly golden undertone. Autumns have reddish highlights in hair ranging from blond to dark chestnut brown. Their eyes are usually brown, but some are green, hazel, or blue-green. Their skin may have freckles. It has a yellow undertone. These people should wear strong, but dusty, colors with orange and yellow undertones. Earthy, muted shades such as the colors of wood and metal are good. Yellows, oranges,

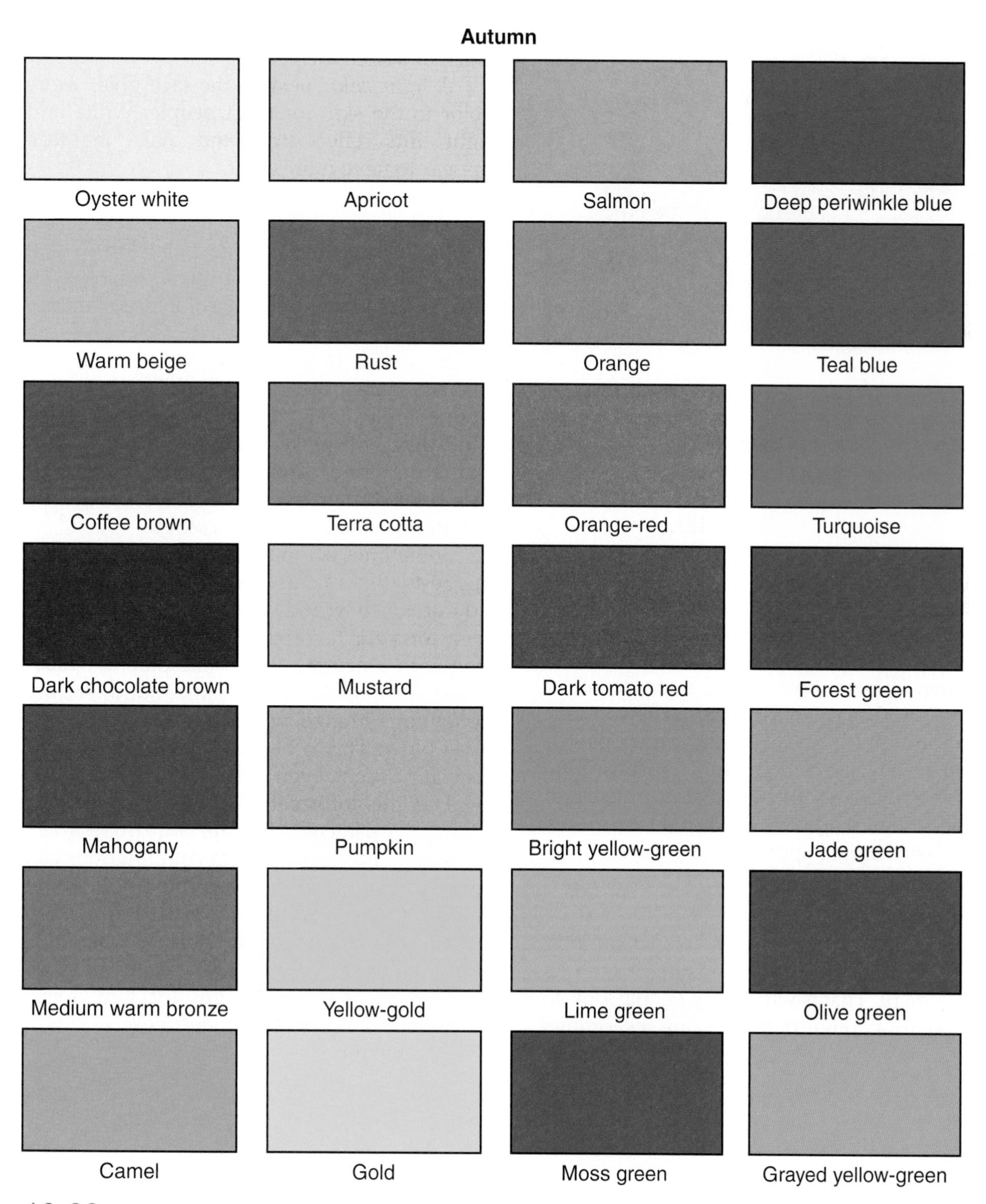

10-23 People of the "autumn" type have yellow undertones and look best when wearing these colors.

Russ Togs, Inc.

10-24 The right colors can enhance a person's natural features and overall appearance. Compare these colors with the text description and chart for "spring."

and browns are accented nicely with gold jewelry. Look at the colors for "autumn" in 10-23.

In summary, the personal coloring categories with blue undertones are winter and summer. These people should wear clothes in cool colors. The categories with yellow undertones are spring and autumn. These people look best in warm colors. See 10-24. Those who wear clear colors the best are winter and spring. Those who look good in dusty, muted colors are summer and autumn.

Wear Your Best Colors

A desirable hue in your personal coloring may be emphasized by contrasting it with its complement. That is why many people look so nice wearing blue; it brings out the orange (blue's complement), or brownness, of their skin. To intensify a faint pink tinge in your cheeks, use a shade of green in your clothes. But don't wear green or murky shades if your cheeks are flushed or sallow.

If you are fond of a color, but it does not flatter your coloring, use it with a hue that is good for you. Wear the flattering color near your face to show off your skin, hair, and eyes. Wear the less attractive color in your slacks, skirt, or accessories.

A light color next to the face gives more color to the skin for most people. White and light tints reflect light and make the face appear to have more color.

Dark colors absorb light and tend to drain the skin of color. If a dark value, such as black, is placed directly next to a pale skin, it will drain color away to make the skin look even lighter. That is why light colored collars are often used on dark shirts or dresses.

It is wise to avoid highly intense colors near the face, except for a little touch, unless your complexion is close to perfect. A subdued blue shirt will make your blue eyes sparkle, but a bright blue shirt may detract from your eye color. Brown eyes are not influenced by color as much as other eye colors.

Sometimes a somber or business occasion suggests that a neutral would be best, even if you prefer to wear a certain color. In this case, look for your favorite color tone in combination with a neutral. If you like blue, try blue-gray. A necktie or pocket handkerchief containing orange will make the gray look even bluer. This is because a color brings out its complement, even in a neutral.

Has this information shown you that the ways you use color in your wardrobe affect your appearance? The right combinations can give you a great look! Keep these pointers in mind. Do not merely wear what happens to be the latest fashionable colors. Use this year's colors to your advantage, and if some of them do not seem right for you, reject them. Use only the good ones; next year will bring some more!

Chapter Review

Summary

Color is the most exciting design element, especially for fashion. Color has a great deal of symbolism, and it affects our feelings and emotions.

Color has hue, value of tints and shades, and intensity. Black, white, gray, and sometimes beige are neutrals. The color wheel shows primary, secondary, and intermediate hue relationships. It has warm and cool sides.

How colors are used together includes monochromatic, analogous, complementary, split-complementary, triad, and accented neutral color schemes. Colors are usually used in apparel according to accepted guidelines. Colors can appear differently under different lights and with different combinations and textures. Use color to create desired illusions of size, feelings, or attention.

To flatter your personal coloring, know if your skin undertone is blue or yellow. Find your best colors through color analysis. All people can be placed within a winter, spring, summer, or autumn season type that offers attractive apparel color options. Wearing the best colors gives a positive effect on appearance.

Fashion in Review

Write your answers on a separate sheet of paper.

Short Answer: Write the correct answer to each of the following questions.

1. What are the three dimensions, or qualities, of color?
2. Name the three primary hues.
3. Name the three secondary hues and explain how they are made.
4. Name the six intermediate hues and explain how they are made.
5. Which colors appear to advance, warm colors or cool colors?
6. Which of the following is an example of a monochromatic color scheme?
   A. Light blue, medium blue, and dark blue.
   B. Red, yellow, and blue.
   C. Red and green.
   D. Yellow, yellow-orange, and orange.
7. Which of the following is an example of an analogous color scheme?
   A. Light blue, medium blue, and dark blue.
   B. Red, yellow, and blue.
   C. Red and green.
   D. Yellow, yellow-orange, and orange.
8. Which of the following is an example of a complementary color scheme?
   A. Light blue, medium blue, and dark blue.
   B. Red, yellow, and blue.
   C. Red and green.
   D. Yellow, yellow-orange, and orange.
9. Everyone's skin color has an undertone of either _____ or _____.
10. The largest number of people in the world are of which personal color season type?

True/False: Write true or false for each of the following statements.

11. Colors appear to change when viewed under different lights.
12. In outfits with contrasting color values, the light garment will look even lighter, and the dark garment will look even darker.
13. Texture has no effect on color.
14. Clothing outfits are generally more attractive if they have equal areas of light and dark.
15. Dark, cool, and dull colors make a form seem smaller.

16. Wearing light or bright colored pants will make large hips look smaller.
17. A single color for an entire outfit makes a person look thinner and taller.
18. A wide belt or waistband of a sharply contrasting color makes a person look taller.
19. A small amount of a light or bright color can be used to draw attention to a person's best physical features.
20. In most cases, no more than one major color should be used in an outfit.

Fashion in Action

1. On white poster board, make a color wheel of twelve hues using only blue, red, and yellow paints. Make a value scale of tints to shades for one of the hues. Show a gradual dulling of the intensity of a different hue by adding varying amounts of its complementary color. Label your work.
2. List several "fashionable" names for the twelve hues of the color wheel. Use names that would affect people's emotions and make them feel good about wearing clothes of those colors. From catalogs or magazines, cut out pictures of clothing that illustrate your descriptive names for colors.
3. Determine the personal color season for someone you know (perhaps yourself). Write a description of the hues, values, and intensities that are best for this person's clothing. Also describe how colors can best be used for this person's personality and body shape. Put together a guide for this person's use in future dressing and shopping.
4. Illustrate at least four of the six different color schemes for clothing described in this chapter. Cut out a picture to show each, from magazines or catalogs. Mount the pictures in a booklet with a color scheme title for each.
5. Conduct research about one of the main subjects of this chapter as it relates to fashion. Write and essay about the that aspect of the element of color. Back up your statements with a bibliography of your sources of information.

11 More Elements of Design

Fashion Terms

shape	decorative lines
line	texture
vertical lines	structural texture
horizontal lines	added visual texture
diagonal lines	motif
structural lines	

After studying this chapter, you will be able to

- describe the effect that clothing shape has on appearance.
- list line types, directions, and applications.
- explain how to use lines to the best advantage in garments to enhance the appearance of body shapes.
- discuss texture and how to use it effectively to improve appearance through clothing.

All four elements of design—color, shape, line, and texture—contribute to the overall design of a garment. A garment of good design is pleasing to the eye. It makes the wearer look his or her best. It has a good combination of design elements.

Fashion designers combine the elements of design in different ways to produce various garment styles. Architects, painters, landscapers, and others use the same design elements in their artistic endeavors.

By observing and experimenting with design, people can develop good taste. Once people know how to use the elements of design effectively, they can dress to look their best.

Shape

The ***shape*** of a garment is its form or silhouette. It is the overall outline. It is created by the cut and construction of the garment. The shape of an outfit is the outline when seen from a distance or in a shadow.

Since shape can be seen at a distance, it is noticed early. Thus, it is a major factor in a viewer's first impression of a person. This is why it is important that your clothes have a becoming shape for you. The cut of a jacket, the fit of a shirt, or the fullness of slacks should flatter your figure or physique.

Clothes can reveal or disguise the natural body contour. The outer shape of a garment can enhance or hide much of the figure underneath. Some silhouettes draw attention to specific parts of the body. Clothing shapes that are most flattering to a person emphasize his or her good features and hide the less attractive ones.

Using Shape in Clothing

Full, wide clothing shapes make people look larger. They look best on people who are slim, as in 11-1. Trim, compact silhouettes make people look smaller.

Straight, tubular shapes seem to add height, as in 11-2. If a person is very tall, a tubular shape would make him or her look even taller. A better choice would be a bell or back fullness silhouette to emphasize width and cut height. A straight silhouette also gives the impression of slimness. To hide a woman's large waistline, a tubular dress could be worn that intentionally falls straight from the shoulder to the hem.

Form-fitting clothes reveal any unattractive contours that a body might have. They are

Orvis

11-1 A full silhouette is best for a slim person who can look good with apparent added width.

American Printed Fabrics Council

11-2 Straight, slim clothing silhouettes make people look taller and more slender than wide silhouettes.

best for people whose shape is close to perfect. If clothes are too tight, they can make people look overweight—as if they've outgrown their clothes! On the other hand, if a waistline is especially small, the bell shape would help to emphasize its smallness.

When choosing clothes, a person should decide where his or her shape is best and select a silhouette that accentuates that feature. At the same time, the person should try to draw attention away from an area that should not be noticed.

The Shape of Fashion

Shape also reveals whether or not apparel is in fashion at any given time. As discussed in Chapter 2, the three basic silhouettes of clothes throughout history have been tubular, bell, and back fullness. However, many variations of shapes are possible within these main categories.

The shapes of fashion are more accentuated in high style designs than in those for the mass market. Sometimes the fashionable silhouette has big, wide (padded) shoulders and a slim look at the hips. At other times, the "in" look has accentuated height with tall, stand-up collars and longer than normal hem lengths. At other times, the silhouette seems to be "cut" by a jacket length below the hips, above the waist, or somewhere in between. The most lasting fashions have been the ones that have shown the natural contour of the figure to some degree.

Review some of the shapes of previous and present fashions. Decide which ones flatter or disguise various parts of the human body. Then decide which silhouettes are best for the good and not-as-good parts of your own body.

Facial Shapes

The shape of your face should also be considered to achieve your best look. To determine your facial shape, hold your hair back. Close one eye and trace the reflection of your face in a mirror with the edge of a piece of soap.

The "ideal" face shape is an oval. Strive to create the illusion of an oval with well-chosen apparel neckline designs. The neckline of apparel frames the face. The neckline should complement and flatter its shape. If the shape of a person's face is extreme in any way, do not repeat that shape in the neckline. The neckline would reinforce or strengthen that shape. A contrasting neckline balances the face.

Someone with an oval face can wear any neckline shape. Someone with a long face should wear wide, horizontal neckline designs. A pointed chin would look even more pointed with a V-neckline. On the other hand, someone with a round face would look attractive wearing a V-neckline.

Line

Line is a distinct, elongated mark as if drawn by a pencil or pen. Lines have direction, width, and length. The element of line can play up a person's good points and play down the bad ones just like color and shape can.

Eyes follow lines. Lines suggest movement or rhythm when they lead the eyes. Lines lead the eyes up and down, side to side, or around.

Lines outline and form outer and inner spaces of garments. They also connect parts. Lines can emphasize or create height, conceal weight, or focus attention to a certain area. Use lines to fashion advantage in apparel.

Lines of garments can be categorized in three ways: by type, by direction, and by application. All garments contain a combination of lines from each of these categories.

Line Types

The three types of lines are straight, curved, and jagged. These line types are shown in 11-3.

Straight lines are bold and severe. They suggest dignity, power, and formality. They give steadiness or stability. If overdone, they can give a stiff look.

All clothes have some straight lines in them. Pants have straight lines in the seams going down the legs. The bottom of a hemmed skirt forms a straight line.

Curved lines can be rounded and circular, or somewhat flattened out. Circles and curves make spaces look larger than they really are. They increase the size and shape of the figure. Circles are closed lines, so they stop the eye entirely. Somewhat flattened curves are the most flattering to the human shape.

Curved lines are less conservative, formal, and powerful than straight ones. They add interest and smoothness. They give a soft, gentle, youthful, charming, graceful, and flowing feeling. They accent the natural curves of the body.

Curved lines are found in round scoop necklines and along scalloped edges. Fabric prints sometimes contain them. The stronger or more abrupt a curve is, the more powerful and moving feeling it creates. Long, sweeping curves seem smooth and pleasant. Notice the curved lines in 11-4.

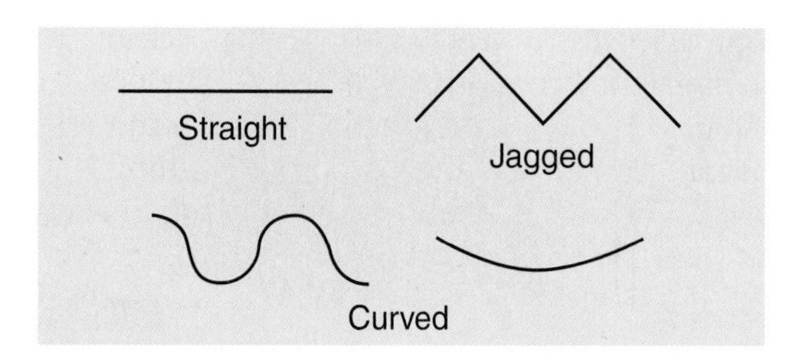

11-3 The three main types of lines are straight, curved, and jagged.

Use curved lines sparingly in apparel. Too many of them used at once can become confusing.

Jagged lines change direction abruptly and with sharp points like zigzags. They are designed into clothing with rickrack, fabric prints, or intentional seaming. The knit design shown in 11-5 has a bold jagged pattern. The outer notch of a suit collar lapel is a mild form of a jagged line.

Overused jagged lines can create a jumpy, confused feeling. Use them sparingly, since they are very noticeable. Wear clothing with dominantly jagged lines only if you have lots of self-confidence. Jagged lines are appropriate for fun-loving people who do not need to create a serious image at the events they will attend while wearing them!

Line Directions

Line direction may be vertical, horizontal, or diagonal. Line directions are shown in 11-6.

Bill Blass by Rubin Toledo

11-4 Curved lines are used in the silhouette of the garment parts as well as the applied fabric decoration of this designer dress.

Vertical lines are those that go up and down. In clothes, vertical lines lead the eye up and down. They give the impression of added height and slimness. In other words, they make the body look taller and thinner. They also give a feeling of dignity, strength, poise, and sophistication.

Vertical lines are found along the front line of a shirt, a center back seam, and in princess seam lines. Notice the vertical lines in the

Sandra Miller Knitwear

11-5 This fashionable sweater uses jagged lines to make a dramatic design statement.

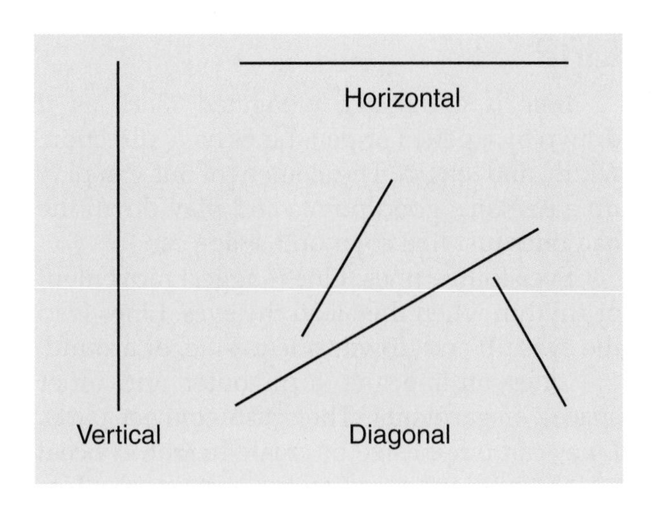

11-6 Lines can go many different directions in apparel.

clothing pictured in 11-7. For the best effect, put vertical lines over an area of the body that you want to look thinner.

Horizontal lines go from side to side like the horizon. In clothes, they carry the eye across from side to side. They give the impression of less height and more width. In other words, they make the body look shorter and wider. Also, horizontal lines tend to give a relaxed and calm feeling. They suggest rest and gentleness.

Put horizontal lines across the body where you want it to look wider. Tall, thin people would use horizontal clothing lines to make themselves look more filled out. If used across the upper body, as in 11-8, the chest will look wider or bigger. Horizontal lines are found where a belt goes around the waistline and at the bottom edge of a jacket or skirt.

Diagonal lines are slanted. The degree of slant determines their visual effect in clothes. If they have a vertical slant, they are slenderizing. If they have a horizontal slant, they add width.

Diagonal lines are versatile and interesting. They are strong and draw attention to the area where they are used. They can give a feeling of action and strength. They can seem dramatic and eccentric.

Diagonal lines are found in V-necklines, along collar lapels, and down the edge of surplice closings. They are seen along the outer edges of flared skirts and bell-bottom trousers. Stripes on a fabric that is cut on diagonal grain (bias) can be matched at seamlines. This creates angles known as *chevrons*. See 11-9.

The Fashion Association / Pronto Uomo

11-7 Vertical lines are present in this fabric design, at the shirt center front, and along the inside and outside edges of the pant's legs.

Liz Claiborne

11-8 One or more horizontal stripes going across the chest and arms of a slender person gives the illusion of more width in the upper body.

Italian Trade Commission/Prada

11-9 Varying degrees of diagonal lines are present in this outfit. They appear in the overlapped surplice front, shoulder halter edges, angled stripes of the top, and chevron in the center of the shorts.

Line Applications

Lines are incorporated into clothing in two basic ways. There are structural lines and decorative lines.

Structural lines are formed when parts of the garment are constructed. They are the seams, darts, pleats, tucks, and edges of the garment that were created when it was made. They are the assembly details that also create visual interest. They can look very decorative but are structural since they are a necessary part of the garment's construction. Structural lines are most noticeable if the fabric of the garment is plain. Also, more of them can be used with a plain fabric.

Decorative lines, or applied lines, are created by adding details to the surface of the clothing. They are added simply to decorate the outfit and make it more interesting. They add style and personality.

Decorative lines can be formed with ruffles, braid, fringe, edgings, top-stitching, lace, tabs, flaps, appliqués, or buttons. Decorative lines can also be created with accessories, such as scarves and necklaces.

The decorative lines of an outfit should be in harmony with its structural lines. Decorative lines often accentuate structural lines through repetition or contrast, as in 11-10. They may accent cuffs, necklines, seams, hemlines, or openings. For instance, a row of top-stitching down the front of a shirtwaist dress emphasizes

The Fashion Association / Mondo Di Marco

11-10 The decorative edging on this sweater closure accentuates the center front opening.

the long straight structural edge of the dress. It adds to the slimming effect.

Too much detail causes competition between the lines and is confusing. It is not pleasing. The more elaborate the structural and decorative lines are, the more attention you will draw to your body.

Creating Illusions with Lines

As mentioned earlier, different lines are mixed in garments. The ways lines are combined produce various, predictable effects. They lead the eyes in certain directions. The dominant line catches the gaze. Skillfully used lines can create various optical illusions.

Figure 11-11 shows how lines can play tricks with our eyes. In A, the straight lines are the same length. However, the one on the left looks shorter because the diagonal lines added to each end bring the eye back toward the center. The line on the right looks longer because the diagonal lines on each end keep the eye moving out.

In B, both rectangular boxes are the same size. However, the box on the left looks taller and narrower than the one on the right. The up-and-down stripe in the left box gives it height and cuts its width by making the eye move up and down through its center. The line in the box on the right moves the eye across the horizontal width of the box, which gives the illusion of more width and less height.

In clothing, lines are often combined into designs that appear to form an arrow, or the letters T, I, or Y. These configurations cause certain optical illusions. Look at the examples of corresponding clothing designs in 11-12. The garments in these examples have only simple drawn lines. However, most real garments have many lines. You will have to learn to pick out the most important lines in garments and to consider the effects created by those lines.

Lines that form an arrow tend to deflect the gaze downward. They shorten, or reduce the height of a person. Lines that form a "T" also stop the upward movement of the eye. The height is again cut, but width is given to the top. An illusion of broader shoulders results. This look is perfect for someone who is tall, with narrow shoulders and wide hips. Attention would be drawn to the shoulder area, which would help create the illusion of smaller hips.

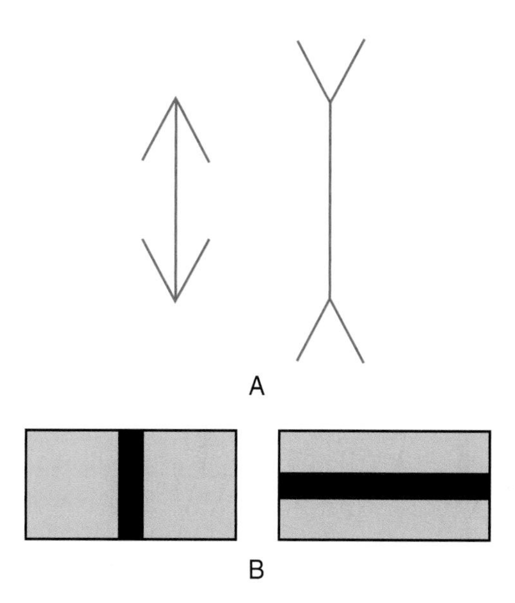

11-11 These are standard examples of how lines can play tricks, or create illusions, when used in certain ways.

Lines that form an "I" tend to give a vertical feeling that is contained at the top and bottom. They carry the gaze upward and make the body look somewhat taller and thinner. Lines that form a "Y" keep the gaze moving upward even further. The appearance of more height is given to the body with a raised collar or a V-neckline.

In summary, for a taller, thinner look, select lines that keep the eye moving up the figure without interruption. For a shorter, wider look, put horizontal lines across the body, especially where width is desired.

Further Use of Lines in Clothing

Apparent size is affected by how the garment lines divide the figure. To create the illusion of more height and slimness, choose a narrow overall shape with vertical lines added. This can be done with a narrow center panel or button placket, vertical trimming, neck-to-hem closing, or princess seamlines. Notice in 11-13 how an accented line down the center front, plus vertical seamlines divide the body lengthwise and give a strong vertical feeling.

Lines spaced far apart make the figure look larger. A wide panel, even though it runs

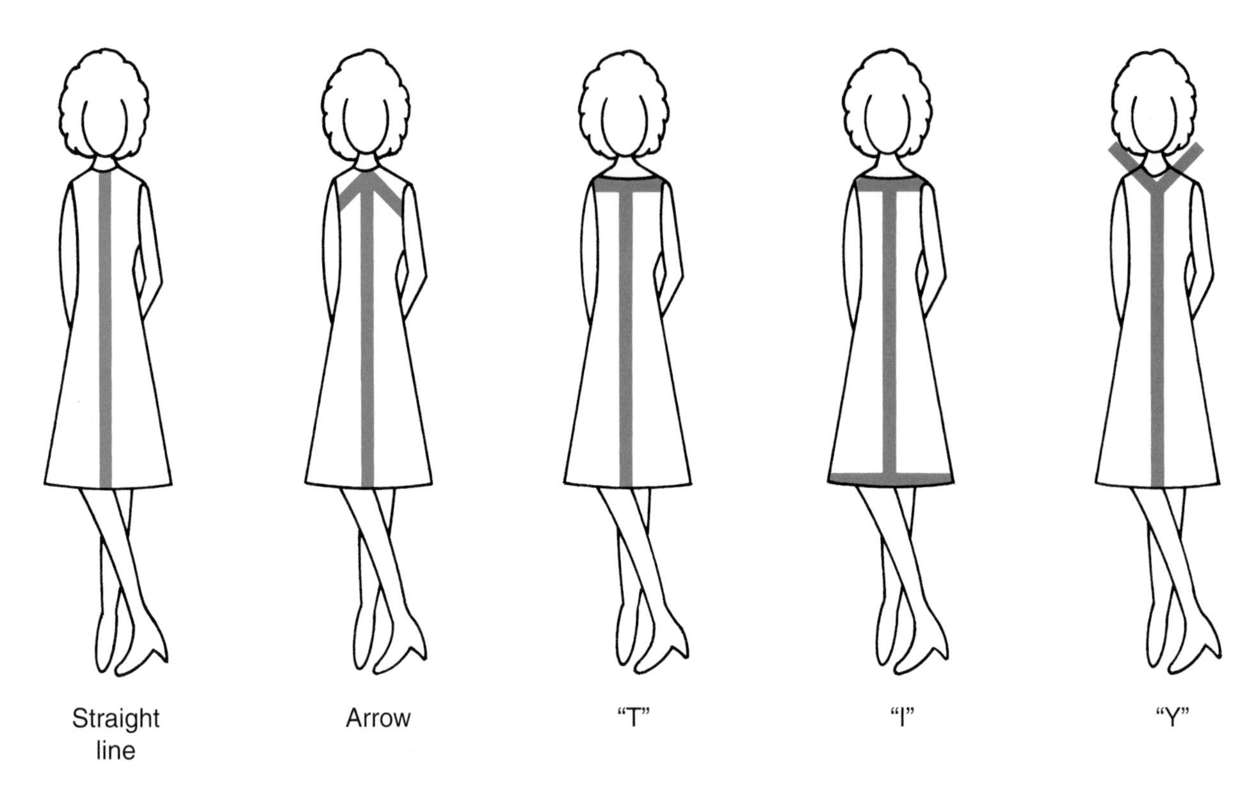

11-12 The ways lines are used in apparel can create optical illusions.

11-13 This dress contains slimming vertical lines, plus a horizontal line at the collar edge providing some shoulder width.

vertically in the garment, can make the body look wider and heavier. This is because the viewer's eye hops back and forth across the panel. Check this out in 11-14 and 11-15.

When lines are used in a series, the space between them determines what illusion is created. Widely spaced vertical lines create the illusion of width, while narrowly spaced lines produce a slimming effect. The spaces play tricks by the way the area is divided by the lines. For instance, the wider center front space on a double-breasted jacket is less slimming than just one row of buttons in the center.

The more elaborate or bold the lines of your garment are, the more attention you will draw to yourself. Large, bold lines make you look larger than you actually are. Wide lines are more noticeable and have more widening effect than narrow ones. Large stripes emphasize the area where they are used. They make you look bigger, even if they are running up and down the garment.

If lines are all the same width, and evenly spaced, the eye will move the opposite way. The equal lines and spaces cause the eye to move easily from one line to another without stopping.

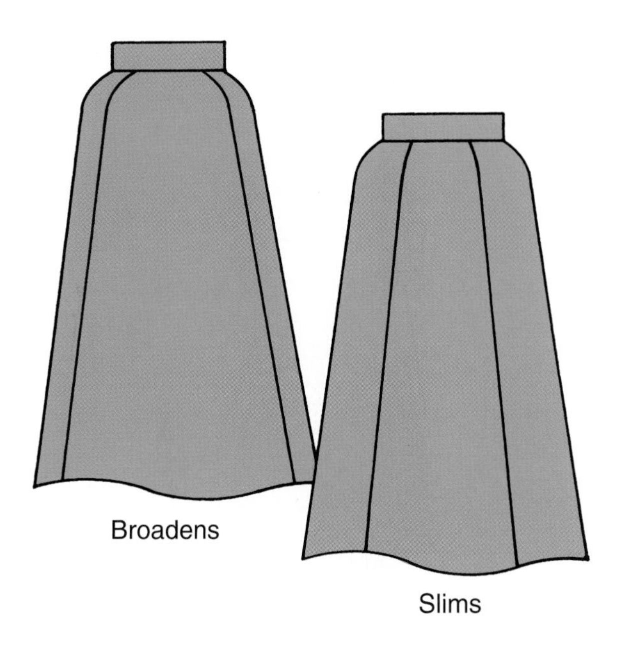

11-14 The skirt shown here with a narrow center panel is more slimming than the one with a wide panel.

11-15 Once again, the wide panel gives a wider look to the body. The narrow panel gives a more slender look.

Long, unbroken lines give the feeling of continuity. Broken lines give a spotty effect compared to continuous lines.

V-shaped lines are slenderizing if they are long and narrow. If V-lines are horizontally flat, they give a widening image. For instance, a wide V-neckline would make the shoulders look broader.

If you want to create the illusion of more body width and less height, use a wide overall shape with horizontal effects added. This can be done with a full garment and uneven horizontal stripes, full or short contrasting sleeves, or pockets or flaps on both sides of the garment. A wide contrasting belt would also give width by forming a line that cuts the body horizontally. Square necklines or wide collars add width, as do yokes.

When lines cross each other, such as at the waistline, hip, or hemline, attention is drawn to that area. A contrast of line direction at the hem of a skirt (possibly created with trim) directs attention to the legs. A mismatched plaid at a seam attracts attention and will make that area look larger. If apparel contains very plain lines, they will not be as noticeable, and the attention will go to the face.

Two-piece outfits add width by cutting the body with a horizontal line. The horizontal line gives the greatest shortening and widening effect if it is placed low in the outfit because the eye is drawn downward, as in 11-16. Thus, tunics are not a good choice to wear unless you are tall. Horizontal lines high on the figure produce less shortening image, such as with an empire waistline.

Texture

Texture is the surface quality of goods. Fabric texture is how the fabric's surface feels and looks. It is the quality of roughness or smoothness, dullness or glossiness, or stiffness or softness. Texture is determined by a fabric's fibers, yarns, and method of construction, such as weaving or knitting. Fabric texture is also affected by mechanical and chemical finishes, including any applied designs that are added. Thus, texture may be described as the "character" of the fabric, as in 11-17.

Some words to describe the texture of fabrics are: rough, smooth, dull, shiny, firm, crisp, fuzzy,

The Fashion Association / Nautica

11-16 A strong horizontal line exists at the bottom of this shirt. The fabric design in the shorts also gives a horizontal feeling.

Sigrid Olsen

11-17 Notice the textures of the ribbed mock turtleneck, bouclé multicolor sweater, woven wool skirt, and patterned hose.

bulky, nubby, soft, shaggy, flat, harsh, sheer, loopy, furry, scratchy, pebbly, delicate, sparkling, and fine. Textures can also be described as lightweight, medium-weight, or heavyweight. You can probably think of a few more descriptive terms for fabric textures. Tweed is usually rough; silk is most often smooth and glossy.

Structural texture is created when fabrics or garments are manufactured, as shown in 11-18. ***Added visual texture*** is printed onto the surface of fabrics or garments, as shown in 11-19.

Using Structural Texture in Clothing

Garments made of bulky fabrics that are heavy, fuzzy, or shaggy add visual size when worn. They make a person's shape look bigger. However, they also disguise figure irregularities. They can overpower a small person. The main advantage of bulky fabrics is the warmth they provide.

Smooth, flat textures make people look smaller. They are suitable for almost all figures and physiques. They can usually hide some

JC Penney

11-18 Notice the bumpy stripes of this shirt, the flat texture of the jeans, and the shininess of the sandals and bracelets. These are all structural textures.

The McCall Pattern Company, McCall's

11-19 The black and gray flowered print on this dress fabric is an example of added visual texture.

figure irregularities because they hold their own shape. Fabrics such as percale, wool crepe, and soft linen have this texture.

Shiny textures make the body look larger because they reflect light. Shiny textures emphasize body contours. They make fabric colors seem lighter and brighter.

Rough textures have the opposite effect. They tend to subdue the colors of fabrics. This is caused by light hitting their uneven surfaces at different angles.

Sheer fabric also subdues colors because the skin of the wearer is seen through the fabric. Sheer fabric reveals the true body shape. It softens the figure when used over a soft lining.

Dull textures make a person look smaller because they absorb light. They do not have highlights, so heavier body areas look thinner. Dull-textured fabrics in medium weights are almost always flattering.

Clinging, soft textures reveal the body's silhouette. They emphasize any heavy areas. They are good for draped designs, especially on a nice body shape.

Stiff, crisp textures make the total shape look bigger, because they stand away from the body. Stiff fabrics do, however, conceal figure faults. They should not be used in draped styles. They require seams and other shape controllers to give form to garments. They are good for bell silhouettes.

Textures chosen for various outfits must be suited to the occasions for which they will be worn. They should be appropriate for the garment and for the wearer. Rich and luxurious velvets and brocades are dressy and look best on people who have poise and dignity. Glossy satin and fine silk are usually for fancy clothes. Heavyweight fabrics are for hard work or for warmth.

Strive for interesting combinations of textures in apparel. Often just one main texture is used with an accessory having an accent texture. For instance, a rough tweed coat might have a smooth leather belt. Try not to combine too many textures in one outfit. An appearance of confusion might result.

Harmony is achieved when related textures are used together, as in 11-20. Variety and added interest is provided with a combination of unlike textures as long as they do not look too "busy" together. Medium textures can be easily combined in clothes with heavyweight or lightweight fabrics.

Use texture to advantage. Use a bulky sweater to make small shoulders look bigger. For large hips, smooth-textured slacks will deemphasize them. Combine your knowledge of texture with the best use of color, shape, and line to achieve the optimum look for your body.

Using Added Visual Texture in Clothing

Added visual texture is "printed-on" surface design. The print is the overall pattern created by design motifs. A ***motif*** is one unit of a design that is usually repeated. Patterned prints applied to the fabric add textural interest. In fact, the applied design is often more noticeable than the structural texture of the fabric.

Added visual texture can affect the apparent size of the wearer just as structural texture does. Added prints can give an overall vertical, horizontal, diagonal, curved, or jagged feeling. If most lines of the added design move in a vertical direction, the body wearing it will look taller. Dominant horizontal lines will shorten the apparent body height. Stand away from the fabric and study the direction its design seems to take.

Added prints can be small, medium, or large. They can be quiet and subtle, or loud and bold. If added designs are large and bold, the structural texture will be secondary.

An added print will diminish the effect of structural and decorative garment design lines. A busy print would diminish the appearance of these lines a great deal. A plain, subdued texture will emphasize apparel design features. The more interesting the texture is, the simpler the garment lines should be.

Large, bold patterns emphasize the area where they are used. They increase the apparent size of the wearer. This is compounded if the print has bright colors or sharp color contrasts. Small, subdued, overall prints tend to make a person look smaller, especially if they are in closely related colors.

What is fashionable is always changing. Good structural and added visual textures must be planned and organized in interesting ways, as in 11-21.

Pendleton Woolen Mills

11-20 The harmonious textures of this sweater knit, woven pleated skirt and box-check jacket are accompanied by an accent of shiny jewelry.

Escada by Margaretha Ley

11-21 The leopard pattern is knitted into this fabric, then combined with coordinating plain edging, shiny leather boots, and a soft furry hat.

Chapter Review

Summary

Besides color, other elements of design are shape, line, and texture. The shape, or silhouette, of a garment can form an outline to enhance or hide body contours. Shape also reveals whether or not apparel is in fashion at any given time. Facial shape should be considered when deciding on necklines and collars.

Lines have direction, width, and length. Line types are straight, curved, and jagged. Line directions can be vertical, horizontal, or diagonal. Line applications are structural and decorative, which should coordinate with each other in garments. The ways lines are combined in outfits can produce predictable effects with optical illusions. The apparent size of the wearer is affected by how apparel lines divide the figure.

Texture, or the surface quality of goods, can be structural or added. Bulky, shiny, and stiff structural textures make people look larger. Smooth, dull surfaces help people look smaller. Printed-on motifs or added visual texture can also affect the apparent size of the wearer.

Fashion in Review

Write your answers on a separate sheet of paper.

True/False: Write true or false for each of the following statements.

1. Full, wide clothing shapes make people look slimmer.
2. A tubular shape makes a person look taller.
3. Clothes that are too tight can make people look overweight.
4. Someone with a pointed chin would look most attractive wearing a garment with a V-neckline.
5. Straight lines provide a feeling of steadiness or stability.
6. Jagged lines provide a gentle, youthful, and graceful feeling.
7. Vertical lines give a feeling of dignity, strength, poise, and sophistication.
8. Horizontal lines make a person look taller and thinner.
9. Diagonal lines give a feeling of action and strength and can seem dramatic.
10. Structural lines can be formed with ruffles, edgings, top-stitching, lace, tabs, flaps, appliqués, or buttons.
11. Decorative lines are formed by seams, darts, pleats, and tucks.
12. To create the illusion of more height and slimness, choose a narrow overall shape with vertical lines added.
13. Lines spaced far apart make the figure look larger.
14. Two-piece outfits add height by cutting the body with a horizontal line.

Short Answer: Write the correct answer to each of the following questions.

15. Name the four elements of design.
16. Texture is determined by a fabric's _____.
    A. fibers and yarns
    B. method of construction
    C. mechanical and chemical finishes
    D. All of the above.
17. Explain the difference between structural texture and added visual texture.
18. How do bulky fabrics affect the wearer's appearance?
19. How do shiny textures affect the wearer's appearance?
20. How do large, bold patterns affect the wearer's appearance?

Fashion in Action

1. Cut out the shape of a garment from the inside of a large piece of cardboard. You will then have cardboard edging around the open silhouette of a shirt, dress, jacket, skirt, or pair of slacks. In front of the class, place it over pieces of fabric with various lines and textures. Include printed motifs. Discuss the effects of the different pieces of fabric on the apparent shape of the garment.

2. Clip at least five pictures of clothing that show straight, curved, jagged, vertical, horizontal, and diagonal lines. Mount the pictures on paper and write descriptions of the effect of the lines. Identify the lines as structural or decorative.
3. Write a report about what shapes, lines, and textures would be most flattering over the various parts of an imaginary person's body. Describe the person in terms of height and body shape. Include drawings or pictures of at least six garments that would have the best elements of design for this person.
4. In an oral report to the class, analyze the elements of design in at least four actual garments or clear pictures of garments. Compare the relationship between outer shape and inner structural and decorative lines. Describe the structural and added visual texture. Tell what illusions might be created by each garment's color, shape, line, and texture.
5. Prepare a sample test of at least 25 questions about information in this chapter. Include matching, essay, fill-in-the-blank, and true and false questions.

12 Principles of Design

Fashion Terms

principles of design	rhythm
design	repetition
balance	gradation
formal balance	transition
informal balance	opposition
proportion	radial arrangement
emphasis	harmony

After studying this chapter, you will be able to

- name the four principles of design.
- give examples of the use of each principle of design.
- explain how the principles of design can be used to produce harmony in clothing.
- describe apparel outfits that have the best design for the assets and liabilities of various body shapes.

The ***principles of design*** are balance, proportion, emphasis, and rhythm. They are guidelines for the use of the elements of design: color, shape, line, and texture. When the elements of design are used effectively according to the principles of design, harmony is created. This is true for buildings, paintings, landscapes, interiors, and sculpture, as well as apparel.

Design is the plan used to put an idea together. The process of designing is the selecting and combining of the design elements according to the principles of design in order to achieve harmony. The design process is illustrated in 12-1.

Balance

Balance implies an equilibrium or steadiness among the parts of a design. It is a visual distribution of "weight" in the way details are grouped. Balance brings overall stability to a design. It produces a feeling of rest or lack of movement.

Clothing should have equality in the amount of visual weight on each side. The parts, or spaces, of a garment should relate well to each other. To achieve proper balance, all sides should be equal in weight. If one side is heavier than the other, the garment or outfit is out of balance.

Balance in garments is produced by structural parts and added decoration. It should also be present in the fabric design.

Color plays an important part in achieving harmony with balance. Warm and dark colors appear heavier than cool and light colors. A small amount of a bright color balances a large amount of a dull color. A small area of a warm color balances a larger amount of a cool color. Large amounts of tints or neutrals balance smaller areas of shades or bright colors.

Line and texture are also important in harmonious balance. For instance, one long line balances two short ones. One wide stripe balances two thin ones. A larger area of fine or soft texture balances a smaller area of heavy or coarse texture.

Types of Balance

There are two main types of design balance. They are formal and informal balance.

Formal balance is symmetrical. Its design details are divided equally to create a centered

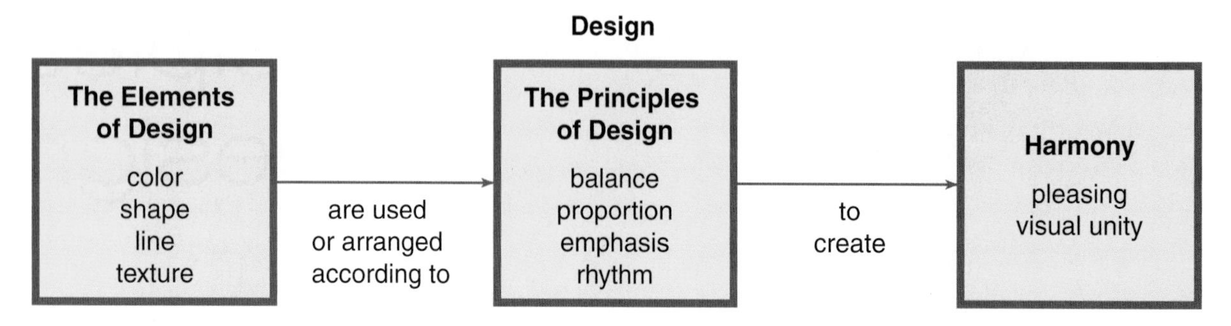

12-1 When the elements of design are used according to the principles of design, harmony is created.

balance. In other words, both sides are the same, as in 12-2. Identical details are arranged the same distance from the center on the right and left sides. One example would be the same number of tucks or pleats on each side of a pair of trousers. Another would be similar pockets on both sides of a dress.

Formal balance is the most common type of balance in clothes. It is the simplest and least expensive to produce.

Informal balance is asymmetrical. Its design details are divided unequally from the center. An arrangement of colors, shapes, lines, and textures on one side balances a different arrangement on the other side, as in 12-3. Informal balance is often achieved with diagonal lines and off-center closings.

Informal balance should not look heavier on one side than the other. If done properly, the design appears to be balanced, even though its two sides are different. It is more unusual and interesting than formal balance. It can be quite dramatic.

The McCall Pattern Company, Vogue® Patterns, Adri

12-2 This outfit has formal, or symmetrical, balance. If it were folded in half lengthwise, the right side would be the same as the left side.

Proportion

Proportion is the spatial, or size, relationship of all of the parts in a design to each other and to the whole. Proportion is sometimes called *scale.* The size of all the parts of an outfit should be related, as in 12-4.

Proportion is determined by how the total space is divided and by how the inner lines are arranged. One part should not be out of scale with the others. When all the parts work well together, the garment is well proportioned rather than out of proportion.

Proportion is not as pleasing when all areas are exactly equal in size. Unequal parts are more interesting. Also, an odd number of parts, such as three, is more interesting than an even number, such as two or four.

Garment designs should be related to the structure and proportion of the human body.

Simplicity Pattern Company, Inc.

12-3 This dress has informal balance, with its right and left sides different from each other. The smaller dark area on one side balances its larger light area on the other side.

J. Jill

12-4 In this example, the flowered print of the dress and size of the striped inserts at the neckline, sleeves, and hem are in proportion to each other and to the size of the wearer.

The ideal body is said to be "8 heads tall" with ⅞ of it below the head, as shown in 12-5. The proportions have ⅜ of the total figure from the waist to the top of the head. The remaining ⅝ of the body is from the waist to the soles of the feet. To coordinate with this, most outfits divide unequally. For instance, a dress usually has a bodice area smaller than its skirt area.

Fashions that flatter the natural figure are pleasing and remain in style. Those that make the body look distorted, or out of proportion, are sometimes popular as fleeting fads. However, styles that complement the body's natural proportions remain in style longer.

Proportion in Apparel

When combining garments and accessories in an outfit, think about the proportion of each variable. For instance, consider the length of a jacket in relation to the length of the pants, to the whole outfit, and to the wearer.

Parts of apparel, such as yokes, collars, and pockets, must be the right size for the total design and for the wearer. A tiny pocket would

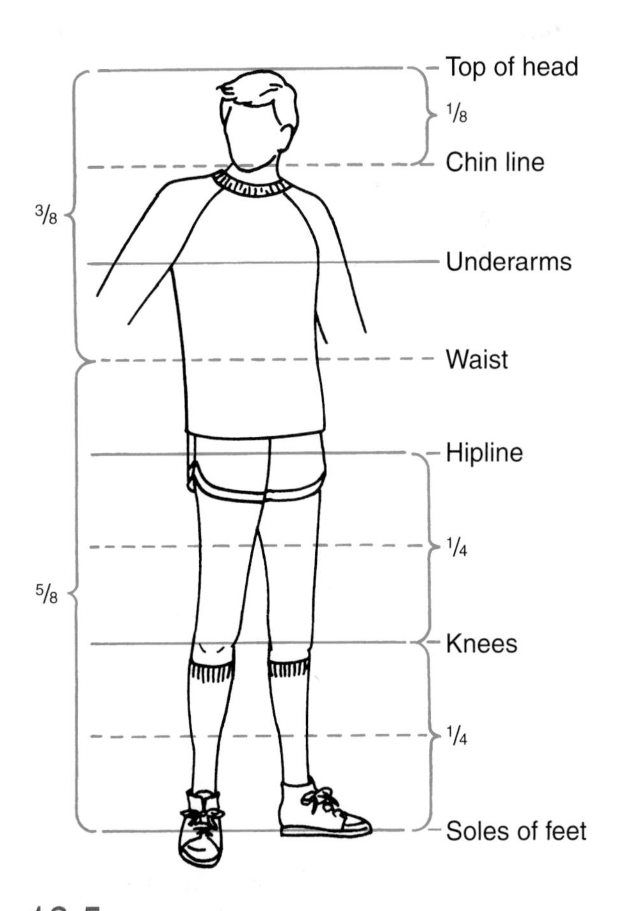

12-5 The ideal male or female body is said to total eight times the height of the head. The waist divides the body naturally into the pleasing proportion of 3:5, with 3/8 above it and 5/8 below it.

look out of proportion on a large, heavy overcoat. Details such as buttons and trimmings should also be related to the overall size of a garment.

Accessories should be in proportion to the garment and body build of the wearer, too. A large belt would look fine on a tall person but would be out of proportion on a small person. A woman with a large frame wouldn't look right carrying a tiny purse at her side.

It is important that the design of the fabric be in proportion to the garment and the wearer. Prints and other textures must be scaled to their use, as in 12-6. Large prints, plaids, or checks would overpower a small person. On the other hand, a tiny print would be lost on a large person. It would be so out of proportion that it would make the wearer look even larger.

In considering the proportions of an outfit, always use the body that is being clothed as your guide. It is most pleasing to divide a garment or outfit at a natural body division such as the chest, waist, or hips.

Emphasis

Emphasis is a concentration of interest in a particular part or area of a design. One part is more important or noticeable than all the others, as in 12-7. The emphasis is the center of attention of an outfit. It is the focal point that attracts the eye first. All other parts of the design are subordinate to the area of emphasis.

Without any center of interest, an outfit looks unplanned and monotonous. It is best to have one main area of emphasis. A secondary

Wrangler/Michael Johnson

12-6 The medium size of the plaid in this blouse, cable of the sweater, and width of the belt are in perfect proportion with the size of the woman wearing this casual outfit.

Simplicity Pattern Company, Inc.

12-7 In the outfit on the left, the primary area of emphasis is the bright pink part of the top. The fabric "twist" is a secondary point of emphasis. In the outfit on the right, the twist is the primary point of emphasis.

emphasis may be included if done carefully. However, too many focal points create a cluttered, confusing design. It is best, for instance, to leave the cuffs, hemline, and other areas of a dress fairly plain if the neckline is being emphasized.

Emphasis can be used in apparel to draw attention to an attractive personal feature. It can also draw the eye away from an undesirable feature. Emphasis is most commonly used to call attention to the face. However, it may spotlight the waist, chest, arms, one shoulder, or another area.

Emphasis can be created with contrasts of colors or textures. Light, bright colors and shiny textures attract attention. Structural lines and decorative trimmings, such as an area of tucks, gathers, ruffles, or buttons can provide emphasis. The unusual shape of an area or a contrasting design outline will also attract attention.

A center of interest can be achieved with one large item or with a group of small ones. A bow, an appliqué, or a monogram on a contrasting background can create emphasis. Accessories, such as belts, scarves, neckties, or jewelry, can add emphasis to otherwise plain outfits.

Rhythm

The elements of design should be arranged to produce a feeling of continuity or easy movement. That is rhythm. ***Rhythm*** is the pleasing arrangement of the design elements so the eye moves well over the apparel. Rhythm directs the flow of the eye movement steadily and smoothly through the lines and spaces of the design. The gaze unconsciously moves from one part to another. Rhythm can be created with repetition, gradation, transition, opposition, or radial arrangement.

Repetition repeats lines, shapes, colors, or textures in a garment. This can be done with all parts having the same shaped edges. They might be squared, rounded, or scalloped. Repetition of colors can create good rhythm, especially if the colors are distributed in an interesting way. Rhythm in an outfit can also be achieved by repeating buttons, tucks, pleats, or trim. An example is shown in 12-8.

Gradation, sometimes called progression, implies a gradual increase or decrease of similar design elements. Colors can go from light to dark as in 12-8, or textures from fine to coarse. Lines may range from thin to thick, and shapes may range from small to large, as in 12-9. The gradual changes provide continuity while giving a feeling of movement.

Transition is a fluid rhythm created when a curved line leads the eye over an angle. The curved lines of transition cause the eye to change direction gradually rather than abruptly. It is found in puff sleeves, cap sleeves, and dropped shoulder designs, as in 12-10. Transition can also be achieved by using scarves, shawls, ascots, jabots, ruffles, and gathers.

McCall Pattern Company, Vogue® Patterns, Tom and Linda Platt

12-8 Rhythm is created in this outfit with repetition of diagonal rippled lines through the length of the skirt.

Butterick Patterns / Butterick Company, Inc.

12-9 The small area of the shoulder and sleeves, medium-sized bodice area in the middle, and the largest area of the long skirt show rhythm through gradation.

Butterick Patterns / DKNY; Butterick Company, Inc.

12-10 Rhythm is created through transition with the diagonal line below the shoulder "short-cutting" the eye past the shoulder angle.

Opposition is rhythm created when lines meet to form right angles. Opposition may be found in fabric designs such as checks and plaids. It is also found in square necklines, square pockets, waistbands, yokes, collars, and cuffs. An example of opposition is shown in 12-11.

Radial arrangement, in which lines emerge from a central point like rays, also produces rhythm. It is created when gathers, tucks, seams, darts, flowing lines, or colors fan out from a central area, as in 12-12.

Rhythm is broken when lines, trimmings, or fabric designs are not matched at the seams or at other construction points. For instance, it is upsetting to the eye when the lines of plaids, checks, or stripes are broken. If patterns cannot be matched at all seam lines, they should at least be matched at center front and center back.

Harmony

Harmony is pleasing visual unity. It is the tasteful relationship among all parts within a whole. It is created when the elements of design are used effectively according to the design principles.

Harmony gives the feeling that all the parts of an outfit belong together and suit the wearer and the occasion. The garments and the accessories work together to help the wearer look his or her best. A value judgment must be made to decide if harmony has been achieved.

The creative use of clothes and accessories is both an art and a science. Through observation and experience, you can learn to use the

Appleseed's

12-11 Horizontal lines in this jacket meet the center front zipper at right angles to create rhythm through opposition.

12-12 Rhythm flows through this tie-dyed design in radial arrangement, or from the center outward.

design elements according to the design principles to bring harmony to your wardrobe. Apparel is harmonious when it is complementary to the person wearing it.

Create the Best Look

The shape of a girl's or woman's body is called her *figure*. The name given to the shape of a boy's or man's body is his *physique*. Body size is usually measured at the chest, waist, and hips. However, the proportion of the parts gives a better indication of shape. This relationship among the different areas of the total human form is called *body build*.

The best qualities, or parts, of a person's body are called *assets*. Physical drawbacks are *liabilities*. Strive to emphasize the assets. Try to minimize or camouflage the liabilities. A well-designed garment enhances the wearer's personal assets. It plays down, or leads the eye away from, the person's liabilities. By using the elements and principles of design effectively in a wardrobe, anyone's appearance can be enhanced.

People come in a variety of sizes, shapes, and proportions. Some changes can be achieved through diet and exercise, but there is no perfect body type. We should try to improve what can be changed and to accept what cannot be changed. The rest of this chapter will describe how to attire each particular body build to advantage.

To determine your build, do a self-evaluation, or analysis, of your body. Study yourself in front of a full-length mirror. Jot down your observations. How would you describe yourself? Are you short, tall, thin, or heavy? Where is most of your weight placed? Are you short-waisted or long-waisted? Do you have a short, thick neck, or a long, slender one? Is your stomach flat or rounded? Note all of your physical assets and liabilities.

You can do this for others, too, such as your friends or relatives. Or maybe you will someday work as a fashion consultant, wardrobe coordinator, or personal shopper. From the results of body evaluations, you can decide what to show off and what to cover up to help people look their best.

The latest fashions may flatter some people but not all. *Personal style* is achieved when design is used to the best advantage. Charts 12-13 and 12-14 summarize ways to utilize design in apparel.

The Total Design for Individuals

A person's figure or physique is the underlying form on which to base clothing selections. The body has height, width, and depth (thickness). Actual measurements are not as important as total appearance. Clothes can help or hurt that appearance.

Before you can solve a problem, you must identify it with an analysis. All of us have some figure or physique assets and liabilities. Once we define them, we can decide what to show off and what to cover up.

Try to balance the shoulders with the chest and hips, as in 12-15. Decide if you want a taller, shorter, broader, or thinner look in various areas. Decide where and how to use color, shape, line, and texture to create the desired illusions. Play up the best features. Choose garment designs and fabrics that disguise or lead the eye away from figure faults.

Current fashions always include enough variations to give a choice of many looks and designs. You, thus, have the opportunity to

Creating Height and Width Illusions	
To Look Taller and Thinner	**To Look Shorter and Wider**
Wear straight silhouettes	Wear wide silhouettes
Use vertical lines	Use horizontal lines
Shape with seams or darts	Shape with gathers or pleats
Use smooth, flat textures	Use bulky, heavy textures
Wear one-color outfits	Contrast colors (top/bottom)
Wear a small, matching belt	Wear wide, contrasting belts
Have fit that skims the body easily	Have tightly fitting clothes
Use dark, dull, cool colors	Use light, bright, warm colors
Wear subtle prints, plaids, etc	Wear bold prints, plaids, etc
Create a simple, uncluttered look	Wear full sleeves, wide pants
Use emphasis to lead the eye upward	Use emphasis to lead the eye downward

12-13 To create body shape illusions with apparel design, follow these suggestions.

Avoiding and Attracting Attention	
To Avoid Attention to Areas	**To Attract Attention to Areas**
Dark, dull colors	Light, bright colors
Cool hues	Warm hues
Minimal structural garment design	Structural accents there
No applied decoration there	Applied decoration there
Flat, dull fabrics	Shiny or textured fabrics
Soft fabrics	Clingy fabrics
Plain, unpatterned fabrics	Large, busy prints

12-14 Use design in clothes to highlight physical assets and camouflage liabilities.

choose what is most flattering. Everyone can look his or her best by selecting the right clothes.

The seven most common body types are discussed in the following sections. Some assets and liabilities may overlap among the groups. Specific apparel recommendations are made for each. Suggestions for what should be avoided are also given.

Tall and Thin

This is the ideal "model" shape for wearing clothes. Choose gathered or pleated full skirts, flared or wide-legged pants, and horizontal stripes and seams. Multicolored outfits with bright, large patterns and prints are also good.

Any part of this person's garments can have shaping with tucks, pleats, gathers, or shirring. Double-breasted jackets and coats can be worn well. Blouson, A-line, and bell silhouettes give width. Long, full sleeves hide boniness. Belted tunics, over-blouses, and wrist-length or waist-length jackets cut the height. This is especially true if the top and bottom are of contrasting colors, as shown in 12-16.

This person can wear almost all fabrics, especially heavier ones. Large weaves and nubby, napped, and pile surfaces can be worn well. He or she can experiment with high-fashion garments and fabrics. Bold accessories and somewhat exaggerated details can be worn. Wide belts and horizontal stripes help to minimize height. Light or brightly colored slacks with a plaid or tweed jacket would be attractive.

Tall, thin people should avoid tight, straight dresses, skirts, or pants. Severely tailored lines are not attractive. Tiny fabric patterns, frilly fashions, and bold verticals should be avoided. Tight, high turtlenecks and clingy jersey are not good. Slacks that are not long enough are poor choices. Sleeveless styles and small accessories should be avoided.

Appleseed's

12-15 Horizontal stripes add width to the shoulders and chest. This balances the hips, which appear smaller, in dark, plain fabric.

Pendleton Woolen Mills

12-16 The contrasting colors and values that cut the height where they meet each other are attractive on this tall, thin woman.

Tall and Heavy

This "statuesque" frame can look stunning, but do not let it become overpowering. Garments should have simple lines and little decoration. This is often called understated design. Choose subtle and muted prints, stripes, and patterns in scale with the body. Easy-fitting A-line skirts that are smooth at the hips should end below the knee. Pants should have straight or only slightly flared legs.

The shaping of garments for this body type should be done with seams or darts. Vertical or diagonal lines should be used. Asymmetrical closings look great, as in 12-17. Flat, firm fabrics are best in grayed or dulled shades. A simple, smooth-fitting tunic or uncluttered suit, like the one on 12-18, can be worn. Details at the neck can be used to draw attention to the face.

Tall, heavy people should avoid loud prints, checks, plaids, and stripes. Fussy details, ruffles, and delicate accessories look out of proportion on large bodies. Clothes that are too tight make these people look bigger. They should also avoid clingy, bulky, or heavy fabrics. Boxy overblouses, untucked shirts, and tent dresses should not be worn. Gathers or excess fullness add visual size and should be avoided.

Short and Thin

"Petite" people who are short and thin should choose soft, fluid, lightweight fabrics. Small scale prints, subtle patterns, and smooth textures are best. Unbroken vertical or diagonal lines are most flattering. Bell, blouson, and flared silhouettes are good if they are not too wide.

Draper's & Damon's

12-17 The easy-fitting asymmetrical and vertical lines for this dark, one-piece dress are perfect for a tall, heavy figure.

Orvis

12-18 Tall, heavy people can minimize their size with well-fitting clothes that have simple design lines. Flat fabrics in dull shades help, too.

Shirtwaist dresses and business suits look nice on short, thin people. Pants should have straight or only slightly flared legs. Shape can be given with tucks, shirring, or gathers. Light or bright one-color outfits are attractive. Construction details (collars, lapels, pockets, etc.) should be limited and small in scale. This creates simple, uncluttered clothes. Hip-length jackets, and tunics and vests of moderate length, can be worn well as in 12-19. Small, neat accessories are desired.

Short, thin people should avoid bulky textures and large prints and plaids. Large pockets, collars, or cuffs should not be used. Heavy looking accessories are overpowering. Wide, contrasting horizontal bands or belts are not good because they reduce height. Layering clothing items can produce an awkward bulkiness. Skirts should not be too long.

Short and Heavy

People who are short and heavy need to choose garments that give illusions of height. Lots of vertical lines give height as well as slimness. Empire, A-line, or narrow and straight silhouettes are good. Long, asymmetric closings are attractive. A sheath dress with long straight sleeves and no belt is slenderizing. If a belt is needed, a thin one of a matching color will not detract from the long vertical image. Pants should be narrow or straight-legged as in 12-20. Construction shaping lines and darts should be vertical. Somewhat high heels can add height to a short person. However, extremely high heels look out of proportion.

Fabric patterns should be subtle with vertical designs. Otherwise, use plain fabrics that are not stiff. Smooth, flat fabrics should be

Draper's & Damon's

12-19 A short, thin person looks great in outfits with subtle patterns and limited construction details.

The McCall Pattern Company; McCall's

12-20 Vertical lines, no belt, and narrow pants help to make a short, heavy person look and feel taller and thinner.

of dark or dull colors without any sharp contrasts. Stockings and shoes look best when they are color-related to the outfit.

It is important that the clothes of short, heavy people fit well. Dark gray suits as in 12-21, and straight-cut dresses are good choices. One-color suits with matching vests are also good. Perk up one-color outfits with scarves, pins, neckties, or other accessories. Trim near the face brings the eye of the beholder up. Open or collarless V- or U-necklines are also good.

Short, heavy people should avoid two-piece or two-colored garments that cut them in half visually. Pants should not have cuffs. Turtlenecks are discouraged. Double-breasted jackets and horizontally striped shirts are out! Clunky shoes or boots, and wide, stiff belts should be avoided. Clingy styles, cluttered fussy looks, sloppy looks, and shiny fabric finishes are not good. Large prints, horizontal lines, extra fullness, and bulky fabrics should not be worn since they give the appearance of added width. Also, clothes that are too tight make short, heavy people look even heavier, as if they have just outgrown their garments.

Top Heavy

This "inverted triangle" figure or physique has a large upper body in proportion to the lower body. Choose simple, slim tailored shirts or long cardigan jackets with a vertical feeling over the top. V-necklines are flattering, as are open collars with lapels. Slim sleeves are good.

On top, use dark, dull, plain fabrics that are smooth or soft. See 12-22. Light, bright, or patterned fabrics are best for the bottom, as in 12-23. Also, napped and bulky textures can be used on the bottom.

The McCall Pattern Company; Vogue® Patterns

12-21 Well-fitting clothes of flat fabrics with subtle patterns are attractive on short, heavy people.

The McCall Pattern Company; Vogue® Patterns

12-22 Top heaviness can be disguised with dark, plain colors over the top and light-colored fullness on the bottom.

Perry Ellis by Hartmarx

12-23 This outfit is great to decrease the size of the upper body with a darker and narrowing horizontal image. Balancing width is given to the lower body with lighter pants, which are also pleated for fullness.

Low waistlines, hipline interest, A-line silhouettes, and full or flaring skirts are good. Pants should have slightly flared legs. Hip-length tunics or vests are good. A belt or band of contrasting color at the hips is very effective, especially if the hips are small.

Do not have any horizontal lines at the chest level. Avoid a low yoke or smocking at the bustline. Shaping above the waist should be done with vertical seams or darts. Try garments that taper gently under the chest. Gathers, pleats, or shirring can be used below the waist.

Top-heavy people should avoid clingy or shiny fabrics on the top. Also avoid high or fussy necklines and large collars. Short full sleeves, breast pockets, and tightly fitted or short tops are not recommended since they give the illusion of width across the chest and upper arms. Ruffles, bows, or trimmings should not be on the upper part of the garment. Large prints and bright colors should not be used above the waist. Use a dark-colored top and light-colored bottom. Tight skirts or pants emphasize the large top.

Thick Middle

This body shape has no well-defined, or indented, waistline. The waist measurement is quite similar to that of the chest and hips. Choose unfitted, but not full, garments, such as overblouses, empire lines, tunics, and long sweaters. Tubular silhouettes are attractive. Garments that hang slightly loose from the shoulders are good, such as the jacket shown in 12-24.

People with this shape should use vertical lines giving an upward direction toward the

JC Penney

12-24 This unbelted dress is perfect to hide a thick middle especially since it is covered by a jacket with mostly vertical lines.

face. Keep fashion details above the waist, and use attractive neckline interest. Skirts can flare a bit at the hemline, and pants can be slightly flared. Shaping should be done with darts or seams. A somewhat loose vest or sweater, as in 12-25, can camouflage bulges.

Choose smooth, lightweight fabrics. Solids or subtle prints are best. Other textures may be used if they can give a streamlined effect. Colors should be of medium shades that flatter the face and hair.

People with a thick middle should avoid clingy styles and clothes with fitted waistlines or tight belts. A baggy or bulky fit around the middle is not good either, since the appearance of even more size there will be given. Styles with trim or contrasting buttons at the center front are not recommended. Also avoid clothes that have bold prints or obvious horizontals.

Lerner New York

12-25 A thick middle can be hidden with a cardigan worn open to form a vertical area leading to the face.

Hip Heavy

This "triangular" figure or physique has broad hips that are large in proportion to the upper body. Sometimes thin arms or narrow shoulders contribute to this appearance. Choose garments that have vertical lines with an upward direction toward neckline interest, as in 12-26. Wear attractive collars. Wide shoulder lines and yokes can be used. Gathered sleeve tops that become narrow toward the wrist are good.

Skirts and pants should fit neither tightly nor with excessive fullness at the hips. Pants should not taper at the bottom. See 12-27. Skirts should have a bit of flare at the hem. A-line styles are good. Skirts and pants should have shaping with darts and seams. Tops can have gathering and other fullness.

Use light, bright, or printed tops with dark, dull-colored skirts or pants. If belts are used, they should match the bottom garments. Use smooth, firm fabrics in solids or subtle prints.

Hip-heavy people should avoid tight-fitting pants or skirts as well as overblouses or shirts that end at the hips. Wide belts and horizontal bands at the hipline are bad. Details such as patch pockets should not be used below the waist. Long full sleeves with big cuffs are not good. Also avoid shiny or clingy fabrics and large prints or plaids below the waist. Tight, skimpy tops make the hips look larger.

General Guidelines

With the knowledge you now have, you should be able to create the illusions you desire in apparel. In addition, think about the following guidelines.

Pockets that have the shape of vertical or slanted slits are slimming. Patch pockets add width. A yoke across the bodice front appears to broaden the shoulders. Wide cuffs at the wrist make arms look shorter.

With broad shoulders, you look taller because your width is high on your body. With broad hips, you look shorter. The closer your width is to the ground, the shorter you look. Also, hips usually look wider in pants than in a skirt.

Use V lines at the top of a vertical line to give sloping shoulders an upward lift. Use the downward slant of raglan sleeve seamlines to soften or modify square shoulders. A raglan line should not be worn by someone with sloping or narrow shoulders because it accentuates them. For narrow shoulders, wear tops with extended or padded shoulders, as

JC Penney

12-26 Hip-heavy people look super in outfits that play down the hip area and draw attention to the upper body.

The McCall Pattern Compnay, Vogue® Patterns, Adri

12-27 This outfit camouflages large hips with top interest, fluid vertical draping, and straight pant legs.

pictured in 12-28. For rounded shoulders, wear set-in sleeves.

Suppose a person's waist produces a pleasing curve and relationship between his or her chest and hips. Then tucked-in shirts, belted garments, or a sheath dress that is fitted from bust to hips will accent those nice proportions. For a short-waisted person, use beltless styles combined with lengthening tricks on the top. Choose a cardigan, dropped waistline style, or tunic, as in 12-29. If legs seem too long for the rest of a body, use shortening tricks there with color, shape, line, and texture.

If someone is long-waisted with short legs, disguise or hide the natural waistline. Try to visually elongate the lower body to appear to have ⅝ of the total height from waist to floor. Chemise silhouettes, empire style dresses, or

The McCall Pattern Compnay, Vogue® Patterns, Issey Miyake

12-28 The appearance of added width at the shoulders is provided by this dress top.

soft skirts or pants are good. Use stockings and shoes in colors that match the skirt or pants.

If a person's legs are too thin, textured hose or pants of bright and heavy fabrics help to round them out. Corduroy or tweed are good choices. On the other hand, with heavy legs, pants in medium or dark tones and longer skirts can be good disguises. Keep fashion interest above the waist, and wear simple shoes with enough bulk to balance the leg.

Heavy thighs and a large seat can be covered with easy-fitting (but not full) pants or with bias-cut or A-line skirts. Put fashion interest near the face.

To lengthen and slim the appearance of a short, thick neck, V-necklines are good, as shown in 12-30. Avoid turtlenecks, high collars, and scarves or ascots at the neckline.

Extra Tips

Make sure the design of the fabric is related to the structural design of each garment. Straight lines in the fabric design are best when used with a garment that has straight structural lines. Medium and large prints should be used in garments with large unbroken areas and a minimum of construction details. That is because seams, darts, and pleats break up the design motifs. Tiny prints, such as small polka dots, can be used in garments with more structural details. Circular prints give an illusion of roundness, which is good for thin bodies but bad for heavy ones.

To reinforce the design concepts you have learned in this chapter, go shopping for clothes. (Leave your money or credit cards at home!) Try on garments that have various colors, shapes, lines, and textures. Try many different styles, shapes, lengths, and fabrics. Take note of the best features of your particular body build. Also notice what features do not flatter you. Look for balance, proportion, emphasis, and rhythm in the garment designs, too. Some outfits are more harmonious than others.

Don't forget to consider the visual effects of shoes. Wide feet look best in simple, one-color shoe styles. A buckle or other decoration on a shoe may make the foot look shorter or wider. Ankle straps cut the illusion of body height and call attention to heavy ankles.

In general, do not let your clothes overpower you. Extreme fashions are usually hard to wear unless you are slender, well-proportioned, and outgoing. They distract from your figure and your personality. Wear what is best for you rather than just what is fashionably new.

Liz Claiborne

12-29 This fluid tunic hides the location of the true waist. It gives the illusion of a longer upper body to someone who is short-waisted.

Spiegel; Claude Mougin, Photographer

12-30 This V-neckline dress gives a longer appearance to the neck. It also gives a tall, slender feeling because of its straight silhouette, front vertical closing, and one-color fabric.

Chapter Review

Summary

The principles of design are guidelines for the use of the elements of design. These are used to put ideas together for every design. Harmony, or pleasing visual unity, is created when the elements of design are used effectively according to the principles of design.

Balance visually distributes "weight" in the way details are grouped. Formal balance is symmetrical while informal balance is asymmetrical. Proportion, or scale, should be related to the structure and proportion of the human body. Proportion should relate all parts of a garment and accessories to each other and to the whole. Emphasis concentrates interest in a particular area of a design with a center of interest. Rhythm arranges the design elements for pleasing eye movement over the apparel through repetition, gradation, transition, opposition, or radial arrangement.

To create the best look for a woman's figure or a man's physique, apparel designs should emphasize the assets and camouflage the liabilities. Different designs are best for various individuals. Tall, thin people can wear garments with fullness, contrasting colors, and heavy fabrics. People with tall, heavy body builds should wear clothes with simple lines and little decoration. Short, thin people need soft fabrics, small-scale prints and details, and smooth textures. People with short, heavy shapes should add illusions of height with vertical lines, smooth and dull colors, and no sharp contrasts. People with top-heavy body builds look best with simple, vertical-looking tops and fuller interest below the waist. People with a thick middle should choose slim, unfitted silhouettes. Hip-heavy people should choose garments that have upward vertical lines toward neckline interest.

General guidelines can be followed to create the desired illusions in apparel. Fabric design should be related to the structural design of each garment. Try on and analyze outfits to reinforce the design concepts for your particular body shape.

Fashion in Review

Write your responses on a separate sheet of paper.

Matching: Match the terms listed below.

1. Gradation.
2. Formal.
3. Repetition.
4. Asymmetrical.
5. Scale.
6. Transition.
7. Focal point.

A. balance
B. proportion
C. emphasis
D. rhythm

True/False: Write true or false for each of the following statements.

8. The principles of design are guidelines for the use of the elements of design.
9. Proportion is a visual distribution of "weight" in the way details are grouped.
10. A large area of a bright color balances a small area of a dull color.
11. Formal balance is the most common type of balance in clothes.
12. The ideal body proportions have ⅜ of the total figure from the waist to the top of the head.
13. Yokes, collars, pockets, and other apparel parts must be the right size for the total design and for the wearer.
14. Opposition is a concentration of interest in a particular part of clothing outfits.
15. Emphasis can be created with contrasts of textures or colors.
16. Radial arrangement is one way to achieve rhythm in an outfit.
17. Harmony gives the feeling that all parts of an outfit belong together and suit the wearer and the occasion.
18. The ideal "model" shape for wearing clothes is short and thin.
19. Short, heavy people look best in two-piece or two-colored garments that cut them in half visually.
20. The closer your width is to the ground, the shorter you look.

Fashion in Action

1. Cut out four pictures of complete outfits from catalogs, magazines, or newspapers. Mount the pictures and write descriptions about the elements of design in each, as well as how the principles of design are used. Classify the designs as having good, medium, or poor harmony according to your judgment. Give reasons for your decisions.
2. Clip and mount pictures of garments that would improve the appearance of any five of the seven most common body types. Provide a written explanation of why each garment design would be good for a particular body type.
3. Write a 400-word report that includes a short analysis of a hypothetical figure or physique. Mention the assets and liabilities. Then explain what design features would be best for that person's apparel. Describe or draw sketches of outfits that would be "perfect" for that person.
4. Using one inch for the height of a head, sketch a fashion figure with ideal proportions. Then, using tracing paper, draw at least two fashion sketches using the basic figure as a guide.
5. Collect fabric swatches of different colors, textures, and designs. Examples might be velvet, burlap, satin, tweed, linen, corduroy, and various printed fabrics. In class, discuss how each could be used in apparel for various body builds.

PART FIVE

Consumers of Clothing

Chapter 13
The Best Clothes for You

Chapter 14
Wardrobe Planning

Chapter 15
Being a Smart Shopper

Chapter 16
Making the Right Purchase

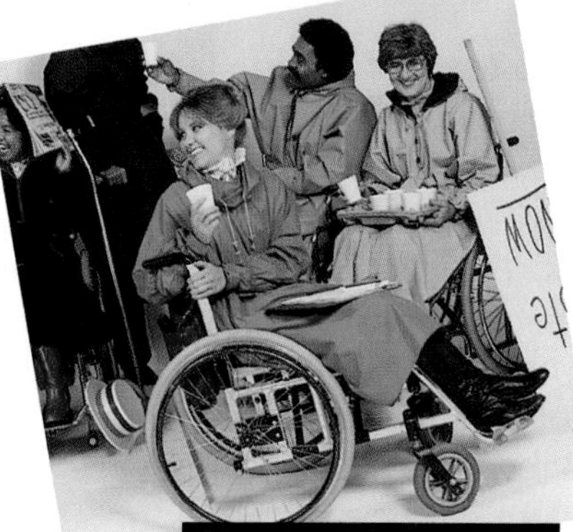

Chapter 17
Apparel for People with Special Needs

Chapter 18
Caring for Clothes

13 The Best Clothes for You

Fashion Terms

image
grooming
communication
personality
cosmology
yin
yang
first impression
lifestyle

After studying this chapter, you will be able to

- give examples of how people use clothes to project images and communicate messages.
- explain the importance of dressing appropriately for your lifestyle, climate, and community standards.
- describe the benefits of a well-planned wardrobe.
- discuss the enjoyment that can be gained from choosing the right apparel.

The best clothes for you are those that help you project a positive image. They allow you to express your personality and make good first impressions. They are appropriate for your lifestyle and for the climate in which you live. By discovering what clothes are best for you, you will be prepared to organize a well-planned wardrobe and be a smart consumer. Then you will be able to purchase and care for your clothes in the best possible ways.

The Image You Project

The ***image*** you project is what people see when they look at you and remember about you later. It is the mental picture that they form of you. The way you dress is a big part of your image. Personal attractiveness, to give the best possible image, requires some effort.

There is something good looking about everyone. To look super, you don't need to be handsome or beautiful. Instead, what you need is good personal grooming plus clothes that fit and flatter.

Good ***grooming*** includes having clean nails, hair, teeth, and body. Try to maintain good health and the right weight through proper habits of eating, exercise, and rest. Sit, stand, and walk with straight posture. Style your hair in a way that complements your face and the shape of your head. Try to develop a personality that is cheerful and alert. Also, remember that pleasant facial expressions and gestures can make the rest of you look better.

To give the best image, your clothes should fit nicely, without being too tight or too loose. They should be the right length. They should flatter your physical assets. They should be neat, clean, and in good repair.

To look your best, you must combine good grooming and clothing sense. Without proper grooming and posture, your clothes will not look very good on you. When you know you look good, you have self-confidence and you automatically feel wonderful!

Clothing as Communication

Communication is the giving and receiving of messages. Not all messages are spoken or written. Some are expressed through facial expressions, hand gestures, body movements, or even clothes! Did you ever think how much your clothes say about you? You speak with a powerful "voice" through what you wear.

You may have heard the saying, "You are what you wear." The clothes you choose, and the ways you wear them, do express visual messages that tell others about you, as in 13-1. Just as a traffic light sends a message without speaking, your appearance communicates in a nonverbal way. Others draw conclusions about you from the way you look. Apparel speaks a silent, but clear, language. The best clothes for you are those that send out the right messages.

JC Penney

13-1 The visual message communicated by these men is one of having self-confidence and caring for themselves and others. Much of this message is conveyed through the clean, neat, and attractive clothes they are wearing.

The colors of the clothes you wear at a particular time may help you express your mood. On happy occasions, you probably choose bright, cheerful clothing. If you are sad or feel down, you probably wear dark, dull clothing.

Clothes let us say different things about ourselves at different times. If we wear dirty and ripped clothes, we send the message, "I don't care what you think of me because I don't think much of myself." If we dress like our peers, we communicate that we are part of that group. If we dress very differently from others, we express our individuality. Conformity and individuality pull in different directions. Most people communicate varying amounts of each. They are telling the world who they are.

Sometimes communication with apparel is not successful. If you were 50 years old and tried to dress like a teenager, your message might be misinterpreted. You might be trying to say, "See how young I am!" However, the message received by others might be, "See how foolish that person looks!" The same thing might happen if you were to dress totally like a popular actor or celebrity. You would probably look silly since you are a different person, and lead a different life from the celebrity. It is important to keep communication in mind when you select and wear your clothing.

Clothing is an important language, or code, that projects the wearer's talents and goals. Apparel can imply that you can hold a responsible job or be reliable. When building your wardrobe, keep these messages in mind so your clothes will speak for the real you.

Your Personality

Your ***personality*** is made up of your own personal thoughts, feelings, and actions. It is the total of the characteristics that distinguish you as an individual, especially your behavioral and emotional tendencies. Everyone's personality is unique. One way to express your personality is through the way you dress.

Clothes are an outward expression of how you feel about the world around you as well as how you feel about yourself. The way you dress is a projection of how you see yourself and how you want others to see you.

Judgments of personality are often based on clothing choices. Your personality and appearance are what "sell" you to others. They can help you be elected to a student government office, get a special date, or achieve other goals. Research has proven that people with high clothing consciousness achieve higher goals in life than others. They have higher levels of education, better incomes, and more responsibility in their communities than those who do not care about their appearance.

Yin and Yang Traits

Yin and yang traits combine personality and physical characteristics. They are described in ancient Chinese cosmology. ***Cosmology*** deals with the order of the universe. Chinese philosophers used yin and yang to describe the opposite, but equally important, forces that exist in the world. Yin and yang, thus, represent opposite complementary qualities. They represent the balance between the contrasts found in the universe. They are pictured by the symbol shown in 13-2.

Yin represents the passive, timid, and delicate elements of personality. Dominantly yin people are submissive, mild, and fragile. They are quiet, shy, sensitive, and introverted. They are followers rather than leaders.

Chinese cosmology pictures yin physical traits as being feminine and short. They include a small bone structure, graceful features, delicate coloring, and soft, gentle speech. Youth and innocence are portrayed by yin people.

Yang represents the active, rugged elements of personality. Dominantly yang people are forceful, aggressive, and strong. They have vigor, dignity, and self-assurance. They are sturdy and not easily offended. They are extroverts and leaders.

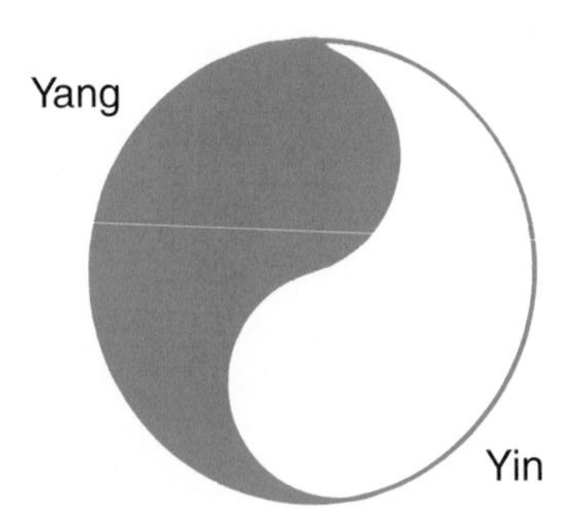

13-2 The yin/yang circle illustrates the balance of opposite forces in the universe.

Chinese cosmology pictures yang physical traits as being masculine and large with strong, prominent features. They include pronounced coloring, a low firm voice, and a tall angular frame. Maturity and sophistication are portrayed by yang people.

No one is completely yin or yang. Every person is a complex blend of personality traits. Usually, either yin or yang is dominant with some characteristics of the other. We all fluctuate with our amounts of yin and yang characteristics as we change our roles and activities daily. As each of us matures, there must be a balance of yin and yang in the personality. There is always a balance in a well-adjusted person. A person may be assertive in business and sports (showing yang characteristics) but gentle and caring in a personal relationship (showing yin traits).

Yin forces are represented in clothing with graceful, flowing lines. Intricate details and curves are used, as well as soft, sheer textures. Fabrics often have small, delicate designs and are in light colors, as in 13-3. Yin people generally wear delicate jewelry. Examples of yin forces in nature are the weeping willow tree, the trailing vine, and the tiny hummingbird.

Yang forces are represented in apparel with bold designs. There is a minimum of detail. Straight lines are used. Fabrics have forceful design motifs and are in strong or neutral colors, as in 13-4. They have heavy, rough textures. Yang people generally wear bold, bulky jewelry. Examples of yang forces in nature are the sturdy oak tree, the rugged mountains, and the fierce eagle.

In apparel, yin and yang qualities are evident in fabrics, garments, accessories, colors, shapes, lines, and textures. Try to balance the yin and yang in your clothing with the yin and yang in your personality and body structure. You will feel most comfortable this way and project the truest image of yourself.

First Impressions

A ***first impression*** is what people think of you when they first see or meet you. An impression is a feeling or a reaction. The

JC Penney

13-3 A dominantly yin person would prefer this outfit with a pastel rose print, small buttons, and pearls with a pink bow at the neck.

JC Penney

13-4 This outfit shows yang influences with its dark colors, straight lines, and bold details. A dominantly yang person would prefer to wear it.

human mind receives and acts on signals very quickly. These signals or impressions work both ways. When other people meet you, they form first impressions of you based on their immediate feelings or reactions. Likewise, you form first impressions of other people when you first meet them. You form opinions on the spot. People you meet may strike you as handsome, friendly, intelligent, sad, sloppy, or shy. Most people agree that it is what is inside a person that matters. However, right or wrong, first impressions are made on the basis of appearance. What would your first impression be of the men pictured in 13-5 and 13-6?

First impressions are determined to a great extent by what a person is wearing. That is because the first view of someone is made up of about 90 percent clothing. Clothing covers the largest area and is easily noticed. What you wear is the first statement you communicate about yourself. Your clothes are an outward expression of your inner identity.

A favorable first impression leads to good feelings. Then others want to get to know you better. A poor first impression has the opposite effect. If the first impression about you is negative, others may not want to get to know you better. You may never get a second chance to prove your worth, especially if you are with someone for only a short time. For instance, you may have an interview with a prospective employer, and you may have excellent qualifications for the job. If your appearance does not make a good first impression, you may lose out to someone who makes a better impression.

The Fashion Association

13-5 Quickly list three adjectives to describe this man. Find out if your classmates formed the same first impression of him.

Tultex Corporation; Swann-Niemann Photography

13-6 Is your first impression of this man different from your first impression of the man pictured in 13-5? Why?

First impressions are very important, especially in new social groups, in new schools, and in business. That is why you should pay attention to the messages your clothes send to others. Once you have made a good first impression, then you can reinforce it with your words and actions.

The "Right" Apparel for You

To be appropriate, your clothes should be comfortable, clean, and stylish. However, they do not necessarily have to be of the very latest fashions. How "right" they are depends on several factors. They should be correct for your lifestyle. They should meet the needs of the seasons of the climate in your region. They should also correspond with your community standards of dress.

Your Lifestyle

Your ***lifestyle*** is made up of all the activities you do and the places you go. It is influenced by your job, sports, travel, friends, income, and attitudes. It includes your school, work, social, and leisure activities. If you take up a new sport, start a new job, or assume a new social responsibility, your lifestyle is altered. Your apparel may have to change accordingly. You may have new clothing requirements in order to be dressed appropriately.

You should be aware of your activities so your apparel is geared to your lifestyle, as in 13-7. To evaluate your lifestyle, write down your usual routine over a normal two-week period. Include evening events, hobbies, sports, and other occasions you attend. What are your specific activities? How many are casual, and how many are dressy? Do you swim, bowl, play tennis, read, watch movies, or go on picnics? Try to make a complete list of your activities.

Next, analyze how you should dress for your activities. Try to develop your sense of what to wear when. Notice how others dress for certain occasions. What kinds of clothes are needed for what you do? Do you have suitable attire for the special events you attend? What clothes can double for school wear as well as for movies, shopping, or hobbies? Fortunately, many garments are suitable for more than one type of activity.

Many sports require special ways of dressing. Sports clothes must be ready for action. You may need special garments that provide protection and safety as well as comfort and freedom of movement. Footwear that provides support, comfort, and durability might be necessary.

How you dress is very important in the working world. Dressing appropriately for a job means wearing the right type of clothes for the work to be done. This might mean sturdy and practical garments for hard physical labor, as in 13-8. Conservative apparel might be best for a law office, while a uniform is required for some types of official duty or schools, as in 13-9. Zany outfits may be best for working in a fashion boutique! Your clothing should conform to the generally accepted standards where you work.

In a job, people will take you as seriously as you present yourself. The "power of positive dressing" causes people to treat you with respect. This, in turn, raises your status and encourages your success, as in 13-10. Dress for the kind of job you seek. Fair or not, job advancement often depends on a well-groomed appearance. If your image is sloppy or unkempt, people may assume that your work and attitude will be that way, too.

Your Climate

Different climates create different clothing needs. Garments in summer and winter wardrobes vary in terms of texture, weight, and color. Your clothes need to feel cool in hot weather. They must provide warmth when cold temperatures hit. Energy conservation has caused houses and offices to be cooler in winter and warmer in summer. This has changed our wardrobes.

To keep warm when the weather is cold, your torso should be covered with several clothing layers. The layered look is perfect for adjusting clothing to meet body needs. Each layer traps a pocket of air. This keeps body

Kellwood Company

13-7 This type of apparel may be needed in your wardrobe if you have an active lifestyle that includes exercising.

Tultex Corporation; Swann-Niemann Photography

13-8 Rugged, physical work requires sturdy apparel that can take hard wear and be laundered frequently.

13-9 Many schools require uniforms, partly so students feel serious about concentrating on their studies.

The Fashion Association/Doncaster

13-10 The appearance of this woman is businesslike. It gives the impression that she is capable of getting the job done.

heat in and lets moisture out. However, do not wear so many layers that you can't move around easily. Also, the outer layer must be big enough to go over all the rest, as in 13-11.

The fabrics of garments should be chosen according to the climate. To gain warmth, plan to have fuzzy, bulky, wool-like garments in your wardrobe. Double knits, quilted fabrics, and laminated fabrics have extra warmth. Down or other fillers are used in garments to give warmth, too. Garments that fit snugly at the neck, wrists, and ankles keep drafts of cold air out. Avoid clothing that is extremely loose or extremely tight.

For hot weather, garments of cotton are good choices. The outfits in 13-12 and 13-13 are both appropriate for warm weather conditions.

When outdoors, you sometimes need protection against wind and rain. You must

JC Penney

13-11 These jackets can help provide different amounts of warmth, depending on what garments are layered under them.

have apparel to keep your head, hands, and feet warm and dry. In cold weather, long underwear worn under clothes can provide a surprising amount of warmth.

Northern, cold climates demand higher spending costs for clothing. Items are needed to protect against the cold, such as coats, boots, hats, scarves, gloves, sweaters, and long

International Male

13-12 This "summer-weight" outfit is especially suited for warm weather.

Simplicity Pattern Company, Inc.

13-13 Loose-fitting, sleeveless, collarless garments of lightweight fabrics are ideal in hot weather.

underwear. People need more variety in their wardrobes when they live in geographic regions that have distinctly different seasons. A mild climate, where temperature variations are less severe, causes fewer clothing needs and is easier on the budget.

Community Standards

Community standards influence the appropriateness of certain clothing. There are different standards of dress in different parts of the country and world. The tempo of life in a community near a resort area, as in 13-14, is different from that of one near a major business or financial center. Tastes are usually more conservative in small towns and rural areas than they are in big cities. Appropriate dress for one activity might differ from place to place.

Particular rules of dress are followed in communities. When worn in the wrong situations, clothes can cause disfavor. For instance, at a rock concert, jeans are probably favored.

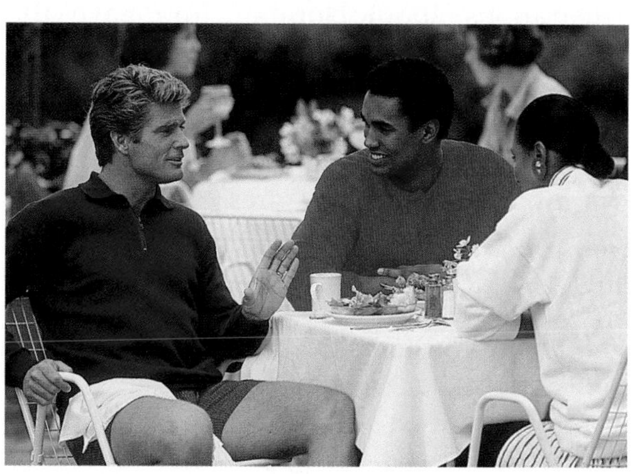

Tultex Corporation; Swann-Niemann Photography

13-14 Appropriate attire for a restaurant in a resort area is much more casual than in the business area of a city.

However, at a symphony performance, jeans would show disrespect in most areas. What is considered to be acceptable dress by your friends, school, or employer? Your clothes should conform to the times and your social group, as well as showing some originality or creativity.

The Benefits of a Well-Planned Wardrobe

As discussed in earlier chapters, your wardrobe includes all the clothes and accessories you have. With a well-planned wardrobe, you have enough of the right clothes for your lifestyle, climate, and community standards.

You may have lots of clothes. Are they in style? Do they fit you? Do they need repair? Do you often say, "I don't have a thing to wear," even though your closet and dresser drawers are full? You can unclutter and gain control of your wardrobe with planning. It will then contain enduring, thrifty, mix-and-match apparel.

You Can Project Your Best Self-Image

With a well-planned wardrobe, you look and feel your best. You have apparel that suits your body shape, coloring, personality, and lifestyle. A well-planned wardrobe reduces frustrations about clothing. You have the clothes you need, and they fit well and are stylish.

A well-planned wardrobe helps you feel comfortable and confident. The ideal wardrobe plan lets you dress to forget your clothing. You can then pay attention to more important matters. When you know you look good, you feel attractive, secure, and self-confident. You are sure that the first impressions you make are always the best they can be. You are then able to concentrate on the activity or job at hand. You can turn your thoughts and attention toward others.

When planning your wardrobe, try to establish the right blend of conformity to peer group pressure and individuality. Individualize your clothes enough to express your own personality. Use your clothes to tell others who you are and how you feel. Identify yourself with your clothes. Let your clothes conform to current fashion to meet your needs to belong. However, they should also be different enough to express your true self.

Dressing should be fun and creative. Your clothes should be able to give you a lift and change your mood. This is good for your psychological well-being. You should enjoy wearing the clothes in your wardrobe. You must feel comfortable about your clothes and in them. Aim for an appearance that says who you are and makes the most of it. Personal satisfaction is gained with a well-planned wardrobe.

You Can Save Money

The least expensive way to dress is to avoid making wardrobe mistakes. Careful planning and wise buying allows you to do this. It is best to plan a small, well-chosen wardrobe. Having too many clothes is confusing. Your closet should contain a well-coordinated, well-edited, complete wardrobe that works for you, yet is economical.

Looking great does not necessarily mean spending lots of money. A well-chosen wardrobe will give you a collection of flattering garments and accessories that keep you suitably dressed for your activities at a reasonable cost. Your goal should be to look your very best with a minimum expenditure of time and money. It is unnecessary to buy a whole new wardrobe for a new look each season.

A well-planned wardrobe also provides a good basis for continuous changes that will keep you up-to-date. Your wardrobe should be in a slow state of change. You must keep working on it, replacing a few items at a time. With planning, you can anticipate your needs and expenditures ahead of time. You can keep your wardrobe up-to-date without sudden costs.

You Can Gain Flexibility

Planning and having mix-and-match garments based on a particular color scheme makes sense. Then you can combine a few garments in different ways to create several outfits, as shown in 13-15. All parts of the wardrobe become harmonious.

Decide on one or more color schemes that look good with your coloring. Plan your wardrobe around those colors. Examples might be khaki and navy, gray and burgundy, or red,

The McCall Pattern Company

13-15 These four outfits are made from different combinations of seven garments. What other outfits are possible?

white, and black. By doing this, your garments can be put to better use. The color schemes become the foundation for your whole wardrobe. Your apparel items will coordinate automatically as they are mixed and matched for flexibility. Bring in brighter or different colors with your accessories. Use prints, plaids, or patterned fabrics containing colors of the solid garments with which they are worn.

You should be able to reassemble and wear the same clothing items for different occasions to fit your lifestyle. A well-planned wardrobe is easy to mix and accessorize for many different looks.

Fewer clothes are needed if each garment is useful to the maximum extent. For instance, a navy blazer can be coordinated with many different sweaters, shirts, and slacks (or blouses and skirts) for great versatility. A pair of blue jeans with a good cut and proper fit can be worn with many different shirts and sweaters. A basic dress can be worn to school, religious events, and an important dinner by changing what is worn with it. For school, a cardigan sweater could be worn over it. For religious events, a jacket might be added. For an evening out, it might have a pretty silk scarf at the neck or fancy jewelry.

Enjoy the Apparel You Choose

You will never stop experimenting with clothes. Fashion is always offering something new. Many people choose to wear styles that they like even if they are not the newest fashions. There are many different fashion statements from which to choose. There is no one look or silhouette being dictated for your wardrobe. With all the options, the final decision is a personal one based on your individual tastes and needs. Besides, you should not wear items that are supposed to be "in" when they are all wrong for you.

Your fashion tastes change as you mature. As you get older, you may become more individual in your wardrobe selections. You will probably develop a personal style that says who you are through your appearance. You should always be able to look and feel your best, as in 13-16. Your wardrobe plan can allow your personal style to evolve with different strengths as your situation changes.

Besides expressing yourself to others, your wardrobe influences the way you feel about yourself. Feeling that you are suitably and

Lands' End Direct Merchants

13-16 As people mature, they develop their own personal styles that help them look and feel their best.

attractively dressed gives you confidence about your worth and ability. Then you can devote your time and efforts to other people or to important activities. Being well-groomed and properly dressed is not an absolute guarantee of instant success in every situation. However, it will definitely help boost your self-confidence so that your chances of success are greatly improved. Self-assurance comes with putting your best foot forward. When you look good, you also tend to feel good. When your wardrobe includes styles that play up your best features and express your personality, your clothes seem to say, "There is a sharp, confident person!"

Chapter Review

Summary

To project the best image when people see you and remember you, it is important to combine good grooming and clothing sense. Clothing communicates messages to others. Your personality is expressed through the way you dress. Yin and yang traits combine personality and physical characteristics through Chinese cosmology, that describes delicate versus rugged images.

First impressions are often determined by what a person is wearing. The "right" apparel for you should be comfortable, clean, and stylish, but does not need to be of the latest fashion. Your apparel should be geared to your lifestyle activities. Different climates create different needs for garments and fabrics. Community standards also influence the appropriateness of certain clothing.

There are many benefits to a well-planned wardrobe. You can project your best self-image with clothes that fit nicely and are stylish. You can save money by avoiding wardrobe mistakes. You can gain flexibility by having mix-and-match garments based on sensible color schemes. You will enjoy the apparel you choose if you wear the styles you like that are best for you.

Fashion in Review

Write your responses on a separate sheet of paper.

True/False: Write true or false for each of the following questions.

1. To look good, you need good personal grooming plus clothes that fit and flatter.
2. Your clothes express visual messages that tell others about you.
3. One way to express your personality is through the way you dress.
4. Research has proven that people with high clothing consciousness achieve higher goals in life than others.
5. Yin and yang represent opposite complementary qualities that balance each other.
6. Yin forces are represented in clothing with bold designs, straight lines, and rough textures.
7. Yang physical traits include a small bone structure, graceful features, delicate coloring, and soft, gentle speech.
8. A first impression is a person's immediate feelings or reactions to you, or what you think of others when you first meet them.
9. Your lifestyle influences the types of clothes you need.
10. Tight-fitting garments of cotton are good choices for cold, winter weather.

Multiple Choice: Write the letter of the best response to each of the following questions.

11. Dressing appropriately for a job means wearing _____.
    A. the right type of clothes for the work you need to do
    B. clothes that must be dry-cleaned
    C. a sport coat or suit jacket every day
    D. apparel that can be accessorized differently for after-work social activities
    E. All of the above.
12. Having too many clothes is _____.
    A. confusing
    B. unnecessary
    C. costly
    D. space-consuming
    E. All of the above.

Short Answer: Write the correct response to each of the following items.

13. Why is it important to make a good first impression?
14. Why is proper apparel so important in the working world?
15. How can a well-planned wardrobe save you money?

Fashion in Action

1. Cut out five pictures from magazines, catalogs, or newspapers of clothes that show yin styling and five that show yang styling. Mount the pictures and write descriptions of their yin and yang characteristics.
2. Write an essay describing experiences you have witnessed concerning first impressions. Have you seen people treated badly because of their appearance? What kinds of reactions have you seen to well-dressed people who made good first impressions? Have you witnessed occasions when first impressions turned out to be inaccurate?
3. Clip out at least five magazine or newspaper photos of people who are sending out messages by the way they are dressed. Mount them and explain what messages you are receiving in each case.
4. Read an article or book on career dressing for achieving success. Present an oral report on it to the class.
5. Make a comprehensive lifestyle chart. Determine what amount of time is spent in a typical week for your particular activities (or for the activities of a hypothetical person). List days of the week down the left side and activity categories across the top. Fill in with hours or fractions of hours. Then analyze what apparel is needed (in general terms) for this person's wardrobe.

14 Wardrobe Planning

Fashion Terms

wardrobe inventory	costume jewelry
appliqué	basic apparel
want	investment dressing
need	extenders
priority	wardrobe plan
fine jewelry	resources
bridge jewelry	

After studying this chapter, you will be able to

- take an inventory of your present wardrobe.
- distinguish between your apparel wants and needs.
- set priorities on additions to your wardrobe now and in the future.
- evaluate your resources to be used for apparel.

Developing a well-planned wardrobe takes careful thought as well as knowledge and skill. To plan a wardrobe, you should have a knowledge of fashion, fabrics, and clothing construction. You must think carefully about what apparel items you already have and what you want to add. You need the skills to shop wisely and to select items that will enhance your appearance. Wardrobe planning takes time and effort, but the results are worthwhile.

Know What Apparel You Already Have

Your present wardrobe is the basis for your future wardrobe. You should evaluate what you already have by taking a ***wardrobe inventory***. An inventory is an itemized list of what you have. When you write it out, you will be able to decide what clothing items to keep, what to eliminate, and what to add for the best possible wardrobe.

Make a wardrobe inventory chart like the one in 14-1. List all garments and accessories from your closet and drawers by category. Don't forget the garments in the dirty clothes hamper or at the cleaners. Add other categories to your chart if needed. Write all pertinent remarks in the spaces.

As you complete your wardrobe inventory chart, try to visualize your separates being put together as outfits. Keep in mind what clothes can be worn for each of your activities. You may find that some items can be mixed and matched in new ways. You may even discover some items in the back of your closet that you forgot you had!

You may want to analyze your wardrobe just once or twice a year. You may want to do this before the beginning of each season. Keep your wardrobe inventory chart posted inside your closet door. Then you will always know where you are and where you are going with your wardrobe.

Group Your Clothes

Once you have made a list of everything in your wardrobe, consider which items are most suitable and usable for you. To do this, sort your apparel into four groups:

- Clothing you wear often.
- Clothing you wear occasionally.
- Clothing needing repairs or cleaning.
- Clothing not worn in the past year.

Wardrobe Inventory Chart			Date ____________	
Clothes I Have	**Description**	**Condition**	**How I Like It**	**Action to Take**
Dresses (girls)				
Skirts (girls)				
Suits	———			Buy a gray suit?
Blazers/Jackets	1 Navy 1 Gray	Good Too small	OK Don't like	Save for brother.
Dress shirts/Blouses	2 Striped 1 Blue 1 White	Good (missing button) Like new Frayed collar	Like striped ones best	Sew on missing button Buy 1 white shirt.
Pants	2 Pr. cords 3 Pr. jeans	New 2 Wearing out	Good Love them!	Buy 1 pr. dress slacks + 1 pr. jeans.
Shorts	1 Pr. jogging	Old	Comfy	
Casual shirts	3 T-shirts 1 Turtleneck 2 Knit shirts	OK Good 1 New	Super! Good Fit my needs	
Sweaters	1 V-neck 2 Crew-neck	Old but OK Good	Don't like Good	Trade? Give away?
Coats	1 Tan trench coat 1 Down ski coat	Old Good	Don't like Neat!	Keep for emergencies.
Shoes/Boots	1 Pr. sneakers 1 Pr. cowboy boots	Poor Good	Terrific Love!	Replace sneakers with new pair.
Athletic attire	2 Swimsuits 1 Sweat suit	1 Seam open Good	Like Like	Resew seam.
Underwear	9 Briefs	3 Getting old	Fine	Buy a pkg. of 3.
Socks	6 Pr. white 2 Pr. dress	3 Pr. getting old	Fine Fine	Buy 3 pr. soon.
Sleepwear	2 Cut-off sweats	Old	Terrific	
Hats/Gloves	Knit hat Leather gloves	Good Good	OK Good	
Neckwear	Navy print tie Gray/red tie	Good Good	OK OK	
Belts	1 Brown leather	Good	OK	
Jewelry	1 ID bracelet Cuff links	Broken latch Like new	Like Never wear	Fix latch. Give to Dad?
Other (purses, formal wear, work clothes)	1 Sweatshirt	Old + faded	Love it!	

14-1 Filling out a wardrobe inventory chart like this one is the first step in successful wardrobe planning.

Clothing You Wear Often

Look at the apparel you have placed in the first group. These are clothes you really like to wear. They are probably well-suited to your personality, lifestyle, and geographic location. They are probably the most flattering to your body type and coloring.

Note the styles, textures, and colors of your wardrobe favorites. What makes them pleasing, practical, and appropriate for you? This information will help you make wise purchases in the future. You will know that you should look for similar designs when you buy new apparel items. Double-check the condition of your favorite clothes for signs of wear. Then return these items to their places after wiping the dust from your closet and drawers.

J. Jill

14-2 Separates that you wear only occasionally might be combined with favorite items to create wonderful new combinations.

Clothing You Wear Occasionally

Now look at the clothing in the second group. As you try on the garments, study yourself in the mirror. Is there something not quite right about the colors, textures, lines, or fit of the clothes? Do you have some items that are fine, but don't match anything else in your wardrobe? Do some look out-of-date?

Try to create new outfits by combining separates in this group with some of your favorite items in the first group. Wear jackets, shirts, and sweaters with different pants or skirts, as in 14-2. Try to identify the reasons why some of your clothes are just okay rather than great. Once again, that will help you be a wise shopper in the future.

Determine if there are some creative ways to perk up the clothes in the second group. Would you wear these clothes more if you had new accessories? Do you need new tops or a new belt to wear with your slacks or skirts?

Recycle Garments

Figure out if some items that are worn only occasionally need to be altered or restyled. Should the width or length of pants legs be changed? Should a wide necktie be made narrow? Would new buttons perk up a drab blazer? Should extra fullness of a skirt be removed to make it straighter? Should shoulder pads be added to, or eliminated from, a garment? Could some trim add interest as shown in 14-3?

Most outdated garments can be modified to reflect current trends. Even with minimal sewing skills, you can repair, alter, or update existing apparel. Remember that it is almost always easier to take in or shorten a garment than it is to let out or lengthen it. Also, the simpler a garment design is, the easier it is to change.

Some clothes can be recycled by changing them into different types of garments or wearing them in different ways. Pants that have worn-out knees can be made into shorts. Old shirts can be dyed to freshen their colors or to add decorative touches. Long sleeves can be changed to short sleeves. A comfortable summer skirt that is too short may be shortened even more and worn as a tennis skirt. A jumper that is too short can be restyled as a tunic. Jeans that are too small can be made into a tote bag. Be creative!

Clothing Needing Repairs or Cleaning

Now look at the clothes in the third group. They need repairs or cleaning. They

Butterick Patterns / Butterick Company, Inc.

14-3 The addition of studs or fringe to otherwise plain garments is a fun way to make a different fashion statement.

are probably worth saving but need attention before they can be worn. Fixing these will give you more clothes for less money. A little time and effort can work miracles to help these garments look new again.

Clothing should be repaired as soon as a problem becomes apparent. A small rip will become a big rip if wear is continued. Laundering a garment before it is repaired may cause further damage. Thus, delaying a repair job just makes the job harder, or impossible, to do.

You should resew loose buttons before they fall off. Otherwise, you might lose the buttons, and you may not be able to find identical ones as replacements. Also reattach loose hooks and eyes before they no longer fasten. Tape and safety pins should be used to hold clothing together only temporarily and only in emergency situations!

Figure out what must be done to make each article of clothing wearable. Put the garments in a special place so you will work on them. Then take the time to do it! Fix that hem. Resew and reinforce ripped seams. Attach all missing buttons. Patch holes with matching pieces of fabric, iron-on patches, or appliqués. ***Appliqués*** are cutout decorations. New garments sometimes have them. For repair purposes, appliqués and iron-on patches can be bought ready-made at notions counters of fabric or variety stores. You can also design and make your own with fabric scraps to express your creativity.

You may want to keep all of your supplies in a sewing box or repair box. It should include thread in several colors, needles, a thimble, buttons, snaps, hooks and eyes, scissors, and iron-on patches. Tears in expensive clothes can be rewoven by specialists. This may be costly, but it is almost invisible.

Launder dirty items or take them to the dry cleaner. Spots and stains should be removed immediately so they don't become permanent. Your clothes should always be kept clean and pressed. Then they are ready to be worn whenever they are needed.

Clothing Not Worn in the Past Year

The apparel in this group is not right for you at all! Maybe the garments aren't comfortable. Maybe they are just wrong for your lifestyle. Maybe they don't fit you anymore or are very outdated.

If there are some specialty items in this group that you like, but have not had a chance to wear, such as ski pants or an evening gown, put those back into your closet. Dispose of the rest of the apparel items in this group unless you can think of three good reasons to keep any of them.

Clothes you don't wear are not worth the space they take in your closet or drawers. They can be better used by someone else. Maybe some items can be traded with a friend who likes them and has some apparel that you like. Some can be donated to charitable organizations that will give them to people in need. In that case, you may be able to receive a tax deduction. Some clothes may be sold at a yard sale or by consignment through a resale shop. Then you would get some money from them. Also, some garments can be saved and handed down to a younger relative.

Items that no one would want to wear may be used for rags after the buttons and zippers have been removed from them. Any jewelry of good quality that is out of style may be stored in a plastic bag. It might return to fashion.

After you have grouped your clothes and made them all as wearable as possible, you may have to revise your wardrobe inventory chart. Do your best to keep your chart up-to-date.

Distinguish Wants from Needs

A ***want*** is a desire for something that gives you satisfaction. The item would be wonderful to have, but you can get along without it.

A ***need*** is something you must have for your continued existence or survival. It is necessary for your basic protection, modesty, comfort, or livelihood. For instance, if you live in a cold climate and have no heavy coat, you need a coat so you can stay warm in winter weather. If you have a suitable coat that is a bit out of style, you don't really need a new one. The other one can keep you warm. However, you probably want a new one!

Sometimes it is hard to distinguish between wants and needs. When someone craves something, obtaining that item seems to be a need in that person's mind rather than a want. Also, there are varying degrees of need. The most basic ones give you protection from the environment. Others are determined by your values, culture, lifestyle, resources, peer group, and standard of living. A college graduate who is beginning a new career may need a business suit for the job, even though he or she could live without it. You may find that you need a certain uniform to participate in a sport. You cannot function in that activity without it.

You probably have only a few real needs but many wants. Wants are endless, as in 14-4. They add to your good looks, self-confidence, and enjoyment. Maybe other people have certain clothes, so you want them, too, but you don't really need them. To persuade you to want certain items, advertisements heighten the appeal of products and encourage you to discard the old for the new. Store displays try to convince you that you should have particular garments in the latest colors. The clothes worn by television performers create desires for certain apparel, too.

Wants are also influenced by your personal philosophy. Practical people want their apparel to be comfortable and durable. Others want expensive clothes that will give them high status. Others like the attention received with many clothes of the latest fads. Evaluate your philosophy. If it does not result in smart wardrobe planning, it may be best to consider changing it.

Organize your wardrobe planning so it can be carried out in order of need. When doing the planning, think about needs first. They have top ***priority***. In other words, they are most important and should be considered first. After that, wardrobe additions should be made according to the degree of want. One garment may even fill both a need and a want. For instance, an all-weather coat with a zip-in lining can serve as a winter coat that you might need as well as an extra spring and fall coat that you might want. A new pair of jeans might be the pair of pants that you need for your part-time job. They can also double as those wanted extra school pants.

Spiegel

14-4 It is easy to want all of these colorful items that would be comfortable as well as easy to mix and match.

Choosing Accessories

Accessories, as in 14-5, are the items you wear with your garments to create complete outfits. When choosing accessories, try to achieve a coordinated, yet interesting, look. Show your personality or mood. Your individual touch with accessories will keep your outfits from being plain and uninteresting.

Accessories can work fashion magic. Some are bright; some are subdued. They help you put the finishing touch on an outfit for a fashionable "total look." Often the little details make the difference between a mediocre wardrobe and a marvelous one!

The right accessories can make an old outfit look new and up-to-date. Accessories can extend your mix-and-match wardrobe. They offer an easy, and often inexpensive, way to pull different garments together. Accessories can also add variety to a wardrobe. They can make the same outfit look either dressy or casual. They make it seem as if you own more clothes than you really do.

Spiegel; Lenny Prince, Photographer

14-5 Hats, socks, and shoes are accessories that can be coordinated with your garments for different fashion looks.

Using Accessories to Advantage

Some accessories are functional as well as decorative. Examples of these are watches and handbags. Other accessories, such as bracelets and neckties, are purely decorative.

To choose and use accessories with flair, make sure they are appropriate for the outfit and occasion. The entire ensemble or outfit might be tailored, sporty, or dressy. Outfits to be accessorized might be heavy in feeling, light, or just plain fun. When "big" looks are fashionable, accessories generally become bolder and larger in scale. Light, slender fashion silhouettes call for smaller-scale accessories.

When thinking about the size and style of accessories, also consider your personal size. The accessories you wear should be in proportion to your body. A small person should wear small accessories, since large items would overpower him or her. Likewise, a large person looks best with large accessories.

Accessories can create optical illusions to improve your appearance just like garment designs can. Bright, shiny jewelry will draw attention to a good area and away from a less desirable part of the body. Add height with a long necklace or tie. Decrease your height with a wide, contrasting belt.

Experiment by using accessories in new ways. A pin might be used on a fabric purse. Try accessories with different outfits and in new combinations. Express your individuality. This is one way to make a statement about yourself.

Accessories should never detract from your clothing or from your overall appearance. They should not draw too much attention to themselves. All parts of the design are necessary to the whole. Accessorizing is an elaboration on the art of dressing.

Keep accessories simple; don't overdo. If you aren't sure an accessory looks right, leave it off. Avoid too many accessories; you don't want to look cluttered. A lot of small accents give a spotty or confusing look. A woman would probably look overdone with gold earrings, several gold necklaces, a gold belt, a gold bag, and gold shoes. Use good taste to recognize what is appropriate and socially proper.

Plan your accessories just as carefully as you plan the rest of your wardrobe. Note how accessories are used in fashion magazines, advertisements, and store displays. It takes time and practice to learn to accessorize well. If your garment is intended as the center of interest, play down the accessories. If your garment is plain, add flair with accessories.

Since accessories usually do not cost as much as clothing items, you can buy them more often to be fashionable. There are endless choices of accessories. When selecting them, consider how often and with which garments you will be able to use them.

Footwear

Your footwear wardrobe might include dress shoes, casual shoes, boots, slippers, and athletic shoes. Shoes are important accessories. They are quite expensive, so choose them with care. Try to buy shoes that provide support as well as comfort and durability.

Plan footwear selections in your neutral or good basic colors first. Shoes should not dominate your outfits. Depending on what is fashionable at the moment, your shoes should probably blend with the color of your pants or skirts. Generally, they should be the same color or value as, or darker than, the bottom of your clothes. If they are lighter, attention will be drawn to your feet. Bulky footwear makes your feet look bigger. Men's shoes should usually coordinate with the color of their belts.

Handbags

Handbags should coordinate with shoes, as in 14-6, although they do not need to match. If not the same color as your shoes, your bag should be lighter. Plan to have a handbag in a neutral or in one of your good basic colors. Its size should be scaled to your body size. Purses and briefcases should have simple designs. Straw bags are usually just for warm weather. Canvas bags are sporty. Smaller purses and clutch bags are used with dressy attire.

Headwear

A hat can complete an outfit in a special, chic way. It can also keep your head warm in winter or offer protection from the sun in summer. A hat should fit the shape and size of your head, as the one shown in 14-7 does. Your headwear wardrobe might include baseball caps, knit ski caps, sports visors, and straw hats for summer.

Belts

Belts are available in various widths, colors, and materials, as in 14-8. They may be made of metal, leather, fabric, macramé cord,

Spiegel; Lenny Prince, Photographer

14-6 Accessories look good together when they are coordinated in terms of color, scale, and style.

Lands' End Direct Merchants

14-7 Hats can look fashionable while providing warmth and protection to the head.

The Fashion Association / Canterbury

14-8 Belts can be interesting accessories, especially if they have unusual colors or patterns, or are worn in a creative manner.

or other materials. Buckles may be plain or elaborate. They may give a Western flair or an Egyptian touch. They may even display trademarks or designers' logos.

Some buckles are detachable, so different strips of belting can be inserted into them for great versatility. Some belts are reversible with combinations such as a black side and a brown side. Remember that color contrast at the waist will draw attention there. People with thick waistlines look best when their belts match the color of their outfits. Wide belts look best on tall people.

Scarves

Scarves are versatile accessories. They come in all sizes, colors, designs, and fabrics. They can be used to give warmth, or they can be worn strictly for decoration. They can be worn around the neck, on the head, or other ways as shown in 14-9. Fashion changes the sizes of scarves and how they are worn. Scarves can add color near the face. Multicolored scarves can tie together all the separate parts of an outfit.

Neckties

Neckties can express the personality of the wearer with conservative or bold patterns, textures, and colors. The width of neckties swings from narrow to wide and back to narrow as fashion changes. Sometimes bow ties are in style.

Neckties come in many fabrics and surface designs. They can be complemented with a tie tack or clasp when those items are fashionable.

Handkerchiefs

A handkerchief peeking from the pocket of a suit can perk up your appearance, as in 14-10. Some handkerchiefs are made of linen and are monogrammed. Some are made of silk. Some have border designs.

Jewelry

Fine jewelry is expensive. It is usually made of gold, silver, or platinum and may contain precious or semiprecious stones. It should be bought from a reputable jeweler and be of a simple style that will not become dated.

Bridge jewelry is made to look like fine jewelry, but it is less expensive. It is made of good metals and may have semiprecious gems.

Costume jewelry is less expensive. Sometimes it is plated with gold or silver. Sometimes it is made of plastic, shells, wood, or unusual materials. It can be fun, showy, inexpensive, trendy, and obviously fake.

Certain classic pieces of jewelry are wise choices for most people's budgets and don't seem to go out of style. These include simple chains, button or loop earrings, ID bracelets, medium-width rings, pearl necklaces, and circle bracelets.

Choose jewelry that is suitable to your personality and age. Some is dignified; some is striking or bold; some is youthful and fun. Like other accessories, you should scale jewelry to your size. Earrings should be scaled to the size of the face. Rings should be in scale with the hand. A tall person can wear longer necklaces. A heavy bracelet belongs on a large arm since it would make a thin arm look even skinnier.

Optical illusions can be created with jewelry. To make a slender neck look wider, wear a necklace of round beads around the throat. To slenderize a thick neck, choose a chain or necklace of medium length, so there is space between the chin and the jewelry. To widen a long, thin face, wear round earrings.

To be tasteful, the use of jewelry should not be overdone. Jewelry must be appropriate for the occasion to which it is worn. Brushed metal is dressy, especially if used with sparkling stones or pearls. Shiny metal is better for sport or casual wear. Also, the texture and weight of your jewelry should complement the texture and weight of your garment and its fabric.

Other jewelry accessories that come into fashion periodically, or for specific occasions, are lapel pins and cuff links. A *lapel pin* is worn in the buttonhole on the lapel of a suit. It often signifies membership in an organization or an

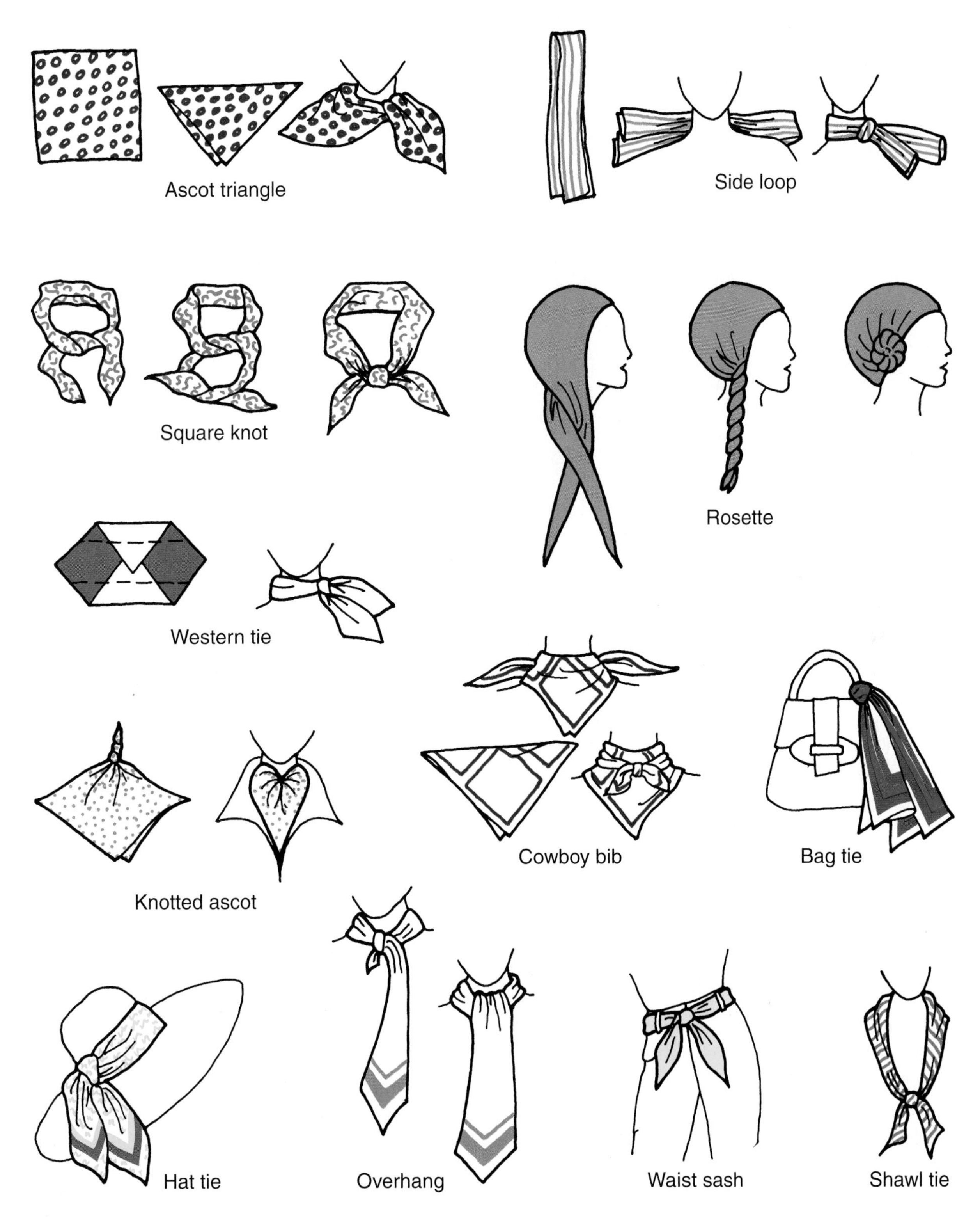

14-9 These are only a few of the ways that scarves can be tied and used.

The Fashion Association / Palm Beach

14-10 The red tie and handkerchief give this outfit a cheerful and stylish look.

award that has been received. *Cuff links* are worn in place of shirt buttons at the wrist. However, the shirt must have special (French) cuffs with buttonholes through which the cuff links can be inserted.

Eyewear

Eyeglasses should harmonize with your eyes, hair, and skin tone. Pick a shape that complements your facial contour and features and that suits your personality and lifestyle. Glasses should not be overly decorative. Eyeglass frames should not be wider than the widest part of your face. Choose a high, slender, or rounded bridge for a small nose. For a long nose, a low, thick, or straight bridge is best. Some people have more than one pair of glasses for insurance against breakage or loss, as well as for fashion interest.

Sunglasses are not just helpful against glare. They are also fashion items, as shown in 14-11. They can be more extreme, large, and fun than regular eyeglasses.

Gloves

Gloves are sometimes a high fashion item. For practical purposes, they are needed for warmth in cold climates. Well-fitting basic gloves of cloth or leather with little or no trim can be worn with almost anything. Extra gloves can be fancy or bright. Gloves with leather or vinyl palms are best for driving a vehicle.

Hosiery

Hosiery includes panty hose, tights, knee-highs, leg warmers, and all other stockings or socks. Stockings of various styles, fibers, and colors are available at different qualities and prices. Most are stretchy for good fit.

The texture and color of a woman's hosiery should be related to her shoes and natural skin color. Darker shades make legs look slimmer. Lighter shades and bright colors add size to the legs. Thick stockings go with low-heeled, sporty shoes. Sheer hose should be worn with dress shoes. Sheer "sandlefoot" hose should be worn with shoes that have open toes or heels.

Socks for men should always be of a dark color for dress. For sports, they are usually white. Argyle, striped, and decorated socks gain fashion popularity periodically.

Spiegel; Giorgio Lari, Photographer

14-11 Sunglasses are fun accessories that allow people to make fashion statements.

Develop a Total Future Plan

Once you have analyzed your current wardrobe, you should know what types of clothes help you look and feel your best. You should be aware of your personal needs and wants. You should also have some ideas about how to accessorize your apparel. The next step is for you to develop a future wardrobe plan.

Check with the right-hand column ("Action to Take") of your wardrobe inventory chart. You will want to build up weak areas of your wardrobe. Make sure to consider any future changes that are likely to affect your lifestyle. If you will soon go to college or begin a job, start planning ahead for your new wardrobe needs. As you develop your wardrobe plan, consider the following points.

Your Basic Garments

Your ***basic apparel*** provides the core of your wardrobe. These garments are the ones that are worn most often. They should be classic garments of top quality. They should not have faddish details or extreme silhouettes. Well-tailored, traditional styling is best for versatility and long-lasting fashion life. They should be stylish items that have general appeal. Their look can be dressed up or down with a variety of different accessories and by combining them with other garments. The leather jacket in 14-12 is a basic garment.

When developing a plan for your wardrobe, first decide what your basic apparel items are. A good wardrobe revolves around these few essential garments. They should be of the highest quality you can afford. They should be of good fabric. They should have simple structural lines without many frills or much applied trim. The simpler the design of a basic garment is, the more you can do with it.

Having several good garments, such as jackets, slacks, or dresses that will last a long time without going out of style, is called ***investment dressing***. Basic, good quality garments are good investments. They can be worn for many years in various outfits and with different accessories. Investment dressing is especially smart in the business world where fashion changes are slower and more subtle.

The traditional, classic styles of your basic apparel will stay fashionable for many seasons. You can wear them for many years and to many different occasions. You only have to adjust the hem and change the accessories over the years. An inexpensive garment of poor quality would lose its shape after several wearings. It would soon look cheap and would have to be replaced. It would cost you more money in the long run. Skimping on the basic apparel items is a poor clothing investment. Spend more of your clothing dollars for styles that will last.

Your basic, more expensive apparel items should be in neutral hues or low values of your best colors. Bright colors would probably be too noticeable, so the garments might lose their fashion interest. Also, you might tire of them quickly. With a planned color scheme, only a few accessories and pairs of shoes are needed, since they will go with many outfits.

Saxony Sportswear Company

14-12 This leather jacket is a basic garment that can be coordinated with many different shirts, sweaters, slacks, and jeans to create different fashion looks. It should remain stylish for a long time.

Extending Your Wardrobe

Once you have decided on your basic apparel, consider the less expensive garments and accessories that can expand your wardrobe. These are your ***extenders*** or "multipliers." Plan for versatile pieces that you can mix and match to make more combinations of outfits from the clothes you own. An example is pictured in 14-13.

Extenders add individuality and flair to your wardrobe. If extenders are coordinated, they can be combined in many fashionable ways.

Extenders might include slacks of an unusual texture, a shirt with interesting trim, a patterned or quilted vest, or a brightly colored turtleneck. They can make an outfit look Western, preppy, athletic, or whatever is "in" at the moment. If each one can be coordinated with several items in your wardrobe, your number of total outfit combinations multiplies.

You will be able to combine more wardrobe items if many of your extenders are solid-colored garments rather than stripes, plaids, or figured designs. A minimum number of shirts, sweaters, pants, skirts, and vests in coordinated colors can be mixed into many outfits.

If your budget permits, you can perk up your wardrobe with the latest accessory or fad. Such accents make your wardrobe interesting and even exciting! By adapting new trends to your wardrobe, you'll be able to look up-to-date while achieving an individual style of dressing.

Plan for *seasonless clothes* whenever it is possible. They can be lightweight woolens, knits, or good quality corduroys. They can be worn during most of the year: fall, winter, and spring.

Appleseed's

14-13 This scarf and sweater can be worn together or separately with various skirts, slacks, and jackets to extend basic garments into many different outfits.

Write Your Plan

Now you can write your ***wardrobe plan*** for the coming seasons and years. This is a "blueprint" of action to be taken to update or complete your wardrobe in the best way. It might change as you consider the resources available to achieve it, but it will guide you in your purchasing. Don't plan garment additions one at a time, even if you will buy them that way. Plan for several garments that can pull together your existing wardrobe. Identify gaps in your inventory. Then include only clothes and accessories that will fill those gaps. Remember, a few carefully chosen additions can make a big change by creating many new outfits.

Make a wardrobe planning chart like the one shown in 14-14. It has been partially filled in for use as an example. To complete your own chart, first list your specific activities. Then for each of your activities, select wardrobe inventory items that can be worn and combined with others. Write your present apparel (from your inventory chart) under the proper categories with a black pen or a pencil.

In the "needs" column of the wardrobe planning chart, write major future expenditures in red. These are such items as shoes, coats, suits, and other basic apparel that should be of good quality and might be expensive. The purchases of these can be spread over a period of time. Then list the extenders that you need. Write them in blue pencil or pen in the "needs" column. They should coordinate with the basics to complete outfits for your activities. List occasionally used items only if it is very important to have them. Your "needs" also include replacement of favorite items that are worn out.

In your wardrobe planning chart, put a number next to each "need" to show its

Wardrobe Planning Chart

Season and year: ________ **Basic color schemes:** ________

Activities	Basic Apparel	Extenders	Accessories	Needs & Priorities
School	2 Pr. cords 1 Pr. nice jeans 2 Crewneck sweaters	2 Striped shirts 2 Knit shirts	Sneakers Cowboy boots	1 Pair jeans (3) 1 Casual shirt (8)
Job: Part-time salesclerk in clothing store	Same 2 cords Same 2 crew-neck sweaters	Turtleneck Same 2 striped shirts 1 Blue shirt	Leather belt Billfold	Loafers (2) Dress slacks (5) 2 Pr. dark socks (7)
Religious and special events	Navy blazer Same 2 pair cords	Same turtleneck Same blue shirt 2 neckties		White shirt (9) Gray suit (10)
Social activities: Movies	Same 2 pair Cords	Same 2 knit shirts Same 2 striped shirts	Cowboy boots I.D. bracelet	Same as for school + job
Sports activities: Swimming	2 Swimsuits 1 Sweat suit	3 T-shirts		
Leisure: Bike riding Shooting Baskets	2 Pr. old jeans	Same 3 T-shirts Old sweatshirt	Sneakers	Sneakers (1) 3 Pr. white socks (4)
Lounging and sleeping		2 Old sweat suits Old jogging shorts		
Outer clothing	Tan trench coat Down ski jacket		Knit hat Leather gloves	
Other: Prom				Rent formal wear when needed Could use some new underwear (6)

14-14 Your wardrobe planning chart should include the basic apparel, extenders, and accessories you already own as well as items you need to fill in any gaps.

priority. As you set priorities, you decide which additions are most important. Then you know which items to buy first.

Your wardrobe plan is like a blueprint to build a house. It is rewarding to follow the plan to reach your goal. If you plan well, your new purchases will enhance the clothes you already own. It may take several seasons to finally get your wardrobe compact and well-coordinated. The result will be well worth the effort!

Consider Your Resources

Your available ***resources***, or the money, time, and skills you have to make wardrobe improvements, should be evaluated. You must combine your wardrobe plan with a sound money management plan, so you know what you can afford.

If you don't have your own source of income, the needs of others in your family must be considered. You must take into account your family's size, income, other expenses, and attitudes about clothing. Clothing is only one of many needs and wants. Housing, food, medical care, transportation, and recreational activities cost money, too. An unexpected emergency or new responsibilities may also limit the amount of money and time available to acquire clothes.

Most people cannot afford to have all the clothes they want. You will want to save toward your planned purchases according to the priorities noted in your wardrobe planning chart.

Small clothing expenses are continuous. They are for clothing upkeep, such as dry cleaning, and for purchases of low cost extenders and accessories. Large clothing expenses occur only periodically. They are for higher cost basic apparel items. Try to plan your resources accordingly. You may want to write a spending plan in order to stay on track to reach your goals.

The total amount you are able to spend on wardrobe items will probably have more effect on the quality of the garments than on the styles you buy. Also, your budget and where you shop will affect how fast you make the additions to your wardrobe.

The philosophy about quantity versus quality of clothes is a personal judgment. Young people tend to favor quantity because they like to have a variety of the latest fashions. This may be a good strategy for quickly growing teens. However, as people get older and stop growing, they seem to enjoy having fewer clothes of better quality. Your own buying decisions will depend on your situation, guided by your wardrobe plan.

Sewing as an Option

Sewing is one of the best and most economical ways to build up a wardrobe. If you have sewing skills, using them as a resource is a super way to stretch a fashion budget, as in 14-15. Sewing allows you to get more clothes for your money.

The new sewing machines have a great variety of useful features. Patterns and fabrics are available in most areas. Sewing lessons can be taken. If you develop and use sewing skills, you can make clothes for less than half the cost of buying them ready-made. You will also be able to select the colors and styles that are best for you, achieve a good fit, and include creative expression in your apparel.

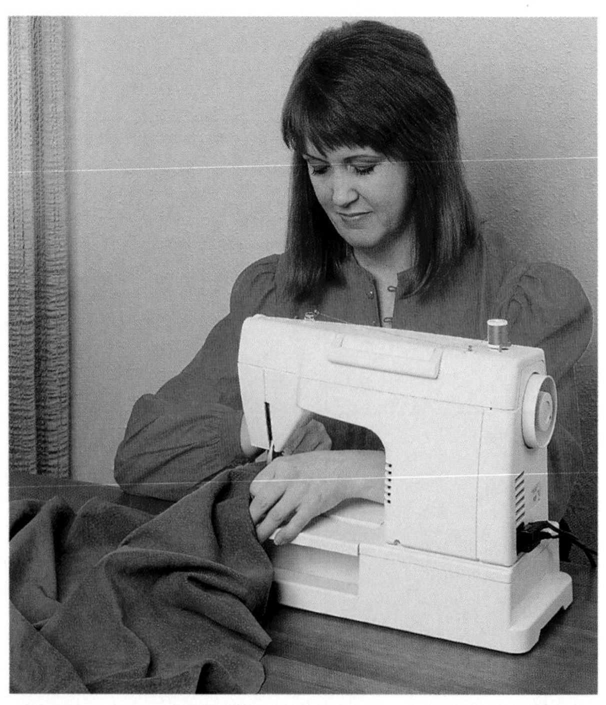

Tandy Leather Company

14-15 Sewing skills are valuable resources that can help people build and maintain their wardrobes.

Chapter Review

Summary

Developing a well-planned wardrobe takes time, effort, knowledge, and skill. Start by taking a wardrobe inventory to evaluate the apparel you already have. Group your clothing, especially noting why you like the clothing you wear a lot. Analyze the clothing you wear occasionally as to what modifications or combinations are needed to wear them more. Take action to do clothing repairs or cleaning. Properly dispose of clothing not worn in the past year. Also, distinguish your wants from needs so you can prioritize wardrobe additions.

Choose accessories that are appropriate for your outfits and for the occasions to which they are worn. Consider proper footwear, hosiery, handbags, belts, headwear, and gloves. Express yourself through scarves, neckties, and decorative handkerchiefs. Evaluate your needs for fine, bridge, and costume jewelry. Also consider harmonious eyewear.

The next step is to develop a future wardrobe plan. The basic apparel for the core of your wardrobe should be top quality, classic garments. Investment dressing and seasonless clothes are often wise choices. Use less expensive garments and accessories to extend your wardrobe with many combinations. Write down your wardrobe plan as a blueprint to build the best wardrobe over time.

Finally, evaluate your resources to know the money, time, and skills you have to make wardrobe improvements. Think about quality versus quantity, and consider sewing some of your clothes.

Fashion in Review

Write your responses on a separate sheet of paper.

Short Answer: Write the correct response to each of the following items.

1. It is a good idea to analyze your wardrobe by taking a wardrobe inventory _____.
   A. at least twice each season
   B. every three to five years
   C. once or twice a year
   D. None of the above.
2. Describe two courses of action to take concerning the clothes in your wardrobe that you wear only occasionally.
3. Name three useful ways of disposing of clothes that you never wear.
4. Give two examples of choosing accessories that are in proportion to a person's body.
5. Name two ways people can use accessories to create optical illusions to improve their appearance.
6. Distinguish between fine jewelry and costume jewelry.
7. Describe the kinds of garments that people should choose as their basic apparel, or as the core of their wardrobes.
8. Why is investment dressing especially smart in the business world?
9. Give three examples of wardrobe extenders.
10. List three advantages of having sewing skills.
11. Your wardrobe should _____.
    A. be as large as you can afford
    B. be in a continuous slow state of change
    C. include lots of shoes, since footwear is important
    D. be just like your best friend's wardrobe
12. Clothing expenses _____.
    A. are continuous for upkeep and for low cost extenders and accessories
    B. are only periodic for expensive basic items
    C. should be part of a sound money management plan
    D. All of the above.

Fashion in Action

1. Invite a wardrobe consultant to speak to your class on the topic of wardrobe planning. Be prepared to ask questions.

2. Plan two wardrobes with basic items, extenders, and accessories. Plan one for a high school or college student and one for a young business person.
3. Make a list of organizations in your community that accept donations of used, but wearable, clothing. Also list resale or "thrift" shops for clothing. Use these lists as part of an informative bulletin board display on wardrobe planning.
4. Discuss the latest fads for your age group. What is the approximate cost of each item? How long do you think the fads will stay popular? Do you think any will become lasting fashions? Explain your views.
5. Bring some scarves and neckties to class. Demonstrate various ways of tying and wearing them.

15 Being a Smart Shopper

Fashion Terms

fashion conscious
money conscious
impulse buying
sales resistance
shoplifting
hangtags
labels
packaging
Textile Fiber Products Identification Act
Permanent Care Labeling Rule
Wool Products Labeling Act
Fur Products Labeling Act
Flammable Fabrics Act
standards

After studying this chapter, you will be able to

- plan ahead as a well-informed shopper.
- decide where and when to shop.
- evaluate the information on hangtags, labels, and packaging.
- name and describe five laws related to apparel.
- identify the rights and responsibilities of consumers.

It is worth your while to become a competent and well-informed consumer. Making smart clothing buys is a skill you can learn. If you develop good shopping habits, you will be better dressed for less money. You will make intelligent buying decisions and spend your clothes money carefully. You will go past the stage of the beginner who picks apparel on the basis of "I like it; I'll buy it!" No matter how attractive an item of clothing is and how much you like it, it has no value for you if it doesn't fit your body shape, coloring, lifestyle, and wardrobe. Your shopping mistakes will dwindle in number as you gather knowledge and experience.

To improve your shopping techniques, you must learn how to plan your purchases and shop wisely. Informed people shop with a purpose. They think about their needs in advance and evaluate what is available. After they buy, they get great satisfaction from their purchases.

Clothes consume a sizable portion of your budget. Because you are going to buy them anyway, why not make the best choices? With good shopping skills, you can buy much more merchandise than an unskilled buyer with the same amount of money. Wise consumers use their resources to the best advantage. They make the decision to buy when they find the best item at the best price.

Prepare Ahead

Apparel is purchased not only as a necessity but also as a popular form of luxury. To enjoy the luxury of a super wardrobe, you must know how to build it intelligently. That starts with good planning before going out shopping. When you select and buy apparel wisely, you are more likely to be happy with your purchases.

If you have prepared a wardrobe planning chart, as described in the previous chapter, you are aware of what apparel items you need. You should plan to buy only those items in the right colors, textures, and styles for you. Then you can be confident that your purchases will fit into your wardrobe and will be put to good use.

Planning what to buy ahead of time is just as important as doing the actual purchasing. Shopping is a decision-making process. Make a list and follow it. Gather information and use it. With practice, your shopping should become easier and more enjoyable. A little time spent planning before shopping will save lots of time and mistakes later.

Make a List

On a piece of paper, list the clothing items and accessories that are your priority needs. Be specific about the color and style of each item. This will become your shopping list after it has been reviewed further.

Estimate the cost of each item on your list of apparel needs. Study catalogs and read advertisements to get an idea of the costs. Write down the approximate purchase price next to each item. Will there be extra costs for alterations, unusual upkeep, or credit? If an item needs to be dry-cleaned, how much will that cost over the time you expect to wear the garment? The cost of transportation and parking to buy the apparel might also have to be considered. A sample list is shown in 15-1 to help you get started.

Decide how much money you have to spend on your clothing needs. Then try to spend your money wisely and tastefully. Plan ahead so expensive items do not have to be purchased all at once. You may discover that to attain your goals, you have to delay some purchases. It might pay to make some of the items. But even if you sew well, you must know how to buy fabrics and ready-to-wear items intelligently.

You may have to assign different priorities to your clothing needs if your budget can't handle certain costs at this time. What garments do you need right away? Which purchases can be delayed? Are there some that can be omitted? Doing without unimportant items means having more resources for important items or for planned purchases in the future.

Try to put sufficient money into clothes and accessories that will take hard wear and will be worn for several seasons. Consider how many times you will wear each item. Think before you spend a lot on a party outfit that you will wear only once or twice. Such a purchase may force you to skimp on an outfit for school or work that you would wear often. Also, estimate how long the style of each garment will last. Does it have built-in fashion obsolescence or fad qualities that will soon date it? Try to recognize what fashions will probably be in style for several more seasons.

Plan to invest in seasonless clothes when you can. For instance, an all-weather coat with a zip-out lining is versatile. A lightweight wool suit or blazer can be worn during most seasons. Items like these will help you stay within your budget in the future. Do your plans include basic styles that you can wear to many places? Do the items mix and match well with what you already have? Buying clothes that will not be worn is a waste of money.

Sample Shopping List

Needs	Specifics	Estimated Costs	Where Available
Athletic shoes	light-colored all-purpose	$50	Mall
Loafers	brown leather good quality	$58 + credit costs	Bud's Shoes, Main Street
Blue jeans	tapered legs	$30	Mall
Dress slacks	classic lines, gray	$36 + future dry cleaning	put off until next month?

15-1 Your shopping list should include enough information to allow you to plan your expenditures carefully.

When planning future purchases, wise consumers evaluate past purchases in terms of price, amount of care required and wearing qualities. They consider pros and cons of brands they have previously bought. They also consider how pleased they have been with various stores they have patronized.

Gather Information

In order to plan ahead, you must be well informed. This prevents you from being gullible or suspicious while shopping. It makes you confident and reduces shopping errors. Doing preliminary "armchair shopping" homework saves time and travel costs later. The more you know before you go out to buy, the better your purchasing decisions will be.

To gather information, read fashion magazines, newspapers, and mail-order catalogs to learn about new silhouettes, colors, and fabrics. Fashion photographs illustrate how garments and accessories are being combined into outfits. Feature articles often give pointers on clothing selection in relation to the use and care of particular apparel items. Fashion writers tell what is coming months before the clothes are shipped to stores. Knowing a little about what is ahead will enable you to adapt your current wardrobe more effectively and to plan future additions.

Browse through departments of stores you like. Study the windows, interior displays, and mannequins of the stores, as in 15-2. See how new outfits are being "assembled." Notice how accessories are being used. Evaluate the new looks and decide if they suit your personality, body build, and lifestyle. Try to distinguish classic, lasting fashions from passing fads. Although "window shopping" is sometimes done as a leisure pastime, try not to be tempted so it causes unplanned buying. Leave your cash, credit cards, and checkbook at home if you think you will be tempted to buy when you shouldn't.

Study catalogs and advertisements to find out the cost of items and where they are available. There are many attractive clothing catalogs from stores and mail-order establishments. Look through them to get ideas of styles as well as prices. They have written descriptions and illustrations to show the features of garments and accessories. They also have information about sizes, fabrics, and care of garments.

15-2 Store displays can give you interesting ideas about combining items into coordinated outfits.

Take advantage of consumer information resources. Note what well-dressed people are wearing on television and in your community. Friends or relatives, who have had good or bad experiences with clothing purchases, may also offer advice.

Your school or town library has books and magazines to help you plan your shopping. Consumer reports are available on a wide range of consumer goods, including some types of clothing. The telephone book can help you locate where certain items are available, but it does not evaluate the retail establishments or the quality of merchandise. A large ad in the Yellow Pages indicates that a business has paid the fee for that amount of space in the phone book. Its products or prices are not necessarily better than those of a store with a tiny listing.

Consumer aids also offer information. They are distributed by fiber producers, pattern companies, fabric stores, and trade group associations. Educational pamphlets, booklets, and fact sheets describe the properties of textile products and give directions for their use and care. Filmstrips, films, and videos are also available to schools and groups. Most of these consumer aids are free.

Some help with clothing can be obtained from community classes and workshops. More help can be received by calling your county or state Cooperative Extension Service. As well as providing personal assistance on specific problems, extension agents might have useful printed information.

Evaluate Advertising

A good source of up-to-date shopping information is found in advertisements. Advertising has two purposes. The primary purpose is to sell. The secondary purpose is to inform. Advertising can be valuable to consumers if it is used with this understanding.

Retailers and manufacturers spend millions of dollars each year to persuade people to buy their products, whether they need them or not. They use printed media, radio, television, the Internet, displays, demonstrations, fashion shows, and billboards. They catch your attention with clever slogans, songs, and phrases. They describe merchandise in glowing, sometimes exaggerated, terms. They play upon the emotions of consumers. They imply that if you wear what the fashion model is wearing, you will look like the model! Advertisements are meant to entice you into buying by making you want the items shown and described.

Advertisements can mislead consumers who have poor planning and shopping habits. Ads can cause them to buy items they don't need. However, advertising can also provide a great deal of useful information.

You must learn to recognize the helpful information in advertisements. Learn to read "between the lines" and notice the small print. Look for ads that present complete views and descriptions of garments rather than just exciting, glamorous ones. Pay attention to information about fibers, fabrics, and garment care.

Advertisements allow you to compare similar merchandise from different manufacturers or retail stores. They tell you about special sales. They give you lots of needed information about what is fashionable and where it is available before you go shopping.

Reputable media companies will not accept advertisements they know are false. However, information may be exaggerated. You must be careful to judge the content of advertisements. Use what you need to prepare for making your purchases.

Where to Shop

Decide what stores, shopping centers, or malls you will visit before you leave home. Map out a route for the most efficient use of your time and travel. This eliminates wasteful backtracking.

Plan to make your large, basic purchases first, as in 15-3, while you are still fresh and alert. After that, if you still have available time and money, you can shop for smaller items.

You may decide to shop from different sources for different items. There are advantages and disadvantages to shopping at each

15-3 Purchases of major items, such as sport coats and slacks, should be made before smaller items, such as neckties, are bought.

type of apparel outlet. You may want to refer to Chapter 7 and review the types of retail establishments. Then decide for yourself what features are important to you.

Price Versus Quality and Services

You may like the personal attention and pleasant atmosphere of a small shop. Some people are willing to pay more to shop in stores that are known for quality merchandise and service. They may want to charge their purchases or perhaps have them gift wrapped. They like the privilege of getting their money back if they return items. These factors are worth the higher prices to them.

On the other hand, the economical prices of discount stores and manufacturers' outlets may be what you want. Stores that order goods in volume offer good buys on basic necessities. Some merchandise may have well-known brand names. However, such stores often have crowds of people, long checkout lines, and sparse interiors.

In general, the stores that offer the least amount of personal service have the lowest prices. However, the quality of the apparel may be low, too. You must rely on your own consumer knowledge and awareness to find the right merchandise to fill your needs. Evaluate your expectations of service and quality in relation to what you are willing to pay. Where to shop and how much to spend are your own decisions.

Used clothing may be available at low prices at clothing exchanges, fairs, yard or garage sales, and thrift shops, as in 15-4. These sources are sometimes good for seldom worn party clothes and for children's apparel. You may turn up some interesting "finds" that you can use to create fashionable outfits!

Types of Merchandise

The specific items you are seeking will influence your choice of retailers. For unusual merchandise, you may need to shop in a boutique. Items in a certain category may be stocked only at a specialty shop. If you are shopping for several types of apparel at one time, you may choose to go to a big department store or to several different stores.

15-4 One way to stretch a clothing budget is to shop at resale stores for used apparel.

Location and Store Hours

You may base your decision of where to shop on the location of stores. It may not pay to travel a long distance just to save a little money on a few inexpensive items. However, if you plan to buy several expensive items, it may be worth your while to travel to specific stores that will give you better selections and prices.

If you go to a department store or mall, you may not have to travel to more than that one location. It may offer "one-stop shopping" with a wide selection of all the items on your list. See 15-5. Also, convenient public transportation or free parking may be available.

If your shopping time is limited, the hours stores are open will be important to you. If you work afternoons and evenings, you may favor a store that opens at 9:30 a.m. rather than at 11:00 a.m.

It is not convenient for some people to go out to shop at all. If that is the case, or if you do not like crowds or dealing with salespeople, you may decide to do mail-order or Internet shopping from your home.

Mall of America

15-5 If your shopping list contains many different items, a large mall such as this one might be the best place to find all of them.

Mail-order and computer shopping is easiest for people who do not have size or fit problems. There is no opportunity to try on or inspect merchandise before buying. Also, you have to wait several days or weeks for the items to come. There are usually shipping charges involved, and someone may have to sign for the items when they arrive. The items can almost always be returned, but some effort and costs are involved in repackaging and sending them back to retailers.

Advertising or Loyalty

Sometimes you will shop at the store that has caught your eye with its advertising. The store may be featuring an item you happen to need, or its advertised prices may be better than those at other stores.

Some people prefer to always patronize the same store because of personal loyalty. A store may have the reputation of providing fair policies, truthful advertising, and extra services. This is very important to some people. The store employees may get to know their loyal customers and cater to their needs and tastes. Also, the customers may have charge accounts at that store, which can simplify their shopping.

Your Decision

Reports from retailers indicate that most customers are ***fashion conscious***. They are aware of, and want, what is new. They usually look in many stores and are loyal to their own tastes rather than to any one retailer. They want to express themselves and gain approval from their peers. When they find what they want, price may not be important.

The wisest shoppers are not only fashion conscious, but also ***money conscious***. They stay aware of how much each item costs and of their monetary resources. They go where they can get the best buys and service for their money. They find the stores that specialize in the kinds of merchandise they need. They know where the buyers' tastes are the same as theirs, so the merchandise will be to their liking.

Consumers, as a group, have "voting power" with their money and purchase choices. Consumer acceptance or rejection of fashion innovations influences the cost, quality, and availability of those specific goods. The consumers' patronage at certain stores gives a vote of confidence to the way those retailers run their businesses. The profits of manufacturers and retailers dwindle if consumers do not purchase their items.

When to Shop

You may want to shop at different times of the season for different items. Fashion timing can be used in a practical way, as in 15-6.

For a wide choice of styles, colors, and sizes, shop early in the season. The new fashions will have just come into the stores. There will be a large selection of new clothes, which is especially helpful if you are hard to fit. Buying early also allows maximum wearing time while the garments are in style and before the season ends. However, clothes are usually most expensive early in the season.

For lower prices, shop late in the season. There will be a smaller selection, but the items may be on sale to clear the stores for new merchandise of the next season.

Shop ahead of when you will need an item. Doing so will give you time to watch for sales and to shop intelligently. Try not to buy emotionally or under pressure. If you have to shop at the last minute, you will make a hurried purchase. You may be forced to take something in the wrong color or style. If you need something to wear that evening, you will

JC Penney

15-6 Shopping early in the season assured this man of a large selection of the latest fashions in outdoor wear, to be worn throughout the fall and winter.

take whatever you can find, even though it may not be what you really want.

Shop well in advance of a holiday to get the best service and selection. Try to avoid shopping just before a store's closing time. If you are tired or rushed, you are less likely to make smart shopping decisions.

Stores are least crowded during bad weather. If you are hardy, that is a good time to shop so you can browse through the merchandise and get help from salespeople. Other good times are during the morning hours and the evening dinnertime. In late morning, the clerks have put out the new merchandise. They have marked sale items, and the lunch crowd has not come yet. Many stores are almost empty from 6:00 to 7:00 p.m. They are busiest on weekends and near holidays. Lunchtime and just after office hours are busy times in business areas.

The best time to shop for shoes is after walking awhile or in the afternoon. This allows for a good fit over normal foot expansion. Your feet are a little bigger after you have been on them, especially in the summer.

The Actual Shopping

When you want to do serious shopping, it may be best to do it alone. It can be more efficient, especially if you know what you want to buy. However, if you feel an objective, critical opinion is necessary, take a knowledgeable relative or close friend with you. This should be someone whose advice you respect. Salespeople may not be reliable or frank with their help. They will often advise you to buy something that is wrong for you since their job is to sell merchandise.

Stay within your list of preplanned priority shopping needs. Control ***impulse buying***. This is sudden and not carefully thought-out purchasing. Reevaluate the buying of fads or sale items if they aren't needed. Don't be talked into or tempted into making unplanned purchases or ones that you know aren't right for you. In other words, develop ***sales resistance*** or shopping self-control. You have a right to browse and try on garments. You are entitled to shop without being pressured to buy. Salespeople are paid to help you. You do not owe them a purchase. Have confidence to make only the right decisions. If in doubt, don't buy!

Stick to the styles that flatter your body build. Consider whether the lines and colors of garments are becoming to you and go with other apparel you own. Don't guess when matching colors. Carry fabric and color swatches with you when shopping. You can snip small fabric pieces from the seams or hems of garments at home. You can put the swatches on safety pins or glue them onto cards. They will help you coordinate your new purchases. If you were to take your actual garments along, they would be bulky to carry. Also, they might get soiled or lost.

At the store, try to match colors under natural light at a window or doorway if possible. Store lighting can make colors look different than they are in natural light.

Be well-groomed and nicely dressed while shopping. Your neat appearance will gain you the respect of salespeople. Also, wear clothes that are easy to put on and take off. Then you will be able to try on garments easily at the store. Wear or take along appropriate underclothes, accessories, and shoes for what you plan to buy. They should be the right type and color, if possible. For instance, you might take or wear brown dress slacks if they will go with a shirt or sweater you intend to buy. Then you will be able to judge how your purchase will really look.

Try on clothes and shoes to judge the fit. Walk around in shoes for several minutes to check their comfort. When trying on a garment, move in it the way you will when you wear it in the future. Walk, bend, sit, and move your arms to see if it has ample room for movement, as in 15-7. Is it comfortable without any pulling or wrinkling? Look at all views in a full-length mirror. Are the sleeves long enough? Is the hem even? Does the garment improve your appearance? Are alterations necessary? If so, will they change the appearance of the garment or cost a lot?

Burlington Industries, Inc.

15-7 This fun outfit is comfortable even when the wearer is bent down near the floor.

Shopping Manners

Use good shopping manners. By behaving properly, you will have more fun shopping, and the store personnel will be more helpful. Always be courteous and polite to salespeople and other shoppers. Wait your turn. State your needs clearly to salespeople. Then you will receive good service and attention. If you do not need help, say, "I'm just looking, thank you."

Being a responsible consumer requires you to handle merchandise carefully. Do not mistreat or damage items. Refold items after you have looked at them. Pick up merchandise that has fallen. When you try on clothes, make sure they don't get stained from your makeup or dirty from your shoes. Be careful not to tear openings or break fasteners. After you have tried on garments, put them back on the hangers as you originally found them.

Cooperate with store policies. Policies and rules are often for your safety and protection. Notice signs that are posted and abide by them. Some garments, such as swimsuits and intimate apparel, have public health requirements.

Some stores do not allow merchandise to be returned for your money back. Sometimes you will only be given exchange credit toward the purchase of other items in the store. Understand a store's return policies before purchasing goods there.

The lack of consideration of shoppers raises prices for everyone. If an item has been torn, soiled, or returned to the store after being worn, it is damaged merchandise. It cannot be sold at full price. In fact, the store may not be able to sell it at all. Any loss in the selling price of retail goods means an increase in operating expenses for the store. Prices of goods then go up for everyone.

The practice of ***shoplifting***, or stealing, also adds to the price of merchandise. When items for sale in a store are stolen, not only must the cost of those goods be recovered by the store, but security measures must be taken. Some stores hire security guards, and some buy special detection equipment and tags. All of these provisions create extra costs. The markup on merchandise must be increased to cover the actual losses as well as the added costs. You should report shoplifting if you see it taking place. It is a serious crime punishable by fine or imprisonment.

Hangtags, Labels, and Packaging

When you purchase clothing items, hangtags and labels are attached to them. They give information about products. This might include the size, price, special features, brand name, fiber content, finishes, manufacturer, and care instructions. Hangtags and labels exist to identify products, to help sell them, to help consumers make decisions about them, and to explain proper care.

Hangtags, labels, and packaging information give consumers important facts. They have been developed through joint efforts of the federal government, fiber producers, fabric and garment manufacturers, the clothing care industry, and consumer groups. As a consumer, you should read them carefully as you make decisions about buying apparel.

Hangtags

Hangtags are detachable "signs," as shown in 15-8. They are usually affixed to the outside of garments by strings, plastic bands, pins, staples, or adhesives. They are made of heavy paper or cardboard. They are often hung from buttons, buttonholes, zippers, belt loops, or underarm seams. They are removed before garments are worn.

In general, hangtags tell consumers what manufacturers want to say about their products. Much of that information is voluntary (not required by law). Hangtags are a form of advertising or promotion to help sell products. Information about performance features, such as special finishes, reinforced pockets, adjustable button cuffs, and reversibility, is usually provided on hangtags. Symbols and logos are often displayed to identify the designers, manufacturers, or sellers. Sometimes a certification or seal of approval tells of good test results from a lab. Guarantees may assure replacement or money refunded if items do not perform satisfactorily.

15-8 Manufacturers use hangtags to give promotional information about their products to consumers. Hangtags are removed before garments are worn.

Labels

Labels, as in 15-9, are permanently attached to garments on the inside where they do not show during wear. They are usually made of ribbon or cloth. They may be attached at the back neckline or waistline, on a facing, or at a sideseam. They may be any color or style as long as they do not ravel. They must give certain types of information. The information may be written on both sides of a label if attached to the garment so it can be turned over. Sometimes label information is stamped onto shirttails or collars with indelible ink, or glued or fused onto fabrics.

A label gives the fiber content of a garment and names its country of origin. It describes the care required to maintain the longest wear and best appearance of the garment. It also identifies the manufacturer or distributor. Most of the information on labels is required by law, as explained later in this chapter.

15-9 Labels are permanently attached to garments and provide specific types of information.

Packaging

Packaging is the covering, wrapper, or container in which some merchandise is placed. Usually, much of the information that is on a garment's inside label is repeated on the outside of the package. The sleeve length, waist, and hip measurements, or chest size may also be printed on the bag, box, or wrapper. There is usually other printed promotional information on the package, too. Sometimes there is additional printed material loose inside the package.

Government Legislation

Laws require that all textile products sold in the United States carry labels showing fiber content, the country of manufacture, the identifying number of the producer, and general care instructions, as shown in 15-10. The

Information Required on Sewn-In Labels	
Fiber content by percentage	The generic names of fibers that compose five percent or more of the item must be listed in order or predominance by weight. A fiber under five percent of the total weight must be listed if it affects the fabric characteristics.
Country of origin	The country where the item was assembled, including the United States, must be named. If the materials were imported, but the garment was assembled in the United States, this must be stated on the label.
Identity of producer	The registered number of the manufacturing plant must be given.
Permanent care requirements	Clear and complete instructions and warnings about care and maintenance of the item must be given.

15-10 Government legislation has specified what information must be included on apparel labels for the protection and benefit of consumers.

generic names of fibers must be given, since so many trade names now exist. Laws require manufacturers and retailers to attach permanent care and performance labels to almost all textile products. Mail-order catalog descriptions must also state the information.

Several specific laws have been passed and amended since the early 1960s to regulate the truthful labeling of clothing, other textile products, and furs. They encourage ethical practices to protect consumers against deceptive labeling and advertising. Most are enforced by the Federal Trade Commission (FTC). The FTC can bring complaints against violators of these trade practice rules. It seeks the cooperation of representatives of the textile industry, retail stores, apparel care businesses, and consumer groups. These people help in developing the rules and regulations that interpret the laws and tell how to apply them in practice.

The laws relating to textiles and apparel give advantages to consumers. They encourage better quality, product safety, reliable service, and honest advertising. They have added slightly to the costs of clothing production, but the benefits are worthwhile. Permanent labeling has become a part of garment manufacturing, because business, government, and consumers know it is necessary.

Textile Fiber Products Identification Act

The ***Textile Fiber Products Identification Act*** (which also specifies generic names for fibers) requires labels to tell what fibers are in textile products. Fibers must be listed in order of predominance by weight. Both natural and manufactured fibers are listed by their generic group names. Trade names may also be listed if manufacturers choose to do so.

The percentage of each fiber must be given. If a fiber makes up less than five percent of the total fiber weight, it can be listed as "other fiber or fibers" unless it affects the fabric characteristics. If it does affect the characteristics, then it must be listed by name and its contribution explained. An example would be "4% spandex for elasticity." Listing the composition of the fabric helps consumers know how the fabric will behave in use and care.

The name or registered number of the manufacturer or persons marketing or handling the product must be given. Registered numbers are on file with the FTC and usually appear as "RN" on labels. They are used to identify the "responsible parties."

If an item is imported, the country where it was assembled must be listed. This is required for wearing apparel, accessories, and textile products used in the home, such as draperies, upholstery fabrics, and linens. When not imported, garment labels sometimes say "MADE IN THE USA" as in 15-11.

Permanent Care Labeling Rule

The ***Permanent Care Labeling Rule*** requires manufacturers to attach clear and complete permanent care labels to domestic and imported apparel. Care labels must be durable to remain readable in garments. They must give

Burlington Industries, Inc.

15-11 The country of origin must appear on the labels of textile products. American-made items might have "MADE IN THE USA" labels.

clear and complete instructions for care and maintenance of the items. They must withstand wear and cleaning for the normal life of garments.

Care labels specify how to dry-clean, launder, dry, and iron apparel items. They give instructions regarding the use of soap, detergent, and bleach. They recommend the water temperature range (hot, warm, cold). They must be specific about hand or machine washing and the method of drying (tumble dry, drip dry, line dry, lie flat to dry, and so on).

Care labels also give specific warnings, such as avoiding bleaching, ironing, wringing, or twisting, when the appearance or performance of the item would be hurt. If the label does not say that you cannot do something, you can assume you can do it as long as you follow normal practice. "Dry-clean only" means the item should not be washed, even by hand. However, washable items can almost always be dry-cleaned.

If an item is packaged when sold, its care label must be positioned so it is easy to read through the package. If that is not possible, the same instructions must be easy to read on the package itself.

Yard goods must have care labels, too. The information, or a care instruction code from 1 to 9, should be on the end of each fabric bolt available to home sewers. Sometimes the fiber content is printed on the selvage. When fabric is purchased, you should be given a free care label by the salesclerk for each piece of cloth you buy. You may have to ask for it. The labels should be sewn into the garments made from the fabrics. See 15-12 to view the labels that are given with fabric purchases.

A few apparel items are not covered by the Permanent Care Labeling Rule. Those not needing care labels include some hosiery, headwear, handwear, and footwear, as well as fur goods. Items that are disposable or that need no maintenance are also exempt from the labels. Other exemptions are for items that are purely decorative or ornamental and for remnants (mill ends) that are cut and shipped by the manufacturer. The FTC may exempt by petition completely washable garments intended to retail for three dollars or less. They may also exempt apparel that would be substantially impaired by care labels, such as sheer or see-through items.

Method Code	Wording
△1	Machine wash, warm
△2	Machine wash, warm; line dry
△3	Machine wash, warm; tumble dry; remove promptly
△4	Machine wash, warm; delicate cycle; tumble dry, low; use cool iron
△5	Machine wash, warm; do not dry clean
△6	Hand wash separately; use cool iron
△7	Dry-clean only
△8	Dry-clean; pile fabric method only
△9	Wipe with damp cloth only
Note:	
△3	This applies to most wash-and-wear or permanent-press fabrics. To minimize ironing, remove from dryer immediately after it stops. If touch-up ironing is required, use cool iron.
△4 & △6	Certain synthetic fabrics, although rugged, should not be pressed with a hot iron as they may fuse.
△5	Dyes used on certain prints, although ordinarily colorfast to washing, do not perform well when dry-cleaned.
△7 & △8	Professional dry cleaning is always safest.

Developed by The National Retail Federation

15-12 Each piece of sewing fabric that is purchased should be accompanied by an appropriate care label.

Wool Products Labeling Act

The ***Wool Products Labeling Act*** defines the terms "wool" or "virgin wool" and "recycled wool" as described in Chapter 8. It also defines "lamb's wool" as wool from a lamb up to seven months old. The law protects producers and consumers from misbranding and false advertising.

The law also requires labels to specify the percentage of each type of wool in the fabric. If fibers other than wool constitute five percent or more of the product, they must be named and listed by percentage. If the wool is imported, the label must say so and give the country of origin. There is no provision in the law, however, requiring disclosure of the quality of the fibers.

Fabrics cannot be given trade names to imply they are made up of hair from a certain animal unless they really are. For instance, a trade name like "Lambsoft" implies that the yarns in a garment contain lamb's wool. The yarns may be soft, but if they do not contain lamb's wool, that would give a false impression and would be illegal.

Fur Products Labeling Act

The ***Fur Products Labeling Act*** protects consumers against deceptive information on labels and in advertisements of fur products. Fur is animal skin with hair, fleece, or fur fibers attached. The label must list the animal or animals that produced the fur. It must tell if the fur has been bleached, dyed, or otherwise artificially colored. It must tell if it is waste fur, such as from the ears, throats, paws, tails, or bellies of animals, or scrap pieces. It must also list the country of origin of imported furs. This should be done even if a garment merely has some fur trim attached to it.

Flammable Fabrics Act

The ***Flammable Fabrics Act*** has flammability, or burning, standards for fabrics in clothing. It is especially concerned with children's sleepwear. It also covers many household textile products, such as rugs, mattresses, and upholstery fabrics. It has caused manufacturers to develop new fibers and finishes to make more flame-resistant fabrics. It helps assure safe products for consumers.

Consumer Rights and Responsibilities

You have certain rights, or privileges, as a consumer. You should be able to have certain expectations from the goods and services you buy. You also have consumer responsibilities that go with the rights. You must make the effort to carry out these responsibilities when you do your shopping. You must act responsibly. Consumer rights and responsibilities go together.

There has been strong interest in consumer protection during the past few decades. The rights of consumers were defined by the federal government in the 1960s. They were described as the right to safety, the right to be informed, the right to choose, and the right to be heard.

Standards have been established in many categories. ***Standards*** are criteria set by authorities, such as those at testing labs or government agencies, who judge products. The standards verify certain levels of quality.

Right and Responsibility for Safety

Your right to safety is to expect textile products to perform in a safe way under normal use. The Flammable Fabrics Act is one law that provides for this by prohibiting the sale of highly flammable fabrics and wearing apparel. The government cooperates with public and private agencies to conduct research into the flammability of fabrics and textile products. Also, textile firms have been doing research to develop fibers, blends, and finishes that are more resistant to fire.

Most fabrics are somewhat flammable. A flammability standard is a measure of flame resistance. Special standards apply to children's sleepwear, underwear, and dresses. If a label says "flame resistant," it means the garment will not support flame once the source of the fire is eliminated. Other aspects of safety are concerned with fibers and finishes developed from chemicals that could irritate the skin or eyes. Some might give off unpleasant odors or activate allergies. Health irritants must be controlled under the consumer's right to safety.

Your responsibility is to use products safely. Guard against personal carelessness with fire and other hazards. Be cautious with matches and around ranges, fireplaces, and candles. Keep clothing, such as loose sleeves, and household textiles out of contact with these flame sources. Supervise children when they are near fire-producing products.

Right and Responsibility to Be Informed

Your right to be informed includes having access to the facts needed to make the best purchasing choices. Hangtags and labels give specific information about the products being considered for purchase. Advertisements and other sources also provide information.

It is your responsibility to seek out and understand the information. Be certain you understand the claims made for performance and wear. Try to use the information wisely to make good decisions when buying. The more you know, the better choices you can make for your purposes. You need to know the fibers, fabrics, finishes, and features of each garment. It is good to know where the product was made and who made it. You also must know what care is required to maintain the garment.

Right and Responsibility to Choose

Your right to choose enables you to look at merchandise in a broad range of styles, qualities, and prices. Advanced technology has created new standards of performance for textile products. You can suit your own taste and budget from what is available. It is also your right to shop for products and services without being pressured into buying.

Along with your right to choose, it is your responsibility to select carefully and wisely, as in 15-13. Keep desired performance characteristics in mind when you shop. Read hangtags, labels, and packaging to find out about special features. Consider how practical the garment is before you buy it. Save labels, hangtags, and sales slips after you buy. You will then know any special directions for care. Also, you will know brand names in case you want to buy more of the same product. Saved sales slips will provide the documentation for you to return or exchange items if needed.

Right and Responsibility to Be Heard

Finally, you have the right to be heard about consumer matters. You can expect the clothing you buy to meet certain standards of performance. Manufacturers make promises about their products on hangtags, on labels, and in advertisements. If the products do not live up to those promises, you have the right to complain. You may speak up and let your likes and dislikes be known.

15-13 Consumers have the right to choose from a wide range of products and the responsibility to select thoughtfully.

Most manufacturers test samples of their products in specially equipped quality control laboratories before putting them on the market. They want consumer satisfaction and repeat purchases that depend on the successful performance of their products. Merchandise is judged according to standards set from other products of the same type. Garments are judged according to their features and quality of construction.

Besides manufacturers, many large department stores and mail-order businesses have quality control laboratories with trained technicians. There are also independent, commercial labs that conduct tests of various products. Stores usually order merchandise with basic specifications required.

There are nonprofit consumer testing organizations that publish test results about many different products. They are supported by memberships and the sale of publications that give the results of their tests. Some testing is also done by service industry organizations, such as the International Fabricare Institute. Government consumer protection agencies include the National Bureau of Standards and the American Society for Testing and Materials. They work to improve standards and create new and better products for consumers. They run quality and durability tests on fabrics, and they determine the wearability of garments.

Manufacturers' products should live up to the stated or understood promises of their trademarks. There are actions you can take when a textile product fails to live up to its claims or what you have reasonably expected from it. However, if you have not fulfilled your responsibility of using it and caring for it properly, you cannot expect to return the item. You must be fair and honest when you complain or return items.

If you must return merchandise, do it as soon as feasible. Return clothes clean and unworn, if possible, and in their original packaging. Have your sales slip as proof of date of purchase and price paid. Return garments to the salesclerk, customer service representative, department head, or store manager. Explain why the merchandise is defective or unsatisfactory. Be pleasant and businesslike. Describe the problem in a specific and factual way. Have hangtags or printed matter with you that support your viewpoint. Manufacturers, retailers, and cleaning establishments are usually reasonable about such claims. You should expect to receive a refund, a charge credit, or an exchange for another item that is satisfactory.

It is your responsibility to let legitimate dissatisfaction be known. If a product does not live up to its claims or to your expectations, a complaint may prevent the problem from happening again. The store will inform the manufacturer if it is at fault. When buyers and sellers work together, products can be improved. By being aware of a product's shortcomings, corrections can be made in the merchandise to please consumers in the future. Your com plaints and those of other responsible consumers help to bring about improvements in textile and clothing performance.

There may be a time when you do not receive satisfaction from a store and still feel that your complaint is reasonable. In that case, send a letter to the retail store's president or person responsible for consumer complaints, or to the manufacturer of the item. Write your letter carefully, stating exactly what is wrong. Be businesslike and pleasantly firm, as in 15-14. Type or write legibly, and keep a photocopy of the letter. Such a letter may not get to the specific individual to whom it is addressed, but chances are good that your complaint will get attention. If you receive no response, write another letter in a month.

For serious consumer grievances, seek guidance from your local Better Business Bureau. You may also contact your local Chamber of Commerce, the state attorney general's office, or a state agency dealing with consumer affairs. Another option is the Bureau of Consumer Protection, which is one of the branches of the Federal Trade Commission.

As a responsible consumer, you should report false or misleading advertising. Do not patronize dishonest businesses. Spend your consumer dollars at trustworthy establishments. Reputable manufacturers and retailers want satisfied customers. They deserve repeat business. Through continued purchases, refusals to purchase, or justifiable complaints, consumers can apply pressure on retailers. They, in turn, influence the textile and apparel manufacturing segments of the industry.

902 Loden Grove Road
West Haven, CA 90169
June 4, 20xx

Mr. John Doe, President
Smedley Group Retailers
3100 Washington Avenue
Los Angeles, CA 90075

Dear Mr. Doe:

Please act on this complaint as soon as possible.

On May 1, 20xx I purchased a cotton sweater from your branch store in the West Haven Mall. The sweater is white with a red checkerboard design. The label states that it is part of the "Doberman Collection" by Parker Fashions. A photocopy of my sales slip is enclosed.

After wearing the sweater for a short time, I washed it by hand in cold water according to the directions on the label. After rolling it in a towel, I laid it flat to dry as stated.

However, by the time it was dry, color from the red areas had run into the white areas. I can no longer wear the sweater because it looks so awful.

When I tried to return the sweater to the store, they said they could not take it back because it had been worn. They also implied that I had ruined it by washing it improperly. As you can imagine, I am very upset!

Your attention to this problem will be greatly appreciated. I will await your reply.

Sincerely,

Susan A. Smith

Susan A. Smith

(Telephone: 213-555-2489)

15-14 A letter similar to this one should bring fair results when you have a legitimate consumer complaint.

Chapter Review

Summary

Smart shoppers are better dressed for less money. Preparing ahead to buy only needed items in the right colors, textures, and styles is just as important as the actual purchasing. Make a specific list of your priority needs, including cost estimates. Gather fashion information from magazines, newspapers, catalogs, and stores. Try to understand advertisements without being misled about merchandise or prices.

Decide where you will shop before leaving home. Plan to make your large, basic purchases first. Evaluate low-priced stores versus the quality and services of higher-priced retailers. Also, the specific items you are seeking will influence where you shop. Know locations and store hours of retailers you plan to visit. You may choose to be loyal to a store you trust. Your choice of retailers will let you be both fashion conscious and money conscious.

You may decide to shop early in the season for a wide choice of merchandise and maximum wearing time, or late in the season for lower prices. When shopping, try to have sales resistance and control impulse buying. Stick to your planned purchases, be well-groomed, and try on items to judge fit. Use good shopping manners and cooperate with store policies.

Hangtags, labels, and outside packaging identify and give information about products. Hangtags tell consumers what manufacturers want to say about their products. Labels give information required by law. Packaging might repeat information that is on the inside, such as size.

U.S. government legislation requires that textile product labels contain certain information. The Textile Fiber Products Identification Act, the Permanent Care Labeling Rule, the Wool Products Labeling Act, and the Fur Products Labeling Act are examples of these laws.

Consumers should also be aware of their rights and responsibilities.

Fashion in Review

Write your responses on a separate sheet of paper.

Short Answer: Write the correct response to each of the following items.

1. Why is it smart to invest in seasonless clothes?
2. Name three advantages of planning and gathering information before going shopping.
3. Describe how reading fashion magazines and fashion sections of newspapers can help you prepare to shop.
4. What are the two purposes of advertising?
5. Advertisements are good sources of up-to-date shopping information because they _____.
   A. have clever songs and slogans
   B. play on the emotions of consumers
   C. show and describe new apparel items
   D. can entice you into buying
6. Name two times of the day when stores are not likely to be crowded.
7. Distinguish between hangtags and labels.
8. What four kinds of information must appear on the labels of all textile products sold in the United States?
9. Most laws that protect consumers against deceptive labeling and advertising are enforced by what government agency?
10. What law specifies generic names for fibers and requires labels to list the fibers that are in textile products in order of predominance by weight?
11. What law requires domestic and imported apparel to have labels that give instructions for care and maintenance?
12. What law protects consumers against deceptive information on labels and in advertisements of fur products?
13. Describe the consumer's right and responsibility for safety.

14. Describe the consumer's right and responsibility to choose.
15. List five people or agencies consumers can contact for help with their complaints.

Fashion in Action

1. Find at least five clothing ads from magazines, newspapers, or mail-order catalogs that give written information as well as illustrations. In class, distinguish between the useful information and the sales pitches. Discuss any ads or statements that may be misleading or exaggerated.
2. Shop for a garment of your choice in a department store, a discount store, and a specialty shop. Write a report comparing the prices, quality of merchandise, size of selection, service, store displays, and general atmosphere.
3. Look at the labels and hangtags from several articles of clothing. Note the information on each. What information is required by law? What extra information do some contain?
4. Make a display of consumer information sources. Include articles and ads from newspapers and fashion magazines. Sketch store windows and displays. Include catalogs. List library books that are available. Get some consumer aids from sources such as fiber producers, fabric stores, and your local extension agent.
5. Interview a store manager about the behavior of shoppers. Ask him or her about merchandise that is damaged when it is tried on and about returned items. What is the store's estimated yearly loss from shoplifting? Write a report about your findings.

16 Making the Right Purchase

Fashion Terms

comparison shopping
value
double ticketing
wearing ease
trademark
overdrawn
debit card
layaway purchase
credit rating
thirty-day charge account
revolving charge account
installment plan
Truth-in-Lending Law

After studying this chapter, you will be able to

- comparison shop to make the best purchases.
- judge the value, quality, and fit of garments.
- evaluate bargains and sales.
- choose the best method of payment for your purchases.
- describe home shopping options.

After studying the previous chapters of this book, you should know how to prepare yourself to shop wisely. The actual process of making the right choices in the marketplace will depend on your level-headed judgments at the time of purchase. You will have to compare and evaluate the products that are available. Then you will have to decide which method to use for payment of your purchases.

Comparison Shopping

Comparison shopping means comparing the qualities and prices of the same or similar items in different stores before buying. Be a "looker" before you buy. Check out the fabric and construction of garments. Compare similar garments at different stores and departments. Know what items are worth the prices being charged for them. Read and compare labels and hangtags. Evaluate the details of merchandise to be certain you find what you need and want.

You may also compare merchandise by reading mail-order catalogs and newspaper advertisements. Consider the comfort, appearance, durability, and care of each garment. Weigh the pros and cons of alternate choices. Choose the store and the item that will give you the greatest satisfaction and suit your needs.

Some people like to browse in designer departments or expensive stores every now and then to see how quality clothes look. They check the construction details, fabrics, colors, and fit. Then they search out similar items in less expensive departments or stores. In this way, they learn what characteristics are desirable before they buy.

Comparison shopping takes some time, but it is worthwhile. If you comparison shop, you will be happier with your purchases. Your final choices will be what you really need and want. You will know they are the best available items for the price. You will hardly ever need to return items. You will save money and have the right merchandise. You will also have a good feeling of satisfaction!

Judging Value and Quality

Value is the degree of worth or benefit of something. In clothing, the "best value" is the highest quality of materials, construction, and fashion for the lowest price.

Evaluate apparel purchases in terms of good design, construction, lasting fashion, durability of fabric, ease of care, and suitability to you and your lifestyle. See 16-1. When shopping, look at the inside and outside of garments. Don't settle for something that is not right. To get the most for your money, become familiar with the marks of quality merchandise.

General Quality

The overall quality of a garment can be poor, good, better, or best. Sometimes it is described as low, medium, or high. A garment of high quality should perform better than one of low quality. Quality has a lot to do with long-term performance of appearance, fit, and wear.

Liz Claiborne

16-1 Value in clothes depends on such factors as high quality, personal suitability, and good price.

High-quality garments have the best construction, materials, and design. They emphasize cut, line, and fabric. Attention is given to construction details, so these garments often have extra built-in features and a degree of luxury. See 16-2.

High-quality garments have matched plaids and stripes at the seams. They have generous hems and linings. The linings are color-coordinated with the garments, as are the buttons. Seams are straight, with even and secure stitches. Fasteners are secure and are located so no gapping or pulling occurs. In knits, there is full-fashioned shaping in many areas.

Garments of *medium quality* have reliably good construction, materials, and design. They are usually quite durable. They generally cost less than high-quality apparel but more than low-quality clothing.

Low-quality garments have only fair standards of construction, materials, and design. Poorly made garments become worn and unattractive quickly. Low-quality garments may be okay for fad items that will be outdated soon or for items that will be worn only a few times. For some merchandise, you may not be able to, or want to, spend more money for better quality.

Quality is often, but not always, related to price. Sometimes merchandise of better quality is on sale, or it might be offered in a store with low overhead costs. Thus, a low price does not always mean that a garment is of low quality. An alert shopper can often find good quality products at low prices.

On the other hand, not all high-priced garments are of the best quality. Some merchandise prices are raised to help pay for extra store services or high advertising costs. Some high prices are caused by fur trim on an item, an unusual braid, or a famous designer's name or logo. You must decide if such details are important to you and worth the extra money.

Value is gained when you pay only for the quality you want and need. You will probably want different levels of quality for different garments. Decide what quality is important to you for specific apparel items. Consider the intended use of each garment and your purpose in buying it.

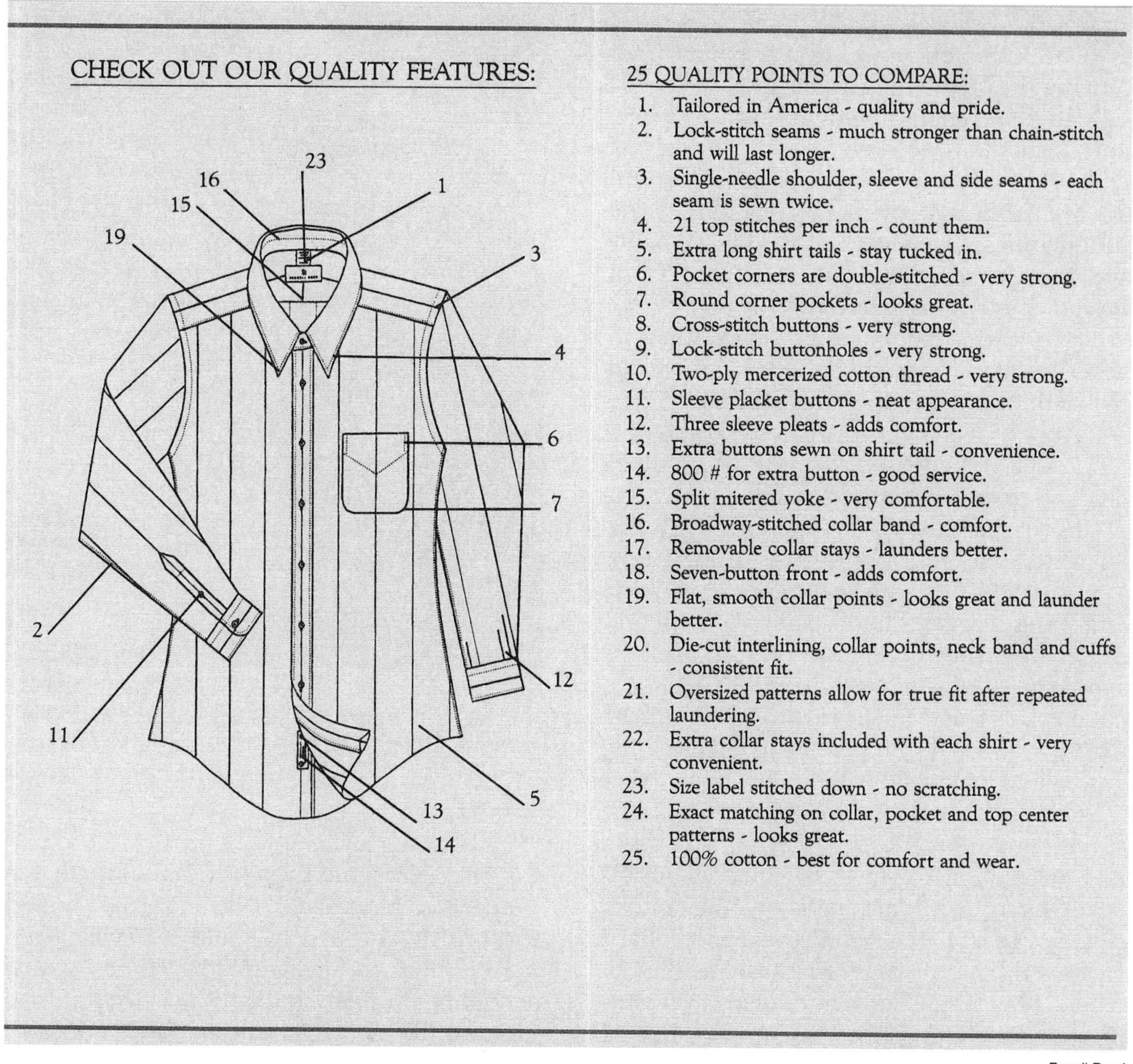

Ferrell Reed

16-2 This hangtag from a high-quality man's dress shirt points out all the quality features that the manufacturer feels are important for consumers to notice.

Buying the best quality you can afford for a basic garment is usually a good investment. Consider the cost-per-wearing rather than the actual purchase price. Paying a higher price for a good quality coat or jacket that you will wear a great deal for several seasons is usually economical. It will be able to withstand lots of hard use.

For extenders, occasional wear garments and accessories, items of lower quality with lower prices may be appropriate. This same philosophy might provide the best value for a fad item, too. If you don't pay much for it, you won't mind discarding it when it is out of date. A high quality fad item might last longer than it will stay in fashion!

Consumers should develop quality awareness. Clothes are available in various qualities and prices. Well-made, good-looking clothes can be found at many price levels. Smart consumers decide what quality suits their needs. They shop for fashion at prices they can afford. They consider a "good buy" to be something that meets a need at a price they want to pay.

Specific Points of Quality

Check the following specific points when judging the quality of garments.

The *fabric* should be of an even weave or knit. It should have even color or pattern, and it should be free from spots or streaks on the surface. It should not contain snags, flaws, or irregularities. Its weight and flexibility should suit the garment type and style. The fabric design should be interesting, flattering for your coloring, and in proportion with your size. The fabric should have the care requirements that suit your lifestyle and budget.

The *cut* of the garment should use ample fabric so it does not look skimpy. It should be cut "on-grain." This means that the center line of the garment parts are parallel with the lengthwise or crosswise threads of the fabric. The fabric threads at the garment's front, back, and outer sleeves should be straight up and down. The threads across the chest and hips should be parallel to the floor. If the garment is cut on "true bias," the fabric threads should be exactly diagonal. Proper grain enables the garment to hang well without twisting.

Garments should also be cut so plaids, stripes, and other fabric patterns are on-grain and matched at the most obvious seams and edges, as in 16-3. Matching should be done at closures, seams, pockets, sleeves, and collars. However, patterns cannot be matched everywhere. Matching is most important where it is most noticeable.

The *seams* of a garment should be smooth, without ripples or puckers. Seam allowances should be of even widths. They should be generous enough to hold firmly if there is strain on the garment. Reinforcing or double stitching may be needed at areas of strain, such as the underarms or crotch. Extra width is required if alterations will be needed. Raw edges should be finished, for instance with overcasting, so the fabric will not ravel. Seam edges and corners should lay flat. In knit garments, seams should have as much stretch as the fabric.

Stitching should be straight and neat. The stitches should be short and even. They should be sturdy with no broken stitches. The thread should match the fabric color and weight. Thread ends should be securely fastened. Topstitching should be smooth and straight.

Pendleton Woolen Mills

16-3 The nicely matched lines of this plaid go across the sleeves and bodice front, continuing across the front closure.

The *hem* should be of an even width that is proper for the garment style and deep enough to let down if needed. It should lie flat and have a finish that is suitable to the fabric. A hem should be finished, so it does not ravel or fray. It should be securely stitched so it is nearly invisible when viewed from the right side.

Reinforcements should be included at points of strain. These include the top corners of patch pockets, the ends of bound pockets, and the sides of shirttails. Reinforcing might be done with extra stitching, bar tacks, or metal rivets (as on the blue jeans in 16-4). Shoulder seams of knit shirts are often sewn with ribbon or binding reinforcement to prevent stretching. Knees of children's play clothes should also be reinforced.

A *zipper* should be long enough and strong enough for the garment and located for easy use. It should be of a suitable weight and matching color for the garment. A zipper should open and close easily without catching fabric in its teeth or coils. The top should have locking tabs to stay closed when stress is applied. A zipper should be smooth and flat,

16-4 Some blue jeans that are manufactured for hard wear are reinforced with metal rivets at both ends of the pockets.

and free from puckers. Stitching should be secure, even, and inconspicuous in thread that matches the fabric color.

Fasteners such as buttons, snaps, hooks and eyes, and hook and loop tape should be of a suitable size, color, and type for the garment. They should require the same care as the garment. They should be appropriately placed, evenly spaced, and firmly attached. There should be enough of them to keep the garment closed without gaps or pulling. They should open and close easily. Also, they should lie flat and look neat.

Quality garments have an extra button of each size used, as replacements, sewn inconspicuously inside the garment.

Buttonholes should look neat. They should be evenly and properly spaced to exactly match the placement of buttons. They should be cut with the grain of the fabric unless they are strictly decorative. The size and type of buttonholes should be suitable for the buttons and for the type of garment. Buttonholes should be firmly stitched with reinforced ends. They should lie smooth and flat on the garment. They should not have loose threads or frayed edges. Hand-sewn buttonholes in tailored garments are a sign of high quality.

Collars should be cut on-grain and centered on the design of the fabric. Backs of collars should lie flat and close to the neck. The outer part of the collar should turn over far enough to cover the neckline seam at the back. Also, the underside of the collar should not show along the outer edge of the collar. Collar points should be neatly finished and should not curl under or up.

Lapels should lie flat and be stiff enough to hold their shape. They should have a graceful roll and a clean-cut edge, as in 16-5. They should end at the top button of the garment.

JC Penney

16-5 Classic lapels, such as on men's suit jackets as well as this vest, should lie flat. They should also hold their shape and have a precise, smooth edge.

Waistbands should be constructed to prevent stretching and rolling. Belt loops should be well placed and sewn securely.

Pockets should be flat, smooth, and well matched. Corners should be reinforced. Pocket linings should be firmly woven.

A garment's *lining* adds body, helps to retain the garment's shape, and adds to wearing comfort. Apparel items can be completely or partially lined. A lining should have the same care requirements as the outer fabric. It should fit smoothly into the garment. It should be of the appropriate color, weight, and texture to enhance the garment. It should never droop or shrink. It should be securely attached, but not so tight that it will rip or come loose when you reach or stretch. In a jacket, the lining should be attached along the neckline, at the top of the armholes, along the front facings, and at the hems of the sleeves and jacket bottom.

Interfacings and *paddings* are extra layers inside garments. They support certain parts of garments, such as collars and cuffs, and hold the style lines of the design. They provide extra strength, body, and a smooth finish. Higher-quality garments use more of them. They should be properly placed, hidden, and securely attached. They should have the same care and be of a similar color as the garment.

Trimmings and *decorations* should be able to withstand the same care as the garment. Otherwise, they must be easily removable. They should be suitable for the garment, well placed, and neatly and firmly attached. Topstitching should be even and attractive.

Hong Kong Trade Development Council / Virginia Lau, Designer

16-6 High-fashion garments such as this one, as well as inexpensive apparel items, are imported into the U.S. from foreign countries.

The Dilemma of Imports

Many garments are imported, such as the one shown in 16-6. Imports have increased sharply over the last few decades. They now account for over half the merchandise available to, and bought by, consumers in the United States. There are several different views about buying or not buying imported clothing.

As discussed earlier in this text, labor costs are lower in less developed countries. Many garments made in other countries can be offered at a lower price to American consumers, even after shipping costs and duties (tariffs) have been added. These products are, therefore, economically attractive to some consumers.

Some people say that low-cost imports enable stores to put bigger markups (and make bigger profits) on apparel made overseas. Retailers like to sell identical or equivalent products at similar prices. Their domestic and imported items have similar prices, even though the imported ones may cost them less to acquire.

Many Americans think the federal government should pass stronger laws to limit the amount of clothing imports. They cite the fact that we have a trade deficit on apparel. That means we bring more clothing into the country to sell than we ship out to other countries.

Other citizens want the widest possible choice of goods, which includes imports. Some consider that international trade fosters goodwill in America's worldwide diplomatic relations in other areas besides apparel. Some believe that apparel jobs are essential to the basic survival of workers in many developing nations.

In some respects, consumers have benefited from the lower prices of imports. Not only do the imported items cost less, but because of competition, prices on American-produced apparel have been kept lower, too. The competition has caused manufacturers to operate more efficiently, develop new technology, and improve their management.

Some consumers choose to buy items made in the United States whenever possible. This helps to maintain our strong industrial base and to support our workers. Citizens who become unemployed may have to accept public assistance. That costs everyone money through taxes.

As you can see, there are several sides to the dilemma of imports. There are no easy, clear-cut answers. When you are shopping, read labels to find the countries of origin of the items that interest you. Evaluate the quality, the price, and your own personal viewpoint about where the merchandise was made.

Evaluate Proper Fit

Clothes don't look their best if they don't fit well. Clothes should not be too tight, too loose, too long, or too short. A good fit has a look of natural ease. If garments have a good fit, they will feel comfortable, look good, and wear well. Improper fit takes away the fashion attractiveness in garments. Clothes that don't fit well or feel right will stay in the closet unworn and will be a waste of money. A well-fitting garment will last longer and require less care than one with poor fit.

Determine Your Size Category

Know what size garments you wear. To determine your size, first be aware of your body structure. This depends on your stage of physical development. It is the relationship of your height to your weight and bone structure. It also is keyed to the relationship among your chest, waist, and hip measurements.

Sizes are in categories by figure types. Study the size categories described in 16-7 and find the one that best suits you.

Size Range Categories

(Infant and children's sizes are listed in Chapter 18)

Figure Types	Sizes	Proportions
Subteen girls	7-16	For girls whose figures are average; undeveloped in bust and hips. Height approximately 4′2″ to 5′.
Slim girls	7S-16S	For girls whose figures are slim; undeveloped in bust and hips.
Chubby girls	8 1/2C-18 1/2C	For girls whose figures are heavier; undeveloped in bust and hips.
Teen girls	7-14	For girls whose figures are beginning to mature and be longer waisted. Height approximately 4′9″ to 5′5″.
Junior	3-17	For girls and women who are short-waisted, have fully developed bust and hips, and small waist. Height approximately 5′3″ to 5′6 1/2″.
Junior petite	3P-17P	Junior figure with height under 5′2″.

(continued)

16-7 Specific apparel sizes are listed within the various figure types.

Size Range Categories *(continued)*		
(Infant and children's sizes are listed in Chapter 18)		
Figure Types	**Sizes**	**Proportions**
Tall junior	3T-17T	Junior figure with height 5′6″ to 5′11″.
Misses	6-20	For girls and women, fully developed, with average figures and proportions. Height approximately 5′4″ to 5′8″.
(Sometimes Misses sizes are: Petite, sizes 2-6; Small, sizes 8-10; Medium, sizes 12-14; Large, sizes 16-18; Extra large, sizes 20-22; XX Large, size 24.)		
Misses petite	6P-20P	For short and slim girls and women. Height under 5′4″.
Tall/misses	6T-20T	Height over 5′8″.
Women's	18-60	Large proportions. Height 5′5″ to 5′9″.
Half sizes	12 1/2-24 1/2	Heavier, short-waisted girls and women. Height under 5′5″.
Male Size Categories		
Subteen boys	7-12	For undeveloped boys of average build.
Slim boys	7S-12S	For undeveloped, slender boys.
Husky boys	7H-12H	For undeveloped boys who are heavier than average.
Teen boys	14-20	For boys of average build who are beginning to develop.
Slim teen boys	14S-20S	For boys of slender build who are beginning to develop.
Husky teen boys	14H-20H	For boys of heavier build who are beginning to develop.
Men's (related to chest measurement)	30-48	For boys and men who are fully developed. Height 5′7″ to 5′11″.
(Sometimes Men's sizes are: Extra small, sizes 30-32; Small, sizes 34-36; Medium, sizes 38-40; Large, sizes 42-44; Extra large, sizes 46-48.)		
Short men	30S-48S	For men of heights 5′3″ to 5′7″.
Tall men	34T-48T	For men of heights 5′11″ to 6′3″.
Extra tall men	40-46	For men of heights over 6′3″.

16-7 *Continued*

Know and Use Your Measurements

After finding your general size category or figure type, take and record your body measurements to discover your exact size. Have someone else help you with this. Put the tape measure close around you but not tight. Wear whatever undergarments you wear under your nice clothes. Record the measurements.

Measure your *height* while standing without wearing shoes. Measure from the floor to a ruler placed horizontally across the top of the head.

For the *chest* or *bust*, measure around the largest part in front and straight across the back. Have your arms hanging down at your sides.

Measure around the *waist* at the smallest part for female figures. That is the natural waistline. For male physiques, the waist may be measured at the position where pants are usually worn. Sometimes that is below the natural waistline.

Measure around the *hips* at the fullest part. The tape should go straight around the body.

For the *neck*, measure the band of a shirt collar that is comfortable. Lay the collar flat on a table. Measure it from the middle of the front button to the far end of the buttonhole at the other front edge. You may also measure the actual circumference of the neck. Sizes are in half inch increments.

A *shirt sleeve measurement* is from the back center base of the neck, across the shoulder, and down the arm around the bent elbow to the wrist bone. This is shown in 16-8.

The *inseam* of pants can be measured from some well-fitting slacks of the right length. Measure along the inside leg seam from the crotch to the hem.

Look at the size chart in a mail-order catalog or pattern catalog. Find your measurements within your size range category. If your measurements are not an exact size, choose the size that matches the closest.

If your top and bottom are different sizes, it is probably best to wear separates most of the time. You can buy one size for the top and a different size for the bottom. For one-piece outfits, choose clothes that fit your larger part and have the other part taken in. It is easier to make garments smaller than to make them bigger.

16-8 When measuring for a shirt sleeve size, start at the center back of the neck (not at the top of the sleeve) and measure to the wrist bone as shown.

Generally, chest measurements determine the size for suits, tops, and dresses. Waist and hip measurements are used for skirts and women's pants. Men's pants sizes list the waist measurement followed by the inseam length, for instance, 34-33. Some are sized by the waist measurement in short, medium, or long lengths, such as 36L. Men's shirts are listed by a neck measurement followed by sleeve length, for instance 15 1/2 - 34.

Double ticketing combines two or more specific sizes into more general categories such as small, medium, and large. This is fine for sweaters or other loosely fitting styles, but is usually unsatisfactory for fitted garments. Check the measurements for the size categories listed on the label or package before buying. It's also a good idea to try on such garments before buying them.

Try on Garments

Even if you regularly wear a certain size, individual garments vary. Sizes are only a guide to how clothes fit, as in 16-9. There is only one way to know whether or not a garment is right for you: try it on. Wear the underclothes and shoes that you will wear with the garment. Look at all views in the dressing room mirror. Move around to check out the feel and the look. Can you reach and stretch without straining the fabric and the seams? Does the garment fit smoothly without wrinkling or binding anywhere?

Make sure the shoulder length is comfortable and the neckline fits smoothly. When the neckline is buttoned, you should be able to slide the index finger easily between the collar and the neck. There should be enough ease across the chest and back to prevent pulling or stretching.

Depending on fashion, the armholes might follow the natural armline. The sleeves should be wide enough not to bind. They should be a good length. Sleeves of blazers and suit coats should end 1/2 inch above the lower edge of the shirt cuff. There should be proper ease through the hips and seat. Lower garments should hang straight from waist to hem. Outer garments should be large enough to go over apparel worn underneath.

16-9 Sizes on apparel in retail stores are indications of which clothes might fit. However, to judge actual fit, it is always wise to try them on.

JC Penney

16-10 Wearing ease allows a person to move comfortably in a garment. Additional fullness may be included as part of the garment's design.

Darts, if present, should go to the proper areas of fitting (shoulder blade, chest or bust, hips, elbows). The center front and center back seams of garments should be the right length. The waistline should be at the right spot. The crotch length of pants should be comfortable when sitting or bending, but not so long that it sags when standing. The length of the garment should be attractive for your height and for current fashion.

The fit of garments changes with different fabrics. Some are stiff, some are fluid, and some have a great deal of stretch. Fit also changes with different styles because different designs have varying amounts of built-in ease, as in 16-10. ***Wearing ease*** is the room you need to move comfortably in a garment. It is added to garments beyond the actual body measurements. Size and fit also vary among different manufacturers.

The garment industry uses uniform sizing. All American apparel manufacturers use the same measurements for their sizes. However, they interpret them differently. They use different body proportions and amounts of ease. Each apparel manufacturer cuts its garments for its "ideal person." If a certain size from a particular manufacturer fits you, all other garments from that manufacturer in that size will probably fit you the same way.

Sometimes manufacturers size clothes to please their clientele. For instance, women's expensive clothes are cut bigger than inexpensive clothes. The women who buy them fit into a smaller size. This gives the customers a psychological lift!

Inexpensive garments are skimpier than more expensive ones. Less fabric is used. Also, inexpensive fabric may sag or bag when worn. However, you should now be knowledgeable enough to judge the fit of garments regardless of the sizes listed on them. Select them according to how they look and feel on you.

Try to find ready-made clothes that fit. Avoid buying garments that require alterations other than simple ones, such as a change of length. Extensive or poorly done alterations can change the lines of the garment. Be skeptical of a high-pressure salesperson who tells you that a tricky alteration can be accomplished easily. He or she may only be trying to make the sale. Once the alterations are completed, you will not be able to return the garment.

Know Trademarks

A ***trademark*** is a distinctive symbol or name that identifies the goods of a particular seller or manufacturer. See 16-11. An awareness of trademarks offers information about the quality of merchandise. The quality of various brands may be high, low, or in the middle. However, it will almost always be uniform from one article to the next within the same brand. Often trademarks have performance guarantees. The reputation of various brands causes shoppers to choose or reject items.

16-11 Manufacturers promote their trademarks and try to have their trademarks instantly recognized by consumers.

Get to know which manufacturers produce the clothing that fits your body shape, taste, and quality standards. If you find a brand that you like, you may want to look for the same brand when you shop in the future. If it has served you well in the past, it will probably be of the same quality and fit again. You can then be assured of getting the type of clothes you want. You give a vote of confidence when you find satisfaction with textile and apparel products and continue to buy them. Your loyalty to brands of merchandise that suit your standards of style and performance shows by your repeat purchases.

Every now and then, companies change their production specifications. Sometimes companies are bought by other manufacturers. Their apparel may then be either of lower quality than before or "new and improved." A little comparison shopping may have to be done in this case.

Compare the quality of big national brands with lesser-known brands. Sometimes store brands (private labels) are offered at lower prices than national brands. This merchandise is often made by large manufacturers. It can sell for less because stores buy it in quantity and it is not advertised nationally. Many items may be quality products and good buys. On the other hand, some may be inferior. If the quality is equal between the national and store brands, the store brand is usually a better buy because the price is often lower. However, a well-known brand name usually assures you of dependably good quality.

Designer Labels

Designer labels are trademarks that give status to the wearer. A designer's logo is often placed on the outside of a garment as a display for all to see when the garment is worn. Such a status symbol is important to some consumers and gives motivation for buying such products. It also adds to the retail prices of such items. When you buy designer fashions, you pay extra for the labels.

Evaluate Bargains

It is human nature to want to get a good bargain. A *bargain* is a favorable purchase. Getting a bargain means getting a high value of merchandise for a small amount of money. Sometimes a low price is the sign of a real bargain. On the other hand, sometimes a low price is the sign of poor quality that is probably not a bargain.

Bargains can save you money and give you great satisfaction. You should look for good bargains, but don't forget your original shopping plan, priorities, and quality comparisons. Judging bargains is part of the decision-making process when you shop for clothes. Buy carefully to get a real bargain.

Don't buy something just because it is on sale, 16-12. A sale is a bargain only when you buy a garment you need. Buying something just because it is on sale is a waste of money. You will essentially lose the amount of money spent on a poor choice if the item does not fit into your wardrobe and if you never wear it. Buy an item because it is right for you and your wardrobe plan, not because the price is low.

Shopping at Sales

Evaluate sale merchandise in terms of whether it fills a genuine need for you. Consider whether it has the quality and fit you desire and whether it is damaged beyond easy repair. Judge sale items the same way you judge regularly priced clothes. A sale item is a bargain if it is of better quality for the same amount of money you would have paid, or if you get what you want for less money. A sale item that is unusable to you is not a bargain at any price. If it doesn't fit, isn't what you need, and can't be returned, it doesn't matter how cheap it is!

Some sale items cannot be returned if you are not pleased with them. Ask about the return policy before you buy. Check sale garments carefully before purchasing. Look for quality fabrics and construction details. Check for flaws.

When buying at sales, you may find that there is not much personalized service. There will probably be crowds. It is hard to think clearly and make wise decisions when you are surrounded by people. There may be confusion, and you may feel hassled. You may have to dig through piles of merchandise to find the right item. You may get caught up in the frenzy and buy too many items. If possible, shop early before the good items are gone. Getting what you want at sales may take more time, but it is often financially worthwhile.

Types of Sales

Learn to discriminate between real bargains and sales gimmicks. Learn to judge the difference between good sales of quality

16-12 Consumers should try not to be tempted by sales if they don't need the items that have been marked down.

apparel, and promotional sales of cheap garments designed to attract business. Clothes are put on sale for many reasons.

General sales are often held throughout stores during holiday times. They lure people into the stores. Retailers have sales for almost every holiday and special event. Other sales are held during slow seasons. Examples are "January white sales" and promotional sales during August.

Stores want to sell inventory that has been left over at the end of the season. This fashion merchandise is no longer selling well at full price, and the store does not want to keep it for the next season. "Clearance," "inventory," and "end-of-season" sales clear out old stock to make room for new merchandise. See 16-13. Items can sometimes be picked up at great savings.

Sometimes leftover garments are in unusual sizes or colors. They are often of good quality and thus good bargains if you need the items anyway. By waiting until after a fashion has reached its peak, you can find a good quality item at a fraction of its original price. This is especially good if you think the style will be popular for a long time.

16-13 Sales that are held at the end of a fashion season encourage consumers to buy items at lower "clearance" prices. Shelf and hanging space then becomes available for new merchandise.

New items are sometimes put on sale to encourage customers to become acquainted with them. Such sales might offer good bargains. Other times, stores buy quality garments at lower than regular prices for "special purchase" sales. They can offer lower prices because they have bought large amounts, but certain sizes and colors may be limited. On the other hand, sometimes great quantities of cheap or inferior merchandise are brought into stores just to be sold at low prices.

Sometimes garments are marked up artificially and then marked down to a regular level and said to be on sale. They are preticketed with high prices that are implied to be the retail selling prices. Those are then crossed off, and lower prices are given to make consumers feel that they are getting bargains. Some of these methods seem unethical when you know about them. However, they are widespread in the retailing world. Always compare quality and price.

A store might have a "fire sale" when part of the store has burned. The sale goods might have water damage or smell like smoke. They could, however, be very good items.

A store might also have a "going out of business" sale. It might have lost its lease. It might have gone bankrupt. In any case, the owners must sell all their merchandise. Bargains may be available, but look closely before buying. Sometimes cheap merchandise is brought in to be sold with the items already on hand.

Some stores mail coupons to customers or include them in newspaper advertisements, such as for 20% off any one merchandise purchase. The coupons are only good on certain days or expire after a specified date. This could result in savings for you if you would have bought the item anyway. If merchandise is not on your list or not needed, getting it at a discount may be an unwise purchase.

Impulse Buying

Sales may tempt you to buy impulsively, as in 16-14. *Impulse buying* is buying something as

16-14 Sometimes you see interesting items when you aren't really looking for them. Unusual designs, your favorite colors, or "special" prices might tempt you to buy impulsively.

Questions to Ask When Buying Apparel

Is it the correct style for my size, shape, and personality?

Is it functional and appropriate for my lifestyle?

Does it fit into my color schemes?

Does it fill a need listed in my wardrobe plan?

Does it provide comfort and easy care?

Have I gathered information and comparison shopped?

Does the quality meet my requirements?

Is it my size?

Have I tried it on to evaluate its fit and look?

Is it a brand name that I like?

Is it a bargain with a reasonable price for the quality and for the amount of use I expect to get from it?

Am I buying impulsively (emotionally) or logically?

16-15 By asking yourself these questions, you can avoid purchasing mistakes that you will regret in the future.

soon as you see it, without thinking carefully. This type of purchase is not part of your plan, budget, or needs. It is an emotional purchase rather than a logical one. You do not consider your needs or other wardrobe items. You just buy on a whim.

Fads are often bought impulsively. Sometimes you are influenced by advertising displays, salespeople, or a friend with whom you are shopping. One impulsive splurge could ruin your whole shopping plan. You may spend more than you had budgeted. You might end up with garments that you don't need and won't wear.

To guard against such impulsive buying, ask yourself, "Do I like this well enough that I would be willing to make a special trip on another day to buy it?" If it is on sale, ask, "Do I like it well enough that I might have been willing to pay full price for it?" If your answers are "yes," then you are probably not making an unwise purchase.

Check yourself by asking the questions in 16-15 when buying apparel. Then make wise, planned purchases. It is nice to know that the items being purchased are what you want and need and that they will fit into your total plan. If you are happy and confident about your purchases, and if you can afford them, then buy them and enjoy them!

How to Pay

Buying is the exchange of money or credit for goods or services. Buying carefully avoids overspending. Unplanned purchases leave you with no money for emergencies.

Cash Purchases

When you make a *cash purchase*, you use money to pay the full cost of what you buy. The transaction is then complete. You are not in debt for that purchase. You will not be sent a bill or have to pay credit charges.

Cash purchases help you discipline yourself to save your money until you have the funds to buy the goods. With cash, you will avoid spending more money than you have. It is easy for you to control your expenditures. Although paying cash is a good idea, you may not be able to afford to pay cash for some expensive items.

On the other hand, if you always make cash purchases, and if you happen to spend money on the wrong things, you may not have enough for your real needs. When you carry cash with you, you take the chance of it being stolen or losing your money. Also, cash purchases do not allow you to establish a credit rating, and you have no record of your spending other than your sales slips.

Writing a Check

You may make a cash purchase by writing a check, as in 16-16. This gives you more evidence of your spending, since records of checks are good receipts. Paying by check eliminates the need to carry large amounts of money with you.

When you write a check, you must show proper identification. You must be careful to always record the amounts of the checks you write, so your checking account does not become ***overdrawn***. That means you have written checks for more money than you have in the account.

Using a Debit Card

Cash purchases can also be made electronically with a ***debit card***. The card looks similar to a credit card. When you make a purchase, the money is deducted from your bank account without you giving actual cash to the store or writing a check. This enables you to buy items without carrying cash and without paying bills and interest later. However, when the balance sheet is sent to you from the bank, always check to see that it is accurate. Keep your own records of purchases and deductions from your account.

A newer version of the debit card is a *smart card*, which has electronic stored cash value built in. This "cash card" also looks like a credit card but has a computer chip embedded in the plastic. The computer chip might be programmed to hold a certain amount of money after the consumer gives that sum to the bank issuing the card. Retail checkout terminals read the information stored on the card's chip. After the cashier rings up the purchase, the terminal requests verification of the purchase total. The consumer pushes an "okay" or "yes" button, and the cash amount is electronically deducted from the value of the card. The purchase amount is stored in the retailer's terminal for settlement with the issuing bank.

Most smart cards can be reloaded at the bank after their value has been depleted. Those that also contain a credit card magnetic stripe can be reloaded through automated teller machines (ATMs), as well as being used

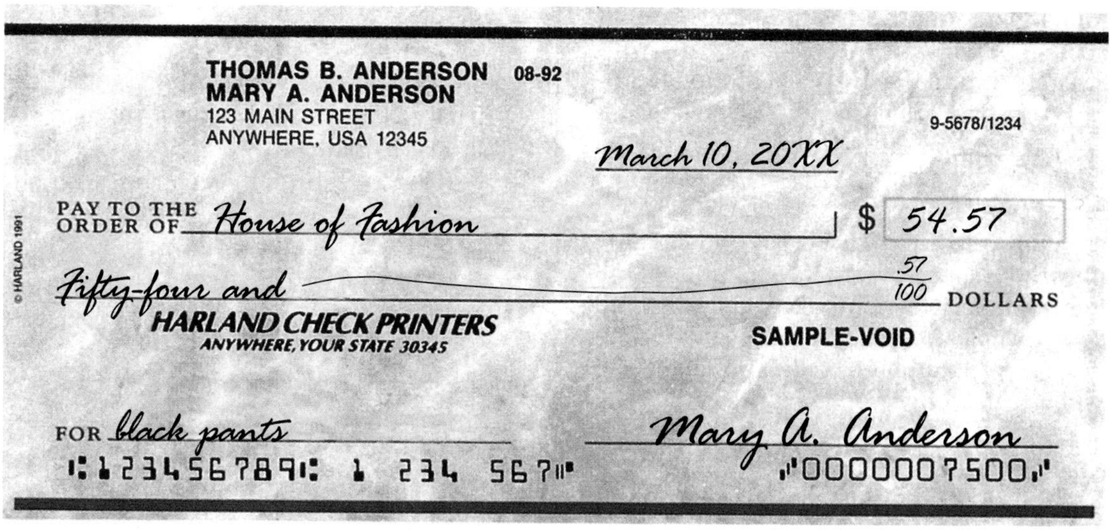

THOMAS B. ANDERSON 08-92
MARY A. ANDERSON
123 MAIN STREET
ANYWHERE, USA 12345
9-5678/1234
March 10, 20XX
PAY TO THE ORDER OF House of Fashion $ 54.57
Fifty-four and 57/100 DOLLARS
HARLAND CHECK PRINTERS
ANYWHERE, YOUR STATE 30345
SAMPLE-VOID
FOR black pants
Mary A. Anderson
⑆123456789⑆ 1 234 567⑈ ⑈0000007500⑈
© HARLAND 1991

16-16 After you pay for merchandise with a check, it is sent to the retailer's bank for the store to receive payment. It then goes to your bank that deducts the funds from your account. Sometimes the "canceled" checks are returned to you.

as credit or debit cards when making a purchase. Consumers find this payment fast and easy for purchases. There are no personal identification numbers to enter, no paper, no signatures, and no on-line authorization. It's faster than a cash purchase with money!

From the merchant's viewpoint, retailers like the reduced risk of accepting debit cards. As retailers replace their checkout terminals with new equipment, they will install versions that can accept credit, debit, and smart cards.

Layaway Purchases

When you make a ***layaway purchase***, a store puts the item away for you for a certain length of time if you make a deposit toward buying it. You can't have the item until you have paid for it in full by making payments. When the total amount is paid, the store gives you the exact piece of merchandise you originally picked. By saving it for you, no one else has been able to buy it. However, if you change your mind or cannot finish paying for it in time, you might have to forfeit the money you have already paid, or get only part of the money back. This is because the store kept the merchandise off the selling floor for a period of time.

Layaway may be a good method of buying items you do not need immediately. It encourages budgeting. You can also establish a credit record by doing this. You do not pay interest and do not go into debt. However, the store has your money ahead of time. When buying this way, always be sure you have a written agreement with the store.

Credit Purchases

A *credit purchase* is a promise to pay for goods or services in a certain, specified way at a later date. It is a "buy now, pay later" arrangement. Cash is not used. A credit purchase can be made with a specific store charge account or with a general credit card.

When you buy something with credit, you sign a paper, called a *charge slip*. It tells what you bought, where and when you bought it, and the amount of the purchase. A special charge card is usually needed.

When you apply to get a credit card or charge account, always give truthful information on your application. Carefully read the credit application, as in 16-17. Know what you are signing. You either agree to pay for your purchases in full when the bill comes, or you agree to pay in monthly payments.

In order to be approved to use credit, you must have a good financial standing or ***credit rating***. That means you have a good record of paying your bills according to the terms of credit agreements. A poor credit rating results from missing payments. It is also caused by sending payments in late or in amounts that are less than what the contracts specify.

A credit purchase can cost more than a cash purchase if the entire balance of the account is not paid on time. In that case, a finance charge is added. Some percentage finance charges are less than others, so it is wise to shop around for the best credit arrangements.

Store charge accounts are among the oldest and best-known forms of credit. Store charge accounts are convenient. However, a store charge account can be used only in that particular store and its branches. It might limit your choice of purchases. You may feel the need to have charge accounts at several stores.

Stores offer thirty-day and revolving charge accounts. Some also offer installment buying. If you want to open a charge account at a specific store, tell the person at the credit office of the store what kind of an account is best for you according to the following explanations.

Thirty-Day Charge Account

Thirty-day charge accounts must be paid in full by 30 days after the billing date. The bill must be paid by a certain day each month. There is no extra finance charge on the bill. By charging the goods this way, it is essentially like borrowing money interest-free for 30 days. However, if you don't pay what you owe by the due date, you will probably be charged a late fee or interest.

Revolving Charge Account

With a ***revolving charge account***, the total maximum amount you are allowed to charge, called your *credit limit*, is determined by your ability to pay. It is set when your account is opened. You can make any number of

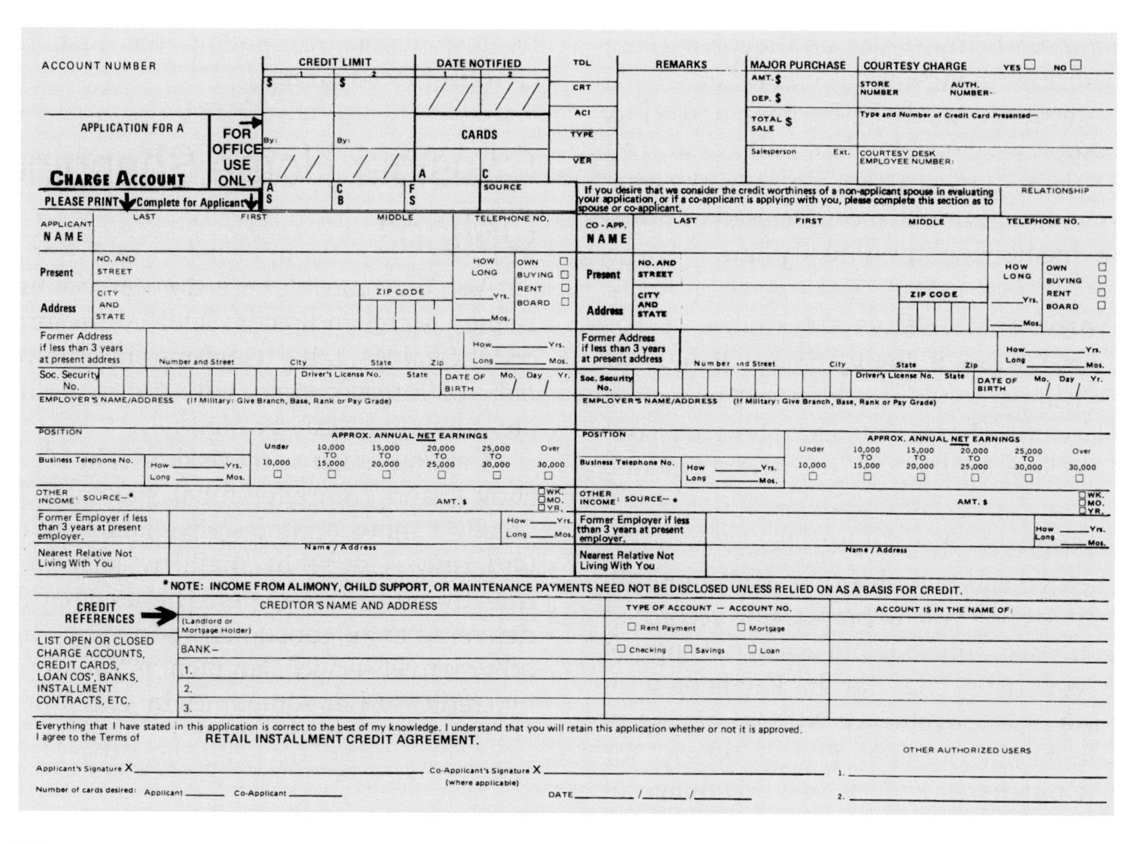

ACCOUNT NUMBER

CREDIT LIMIT 1 $ 2 $ By: By:

DATE NOTIFIED 1 / / 2 / /

CARDS A C

TDL CRT ACI TYPE VER

REMARKS

MAJOR PURCHASE AMT. $ DEP. $ TOTAL SALE $ Salesperson Ext.

COURTESY CHARGE YES ☐ NO ☐

STORE NUMBER AUTH. NUMBER–

Type and Number of Credit Card Presented–

COURTESY DESK EMPLOYEE NUMBER:

APPLICATION FOR A **CHARGE ACCOUNT**

FOR OFFICE USE ONLY

A S C B F S SOURCE

PLEASE PRINT Complete for Applicant

APPLICANT NAME LAST FIRST MIDDLE TELEPHONE NO.

Present Address NO. AND STREET CITY AND STATE ZIP CODE HOW LONG ____ Yrs. ____ Mos. OWN ☐ BUYING ☐ RENT ☐ BOARD ☐

Former Address if less than 3 years at present address Number and Street City State Zip How Long ____ Yrs. ____ Mos.

Soc. Security No. Driver's License No. State DATE OF BIRTH Mo. / Day / Yr.

EMPLOYER'S NAME/ADDRESS (If Military: Give Branch, Base, Rank or Pay Grade)

POSITION

Business Telephone No. How Long ____ Yrs. ____ Mos.

APPROX. ANNUAL NET EARNINGS Under 10,000 ☐ 10,000 TO 15,000 ☐ 15,000 TO 20,000 ☐ 20,000 TO 25,000 ☐ 25,000 TO 30,000 ☐ Over 30,000 ☐

OTHER INCOME: SOURCE–* AMT. $ ☐WK. ☐MO. ☐YR.

Former Employer if less than 3 years at present employer. How Long ____ Yrs. ____ Mos.

Nearest Relative Not Living With You Name / Address

If you desire that we consider the credit worthiness of a non-applicant spouse in evaluating your application, or if a co-applicant is applying with you, please complete this section as to spouse or co-applicant. RELATIONSHIP

CO-APP. NAME LAST FIRST MIDDLE TELEPHONE NO.

Present Address NO. AND STREET CITY AND STATE ZIP CODE HOW LONG ____ Yrs. ____ Mos. OWN ☐ BUYING ☐ RENT ☐ BOARD ☐

Former Address if less than 3 years at present address Number and Street City State Zip How Long ____ Yrs. ____ Mos.

Soc. Security No. Driver's License No. State DATE OF BIRTH Mo. / Day / Yr.

EMPLOYER'S NAME/ADDRESS (If Military: Give Branch, Base, Rank or Pay Grade)

POSITION

Business Telephone No. How Long ____ Yrs. ____ Mos.

APPROX. ANNUAL NET EARNINGS Under 10,000 ☐ 10,000 TO 15,000 ☐ 15,000 TO 20,000 ☐ 20,000 TO 25,000 ☐ 25,000 TO 30,000 ☐ Over 30,000 ☐

OTHER INCOME: SOURCE–* AMT. $ ☐WK. ☐MO. ☐YR.

Former Employer if less tthan 3 years at present employer. How Long ____ Yrs. ____ Mos.

Nearest Relative Not Living With You Name / Address

*NOTE: INCOME FROM ALIMONY, CHILD SUPPORT, OR MAINTENANCE PAYMENTS NEED NOT BE DISCLOSED UNLESS RELIED ON AS A BASIS FOR CREDIT.

CREDIT REFERENCES →	CREDITOR'S NAME AND ADDRESS	TYPE OF ACCOUNT — ACCOUNT NO.	ACCOUNT IS IN THE NAME OF:
	(Landlord or Mortgage Holder)	☐ Rent Payment ☐ Mortgage	
LIST OPEN OR CLOSED CHARGE ACCOUNTS, CREDIT CARDS, LOAN COS', BANKS, INSTALLMENT CONTRACTS, ETC.	BANK–	☐ Checking ☐ Savings ☐ Loan	
	1.		
	2.		
	3.		

Everything that I have stated in this application is correct to the best of my knowledge. I understand that you will retain this application whether or not it is approved.
I agree to the Terms of RETAIL INSTALLMENT CREDIT AGREEMENT.

OTHER AUTHORIZED USERS

Applicant's Signature X ____ Co-Applicant's Signature X ____ (where applicable) 1. ____

Number of cards desired: Applicant ____ Co-Applicant ____ DATE ____ / ____ / ____ 2. ____

16-17 When you apply for credit, read the application carefully and fill it out honestly.

purchases as long as your total does not exceed your credit limit.

A revolving charge account can be paid in full each month to avoid finance charges. Otherwise, it is paid in portions with part of the total carried over. A minimum payment is required each month. It is specified on the bill. You pay a finance charge on the remaining part that you owe since the store has essentially loaned you the money to buy the merchandise.

A 1 1/2% per month rate of interest is commonly charged on the unpaid balance of a revolving charge account. That whopping 18% per year adds almost one-fifth more to the purchase price of apparel bought this way. Sometimes accounts charge 1% per month, which is a 12% annual interest rate. Check out credit and its costs before using it.

Installment Plans

Retail credit is also offered through ***installment plans***. They are used most often for specific large items. You make a down payment when you take the item. A contract or agreement is written for the amount of the purchase. The total cost of the finance charges and the number of periodic (usually monthly) payments are specified. The least expensive way to use installment credit is to make as large a down payment as possible and have a short payment period.

When dealing with an installment plan or any other contract, never sign anything you do not fully understand. Make sure all verbal promises are put into writing. Cross out all blank spaces, adding the date and your initials. Also, always keep a copy of the contract you have signed.

Credit Cards

Credit cards are available from financial institutions and credit card agencies. For many credit cards, you may pay yearly fees for the privilege of having the cards. There is usually no interest added if the bill is paid in full on time. Interest is added on the amount that is not paid. The types of accounts that are available for charge cards (thirty-day and revolving) are also available among various credit cards.

MasterCard, American Express, and VISA are examples of worldwide credit cards. You can use them in most stores. Some banks have their own local cards that can be used in stores that are members of the banks' plans.

Cash Loan

A loan from a bank, credit union, or finance company can also give you money to make large purchases. Interest rates vary, so it is wise to look for the best loan arrangements.

Evaluating the Use of Credit

When you shop with credit, your monthly statements give you a permanent record of your expenses. You have the use of the apparel before you have paid for it. Payment with credit is a nice convenience without having to carry money with you. Use it wisely for planned purchases and to take advantage of sales. Also, it may be easier for you to return goods if you have bought them with credit.

A disadvantage of credit is that people tend to overbuy when they use it. They buy impulsively and charge their purchases. They buy what they really can't afford or don't need. They decide to figure out how to pay for the items later. Note the pros and cons of using credit listed in 16-18.

Never charge more than you can afford. You can end up in trouble if you cannot pay your bills. You could be faced with high penalty charges. You could cause yourself to be denied credit in the future. Also, the finance charges can add up quickly. To reduce interest payments, pay as much as you possibly can toward the bill each month. Remember, every dollar you spend on finance charges is one less dollar to spend on more merchandise. Sometimes it's worth it; sometimes it's not.

The ***Truth-in-Lending Law*** says that consumers must be informed of the credit terms for charge accounts, installment contracts, and cash loans in uniform, easy-to-understand terms. The method of calculating finance charges must be disclosed. Consumers can then compare the terms and choose the type of credit they want.

If a credit card is lost, immediately contact the store, bank, or other institution that issued it. Otherwise, if it is used dishonestly, you may be responsible for paying the first $50 of charges placed on it.

Consider Home Shopping Options

You may consider nonstore shopping from home. Merchandise is selected from mail-order catalogs, cable television shopping channels, or Internet Web sites. You may shop at any time of the day or night.

Catalogs such as those shown in 16-19, TV shows, and computer retail sites and portal "malls" show and describe items in detail. Ordering is done by mail, by telephone, or directly on the computer. The merchandise is delivered to the consumer's address by mail or other parcel service. Payment is usually made by credit card or sometimes by check or debit

Pros and Cons of Using Credit

Pros (Advantages)

- Can pay later and have merchandise now
- Don't have to carry cash
- Can write just one check to pay a bill that covers several purchases
- May be easier to return unwanted purchases
- Can give you a good credit rating if bills are paid on time
- Provides a permanent record of purchases with monthly statements

Cons (Disadvantages)

- Need a special card or loan agreement
- Is easy to buy too much or impulsively
- Involves risk, since card or number can be used fraudulently by others
- May be more expensive to use than cash because of fees and finance charges
- Can give you a poor credit rating if bills are not paid on time
- Might limit where you make your purchases

16-18 The use of credit has both pros and cons. If you use credit, be sure you understand all terms of the agreement and follow them properly.

16-19 Mail-order catalogs such as these, sent to millions of residences, sell all types of garments, accessories, and other items.

card. C.O.D. (cash on delivery) purchases require payment to the delivery service when the merchandise is delivered to the purchaser. C.O.D. deliveries are not very common.

Home shopping is convenient if you are too busy to go out to stores. You can do it whenever you have time. You don't have to find a parking place or pay for public transportation. Unique items are often available, and merchandise prices are sometimes lower because of lower overhead.

On the other hand, with home shopping you can't see, feel, or try on merchandise. If the item is wrong or does not fit, you must arrange for return delivery. Also, once you place an order, your name may be on lists that are sold to other direct purchase merchants. Unless you specify that your name and address not be sold or rented to others, you will probably receive catalogs from other mail-order retailers.

Chapter Review

Summary

Level-headed judgments at the time of purchase will enable you to make the right decisions. Comparing qualities and prices of the same or similar items from different retailers before buying will enable you to know the availability and prices of items.

In clothing, the best value is the highest quality for the lowest price. Quality might be high, medium, or low, depending on the level of construction, materials, and design. Consumers should pay only for the quality wanted and needed for various items. Specific points of quality should be checked carefully on garments. Lower-priced imported garments might be good values, depending on your personal viewpoint about where the merchandise was made.

To evaluate the proper fit of clothes when shopping, first know your general size category. Then, with your measurements, determine your specific size. Check fit by trying on clothes when you shop, wearing the underclothes and shoes you will wear with the garment.

Knowledge of trademarks helps you to evaluate quality and fit, too. Designer labels almost always have higher prices than similar nondesigner items.

A bargain is a high value of merchandise for a small amount of money. Consumers should evaluate sale merchandise in terms of whether it fills a genuine need. Various types of sales are held. Avoid impulse buying of sale items that are not part of your plan, budget, or needs.

Buying is the exchange of money or credit for goods or services. A cash purchase is when the full cost of the purchase is paid with money, check, debit card, or smart card. A layaway purchase involves a down payment and agreement that the item is put away for you until periodic payments are completed. Credit purchases include thirty-day charge accounts, revolving charge accounts, installment plans, credit cards, and loans. You should carefully consider the pros and cons of using credit. Also consider home shopping options, such as from mail-order catalogs, TV shopping networks, and Internet Web sites.

Fashion in Review

Write your responses on a separate sheet of paper.

Short Answer: Write the correct response to each of the following items.

1. How can looking at quality clothes in expensive stores help people do their comparison shopping?
2. Explain what "best value" means in terms of clothing.
3. List five characteristics of high-quality garments.
4. True or False. Quality is always related to price.
5. When judging the quality of a garment, list five points to consider about its seams.
6. List five points to consider when judging the quality of a garment with a lining.
7. When trying on a garment, name five ways to check the fit.
8. True or False. A low price is always the sign of a real bargain.
9. Why do stores have "clearance" or "inventory" sales?
10. What is the cause of an overdrawn checking account?
11. When you make a deposit toward buying an item, and the store puts the item away for you for a certain length of time, what kind of purchase are you making?
    A. Debit card.
    B. Layaway.
    C. Credit.
    D. Mail-order.
12. A credit rating is _____.
    A. the percentage rate of finance charge on credit purchases
    B. a person's financial standing or record of paying bills
    C. an evaluation system that ranks credit companies and banks
    D. the ratio of a consumer's credit purchases to his or her cash purchases

13. What is the major difference between a thirty-day charge account and a revolving charge account?
14. List three advantages and three disadvantages of using credit.
15. C.O.D. means _____.
    A. charge or deposit
    B. clearance of department
    C. cash on delivery
    D. choice of designer

Fashion in Action

1. Comparison shop for two different apparel items in many types of stores. Compare specific points of quality in the garments, as well as prices. Decide which items would be the best buys. Explain your reasoning in an oral report.
2. Select three garments that you will need to buy in the near future. Plan what you will do to get the best value. Figure out what quality each garment should be. Decide where you should shop and when in the fashion season you should buy. Determine the best method to pay for each garment. Write the plan on paper and try to follow your own advice in the future.
3. In class, discuss how people shop for clothes at regular prices and at sale prices. Are shoppers more cautious when they can't return items? Do they settle for lower quality when they buy items at low sale prices?
4. Investigate sources of credit including charge accounts, installment plans, credit cards, and cash loans. Compare eligibility requirements, interest rates, extra finance charges, and payment schedules. Present your findings in a written report.
5. Design a bulletin board display on home shopping.

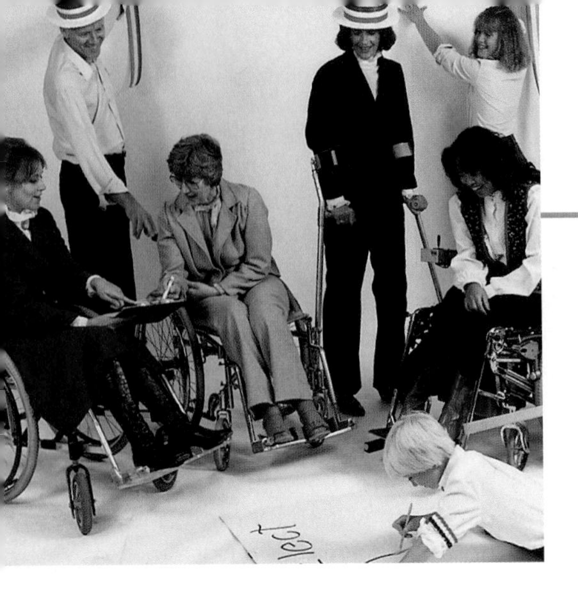

17 Apparel for People with Special Needs

Fashion Terms

grippers
layette
infant kimono
infant gown
self-help features
safety features
growth features
trunk
maternity fashions

After studying this chapter, you will be able to

- evaluate the apparel needs and sizes for infants.
- judge the clothing needs and sizes for children.
- describe wardrobe selection guidelines for the elderly.
- discuss the special apparel needs of people with disabilities.
- put together and pack a wardrobe for travel.

Some people need special features in their apparel because of their age or physical limitations. Enabling these people to dress with ease and to look their best helps them gain self-confidence and pride. Appropriate clothing can enhance their physical, social, and emotional well-being.

Apparel Needs of Infants

Many factors influence the selection of baby clothes. Comfort, practicality, and safety are important. Other considerations include the climate, finances, and availability of laundry facilities.

Infant growth is rapid, and wardrobe needs change quickly. Shopping for slightly larger sizes will best accommodate a fast-growing baby. The garments can then be worn longer. However, garments that are far too big are dangerous, since they can become tangled.

Comfort

Comfort is a must for infant clothing. Tight garments, or too many garments, will make the baby uncomfortable. Soft, durable garments are best for the baby's delicate, sensitive skin. Harsh fabrics can cause a skin rash. Fabrics that are extremely fuzzy can irritate the nose and throat. The baby will be most comfortable when dressed in supple, nonirritating garments.

Infant garments should be constructed of fabrics that are absorbent and let moisture evaporate. Cotton flannelette, as shown in 17-1, challis, batiste, plissé, terry cloth, and jersey are good. Modacrylics have good hand and natural flame resistance. Nylon gives good strength and wearability. Blends of these fabrics, and others, add the needed characteristics to garments. The fabric should be preshrunk or shrink-resistant.

Knit garments are very popular because of the built-in stretch that "grows" with the infant. They also provide warmth and ventilation.

Practicality

Infant clothing should be designed for ease of changing. It is important to keep the baby clean and dry. Most infants dislike clothing that has to be pulled over the head. Dressing and undressing will be easier if the baby's clothing has front openings. Shirts often have open shoulders that overlap, or fronts that overlap and snap near the side. Blanket sleepers and coveralls usually have a front zipper that extends down into one leg, as in 17-2, to allow for easy dressing. Some have ***grippers*** that are heavy-duty types of snaps.

The time of year the baby is born and the climate are important considerations when purchasing clothes. It is best to buy only what is needed for each season as the baby grows.

Warmth is necessary, since new babies lie still and sleep most of the time. Their garments provide their protection.

Fewer garments are needed if they can be laundered frequently. When buying clothes, washability is important. Easy care is necessary for all infant items.

Safety

The flammability of baby clothes has been a concern in recent years. The federal government has taken action to require flame-resistant sleepwear. Flammability standards have been set for all infant sleepwear. Garment packages and tags indicate if a garment is flame-resistant. Another safety concern is the type of fasteners on infant clothing. Lightweight zippers or gripper snaps are best. Buttons and trims that can be pulled off are not practical and may be dangerous. They can be swallowed or poked into a nose or ear by the baby if they become loose. Trimmings should be soft and safe. They should be washable in the same way as the fabric.

The Right Start Catalog, Los Angeles

17-1 These pajamas of 100 percent cotton are absorbent and comfortable.

The Layette

A ***layette*** is the assembled minimal needs of clothing and textile goods for a newborn. It consists mostly of diapers and standard sleeping clothes. Only a few dressy clothes are needed for outings. Use the chart in 17-3 as a guide in selecting what should be included in a basic layette.

Diapers

Disposable diapers are popular. Many people feel they are better for babies, since moisture is drawn away from the skin and fewer rashes occur. They come in different sizes to fit newborns through toddlers. Thicker ones are available for nighttime wear. Disposable diapers are convenient for travel or to use for spares in emergencies. However, disposable diapers are expensive to use and waste natural resources. They also create a great deal of garbage.

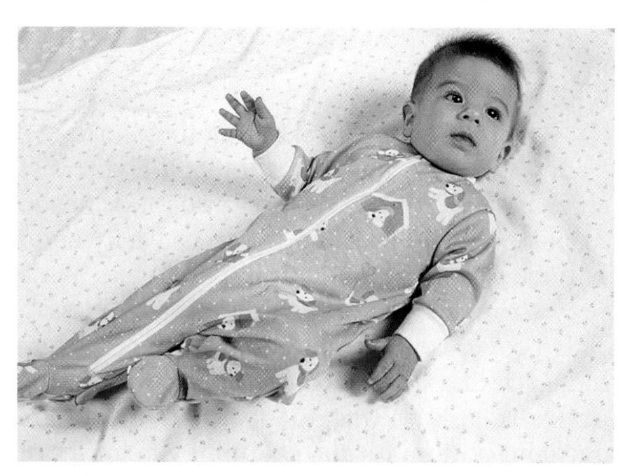

The Right Start Catalog, Los Angeles

17-2 The long zipper from the neckline to one ankle gives a large opening for dressing the baby or for changing diapers.

A Sample Layette

5 Undershirts	2 Blanket sleepers
36 Cloth diapers	3 Bibs
4 Waterproof pants	4 Diaper pins
Disposable diapers	4 Crib sheets
4 Gowns or kimonos	1 Crib blanket
4 Coveralls or stretch suits	3 Receiving blankets
Shawl, bunting, or topper set	2 Waterproof pads
Bonnet or cap	3 Washcloths
Socks	6 Towels
	Sweater and bootie sets

17-3 After getting the items in a basic layette such as this one, fun and cute extras can be added as desired.

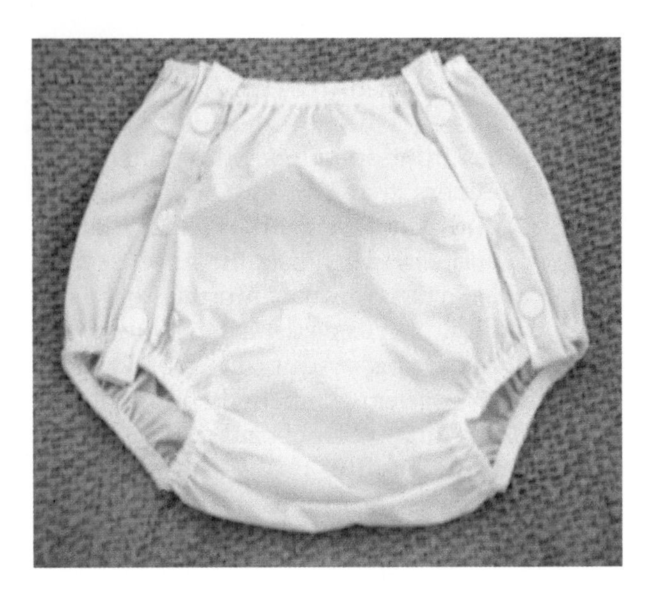

17-4 These waterproof pants snap at the side front seams. They are worn over cloth diapers.

Cloth diapers come in a variety of fabrics and styles. A rectangular shape that folds in several different ways as the baby grows is usually preferred. It is essential that diapers be soft, absorbent, washable at hot temperatures, and quick drying. Disposable diaper inserts are often used inside fabric diapers.

Waterproof Pants

These are used over cloth diapers. They are usually made of soft, sturdy vinyl. They should have sealed seams and soft nylon finished edges on all openings. Waterproof pants either pull on or snap on both sides, as in 17-4. Most are machine washable.

Kimonos and Gowns

A full-length infant garment that looks like a long dress is called an ***infant kimono***. It is closed in the front or back with snaps or ties. It has sleeves and is often made of lightweight flannel. An ***infant gown*** is similar but has a drawstring at the bottom to keep it closed. This provides warmth yet allows the baby to have kicking room inside the garment. The drawstring is untied to open up the entire garment bottom for the changing of clothes or diapers. Cuffs on the sleeves can fold down over the hands to serve as mittens for extra warmth or to stop fingernails from scratching the face. Drawstrings should never be used at the neck since they are dangerous.

Bibs

Bibs for infants are small (drool or teething bibs) or large (mealtime bibs). Bibs usually have absorbent terry cloth on one side and wipe-clean vinyl on the other. The smaller bib is worn most of the day during the baby's teething months. The vinyl side of larger bibs often has a catch-all pocket. The adjustable neck opening of bibs either snaps or ties in the back.

Stretch Suits or Coveralls

These are made of two-way stretch terry fabric in assorted designs. Some have zippers. Some have snaps or grippers along the entire crotch inseam. They have enclosed feet. They are long- or short-sleeved for various climate conditions.

Towels and Washcloths

Towels and washcloths for infants are 100 percent cotton. They are not as thick or heavy as regular towels. Most have a "hood" across one corner to put over the baby's head for warmth after the bath. The towels often come in white or have a small printed design.

Infant Apparel Sizes

Manufacturers base their infant apparel sizes on the average heights and weights of infants.

The clothing sizes are actually listed by age in months or by descriptive terms, such as small, medium, and large. However, clothing should be selected by height and weight rather than by age. Charts 17-5 and 17-6 give the most widely used categories of sizes for infant clothing.

The smallest sizes are not very practical and may be outgrown before being worn. Also, as with other types of apparel, there are differences among the sizing systems of various manufacturers. The garment label and package information can guide you in evaluating a manufacturer's specific size. Sizing charts are also included in mail-order catalogs.

Clothing for Young Children

The selection of clothes for toddlers and preschoolers must emphasize comfort. Young children should be dressed in simple and functional clothes, as in 17-7, rather than in small versions of adult apparel. Their clothing needs are based on growth, constant movement, and the development of their coordination to dress themselves.

Toddlers are children who are actively moving or walking. They are usually between the ages of 1 and 2 1/2 years. They have short bodies, short legs, and a protruding abdomen. Preschoolers are taller and not as round as toddlers. They are starting to have a defined waistline. They are between the ages of 2 1/2 and 5 years.

Appropriateness

Toddlers and preschoolers need clothes built for comfortable, constant action. Their garments should be easy to launder, and they should not be scratchy or itchy. They should have adequate fullness where needed for

Age-Related Clothing Sizes for Infants		
Size	Height (In Inches)	Weight (In Pounds)
3 Months	Up to 24	Up to 13
6 Months	24 1/2 to 26 1/2	13 1/2 to 18
12 Months	27 to 29	18 1/2 to 22
18 Months	29 1/2 to 31 1/2	22 1/2 to 26
24 Months	32 to 34	26 1/2 to 29
36 Months	34 1/2 to 36 1/2	29 1/2 to 32
48 Months	37 to 40	32 1/2 to 38

17-5 Age-related sizes are based on average heights and weights of infants of different ages. However, individual babies vary in their growth rates. Whenever possible, buy infant clothes according to a baby's actual height and weight, not age.

General Clothing Sizes for Infants		
Size	Height (In Inches)	Weight (In Pounds)
NB (Newborn)	Up to 24	Up to 14
S (Small)	24 1/2 to 28	14 to 19
M (Medium)	28 1/2 to 32	20 to 25
L (Large)	32 1/2 to 36	26 to 31
XL (Extra Large)	36 1/2 to 38	32 to 36

17-6 General sizes are also based on both height and weight.

freedom of movement. Tight clothing binds and restricts movement. Extra room should be provided at the seat, armholes, pant legs, and crotch. Elastic inserts should be included to hold garments in place. Soft, unstructured styles are good, such as one-piece playsuits that hang from the shoulders. Shoulder straps should be wide and crisscross in the back. Some shirts have shoulder tabs to keep up straps of pants or skirts. Pants and straps that keep falling off cause frustration and can create a resentment toward dressing.

Clothing for preschoolers must be easy for them to manage. Sometime during the third year, children become aware of clothing and show an interest in dressing and undressing themselves. Children develop greater independence and responsibility if they can get into and out of their clothes by themselves. For this reason, the clothes should have ***self-help features***. Armholes should be large. The backs and fronts of pants, shirts, and undershirts should be different or well marked. Closings should be easy to fasten and located in front, so they can be seen and manipulated. Grippers are the easiest fasteners for children to handle themselves. Zippers and medium-sized flat buttons can be used. Hook and loop fasteners, such as Velcro®, are also good. Elastic waistlines enable children to pull pants on and off by themselves.

Children's garments should be attractive in color and style. Children generally like strong, bright colors, as in 17-8. Prints should be in proportion to the size of the child.

Corduroy, textured fabrics, and prints help hide wrinkles and soil. Pockets are important features because children need places for the treasures they collect. A sense of identity and self-worth is gained by children from compliments received on their clothing or on the special details of their garments.

Safety

Children's clothes should provide ***safety features***. If possible, they should be of flame-retardant fabrics. This is especially required of sleepwear. The designs of children's apparel should not cause the child to trip or catch on corners. Avoid styles with long, flowing skirts,

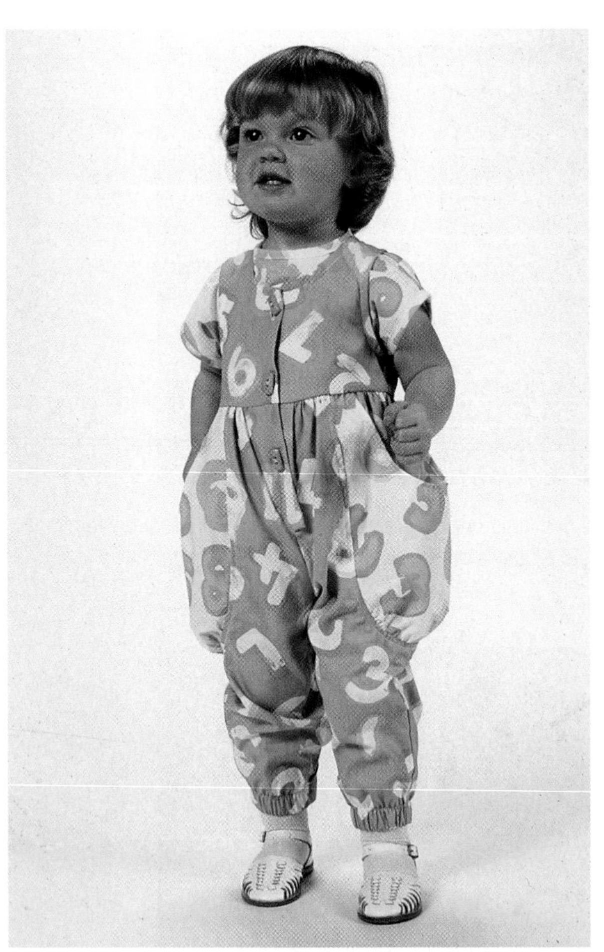

The Right Start Catalog, Los Angeles

17-7 This colorful outfit is comfortable, durable, and machine washable.

JC Penney

17-8 Most children prefer bold, bright colors. These colors are also more visible for safety.

tie belts, drawstrings, or very full sleeves. Trims should be firmly attached and placed where they will not hamper the child.

Children's outerwear must provide protection from severe weather. This means water-repellency as well as warmth without heavy weight. Heavy strain should not be placed on the shoulders. Bulky clothes are tiring and can restrict proper posture. Brightly colored outer garments are readily noticed by motorists.

Practicality

Economy of the wardrobe is also important. Children's garments receive very hard wear. Reinforcements should be included at the points of strain (pocket corners, placket ends, knees, and elbows). For longer wear, the clothes should be of durable, sturdy materials. The best fabrics have soil and stain resistance. They should be made of permanent-press fabric, which provides wrinkle-resistance and easy care.

Ease of care is important for children's clothes, just as it is for infants' apparel. Children are apt to get extremely dirty, and the clothing must be laundered frequently. Children's garments must be durable enough to withstand both the actions of the children and repeated washings. They should be washable without shrinking or losing their color. Even snowsuits and coats for children are available in washable fabrics. Try to avoid purchasing clothing that must be dry-cleaned.

Growth Features

Children's clothes are expensive and are outgrown fast if they do not have built-in ***growth features***. These allow garments to be "expanded" as children grow. This does not mean parents should buy or make clothes that are far too big. Children's clothes should fit well when worn and have features for growth.

Garments with growth features should be of simple designs. They should have large necklines. Raglan or kimono sleeves, or sleeveless armholes, allow for expansion since there is no defined end to the shoulder in those designs. Elastic waistbands and stretch fabrics adapt to growth. Wrap styles are good.

Since children grow taller much faster than they grow wider, length adjustments are the most important. Garments should have wide hems and cuffs that can be let down. They should not have defined waistlines. If clothes have shoulder straps, they should be adjustable so they can be lengthened.

Try to keep a young child's wardrobe down to a minimum amount of apparel. Children often outgrow clothes before they are worn out.

Children's Self-Esteem

In child care, preschool, or school, conformity is important. Children become conscious of what their peers are wearing, and they want to look like others. Approval by their friends gives a sense of belonging and security, as in 17-9. Clothing contributes to a feeling of self-confidence and self-esteem and to sound emotional development. Clothes that make the child look different from others may cause him or her to feel strange, be ridiculed, and become withdrawn or rebellious.

Reinforce interest in appearance by giving children some choice in what they wear. Giving helpful, tactful guidance will help them develop good taste. Clothing selection can be a shared experience between parent and child. Allowing preschoolers to help choose the clothes at the time of purchase ensures that they will wear the garments. The teaching of clothing care can also be started at this age.

JC Penney

17-9 These children are happy being well dressed. Feeling good about themselves will help them do productive schoolwork.

Provide hooks on which to hang outer garments and a hamper in which to deposit dirty clothes.

Children's Apparel Sizes

Sizes of clothes for toddlers are 1T, 2T, 3T, and 4T. They are based on age, but children do not necessarily fit into the size that corresponds with their age. Order or buy by correct measurement, not by age. Garments for toddlers are made to go over diapers or training pants. They fit children who have not lost their baby roundness. They usually have a snap crotch with grippers.

Clothing for preschoolers is sized from 2 to 6X. These sizes are also related to age. However, always buy or make children's apparel in sizes according to height rather than age indicators. Height is the most important measurement for determining size. Children grow faster in height than in width. The length of the arms, legs, and ***trunk*** (measurement from shoulder to crotch) change much more quickly than the width of the shoulders, chest, and hips. Slim, regular, and chubby options are offered within the preschool sizes.

The two charts in 17-10 and 17-11 show the most widely used approximate measurements (in inches) of clothing sizes for toddlers and young children. Toddlers' and children's sizes 2, 3, and 4 are based on the same chest and waist measurements. They differ in the diaper allowance, length, and width through the shoulder and back.

Sometimes preschooler clothing sizes are labeled small, medium, and large. In such cases, "small" relates to sizes 2 and 3 in the chart, "medium" is sizes 4 and 5, and "large" is sizes 6 and 6X. Elastic waist garments will fit waistlines approximately one inch smaller to one inch larger than shown in the size chart.

Clothing Sizes for Toddlers

Toddler Size	Height (In Inches)	Chest (In Inches)	Waist (In Inches)	Approximate Weight
1T	29 1/2 to 32	20 to 20 1/2	19 1/2	25 lb.
2T	32 1/2 to 35	21 to 21 1/2	20	29 1b.
3T	35 1/2 to 38	22 to 22 1/2	20 1/2	33 1b.
4T	38 1/2 to 41	23 to 23 1/2	21	38 lb.

17-10 The measurements shown here are similar to those used by apparel manufacturers and commercial pattern companies to determine clothing sizes for toddlers.

Clothing Sizes for Young Children

Children's Size	Height (In Inches)	Chest (In Inches)	Slim Waist (In Inches)	Regular Waist (In Inches)
2	32 1/2 to 35	21	18	20
3	35 1/2 to 38	22	18 1/2	20 1/2
4	38 1/2 to 41	23	19 1/4	21 1/4
5	41 1/2 to 44	24	20	22
6	44 1/2 to 47	25	20 1/2	22 1/2
6X	47 1/2 to 49	25 1/2	21	23

17- 11 Garments in sizes 2 to 6X are for young children who are taller and more slender than toddlers.

More two-piece outfits are needed at this stage. Interchangeable separates will give the greatest wardrobe mileage for the least cost. Separate tops must be long enough to stay tucked in.

Children's shoes are in sizes 6 to 13 1/2 and then start over at size 1 of a new scale as they get bigger. They then continue up into adult sizes. Be sure to have the child's foot measured by a well-trained salesperson for an accurate fit. Nothing should ever be put on a child's foot that restricts the normal growth of the bones, muscles, or nerves. Children's shoes need to be checked, and often bought, every few months to avoid later deformity or discomfort. Socks must also fit properly.

Sewing Children's Apparel

Children's clothes are fun and easy to sew. You can save a great deal of money by making them. The investment for fabric is small. Cute patterns are available that are quite simple to make. You can add creative touches that will make the child love each garment. If you make a sewing mistake, you can cover it up with an appliqué or some trim. If a tear appears during wearing, you will have fabric scraps to use to put on a matching patch. Leftover fabric can be used for mending, lengthening, and adding decorations.

By adding growth features and using durable sewing techniques, you can make garments that will last a long time. For instance, include several decorative rows of tucks parallel to the hemline of a skirt or dress to be let out later for more length. Choose designs and fasteners that make the garments easy to change. Include flat seams for comfort. Put extra layers or padded appliqués at the knees.

Good fabrics for children's garments are those that are absorbent and either firmly woven or closely knitted. Corduroy, denim, poplin, gingham, seersucker, and broadcloth are good choices. Knit and stretch fabrics are especially good for comfort, action, and growth. Also, children enjoy soft fabrics, such as brushed flannel and terry cloth. Cotton blends are wise choices. Clothing made from fabrics that contain nylon, acrylics, and polyester fibers wear well and have easy care. Strive for comfort without bulkiness and warmth without weight.

Hand-me-downs can help stretch clothing dollars. These can be adjusted to fit. New trim, appliqués, or other personal touches can make them special for the child. Permanent marks, such as lines where the hem has been let down, can be covered with preshrunk ribbon or topstitching that coordinates with the design of the garment.

Clothing for Older People

It is important for older people to maintain their self-esteem by looking attractive. Appropriate and becoming clothing can help them feel better about growing older. Some people lose interest in their appearance as they approach old age. This may be because of poor health, weight changes, or loneliness. Being well-dressed helps older people retain a greater interest in life.

Proper clothing for the aged is receiving more emphasis these days. With declining birth rates and improved medical practices, there are larger numbers of older people all the time. Americans are living longer. Also, they are tending to stay active in recreational and community activities and thus need up-to-date clothes, as in 17-12.

Many older people live in retirement communities where they have frequent social activities. Fashionable apparel is important for these people. It can become a source of compliments, which give the wearer self-esteem and a feeling of well-being.

Physical Changes

As people pass middle age, their body proportions usually change. Gradually their faces thin, their abdomens and hips get larger, their legs get thinner, and their waistlines thicken. Older women often have a large and low bust and stooped shoulders. Older men often have a protruding stomach. These changes become more accentuated with time until the older person can sometimes no longer wear standard-sized, ready-to-wear clothing.

17-12 Many older people lead active lives and need versatile, wardrobes.

The physical changes in body proportions with age result in needed alterations if standard-sized garments are purchased. Sewing garments at this stage in life is a good way to occupy extra time as well as to achieve good fit. However, sometimes eyesight and finger dexterity has diminished.

For people who suffer from arthritis or partial paralysis, getting dressed and undressed is often difficult. Older people may also lack energy or feel weak due to infrequent physical activity. They need large, accessible openings with easy-to-close fasteners. Many older people have trouble with closures in the back of garments since they may not be able to extend their arms over their heads. Outfits that have jackets provide for easier dressing and protection against temperature changes. Cardigan sweaters are also good.

Older people often feel cold temperatures more intensely than the young. Heavy clothes are not comfortable to older people. Lightweight fabrics that provide warmth are needed. Acrylic and modacrylic sweaters and polyester fleece garments can provide needed warmth without weight.

The physical disabilities of older people require their clothing to have many of the same characteristics as those for young children. They need safety features and fabrics that feel soft to the skin. As glandular secretions decrease, their skin often becomes thin, dry, and inelastic. Heavy fabrics and rough textures may irritate delicate skin.

Personal coloring in later years may be different from coloring earlier in life. Gray and white hair enables different colors to be worn attractively. Yellow and brown seldom look good if skin is pale. Colors and styles that focus attention on the face are helpful in hiding figure problems and physical disabilities.

Selecting Apparel for Older People

Certain style features are generally recommended for older people. They include shift-style dresses without fitted waistlines, A-line skirts, and lowered necklines. Men's trousers should be less fitted with more wearing ease. Garments with elasticized waistbands, long center front closures, and large patch pockets are good. Also, raglan sleeves and straight long sleeves without cuffs reduce fitting and buttoning problems. Knit fabrics and blends with polyester are good since they have easy care.

Shopping must be done carefully for older people to satisfy their clothing needs. This may not be easy for an older person who is not used to the complex consumer decisions of today, or who may not have available transportation. Also, the lower incomes of many retired people create extra concerns about the price and durability of their garments.

Apparel for older people should be attractive as well as functional. It should be conservative, yet fashionable. Clothes should have simple care requirements and should be easy to put on and take off. The physical and psychological needs of older people for proper clothing must always be considered.

Apparel for People with Disabilities

Millions of American adults and children have physical or mental disabilities that limit their activities. They need clothes that are comfortable, functional, and attractive, as shown in 17-13. Appropriate apparel can enhance their independence and productivity.

Easy care is an important factor in clothing selection for many people with disabilities. Those who are weak and inactive need clothes that provide warmth without weight. Garment fabrics should be soft and absorbent. Unsuitable or tight clothing can hinder movement, produce discomfort, make the wearer feel unattractive, and thus lower the wearer's self-image and morale.

People with physical limitations need apparel that simplifies the task of dressing. Some disabilities cause reduced manual dexterity and strength. Persons with such handicaps feel better psychologically if they can dress themselves.

Specific Garment Features

For people with disabilities, choose styles without fussy details and with a minimum of fastenings. Self-help features may include easy-to-handle front or side closures. Grippers and zippers are good fasteners. Small loops stitched to the garment near the bottom and top of zippers can be used as leverage when zipping and unzipping. A decorative zipper pull, such as a ring, yarn tassel, or braided ribbon can be added to the existing zipper pull as an aid in grasping it. Velcro® fastening pieces are convenient. However, a long tape of it might be hard to pull apart.

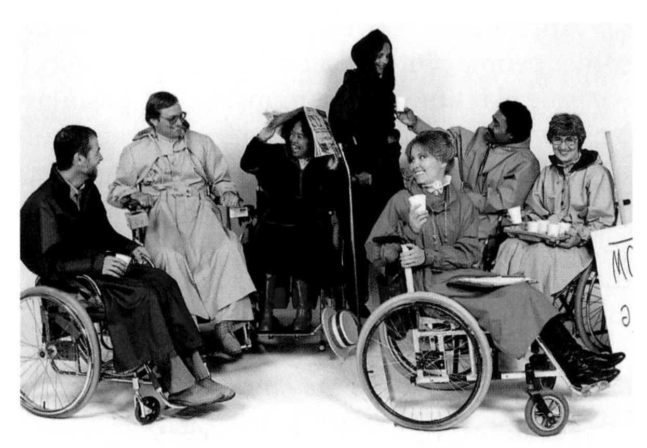

Pirca Fashions of Sacramento

17-13 Apparel for outdoor activities for people with disabilities can be stylish as well as practical.

Buttons may be difficult to fasten. If buttons seem necessary, they should be large and can be sewn on with elastic thread for easy maneuvering. Also, Velcro® squares or dots can be used under existing buttonholes that are sewn closed with the buttons sewn onto them. The "mock button" closing is then actually held with the Velcro® piece. Whatever the method, there should be as few fasteners as possible above mid-chest level.

Raglan and dolman sleeves are easy to slip on and off, and they allow for ease of movement. Large armholes may be needed. Shoulder openings, longer neck openings, and drawstring necklines and waist closures often add convenience. Elastic bands can be used at the waistlines of pants and skirts. Garments will then slip on and off easily, as well as have roomy fullness. Also, a longer back rise can make pants fit better for someone who must remain in a sitting position. This is illustrated in 17-14.

Wrap-around styles are easy for many people with disabilities to handle. Pockets are more usable if they are slightly larger than normal. An air opening in the back of a waterproof rain suit gives ventilation needed for those in wheelchairs. It can be covered with a flap to keep out the rain. Also, a built-in sling can support a weak or paralyzed arm, as shown in 17-15.

Trousers with few or no seams on the back or sides are more comfortable for those who must spend hours every day in a wheelchair. A pocket just below the pants knee can hold tickets, bills, and other items that may fall out of disabled hands. For foot or leg casts, the pants leg seam may have to open its entire length. Pants and underwear may need added seam openings.

Knitted fabrics allow for both ease of wear and ease of dressing. Woven fabrics may be needed if braces or crutches snag knits. Double-knit jersey, velour, and stretch terry are good washable fabric choices. Avoid sheer, flowing fabrics that can catch fire or become caught in wheelchairs or doorways.

Shape and Fashion

Clothing for people with disabilities must often hide figure irregularities. Garments may

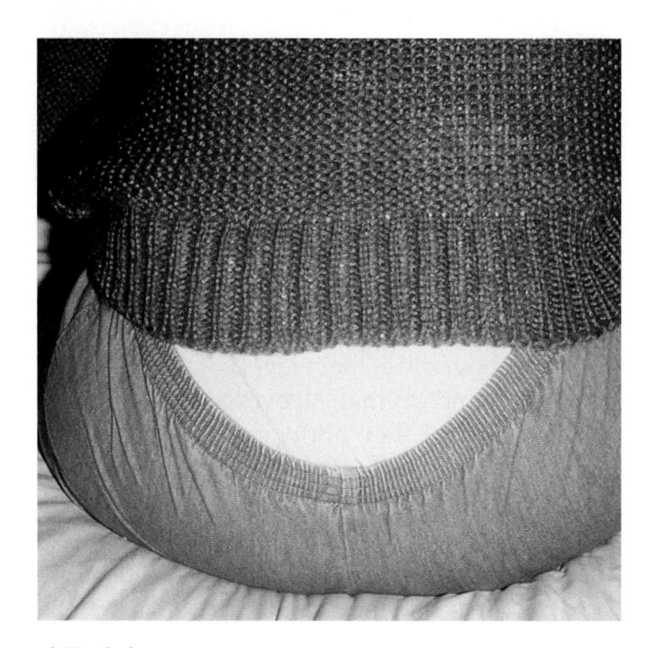

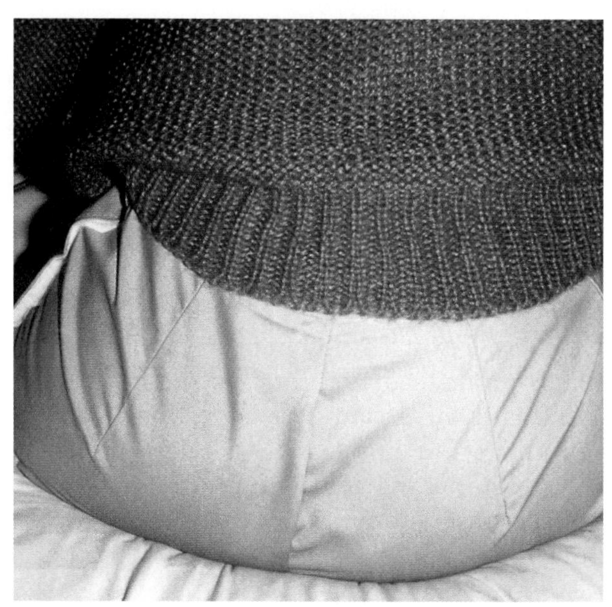

West Virginia University/Laurie Sykes

17-14 The back rise of pants is often too short for someone who sits all the time. With a longer back rise, pants fit nicely up to the waistline at the center back.

West Virginia University/Nora MacDonald

17-15 Specific features for certain physical disabilities can be built into apparel, especially if they are sewn at home.

have to be of different shapes than usual. They may have to go over mechanical devices or body casts. They may have to cover an unusually small limb or an enlarged area. The body may not be symmetrical, so clothing camouflage techniques may be effective. In many cases, separates are a good choice so one size on top can be paired with a different size on the bottom.

Clothing for people with disabilities should help the wearer project a fashionable image, as in 17-16. Most people who have special clothing needs prefer to wear the latest fashions. Looking good gives their spirits a lift.

Apparel can convey a sense of normalcy to some people with disabilities. For instance, wearing daytime clothing, instead of pajamas or loungewear, can help to create a feeling of normal health. Protective cover-ups for students with disabilities who attend a special school have been designed to resemble a turtleneck shirt, a football jersey, overalls, a jumper, and a cobbler's apron. They are functional, durable, attractive, and appropriate in design for the ages of the students.

Fashionable apparel helps the self-image and adjustment of students with disabilities in public schools, too. When these students are mainstreamed, they need clothes that look like those of other students but that are modified to solve some of their dressing problems. Such garments keep them from being different from the others. They want the opportunity to be just like everyone else.

Obtaining Apparel for People with Disabilities

It is difficult to find appropriate and attractive self-help clothing for people with disabilities. There is a definite need for specially designed clothing to be manufactured and promoted for people with disabilities. Although there is a sizable demand, it is not large enough to encourage manufacturers to produce special items on a mass scale. Manufacturers feel that few people need the same adaptation. One possible alternative would be for the manufacturers to offer styles that could also be worn by non-disabled people. Some ready-to-wear apparel is already suitable, such as wraparound skirts and some sweat suits.

A group of Japanese firms has designed and manufactured a line of clothes for people with disabilities. Each company produces one category of garment, such as pajamas, raincoats, shirts, and underwear. The Japanese name of the line means "self independence." The products meet foreign regulations and are being exported to other parts of the world. Retail prices are about 30 percent higher than normal because of the complicated sewing processes. High fashion designers in Rome have also done collections for disabled people. The fashions are shown on the runway by models in wheelchairs.

Special clothing needs can often be met by modifying ready-to-wear clothing. For instance, set-in sleeves can be adapted to include an action pleat that allows more room for arm movement from a seated position. This is shown in 17-17. Instructions for such modifications, and publications about selecting or changing clothing to meet special needs, are

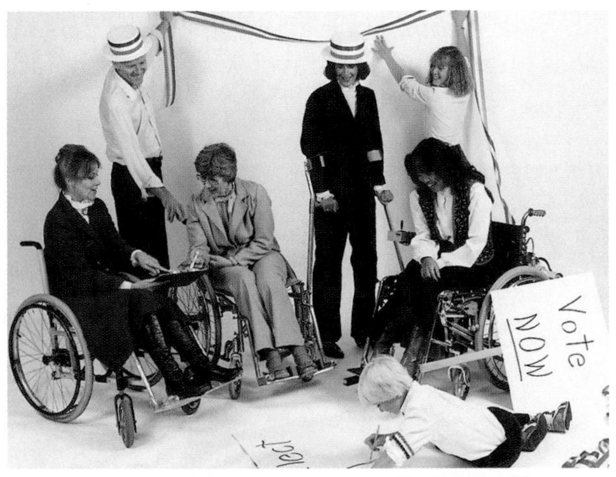

Pirca Fashions of Sacramento

17-16 These people with disabilities look fashionable and professional.

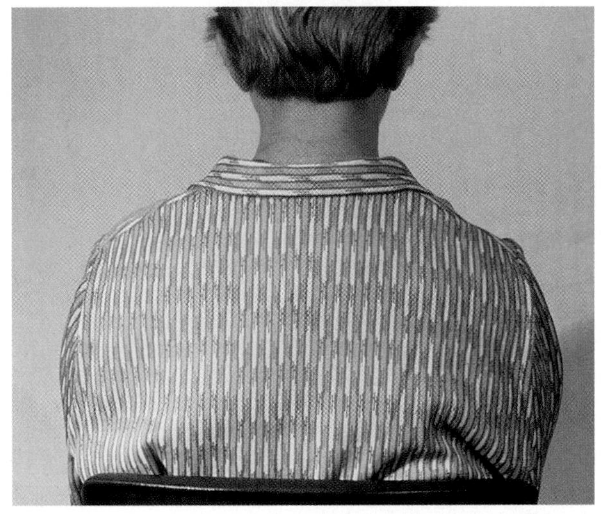

West Virginia University/Nora MacDonald

17-17 This action pleat allows greater freedom of arm movement from a seated position. When the arm is at rest, the action pleat folds inside and does not show.

available through county extension agents. Extension agents may also have other resources, such as videotapes and specialized mail-order catalogs.

It is feasible to successfully adapt home sewing patterns to the individual needs of people with disabilities. Several booklets have been written on how to do this. Helpful information is also available on the Internet.

Knowledge of clothing modifications might give you an opportunity to provide service in your community. You could combine interests in clothing design and social service to develop garments suitable for people with disabilities.

Maternity Fashions

Maternity fashions are for pregnant women. They usually have soft pleats, gathers, and added fullness in front, as shown in 17-18. Maternity skirts and pants have an elasticized panel across the front with an expandable elastic waistband. With so many women in the workforce, there has been a large demand for maternity business apparel that gives a professional look, as in 17-19.

Maternal Instincts/Bob Hennessy, Photographer

17-18 This casual maternity outfit has plenty of fullness for comfort and a stylish appearance throughout a pregnancy.

Entire shops and mail-order catalogs cater to only maternity fashions. Department stores usually have a maternity apparel section that carries fancy attire as well as casual wear. Fashionable home patterns are also offered for pregnant women.

Since maternity clothes are worn for about five months during a pregnancy, they are often only needed for one or two seasons of the year; for instance, summer into fall, or winter into spring. Some maternity dresses can continue to be worn after the baby is born by adding a sash or belt to the waistline.

Travel Wardrobes

A *travel wardrobe* is all the apparel a person takes on a trip. Good travel wardrobes should

Simplicity Pattern Company, Inc.

17-19 Many maternity fashions are appropriate to wear in business settings because of their professional styling.

be lightweight, well coordinated, and wrinkle-resistant. Do not try to pack most of the clothes you own. Much of it will be unused, excess baggage. Some people say that after you have planned what to take, eliminate half of it! Careful planning is the key. Pare down your wardrobe to a few coordinated pieces that will reduce your amount of luggage.

Mix and Match

Clothes for trips must be appropriate as well as good travelers. There is no need to have lots of new clothes. Your present apparel will look new to people elsewhere. Also, you already know how to mix and match your old favorites, and you will be more comfortable in them, as in 17-20.

When you pack, select coordinated separates and simple garments that can be accessorized for many options. With three tops and three bottoms, you can have nine different combinations. Travel light with a small number of clothes that go together well.

The Fashion Association

17-20 By mixing and matching several favorite garments, different outfits that will be comfortable for the various activities of your trip can be created.

Plan your travel wardrobe around your most becoming colors and silhouettes, which complement each other so outfits are interchangeable. Use only one or two basic colors, if possible, for easier coordination. Bright colors can accent neutral garments. Be inventive with belts, scarves, jewelry, neckties, and other accessories, so your wardrobe possibilities will be stretched.

Be Practical

Fabrics with dull finishes and moderate textures show less soil and fewer wrinkles. They are also appropriate for many different places and occasions. Stretchy fabrics and knits are good. Small prints, checks, and tweeds also hide soil and wrinkles. Accessorize with light or bright touches near the face so you won't look tired (which you really might be while traveling).

Comfortable shoes are a must. They should coordinate with all items in your travel wardrobe. Never wear new shoes on a trip. You will probably be doing lots of walking, plus your feet might swell from sitting en route. A pair of dress shoes and a different pair of walking shoes are required for regular travel. Hiking or sports activities will have different footwear requirements. However, since shoes and boots are heavy, only take what you really need.

Determine Your Needs

Your wardrobe needs are determined by where you travel and the type of transportation you take. Try to find out about the trip's activities and dress code ahead of time. Figure out the number of days you will be away. Make a list of what you plan to wear for each occasion. See how many outfits you can make with a few key garments, as in 17-21.

For long distances, you may need an outfit that "sits" well. That is, it doesn't bind while sitting for a long time and looks fresh upon arrival. If going by car, make sure you aren't so casual that you feel out of place in restaurants or at sight-seeing spots. Plan to arrive at your destination dressed presentably. Metropolitan areas may require more sophisticated clothes than rural resorts.

On an extended trip, a larger wardrobe is not needed. You will see different people at

The McCall Pattern Company

17-21 Minimize your luggage by packing only what you will really need. Taking separates that you can mix and match reduces the total amount taken.

your various stops, so the same clothes can be worn over and over again. You will always look good, and no one but you (and your traveling companions) will know you are wearing one of your few, well-chosen outfits. Plus, you will not be exhausted from carrying excess baggage! However, provisions will have to be made for laundering.

You may have to wash garments in a bathroom sink and dry them overnight. You may want to take travel-sized packets of detergent. You can also take a small stretch clothesline with a few clothespins. You may want to hang garments on plastic hangers.

Lightweight travel irons have assorted plugs for different electric currents in foreign countries. Other clothing care items you may want to take include a clothes brush, a small shoe cleaning kit, and a small bottle of cleaning solvent. Large, plastic zipper bags may serve as see-through organizers when starting the trip and come in handy if some items are damp or dirty when they have to be packed for the return trip.

Before you pack, be sure that none of your clothes need mending. Check that buttons are secure, that hems are intact, and that seams are not coming unstitched. You don't want to have to make any clothing repairs while on your trip. However, bring a small sewing kit and some safety pins just in case! Also, be sure all clothes are clean and without stains before packing.

Packing Tips

If you plan to be making many stops, try to organize your suitcase to eliminate unpacking items from the bottom as much as possible. In other words, pack last what you will use first.

Before putting items in the suitcase, stuff shoes with socks, hose, belts, or soft items. Turn the tops of the shoes toward each other with heels at opposite ends. Place the shoes in plastic or fabric bags to prevent them from soiling your clothes. Button jackets and close zippers when folding garments. Match the leg seams on pants to preserve the center front and back creases.

A good deal of wrinkling can be avoided with proper packing techniques. You may either pack by layering, rolling your garments, or a combination of the two.

In a hard-sided suitcase, pack the odd-shaped and heavy items on the bottom, preferably near the edge opposite the handle. Then pack your large apparel. Pants legs and other long items should be laid in alternating directions across the suitcase. One end of each garment should be at the "wall" of the case. The remainder can hang out the other side, to be folded in around all the other items later. Add tissue paper or dry cleaning bags on top and along the edges of the suitcase to prevent creasing when the ends are eventually folded.

Roll up sweaters, lingerie, and other non-wrinkling items. Make a layer of rolled items on top of your pants layer. Stuff pajamas and underwear around the rolls to fill in the spaces. This keeps items from shifting.

Now place dress shirts, jackets, skirts, and dresses flat on top. Alternate the edges of each piece so collars, shoulders, waistbands, or other bulky parts do not pile up in one area. The sleeves, ends, and sides should be left hanging over the edge of the suitcase. Lay any delicate items on top. Then complete the packing job alternately folding in the shirts, dresses, skirts, and pant legs, one over the other. It will give you a neat arrangement that will travel well and protect your garments from wrinkles.

If you have plenty of room in your suitcase when packing, here is another method you can try. Hang pants, skirts, shirts, and dresses on individual hangers. Cover each garment with a dry cleaning bag. Lay the entire pile centered lengthwise across the suitcase, hanging out on both sides. Then fold the pile in thirds, hanger side first. When you arrive, shake out and hang up the garments. The plastic bags are amazing crease fighters since they trap air that provides a buffer cushion for the clothes.

If you are taking only a duffel or sports bag, make one big roll of your soft items. Slacks, skirts, and blouses that wrinkle should be on the outside of the roll. Lay them on a bed or other flat surface first. Then place knits and underwear on one end of them, to be on the inside when you roll everything together. Start rolling from the end with the stack of items toward the flatter end. Then place the roll into the bag as a smooth, round bundle. Shoes and a toiletry pack can fit at one end of the bag.

Other Considerations

If you have packed a large suitcase, you may want a tote or shoulder bag for last-minute items such as a magazine or book, hairbrush, and cosmetics. If going by plane, pack some necessities, such as toiletries and a change of underwear, in a carry-on bag. They will be "lifesavers" if you arrive at your destination without your luggage.

When you get to your destination, hang up your clothes as soon as possible. If some clothes have wrinkled, let the steam of a hot shower relax the wrinkles. If you were able to pack a travel iron, use a fluffy towel placed over a bureau or coffee table as an ironing board. If you stay at a hotel, the laundry service might press clothes for you overnight for a fee. The hotel may also loan you an iron and an ironing board.

If traveling by plane and checking your luggage with the airline, mark your bag with a bright piece of tape on the sides or handle. This provides easy identification when claiming baggage at the other end. If you have room, you might also pack a lightweight folded duffel bag in your suitcase. It can be used to carry home those cherished treasures you buy while away!

You will enjoy your trip more if you are comfortably and appropriately dressed, as in 17-22, and not hassled with unnecessary luggage.

Draper's & Damon's

17-22 Knit garments are good for travel wardrobes. They are comfortable and easy to pack. If they wrinkle, the steam from a hot shower can usually restore their shape.

Chapter Review

Summary

Appropriate apparel for people of certain ages or physical limitations can enhance their physical, social, and emotional well-being. Infants need the comfort of soft, durable, absorbent fabrics. Practicality of infant clothing gives ease of changing and laundering. Safety considerations are flammability and safe fasteners and trimmings. The layette includes diapers, waterproof pants, kimonos and gowns, bibs, stretch suits or coveralls, towels and washcloths, and other items of choice. Infant apparel sizes should be selected by height and weight.

Simple, functional clothes for toddlers and preschoolers must also emphasize comfort. They should be appropriate for constant action and laundering. They should have self-help features, safety features, and growth features. Economical, durable wardrobes for young children should also have styles similar to those worn by friends. Children's apparel in toddler and preschooler sizes should be purchased by height rather than age. Children's clothes can also be fun and easy to sew.

Clothing for older people should enable them to look attractive to maintain high self-esteem. Changes in old age involve different body proportions and coloring, as well as limiting physical conditions. Certain style features can help elderly individuals dress themselves, look attractive, and function well.

People with disabilities need clothes that are comfortable, functional, and attractive. Specific garment features include self-help fasteners, a longer back rise, wrap-around styles, added seam openings, and other considerations. Sometimes clothing must hide figure irregularities or mechanical devices, so camouflage techniques and separates are good choices. Since it may be difficult to find appropriate and attractive clothing for people with disabilities, ready-to-wear clothing or sewing patterns can be modified appropriately.

Maternity fashions, with added fullness in front, are available from department stores, specialty shops, mail-order catalogs, and home sewing patterns. Travel wardrobes should be lightweight, well coordinated, and wrinkle-resistant. Plan to mix and match separates and simple garments in your best colors that can be accessorized for many options. Be practical and comfortable. Determine your needs ahead of time, considering where you will travel, the type of transportation, the trip's activities, laundering facilities, etc. Organize your suitcase to eliminate unpacking from the bottom at interim stops. If you take a large suitcase, have an additional tote bag with necessary items. At your destination unpack as soon as possible to minimize wrinkling.

Fashion in Review

Write your responses on a separate sheet of paper.

Short Answer: Write the correct response to each of the following items.

1. Why are knit garments popular choices for infants?
2. Name two safety concerns related to infants' clothes.
3. Describe two kinds of bibs worn by infants.
4. On what do manufacturers base infant sizes?
5. What factor is of least importance when choosing infant apparel?
   A. Fabric type.
   B. Comfort.
   C. Fashion interest.
   D. Flame-resistance.
6. What factor is of least importance when choosing clothing for young children?
   A. Comfort.
   B. Likeness to adult apparel.
   C. Growth features.
   D. Self-help features.
7. Name three self-help features in children's clothes.
8. Describe three safety features in children's clothes.
9. List three growth features in children's clothes.

10. How can clothes contribute to a child's self-esteem?
11. Name two physical changes that occur as people age, and explain how those changes influence clothing needs.
12. List three style features that are generally recommended for older people.
13. Name two kinds of fasteners that are often used on clothes for people with physical disabilities.
14. Explain how trousers can be constructed to be more comfortable and practical for people who are confined to wheelchairs.
15. What special features are common in maternity clothes?
16. List three tips for selecting items for a travel wardrobe.
17. Why should a person avoid wearing new shoes on a trip?
18. If traveling by plane, what items should be packed in a carry-on bag?

Fashion in Action

1. Hold a class debate on the issue of cloth diapers versus disposable diapers. Discuss convenience, cost, environmental issues, and babies' comfort.
2. Interview parents of preschool children about apparel for the young. What features do the children and parents like or dislike the most? What clothes wear out first, and where do they wear out? What are some good safety, self-help, and growth features? What fabrics are best for certain types of garments? Briefly write down your findings. Then design what you consider to be a good outfit for a preschool boy or girl. Include a sketch, a written description, and (if possible) swatches of its fabrics and trims.
3. Talk with older neighbors or relatives about their clothing likes, dislikes, and problems. Then shop for apparel items for a particular older person you have in mind. Don't actually buy anything. However, write your findings in a short report. If you were going to give that person a gift of clothing, what would it be? Describe its features, cost, colors, and other details. Explain why you think it would be a good choice for that person.
4. Develop a fashionable collection of clothing for a person with a physical disability. Draw or collect pictures of garments that accommodate the person's disabilities. Describe any special features of the garments.
5. Plan a vacation (real or imaginary) and figure out what a suitable travel wardrobe would be. Coordinate specific apparel items and accessories. With pictures or drawings, as well as explanations, tell how the garments and accessories would go together for all the types of vacation activities planned.
6. Use three different methods of packing in three suitcases. (Pack similar types of garments and accessories.) Unpack the suitcases a day later. Compare the condition of the items. Which packing method produced the fewest wrinkles?

18 Caring for Clothes

Fashion Terms

laundering	enzymes
wash load	bleaches
detergents	water softeners
surfactants	fabric softeners
builders	ironing
soaps	pressing
biodegradable	dry cleaning

After studying this chapter, you will be able to

- describe how to provide daily and weekly care for your clothes.
- discuss the proper storage of clothes.
- evaluate the best ways to treat and remove stains.
- explain how to keep your apparel clean and pressed.

Considerable time and effort go into selecting clothes and building your wardrobe. Caring for your clothes properly is just as important. Taking good care of your present clothes will allow you to put off buying replacements, thus saving you time, money, and energy. Most garments today have easy care requirements. You should get into the habit of treating them right with regular cleaning and upkeep. By giving them regular attention, you get the most benefit from your apparel dollars, and you have a well-groomed, neat appearance.

Daily Care of Clothes

A daily pattern of clothing care should be established. Preventive care keeps soiling and damage to a minimum. Simple and immediate procedures are important for prolonging garment life and reducing maintenance costs. Routine care of clothes will keep garments and accessories in good condition and ready to wear, as in 18-1.

While Dressing and Undressing

When dressing and undressing, be careful not to ruin clothes by snagging, ripping, or stretching them. If you fix your hair or put on makeup after you are dressed, put a towel around your shoulders, over your clothes, to catch any hairs, dandruff, or powder. If clothes must be pulled on over makeup or creams, put a large scarf around your hair and face when slipping them over your head. Let a freshly applied deodorant or antiperspirant dry on your skin before putting on your clothes to avoid staining them.

When you dress and undress, open fasteners so garments will slip on or off easily. Extra strain from unopened fasteners can cause rips, broken zippers, and missing buttons. When putting on hose, gather up the leg portion in both hands and slip the foot into the toe. Then gently draw the stockings over the legs. Make sure your hands and fingernails are smooth so runs will not occur.

After Wearing

After you have worn clothes, put them with the dirty clothes, on a hanger, or in a drawer. Wrinkles added from tossing a garment onto a chair or on the floor may not hang out later.

After taking clothes off, check to see if they need to be brushed, laundered, spot-cleaned, dry-cleaned, pressed, or mended. If they need special attention, put them where they will get it. The needed care will be forgotten if the garments are put back into the closet. Only put away those that are ready to be worn again.

Brush good suits, jackets, and pants after each wearing. If dust and lint are left in the fibers, the life of a garment will be shortened, and the garment's appearance will be affected. Use a soft-bristled clothes brush for delicate fabrics and a hard-bristled one for firmly woven fabrics. Brush with the nap or grain of the fabric. The upholstery attachment of a vacuum cleaner is also good for removing loose dirt. A lint roller can be used to remove lint. Turn the pockets inside out to shake or brush out accumulated dirt and lint. Then put the pockets back to their usual positions. If the garment has cuffs, do the same with them.

Put clothes to be hung carefully and squarely onto sturdy hangers. Clothes that are hung correctly on firm hangers will keep their shape and press. Padded hangers, as in 18-2, or broad hangers of shaped wood or plastic are best for most garments. Thin wire hangers may leave ridges across the shoulders or droop under heavy coats and jackets. They may cause garments to lose their shape. Some may even leave rust marks.

When hanging garments, make sure lapels are flat and pocket flaps are pulled out (not tucked in) and lying smooth. Empty pockets so the garments will not be stretched out of shape.

Fasten zippers and buttons of garments when they are on the hanger. This keeps them from twisting, wrinkling, or slipping off. Skirts can be hung on special hangers with clips, or they can be secured on regular hangers with clothespins, safety pins, or straight pins. Do not fold them over a hanger. That way they

The Fashion Association / Mondo Di Marco

18-1 Regular care of clothes keeps them in good condition.

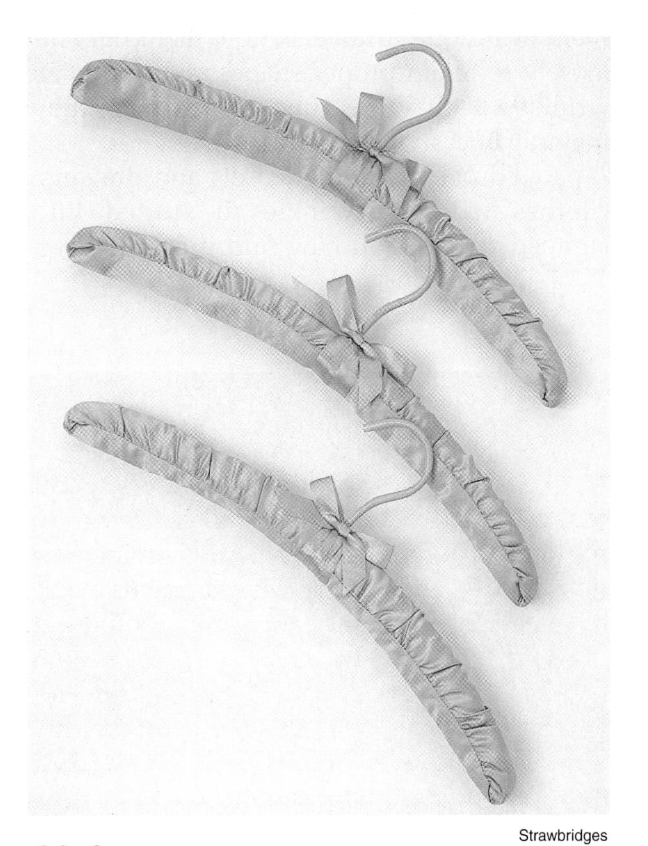

Strawbridges

18-2 Hangers such as these give excellent support to apparel for good care of clothes, plus they look pretty in the closet!

would get wrinkles as well as take up more space in the closet. If your closet space is limited, you might want to try multiple skirt, shirt, or slacks hangers.

Trousers can be hung upside down with the hem area held between the two sides of a pants hanger. See 18-3. They can also be hung over hangers that have a thick or round bottom bar.

Until the Next Wearing

Allow your clothes to air in an open room, outside, or in front of a fan before putting them away in a closet or drawer. Airing helps remove odors and helps wrinkles fall out. By leaving garments where they are exposed to air for a time, they will stay fresher longer. Also, let wool garments and woven stretch fabrics rest for a day between wearings to regain their original shape.

If garments are badly wrinkled, you may want to give them a quick and easy bathroom steam press. Hang them in the bathroom while you take a shower. Be sure the garments are fully dry before you put them away.

Fold sweaters and knit garments and place them neatly into drawers. Store them flat and loosely to retain proper shapes and minimize wrinkles. If they are hung on a hanger, they may sag and stretch out of shape.

Avoid overcrowding closets and drawers. Clothes will get wrinkles if stuffed into jammed spaces. You may find it best to keep seldom-worn garments in a different location. This will allow more space for the clothes you wear often.

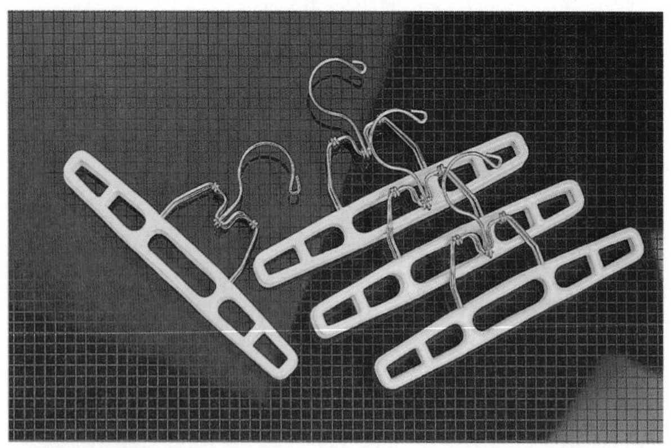

Strawbridges

18-3 These hangers are especially designed for hanging skirts or trousers. The bottoms of the hangers are double, to separate for inserting the garment. Then they lock tightly together to hold the garment for hanging in the closet.

If you get caught in wet weather, hang your clothes up neatly to dry as soon as you take them off. Hang them in a well-ventilated area away from direct heat. With heat, they could start to steam or shrink. Stuff wet pockets lightly with tissue paper to help them dry faster. Prevent woolen gloves from shrinking by inserting wooden clothespins in the fingers. All rainwear should be kept where there is plenty of ventilation.

Stuff wet shoes and leather boots with wads of paper to help retain their original shape. A shoetree might cause damp footwear to change shape or stretch unnaturally. If the leather starts to harden, rub on some oil. Polish as usual when the footwear is dry. The inside of shoes can be dried faster by using the "cool air" setting on a hair dryer.

Weekly Care of Clothes

Upkeep of clothing should be done during a set time each week. If a weekly schedule is established, clothing care will not be put off and will become routine and easier to accomplish.

Sew on loose buttons, hooks and eyes, and snaps. Replace lost ones. Mend small rips and tears before they become large holes. These repairs should be done before garments are laundered or cleaned. Otherwise the broken stitches will pull loose, rips will fray and get bigger, and fasteners will get lost.

Remove spots and stains. Launder the washable clothes that are dirty. Hand wash sweaters and other delicate items. Take soiled garments that can't be washed to a dry cleaner. Take clothes with stains that have not come out with home spot removal methods there, too. Tell the dry cleaner what types of stains are present, and where they are, so the right treatments can be used.

If outerwear has lost its water repellency, put a new coating on it with a water-repellent spray.

Such a spray can also be used to protect parts of garments from perspiration damage. Follow the directions on the container carefully.

If some garments have become pilled, brush them with a stiff bristled brush to remove the pills. Press clothes that are wrinkled. If shoes are dirty and scuffed, clean them with a dry cloth and a bristle brush. Then polish them. If the heels or soles need replacing, take them to a shoe repair shop as in 18-4. Replace old, worn shoelaces with new ones.

Home Storage Areas

Home storage areas, such as closets, shelves, and drawers, should be neat and well organized. You may have heard the old saying, "a place for everything and everything in its place." That saying applies here! Good use of storage will help your home look neat and will keep your apparel in its best condition. Look at the well-organized closet in 18-5.

Storage areas should provide adequate space for both hanging and folded clothes. Store items as close as possible to the place where they will be used or put on. For instance, coats should be near the door. Jewelry should be near the mirror.

Store items together that are used together or that are similar. All underwear items should be in the same, or adjoining drawers. Shirts should all be together, as well as sweaters, slacks, jackets, and so on.

Closet space should be big enough to hold a season's wardrobe without crowding. Traditionally, a sturdy pole is attached from end to end of the closet to hold hanging clothes. The pole should be high enough for long garments. Sometimes a second pole extends only partway across the closet. This gives double space for short items, such as shirts and skirts.

Use all the space you have to good advantage. A shoe rack can be put on the floor of a closet. A shoe bag can be hung on a wall or on the inside of a closet door. A storage box can be put under hanging shirts in a closet since they are not full length. A tie and belt rack can go on

18-4 Shoe repair shops are available in most areas, either in malls or in their own independent locations.

California Closet Company

18-5 All of the apparel items in this closet are clean and organized. They are easily seen and ready to be worn.

the back of a door. Shelves can hold items in the top area of a closet. Can you get any ideas from 18-6?

Having several hooks in a closet provides handy places to hang pajamas, belts, and scarves. Hooks usually screw into a wooden frame area of a closet for strength. Adhesive-backed hooks can be placed at other spots but cannot hold as much weight.

Store items so they are easy to find at a glance. Store frequently used items where they can be easily reached. Cover closet shelves with washable paper so they look nice, have no splinters, and are easy to clean. Shelves are good places to store boxes containing seldom-worn shoes, hats, or purses. Be sure to label the boxes so you can tell at a glance what is in each one. A light in the closet will help you locate these items when they are needed.

The closet should be kept clean and dry to prevent mildew from damaging your clothes and shoes. If you live in a damp climate, a heating rod, a low-watt lightbulb that is on all the time, or a special kind of absorbent material will keep the closet dry from mold and mildew. Large walk-in closets can be kept fresh with a dehumidifier or simply with good ventilation that can circulate air freely.

Drawers should be free of dust, lint, and dirt. Line them with shelf paper or special liner. Do not use newspaper, since the printer's ink can rub off onto clothes. Treated shelf paper absorbs grease and repels moisture.

If your accommodations do not have enough closet or drawer space, create your own arrangement. For an attractive storage unit in an open space, tack or glue fabric to a bookcase. Add a coordinating shade across the front. When the shade is pulled down over the shelving, a unique room divider is created. To carry out the scheme, use matching fabric for the window treatment or as throw pillows for the bed. If you have a small chest of drawers in which you store out-of-season items, put a circle of plywood or particleboard over it. When covered with fabric or a floor-length tablecloth, it can be used as a bedside table.

IKEA

18-6 See-through bins and a covered rolling cart can store items wherever space is available. A mirror in the dressing area is a great convenience.

Seasonal Clothing Storage

Seasonal clothing storage is usually done in the spring and the fall for temperature and weather changes. Apparel that will not be worn for several months should be put away. Your closet and drawers should be cleaned and prepared for the current season.

Never store clothing that is dirty. Dirt will set with time and heat, and may become permanently embedded. Food stains attract insects to fabrics. Even the substances found in deodorants, perfumes, body lotions, and perspiration break down fibers over time. Therefore, always wash or dry-clean garments before storing them. Also, make any needed repairs before storing out-of-season garments. Some dry cleaners provide a clothing repair service for a fee. See 18-7.

To prepare clothes for storage, empty all pockets, and remove belts. Close all fasteners so garments will maintain their shape.

Knitted garments should be folded, not hung. To prevent fold marks, insert layers of flat tissue paper as you fold garments. Folded

18-7 If you cannot make the necessary repairs to clothing, dry cleaners sometimes offer this service.

Spiegel

18-8 Garment storage bags and other closet accessories can be purchased in various sizes and styles.

items can be kept in trunks, suitcases, plastic storage boxes, or cardboard boxes.

Store hanging clothes in zippered garment bags of the right length. These may be purchased, as shown in 18-8, or made. Make sure there is sufficient room between garments to allow the items to breathe. Plastic dry cleaning bags may "suffocate" the fibers in clothing. To store purses and shoes, fill them with wads of tissue paper and protect them from dust.

Thoroughly mothproof all apparel to be stored. Put mothballs into the closet or storage containers. Moth protection also comes commercially in pouches, boxes, and fragrant sachets. Additionally, dry cleaners can mothproof garments after cleaning them.

Put items to be stored in a clean, dry, cool, and dark place. The storage location should be dust free to keep garments clean. It should be dry and cool to prevent mildew, and out of direct sunlight to keep color from fading. You may want to spray storage spaces with insect repellent to prevent the hatching of insects.

Cedar closets or cedar chests provide good seasonal storage, especially if you find the smell of mothballs offensive. Insects avoid garments surrounded by cedar. However, cedar will not kill the eggs of moths or silverfish if they are already in the clothing.

Some dry cleaners provide sealed storage vaults with controlled humidity and temperature for a fee. Furs and leathers can be specially cleaned, reconditioned, and stored. Cold storage keeps the fur hair strong and supple, and it prevents cracking and mildewing of the skins. Furs should not be sprayed with moth or insect repellent. Some repellents contain oil that may harm the furs. Garments are returned from the dry cleaner at the end of the season clean, pressed, and ready to wear. This might be a good storage option if your space at home is limited and if you can afford the cost of professional clothing storage.

Removing Spots and Stains

If a garment gets a spot, do something about it as quickly as possible. Stains are harder to get out the longer they stay in fabrics. If treated correctly, most stains can be removed from washable fabrics at home. The wrong treatment, or no treatment until later, can make the stain permanent.

Treat all stains before you launder or dry clean an article. Pretest the stain removal method on a seam allowance, facing, or other hidden area before using it on a visible part of the garment. You will then be sure the color or appearance of the fabric won't change.

When a spill occurs, immediately absorb excess liquid with a napkin, towel, tissue, or other absorbent material. Gently blot, don't rub. Do not apply pressure. That could force the stain further into the fabric. Also, never iron over a stain or expose the garment to heat or sunlight. That could permanently set the stain.

Identify the Fabric and the Stain

Spots and stains on clothing have many causes and cures. The way to remove them depends on the fiber content of the garment and the source of the stain. Stain removal often takes some time to accomplish successfully.

First, identify the fiber content. Determine if the fabric is washable or bleachable. Check the care label and any hangtags you have saved. Stains on dry-cleanable fabrics should be treated with cleaning fluid or cleaned by reputable dry cleaners. Dry cleaners have special equipment, products, and knowledge to remove them. Do not attempt to remove stains on antique or extremely delicate articles yourself.

Stains on fabrics of manufactured fibers or with permanent-press finishes are often difficult to remove. Most manufactured fibers do not absorb water, so the outside of the fibers are cleaned by water but not the inside. If the fibers absorb an oily stain, special treatment with a prewash soil and stain remover is needed.

If you can identify the stain, you can find out what to use to remove it. If the stain is on a washable fabric, refer to chart 18-9 for the specific removal procedure.

Treat an unidentified stain on a washable garment with a prewash soil and stain remover. Rinse it out and let the garment dry. If the spot remains, treat it with a dry-cleaning solvent. After treating the stain, air or rinse out all spot-removing products. Wash all washable items to make sure the spot-remover and stain are gone.

Stain Removal Supplies

Before using a cleaning product, be sure to read the manufacturer's directions and warnings. Many products are poisonous and flammable. Some treatments damage certain fibers or cause fading or bleeding of dyes. Some cause shrinkage, stretching, or a loss of luster. Some come as a gel and dry to a powder. Use a medicine dropper or glass rod to apply liquid cleaning preparations to very small areas.

Some safety precautions are necessary when working with cleaning products. Always work in a well-ventilated area. Do not breathe any more solvent vapors than necessary. Wear rubber gloves to protect your hands. Do not let cleaning agents come in contact with the skin, eyes, or mouth. Keep them away from electrical outlets and open flames. Never use dry-cleaning solvents in a washing machine. Also, do not put articles in the dryer that are damp with solvent.

When using stain removal products, work in a clean, well-lit area. Have a hard work surface that will not be affected by the chemicals. Never leave any spot or stain removers, or other cleaning products, within reach of children or pets. Store them closed and in their original containers in a cool, dry place away from food products. Wash any spilled products off your skin and the work surface immediately.

Stain removal supplies should be kept together near the laundry area. Absorbent materials should include clean white cloths and white paper towels. You should have a heavy-duty liquid detergent, a bar of white soap, chlorine bleach, and oxygen (all-fabric) bleach. Also stock a prewash soil and stain remover, laundry detergent, spot "lifter," cleaning fluid, and rust stain remover. These and other stain removal supplies can be purchased at most food, drug, and variety stores.

You should have a small flat brush with nylon bristles, a medicine dropper, and a tub or pail for soaking stained articles. Glass or unchipped porcelain containers are best for stain removal treatments. Do not use plastic with solvents. Never use a rusty container.

Stain Removal Guide for Washable Fabrics

Adhesive Tape
Rub with ice; scrape with dull knife; sponge with cleaning fluid; wash.

Ballpoint Ink
Soak with hair spray; rinse; hand scrub with liquid detergent; rinse well. Can try rubbing alcohol, glycerin, or prewash spray.

Blood
Soak in cool water with enzyme presoak; rub with detergent; rinse; can try hydrogen peroxide or ammonia solution; wash.

Candle Wax, Paraffin
Freeze and scrape; place between paper towels or tissues and press with warm iron; place face down on paper towels and sponge with cleaning fluid or rubbing alcohol; wash.

Chewing Gum
Harden with ice; scrape with dull knife; can soften with egg white; sponge face down on paper towels with cleaning fluid; wash.

Chocolate, Cocoa
Soak in club soda or cool water with enzyme presoak; sponge with cleaning fluid, later with detergent; launder in hot water.

Coffee, Tea
Soak with enzyme presoak or oxygen bleach; rub with detergent; wash in hot water.

Cosmetics
Dampen and rub with detergent; rinse; sponge with cleaning fluid; rinse; wash in water as hot as fabric care permits.

Crayon
Scrape (can loosen with cooking oil); spray with prewash or rub with detergent; rinse; sponge with cleaning fluid face down on paper towels; rinse; launder hot with bleach.

Deodorants
Scrub with vinegar or alcohol; rinse; rub with liquid detergent; wash hot.

Egg
If dried, scrape with dull knife; soak in cool water with enzyme presoak; rub with detergent; launder in hot water.

Felt Tip Ink
Rub with strong household wall and counter cleaner; rinse; repeat if needed; launder; this stain may not come out.

Fingernail Polish
Sponge white cotton with polish remover and all other fabrics with amyl acetate (banana oil); scrape with dull knife; wash.

Fruits and Juices
Soak with enzyme presoak; wash; if stain remains, cover with paste of oxygen bleach and a few drops of ammonia for 15-30 minutes; can also try white vinegar; wash as hot as possible.

Grass
Soak in enzyme presoak; rinse; rub with detergent; hot wash with bleach; if stain remains, sponge with alcohol.

Gravy
Scrape with dull knife; soak in enzyme presoak; treat with detergent paste, later cleaning fluid; hot wash with bleach if safe.

Grease
Scrape off excess or apply absorbent powder (talcum or cornstarch) and brush off; pretreat with strong detergent; rinse; sponge with cleaning fluid; hot wash with extra detergent; bleach if safe.

Ice Cream
Soak in enzyme presoak; rinse; rub with detergent; rinse and let dry; sponge with cleaning fluid if needed; hot wash with bleach if safe.

Lipstick
Moisten with glycerin or prewash; wash; bleach if safe.

Margarine
Same as for grease.

Mayonnaise, Salad Dressing
Rub with detergent; rinse and let dry; sponge with cleaning fluid; rinse; hot wash with bleach if safe for fabric.

Mildew
Rub with lemon juice; dry in the sun; rub with detergent; hot wash with bleach; if stain remains, sponge with hydrogen peroxide.

(continued)

18-9 If one of these stains appears on a washable fabric, carefully follow the suggested treatment as soon as possible.

Stain Removal Guide for Washable Fabrics *(continued)*

Milk
Soak in enzyme presoak; rinse; rub with detergent; launder.

Mud
Soak in water with dishwashing detergent and 1 tbsp. vinegar; rinse; sponge with alcohol; rinse; soak in enzyme presoak; wash with bleach if safe.

Mustard
Spray with prewash or rub with bar soap or liquid detergent; rinse; soak in hot water and detergent; launder with bleach if safe.

Oil
Same as for grease.

Paint
Do not let paint dry; sponge oil-based types with turpentine or paint thinner; rub with bar soap; launder.

Peanut Butter
Saturate with mineral oil to dislodge oil particles from fibers; blot; apply cleaning fluid and blot between absorbent mats; rinse and launder.

Perspiration
Soak in salt water, enzyme presoak, or rub with baking soda paste; rinse; rub with detergent; can apply ammonia to fresh stain and white vinegar to old stain; rinse and launder.

Scorch
Soak with enzyme presoak or wet with hydrogen peroxide and a drop of ammonia; let stand 30-60 minutes; rinse well; wash in hot water; rub with suds; bleach; dry in sunshine; may not come out.

Shoe Polish
Scrape off excess; rub with detergent; wash.

Soft Drinks
Dampen with cool water and rubbing alcohol or enzyme presoak; launder with bleach if safe; stain may appear later as a yellow area.

Tomato Products
Sponge with cold water; rub with detergent; launder with bleach if appropriate.

Water Ring
Rub with rounded back of silver (not stainless steel) spoon.

Wine
Same as for fruits; sprinkle a red wine spill immediately with salt.

18-9 *Continued*

Stain Removal Methods

There are many stain removal methods. The most commonly used are sponging, chilling to harden, scraping, and soaking.

For *sponging*, place the stained side down over a clean, dry, absorbent material. Dampen another piece of absorbent material with water or the appropriate stain remover. Then sponge lightly from the center of the stain toward the edge to minimize the formation of rings. By applying stain remover to the underside of the fabric, the stain will be sponged off the garment surface rather than through the fabric. If the stain has hardened, soften it by soaking the spot first. Sponge irregularly around the edges of the stain so there will be no definite line when the fabric dries. Change the under pad and sponge pad frequently so the stain will not be redeposited on the fabric. When the spot is gone, blot up as much excess moisture as possible.

Hardening a substance such as candle wax or gum makes it easier to remove. To do this, gently rub an ice cube across the area. When the material has hardened, lift or scrape it off the fabric. If the garment is small enough, it may be wrapped in a plastic bag and placed in the freezer until the stain hardens.

Scraping the substance off the fabric is done with a dull knife or a spoon. Place the fabric (stain side up) directly on the work surface. After applying stain remover, gently scrape back and forth over the stain. Do not press hard. This procedure should not be used on delicate fabrics because damage could result.

If a garment is washable, *soaking* it in water for about 30 minutes is another stain removal method. An enzyme presoak may be helpful. Prewash sprays work well on the inside of collars and cuffs. If necessary, repeat any treatment until all the stain is removed.

If you use an enzyme presoak product or a prewash soil and stain remover, follow the instructions on the package. When using bleach, do not apply it to just a spot on a garment. It will cause uneven color removal or make thin spots in the fabric. Dissolve any bleach in water first. Then put the garment in the solution. The color of the entire garment may get lighter, but at least the change will be uniform. Do not combine chlorine bleach and any detergent full strength. These two products combine safely when properly diluted.

Do not put chlorine bleach on rust stains. Instead, apply a rust stain remover. White fabrics that have picked up dye from colored fabric that "bled" might be restored by using a fabric color remover.

A spot lifter is sprayed or rubbed into a dry stain. Dampened talcum powder or cornstarch, or a commercial product may be used. After the spot lifter dries, it is brushed away. If successful, the spot is lifted out and brushed away with the dried spot lifter. If a stain cannot be removed from a nonwashable garment, take it to the dry cleaner as soon as possible.

Laundering Clothes

Laundering is washing apparel with water and laundry products. The use of good laundering procedures and equipment, as in 18-10, extends the life of garments and keeps them looking newer longer. Clothes that are allowed to get too dirty may be hard to clean. Also, when soil gets ground into fibers, it can weaken them. With proper laundering, the original size, shape, color, and overall appearance of apparel items should be maintained.

Check the fiber content and washing and drying instructions of garments before laundering them. The chart in 18-11 explains the meanings of terms used on labels. The symbols shown in 18-12 are from an international code for care labels. Read and follow the directions on care labels, laundry products, and appliances.

The Maytag Company

18-10 Automatic washers and dryers make the task of doing laundry fairly easy.

Keep the laundry area well organized, as in 18-13. Store cleaning and laundry supplies where they are handy, but out of the reach of children and pets. A place to hang up garments as they are removed from the dryer will minimize ironing time later.

Prepare clothes for washing by closing fasteners and emptying pockets. A marking pen or facial tissue can affect every item in the wash load with stains or lint. Shake dirt from cuffs and pockets. Loosely tie long belts and sashes to prevent tangling. Turn knits inside out to prevent snagging.

Sort Clothes

Sort clothes carefully to separate items that could damage other items. For instance, if colored dyes run onto white garments, the resulting marks could be permanent.

Separate clothes into piles that are suitably sized wash loads. A ***wash load*** is the bunch of clothes you put into a machine to launder together. Wash loads should be large enough to make good use of water but not too large to prevent thorough cleaning. Washable items should move freely in the wash water. To save energy and money, coordinate the water level adjustment on the washer with the size of the wash load you're using.

Sort by Color

White clothes should be in a separate load. Colorfast prints and pastel solids should be in another load. Dark colors are still another

	When Label Reads:	It Means:
Machine Washable	Machine wash	Wash, bleach, dry, and press by any customary method including commercial laundering and dry cleaning.
	No chlorine bleach	Do not use chlorine bleach. Oxygen bleach may be used.
	No bleach	Do not use any type of bleach.
	Cold wash Cold rinse	Use cold water from tap or cold washing machine setting.
	Warm wash Warm rinse	Use warm water or warm washing machine setting.
	Hot wash	Use hot water or hot washing machine setting.
	No spin	Remove from washer before final machine spin cycle.
	Delicate cycle Gentle cycle	Use appropriate machine setting; otherwise wash by hand.
	Durable press cycle Permanent-press cycle	Use appropriate machine setting; otherwise use warm wash, cold rinse, and short spin cycle.
	Wash separately	Wash alone or with like colors.
None-Machine Washing	Hand wash	Launder only by hand in lukewarm (hand comfortable) water. May be bleached. May be dry-cleaned.
	Hand wash only	Same as above, but do not dry-clean.
	Hand wash separately	Hand wash alone or with like colors.
	No bleach	Do not use bleach.
	Damp wipe	Clean surface with damp cloth or sponge.
Home Drying	Tumble dry	Dry in tumble dryer at specified setting—high, medium, low, or no heat.
	Tumble dry Remove promptly	Same as above, but in absence of cool-down cycle, remove at once when tumbling stops.
	Drip dry	Hang wet and allow to dry with hand shaping only.
	Line dry	Hang damp and allow to dry.
	No wring No twist	Hang dry, drip dry, or dry flat only. Handle carefully to prevent wrinkles and distortion.
	Dry flat	Lay garment on flat surface.
	Block to dry	Maintain original size and shape while drying.
Ironing or Pressing	Cool iron	Set iron at lowest setting.
	Warm iron	Set iron at medium setting.
	Hot iron	Set iron at hot setting.
	Do not iron	Do not iron or press with heat.
	Steam iron	Iron or press with steam.
	Iron damp	Dampen garment before ironing.
Misc.	Dry-clean only	Garment should be dry-cleaned only, including self-service.
	Professionally dry-clean only	Do not use self-service or home dry-cleaning.
	No dry-clean	Use recommended care instructions. No dry-cleaning materials to be used.

The American Apparel Manufacturers Association, Inc.

18-11 This chart explains the meanings of the terms printed on the care labels of apparel.

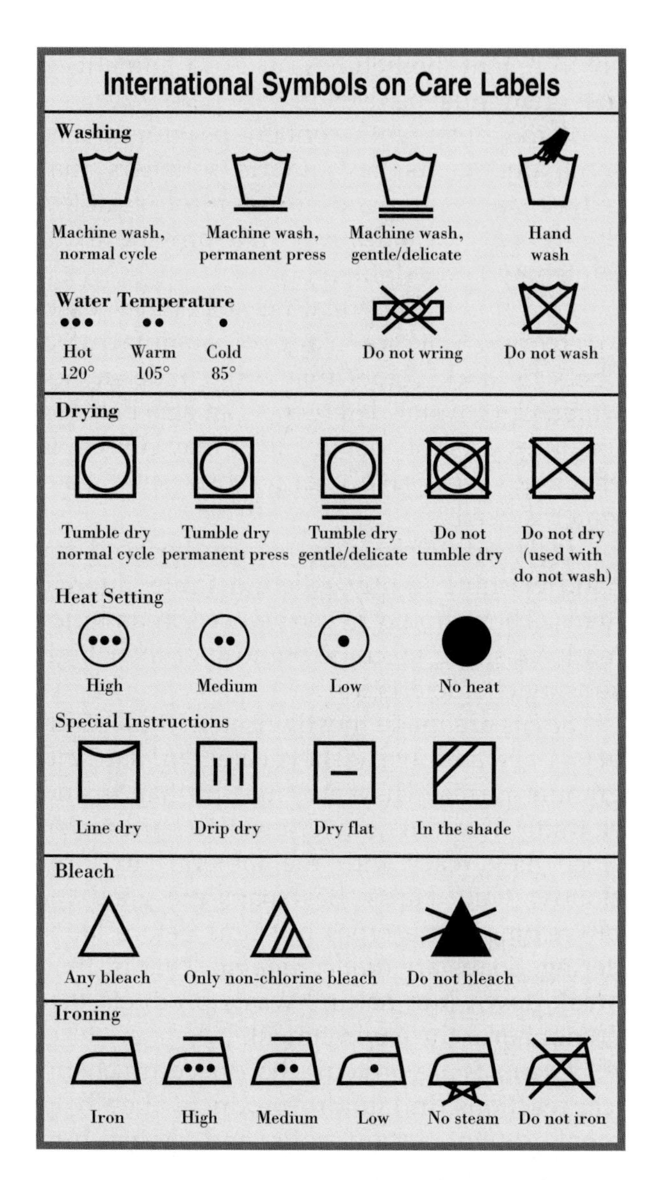

Federal Trade Commission

18-12 These symbols give universal messages without having to write explanations in any language.

category. The dyes in bright or dark-colored fabrics can "bleed" into the wash water to discolor whites or pastels. Manufactured fibers have an especially strong tendency to pick up traces of dyes in the wash water.

Colored clothes are most likely to bleed when they are new. To help items hold their dye, soak them in heavily salted, cool water for 10 to 15 minutes. To test for colorfastness, wash them separately to see whether color appears in the wash or rinse water. If color does appear, wash them separately or only with garments of the same color and intensity.

18-13 A well-organized laundry area can minimize the effort of clothing care and be supplemented with attractive and useful storage areas.

Sort by Type of Fabric and Garment Construction

Delicate items that might snag or tear should not be washed with sturdy items that require stronger washing action.

"Lint-producing" fabrics, such as new towels, terry cloth robes, or chenille spreads, should be washed alone. They will transfer lint onto items washed with them, especially onto "lint attracting" fabrics. Examples are linens, corduroys, dark cottons, fabrics of manufactured fibers, and fabrics with permanent-press finishes. It may also help to turn "lint-attracting" items inside out when washing them.

Sorted groups should contain a mixture of large and small items to provide good washing action. Also, different textures should be combined, if feasible, so more cleaning will result as they rub against each other.

Sort by Kind and Amount of Soil

Extremely dirty items can make lightly soiled garments look dingy after washing. Wash heavily soiled garments separately. Do not wash polyester or nylon fibers with oily or greasy items, since they have a tendency to pick up oily substances.

Choose Correct Products

Choose the correct laundry products for the job. There are several kinds of laundry products with different purposes. Each category has several brands from which to choose. They meet the different needs, preferences, and budgets of consumers. The products you decide to use should be compatible with your laundry equipment, the hardness of the water, and the clothes you are washing. Choose only the products you need. Always read and follow package directions.

Detergents are made synthetically from chemicals that suspend and hold dirt and grease away from clothes. They are granular or liquid. They dissolve readily in water of all temperatures and degrees of hardness. Lately, technology has made them more effective in cooler water (to save energy) and safer for the environment. They can be high- or low-sudsing.

High-sudsing detergents are all-purpose and the most popular. They are used in top-loading washers.

Low-sudsing detergents contain suds control agents and are recommended for washers that are "suds sensitive." They are especially recommended for front-loading, tumbler-type automatic washers. However, they can be used in all washers.

The most important ingredients in detergents are ***surfactants*** (surface active agents). They reduce the surface tension of water to allow it to penetrate soiled fabrics more easily. They loosen and remove soil with the aid of the wash action. They also dissolve or suspend the removed soil in the water until the end of the wash time so it will go down the drain.

Detergents may also contain ***builders*** that inactivate hard water minerals. They prevent the formation of insoluble residues that could be redeposited onto clothes. Some builders are phosphates. Detergents that contain phosphates are banned in some areas to protect the environment. Sometimes an extra amount of detergent must be used in very hard water.

Some detergents contain perfumes, and some detergents have fabric softeners that reduce static cling and help control wrinkles. Bluing aids are additives that provide extra whitening and brightening.

Cold water detergents are sometimes used for fabrics that are made of manufactured fibers or dark colors that could fade in hot water. Cold water detergents can also be used for other wash loads, but they may not be as effective. They save energy costs since cold water is used.

Light-duty dish detergents can be used for hand washing lightly soiled, delicate laundry items. They should never be used in an automatic washing machine because of their high sudsing characteristics.

Soaps are made mostly from natural products (fats and lye). They come in bar and granule forms. They are mild on hands and clothing. They do not work well in hard water, since they react with the minerals to form cloudy white curds. Bar soaps are good for pretreating some types of stains before laundering. Soaps are ***biodegradable***. That is, they break down into natural waste products that do not harm the environment.

Enzymes are proteins that speed up chemical reactions. In laundry products, they help break down certain soils and stains into simpler forms. The soils can then be removed easily in the wash.

Enzymes are found in some all-purpose laundry detergents as well as in presoak products. Look for the words "protease" or "amylase" on the product's label. Enzymes are inactivated by chlorine bleach, so their purpose is defeated if they are used with chlorine bleach.

Bleaches are either of a chlorine type or an oxygen type, as in 18-14. The least expensive is liquid chlorine bleach, used mainly for white cotton fabrics. Oxygen bleach, available in liquid form or boxes of dry powder, is safe for most fabrics. Bleach substances are now often premixed with detergents when purchased. If using such a detergent, do not add more bleach.

Chlorine bleach whitens clothes and helps remove soils and stains. It also disinfects and deodorizes laundry. Do not use it if your water

18-14 Chlorine bleach is used on white cotton fabrics. Oxygen bleach is safe for most fabrics.

supply contains iron. Use it for items that have care labels that say they can be laundered with chlorine bleach. Do not use chlorine bleach on silk, wool, mohair, or spandex. It can interact chemically with the fibers and destroy them. It can also destroy flame-retardant finishes. It may cause polyester fibers to turn yellow, and it reduces the effectiveness of permanent-press finishes. It can shorten the life of a garment by weakening the fibers if too much is used too often. Used correctly, it is a very good laundry aid, especially for white cotton products.

Many automatic washers have dispensers for adding liquid chlorine bleach to the wash load. They dilute the bleach before it goes into the wash water. Some machines delay adding the bleach until the last few minutes of the wash cycle. This enables any other products that may be used in the wash, such as enzymes, to do their work. If you add chlorine bleach at the beginning of the wash cycle, dilute it or mix it into the water before adding the clothes. If poured directly onto fabrics, chlorine bleach can cause holes and ultra-white spots.

Oxygen (all fabric) *bleaches* can be used on all fabrics and, if used properly, for most colors. They are considered to be less effective, or weaker, than chlorine bleaches. You may want to use them in your regular laundry loads to keep clothes looking their best. Just as for chlorine bleach, do not pour oxygen bleaches directly onto clothes.

Packaged ***water softeners*** may be needed if you have hard water. They neutralize the mineral ions found in hard water or hold them in solution. Powdered water softeners can be added directly to the wash water. They enable soaps and detergents to work better. Without them, the clothes could get rust marks or become gray, dingy, and stiff.

A mechanical water softening device can be installed in your home. The water supply runs through it and is softened by a chemical reaction before it comes out of the faucet. This is probably the most effective long-term solution for people with very hard water.

Water hardness varies from region to region. To determine your water hardness, consult your local water company, cooperative extension office, or agricultural university.

Fabric softeners give softness and fluffiness to washable fabrics. They have nothing to do with water softening. They prevent fibers from getting stiff and harsh. They control static electricity, thus preventing "static cling." However, with repeated use they can build up on fibers and reduce the fabric's absorbency.

Some fabric softeners are added directly to the wash water. Sometimes they are put in detergents. Some liquid fabric softeners are added during the rinse cycle of the washing machine. Other fabric softeners are put into the dryer with the laundry. They are in the form of sheets that tumble with the clothes and transfer softness and fragrance to them.

Starches and *fabric finishes* are "sizings." They restore body and crispness to fabrics that have become limp from laundering and wear. They cause the fabrics to have a fresh, smooth appearance after being pressed. Starches are not used in the final rinse water as much these days with the popularity of easy care fabrics. Instead, spray starches and fabric finishes are often applied when clothes are ironed.

Disinfectants are not essential for regular laundering. They are recommended if there is illness in the family or some other reason that clothes must be sanitized. Disinfectants reduce or kill microorganisms to control or eliminate infections. Chlorine bleach does some disinfecting. Also, household liquid disinfectants can be used. Be sure to follow package directions.

Use Laundry Products Correctly

The containers in which detergent, bleach, and other laundry products are sold have instructions for their use. Read them and follow them. For best results, measure amounts of laundry products accurately, as in 18-15. If the cap of the product is not intended for measuring, use a measuring cup that is used only for laundry products. Use separate cups for different types of products so chemicals do not mix with each other. Also, if only one cup is used, rinse it thoroughly between measuring different products.

18-15 To use laundry products properly, read the instructions on the labels and measure accurately.

Package recommendations are based on average washing conditions. However, water hardness, type of laundry equipment, amount of soil on the clothes, and size of the wash load vary. Thus, you must use your own judgment in deciding how much of each product to use for specific wash loads. Use more than the recommended amount if you have very hard water, a large wash load, very dirty clothes, or the extra water volume of a large washing machine. Use less than the recommended amount if you have soft water, a small wash load, very light soil, or the reduced water volume of a small washer or a partial fill. Too little detergent can cause graying, yellowing, or dingy laundry. Too much detergent can cause clothes to become stiff and harsh.

Use the Right Water Temperatures

The wash and rinse water temperatures can directly affect the cleaning, wrinkling, and durability of clothes. Although most detergents work in all water temperatures, they clean best in warm or hot water. However, the heating of water uses energy and creates more cost.

Most automatic washers offer a choice of three wash temperatures (cold, warm, or hot) and a choice of two rinse temperatures (cold or warm). However, some machines have water temperatures programmed into the cycles. In such a case, review the instruction book about the options of the machine. For hand washing, you control the water temperature manually.

The *cold* wash setting is mainly for dark colors that bleed, for lightly soiled loads, and for delicate items. It helps prevent fading and shrinkage. It draws water only from the cold water line as it enters the house. This temperature varies depending on climate areas and seasons of the year. The colder the water, the harder it is to accomplish cleaning.

The *warm* setting gives a mixture of cold and hot water. Warm water is about halfway between the temperatures of the cold and hot water. It varies depending on the temperatures of the incoming hot and cold water. Warm water provides good cleaning for most wash loads. It is good for washing garments made of manufactured fibers and those with permanent press finishes. It cleans moderately soiled items well. Most people wash their laundry in warm water.

The *hot* setting draws water only from the hot water line coming from the water heater. Hot water provides the best soil removal and sanitizing. However, it may set stains, cause shrinkage, fade colors, and encourage wrinkling of some fabrics. It should be used only for white or colorfast fabrics and heavily soiled loads.

For energy reasons, it is most practical to rinse clothes in cold water after washing. It is also better for the clothes and lessens wrinkling of permanent press items.

Follow the Correct Laundering Procedures

Know about and utilize the features of the laundry equipment you use. Controls on many washers allow you to select the best water temperature, water level, and length of wash time. You can choose the agitation and spin speeds. *Washer agitation* is the mechanical action that helps to loosen and remove soils from the clothes during the wash cycle.

Put the water and cleaning agents (detergent, bleach, etc.) into the washer first. Some products may be put into special dispensers in the machine. The cleaning products should dissolve in the wash water before the clothes are added. Water hardness minerals can be tied up before the soil is introduced. Bleach can be dispersed, too, so concentrations of it will not cause spotty color damage on the clothes. If detergent is poured over clothes in the machine, undissolved clumps of it may stick to the clothes and not rinse out. Also, discolored blotches could occur in fabrics.

Be sure the proper water level is selected and the size of the load is right for the washer. Add the clothes, a few at a time, distributing them evenly around the agitator of the machine. This will balance the load. Never wrap large items around the agitator. Remember, overloading reduces cleaning and causes extra wrinkling. It is better to underload than to overload the washer. To get garments clean, they need room to circulate in the water. Soil and lint can then be washed away. There should always be enough wash and rinse water to cover the clothes.

Select the proper washing action cycle on the machine. A regular or normal speed is good for most items. The agitation time should run for a longer time for heavily soiled, sturdy items. A slow or gentle speed should be used for delicate garments such as lingerie. It should run for a short time. There may also be cycles for knits and permanent press items. They automatically provide a "cool-down" period and a cold rinse to minimize wrinkles.

Rinse the clothes thoroughly with clean water. This carries away dissolved laundry products, suds, and loosened soil and lint.

In automatic machines, water is extracted from the clothes after each of the wash and rinse cycles by spinning. Automatic washers have a regular spin speed for most fabrics and a slow speed for permanent press or delicate clothing. On washers having only one speed, the spin time is longer for the regular cycle than for the gentle cycle. After using the washing machine, clean the lint filter if it has one.

Hand Washing

Any clothing that can be washed in a washing machine can also be hand washed in a large pan or sink. Delicate items and wool garments should almost always be done by hand. Woolens can shrink from agitation as well as hot water. Before washing shrinkable garments, trace around them with pencil on a large sheet of heavy paper, as in 18-16. A brown paper sack cut open can be used. Do not use newspaper.

Hand wash in cool or lukewarm water, depending on the fabric. Use a light duty detergent or soap. Several special cold water solutions are available. Swish the water to dissolve the laundry product. Then add the garment and squeeze the suds through it gently. Rub softly only if necessary to remove soil. Soak for up to 15 or 20 minutes. If the fabrics are of dark colors that bleed, do not soak for more than 5 minutes. Never wring or twist delicate articles.

Rinse hand-washed articles in cool water at least twice, or until the water is clear. Then roll them in an absorbent towel to remove as much water as possible. Block (shape) shrinkable garments to their original size according to your paper tracings, as in 18-17, as you lay them flat to dry. Occasionally during drying, reshape the garments by stretching to match their outlines.

Drying Clothes

Clothes can be dried by tumble action in an automatic dryer, drip-dried on a hanger, line dried (outside or inside), or laid flat to dry. Usually, what is washed together is dried together. Dark items should be turned inside out to prevent lint from adhering and showing on the right side.

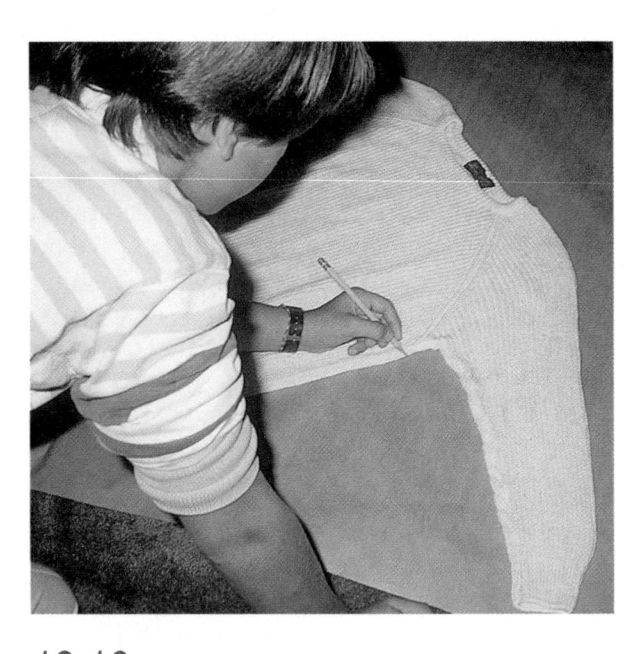

18-16 By tracing around a shrinkable garment before it is washed, its size and shape are recorded for use later.

Automatic dryers usually have regular, permanent press, and air fluff cycles, 18-18. The regular cycle is for items that are not heat sensitive. The permanent press cycle provides a cool-down tumble time with no heat at the end to reduce wrinkles. Remove and hang up permanent press items as soon as the dryer stops. Smooth the seams, collars, and cuffs. The garments should not need to be ironed. If clothes are accidentally left in the dryer after drying, they can be retumbled for a few minutes with a damp towel to restore their permanent press smoothness.

The air fluff cycle of the automatic dryer provides tumbling in unheated air. Some delicate garments may need a cool temperature, gentle tumbling, and a short drying time. Rubber and plastic materials should be dried on the air setting only. This can also be used between washings to fluff items.

Drying clothes in an automatic dryer costs money but saves time and effort. It makes clothes soft and comfortable, helps remove wrinkles, and is not dependent on the weather.

Do not overload the dryer, since the clothes would then not be able to tumble freely. They would take a long time to dry and could become wrinkled and twisted. Do not set the heat too high. A hot temperature can

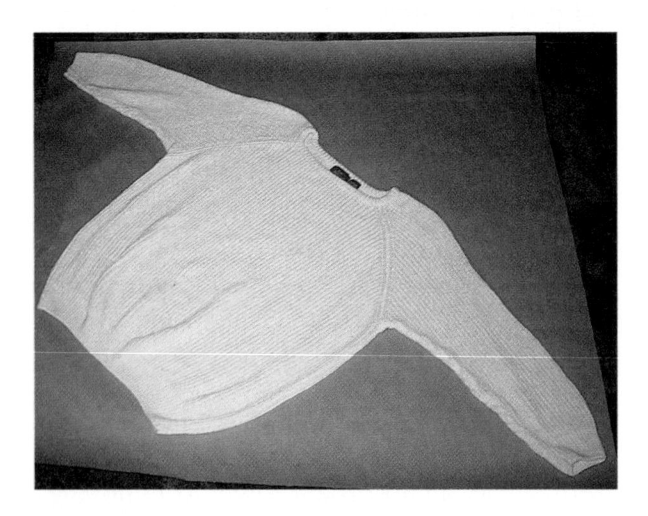

18-17 A shrinkable garment can be placed on the paper drawing and stretched to its original shape to dry.

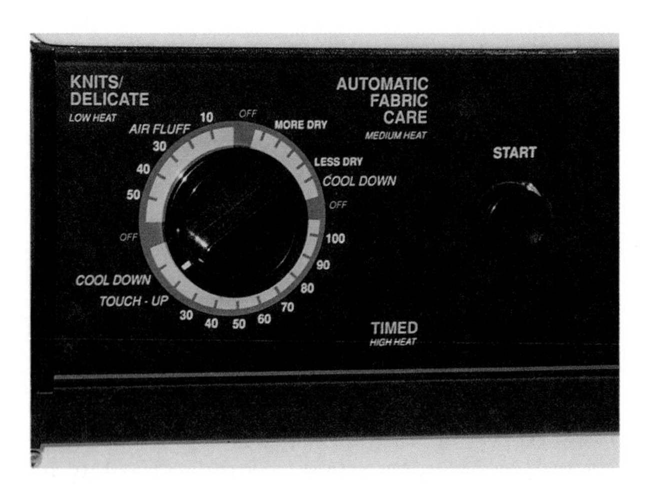

18-18 When using an automatic dryer, choose the setting that is appropriate for the garments that are being dried.

ruin buttons and trims and can cause shrinkage and wrinkles. Also, do not allow clothes to overdry. This makes fabrics harder to handle. Remove clothes immediately, smooth them with your hands, and either hang or fold them. Be sure clothes are completely dry before putting them away in closets and drawers. Also, be sure to clean the dryer's lint filter after each use.

Line drying outside is economical and is especially good for large, flat items. It can give clothes a fresh smell but is not recommended if there is heavy air pollution. Also, some items become stiff. Others fade if they are in direct sunshine.

Before hanging clothes on the line, wipe the line with a damp cloth. Be sure the clothespins are clean. Shake the clothes, smooth out wrinkles, and straighten the seams. Hang items carefully by their firmest parts to avoid creases, stretching, or imprints from the clothespins. Distribute the weight of heavy items over several lines. Dry colored garments in the shade, if possible, and take them down as soon as they are dry to avoid fading. Take items to be ironed off the line while they are a little damp to make ironing easier.

Line drying inside is good for only a few small items. Clothes take longer to dry inside and may become stiff because of the limited air movement.

Few garments require *drip-drying*. However, if this is recommended on the care label, hang the item on a wooden or plastic hanger or over a rack. Metal hangers can cause rust marks. Usually garments that require drip-drying are hung up dripping wet without squeezing or wringing.

Flat drying is good for wool, knit, and leather items, such as gloves. Sweaters are often dried flat. It helps prevent shrinkage but takes up household space. Lay garments away from direct heat on a clean, absorbent surface such as a large towel. Shape the items to their original dimensions.

Ironing and Pressing

Ironing is the process of using an iron to remove wrinkles from damp, washable clothing. Heat and pressure are used to flatten the fabric. Ironing is done with a gliding motion. It is done to entire garments after laundering.

Pressing involves no sliding of the iron. The iron is placed on the fabric and then lifted. Moisture is added from a pressing cloth or steam in the iron. This procedure is good for wool clothing and loose or bulky textures. It is often done to apparel between wearings. It is also done while constructing garments.

A well-ironed or well-pressed garment should be free of all wrinkles. The original garment shape and fabric texture should be preserved. There should be no outline of the edges of seams, facings, hems, or other structural details on the right side.

Ironing and Pressing Equipment

Proper ironing and pressing equipment is needed to do a good job. Basic equipment is needed to iron the laundry. More specific items are necessary for tricky garments and sewing projects.

Many *irons*, like the one in 18-19, have both dry and steam settings. The dry setting is used

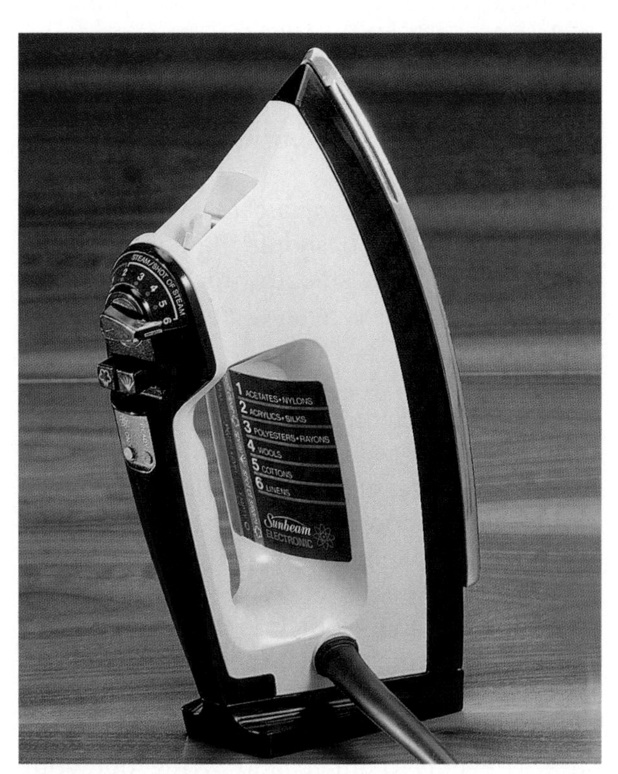

Spiegel

18-19 This iron has both dry and steam settings. It also has a water level gauge, a temperature guide for various fabrics, and an automatic shut-off if left on and not used.

at low temperatures for fabrics made of acetate, acrylic, nylon, polyester, rayon, silk, and some blends. It is also used if the fabric being ironed is damp, or if a damp pressing cloth is used between the iron and garment.

For ironing and pressing with steam, the soleplate of a steam iron has holes that allow steam to escape. Some steam irons have a button that, when pushed, gives either a shot of steam spray or a sprinkling of water when needed. Distilled water should be used in most steam irons, since mineral deposits can build up in the iron. Some can take tap water. Some have an automatic jet steam clean-out feature to help remove lint or solids that can collect and clog the steam holes. Read and follow the use and care directions that came with your iron.

Most irons have a dial with a wide variety of fabric settings. Follow it as a guide to prevent scorching or melting fabrics. When you are not using your iron, store it up on end on the heel rest to protect the ironing surface. To clean the soleplate, use a damp cloth with liquid detergent, powdered cleaner, or silver polish. Make sure you do not scratch the soleplate. If needed, you can attach a special metal soleplate cover ("converter") that prevents sticking, scorching, and shine.

Lightweight travel irons are small, so they fit nicely into suitcases. They press small areas well but are not recommended for complete laundry ironing.

A good, sturdy *ironing board* should not rock as you iron. It should be adjustable to various heights. A smooth, thick, well-fitting pad with a silicone-treated cover gives good results.

A *pressing cloth,* as in 18-20, protects the right side of garments and prevents shine from forming on fabrics. It is placed between the item being pressed and the iron. Pressing cloths can be bought in several weights and sizes. A handkerchief, a cotton/linen dish towel, or a piece of cheesecloth can also be used as a pressing cloth. It is usually used dry with a steam iron, and damp with a dry iron. If the cloth is damp, its moisture makes steam when combined with the heat of the iron. Be sure to evenly dampen and wring out your cloth rather than saturating it with water. A cloth that is too wet can cause spotting. A moistened cloth is recommended for cottons and linens. For silks and woolens, it is best to use two cloths, a damp cloth over a dry cloth.

A *sleeve board,* shown in 18-21, is like a miniature double ironing board. It fits nicely on top of a standard ironing board. It is designed for pressing sleeve seams. It is also good for pressing small, hard-to-reach areas, such as necklines, cuffs, and small flat areas of a garment. It should be well padded and have a clean cover.

Specialized pressing equipment is needed for some jobs. For instance, a *tailor's ham,* 18-22, is used to press curved areas. A *pressing mitt* can be slipped over the hand or onto the end of a *sleeve board* and used instead of a tailor's ham. A *velvet board* has a surface of needles to maintain the plushness of velvet, corduroy and other napped fabrics.

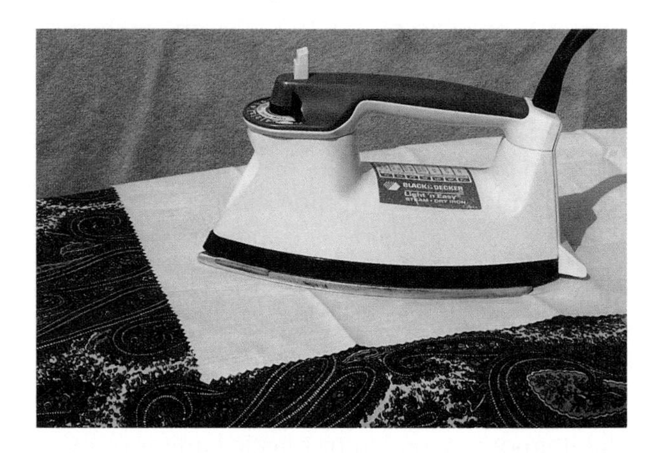

18-20 A pressing cloth is placed between the iron and the fabric to protect the fabric as it is being pressed.

June Tailor, Inc.

18-21 A sleeve board is convenient when pressing sleeves, the legs of pants, or other hard-to-iron areas.

18-22 A tailor's ham has a rounded shape for pressing curved areas. Usually one side of the tailor's ham is covered with a wool fabric, while the other side has a heavy canvas surface.

Ironing Techniques

It is important to use correct ironing techniques. Insert the plug of the iron firmly into the electrical outlet. Unplug it by grasping the plug, not by pulling on the cord. Always iron with the right equipment to maintain the shape of all garment parts.

Iron with the grain of fabrics to prevent stretching garments out of shape. If ridges or impressions result on the right side of the fabric from the edges of seam allowances or facings, place brown paper between the outer garment and inside structural edges.

When ironing a garment, start with the multiple layer areas, such as collars and cuffs. Then do small garment parts, such as yokes and sleeves. Finish by doing the large, flat areas. If the large areas were done first, they would get wrinkled again during the ironing of the small areas.

Iron dark, dull fabrics on the wrong side or use a pressing cloth on the right side to prevent shiny areas. Press wool with moisture. Use steam, but no pressure, on napped fabrics. Iron linens, cottons, and silks damp.

The best use of energy is to do a whole batch of ironing at a time. Start with garments that require a low heat setting on the iron, such as those made of manufactured fibers. Work up to those that require higher heat, such as cottons and linens. An iron heats faster than it cools, so it's quicker to go from low to high heat. You will also prevent scorching items with an iron that has not cooled down from the last piece.

Spray starches and fabric finishes are convenient. They are applied while ironing. They add stiffness and restore body lost in laundering. If desired, they can be used only on certain areas, such as collars and cuffs.

Spray starches provide firm crispness for natural fibers. Fabric finishes give body, but less stiffness, to manufactured fibers, cottons, and blends. Both can be used on either damp or dry clothes. They help the iron glide easier. However, sticking and scorching can result from excess moisture or improper iron temperature.

When using aerosol sprays, shake the contents before using. Line up the markers on the spray button (nozzle) and the can. Hold the container upright about 6 to 12 inches from the fabric. Spray lightly and evenly using a back and forth motion. Too much will lead to flaking and stickiness.

Clean the spray opening of an aerosol can when you are finished ironing. To do this, hold the can upside down and spray out the last residue in the "hose." If this is not done, the spray feature could be clogged the next time you want to use it.

Dry Cleaning

Dry cleaning is the process of cleaning textile items with nonwater liquid solvents or absorbent compounds. Dry-cleaning solvents kill living bacteria. Properly dry-cleaned garments do not smell of the solvents when they are finished. Dry cleaning minimizes shrinkage, preserves tailoring details, and maintains the fine characteristics and finishes of fabrics.

All garments can be dry-cleaned unless the care label states otherwise. When the label says "dry-clean only," the garment should not be washed in water. Clothes can be dry-cleaned professionally, done personally in a self-service, coin-operated machine, or done at home.

With *professional dry cleaning*, you pay for service per garment. You can expect skillful spot and stain removal as well as professional pressing. Cleaners should have trained employees and proper equipment. They should know how to care for various fabrics, dyes, finishes, and tricky garment construction, such as in 18-23.

Tell the dry cleaner if your clothes have stains, what the stains are, and how long they have been there. If requested, fabric pills can often be removed by the dry cleaner's industrial fabric brush. You may want to ask that sizing be added for more body or that a water-repellent finish be restored.

Coin-operated dry cleaning is faster and less expensive than professional dry cleaning. However, you do not have the advantage of professional spot removal, special care for delicate items, or thorough pressing. The machines are similar to coin-operated washing machines.

Prepare your clothes carefully by emptying pockets, repairing loose stitches and buttons, and brushing off dirt and lint. Remove belts or fancy buttons that should not be cleaned. Pretreat spots and stains. Sort the clothes into similar types and colors. Follow the directions for the machine. You will probably have to weigh the load so the proper amount of clothing is put into the machine.

When the clothes are finished, hang them up immediately. Shake them and pat out wrinkles. The fumes from dry-cleaning fluids are strong. Open a car window as you drive home. Shape and press the clothes, and air them, before storing them in a closed space.

Home dry cleaning kits are also now available. They are used in your home dryer, on the "no heat" or "air dry" setting. Such kits include a special bag for the clothes to be cleaned and a chemical packet. Be sure to treat all spots in advance of using the kit and then follow the kit directions carefully. Great care must be taken not to ruin good garments when dry-cleaning them yourself.

Hart Schaffner & Marx/Hartmarx Corporation

18-23 A professional dry cleaner has the skills and equipment to care for special garments of fine fabrics.

Chapter Review

Summary

Taking good care of your clothes gives you the most benefit from your clothing dollars and gives you a well-groomed, neat appearance. Daily care of clothes keeps soiling and damage to a minimum. When dressing and undressing, be careful not to snag, rip, stretch, or stain your garments. After wearing, hang up garments or put them where they will be washed or otherwise treated properly. Let clothes get air circulation and avoid wrinkling. Clothing repairs and laundering are easiest done during a set time each week.

Home storage areas, such as closets, shelves, and drawers, should be neat, handy, and well-organized. Seasonal wardrobe storage is usually done in the spring and fall for temperature and weather changes.

Spots and stains should be removed from garments as quickly as possible. First, identify both the fiber content of the fabric and the source of the stain. Have stain removal supplies on hand and know how to use them properly.

Laundering with good procedures and equipment extends the life and appearance of garments. Before washing, sort clothes by color, type of fabric and garment construction, and by kind and amount of soil. Choose the right laundry products for the job, such as proper detergents, enzymes, bleaches, water softeners, fabric softeners, disinfectants, and starches or fabric finishes. Use laundry products correctly and use the right wash and rinse water temperatures. Knowing about and utilizing the features of laundry equipment and following correct procedures gives the best results. Also, know how to hand wash delicate items and wool garments.

Clothes can be dried by tumble action in an automatic dryer, drip-dried on a hanger, line-dried, or laid flat to dry. Ironing and pressing should remove all wrinkles from garments. Correct ironing and pressing techniques are important for different fabrics and types of garments.

Dry cleaning jobs can be given to professionals, done in coin-operated machines, or done at home with dry cleaning kits.

Fashion in Review

Write your responses on a separate sheet of paper.

Short Answer: Write the correct response to each of the following items.

1. List five guidelines for the daily care of clothes.
2. List five guidelines for the weekly care of clothes.
3. List five guidelines for clothing storage.
4. Name three of the most commonly used stain removal methods.
5. Name three guidelines for sorting garments into wash loads.
6. Explain the difference between water softeners and fabric softeners.
7. Name three ways of drying clothes.
8. Name and describe two pieces of specialized pressing equipment.

True/False: Write true or false for each of the following statements.

9. It is best to hand wash knit garments so they don't wrinkle.
10. It is wise to line drawers with newspapers so no splinters will harm your clothes.
11. Stains often fade or disappear if left alone.
12. To save energy and money, coordinate the water level selector on the washer with the size of the wash load.
13. It is best to use a soap rather than a detergent for laundering if you have hard water.
14. Chlorine bleach is safe for use on all fabrics.
15. Starches and fabric finishes are sizings that restore body and crispness to fabrics that have become limp from laundering and wear.
16. Cold water should be used to wash white and colorfast fabrics and heavily soiled loads.

17. Hot water should be used to wash most wash loads.
18. Pour the detergent over the clothes in the washing machine when starting a load of laundry.
19. The lint filter of an automatic dryer should be cleaned after each use.
20. When ironing a garment, start with the large, flat areas.

Fashion in Action

1. On ten large index cards, write a definition for each of the following terms: all-purpose (high-sudsing) detergent, low-sudsing detergent, cold water detergent, soap, chlorine bleach, oxygen bleach, water softener, fabric softener, and sizings. Then take the cards to a store. On each card, write down the brand names of products that belong in that category. List the ingredients in them. Also write a paragraph about the directions for use after reading the container's labels.
2. Design what you would consider "ideal" storage for all of a person's wardrobe items. Include shelf space, drawer space, and hanging space. Express your ideas through drawings and written descriptions.
3. Compare costs of clothing care. Call or visit an appliance store to find out the approximate purchase price of a basic washer and dryer. Also ask about the "life expectancy" of these appliances. Consult local utility companies to find out the estimated energy costs for running them. Also find out the cost of using washers and dryers at a laundromat, as well as transportation getting there and back. With the number of loads of wash done in an average household each week, is it more economical to do laundry at home or out? Which way is more convenient?
4. Test various stain removal products and techniques. Cut some light-colored washable fabrics of natural and manufactured fibers into ten-inch squares. Make a stain on each with some of the "sources of stains" that are listed in chart 18-9. Keep track of what stain is on which fabric scrap. Then try appropriate stain removal treatments on each. Document the procedures and results.
5. Make a list of at least five apparel items in many people's wardrobes that need unusual care or storage. Describe the specifics of what should be done for them and why.

PART SIX

Apparel Industry Careers

Chapter 19
Careers in the Textile Industry

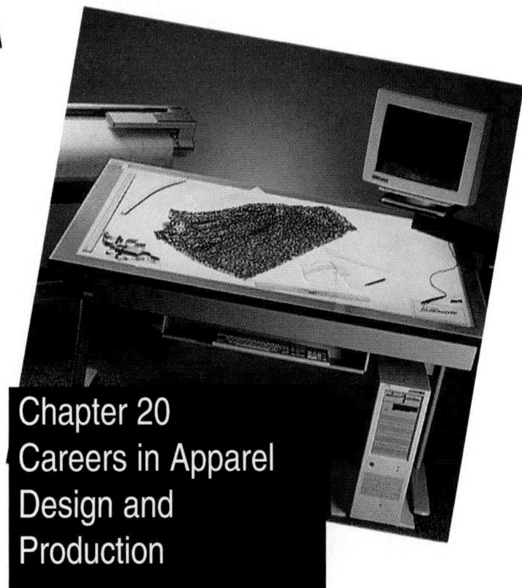

Chapter 20
Careers in Apparel Design and Production

Chapter 21
Fashion Merchandising and Other Retail Industry Careers

Chapter 22
Careers in Fashion Promotion

Chapter 23
Other Careers and Entrepreneurial Opportunities

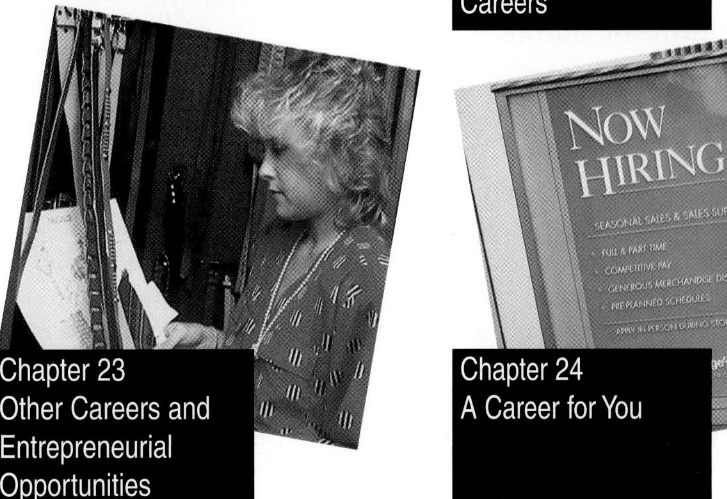

Chapter 24
A Career for You

19 Careers in the Textile Industry

Fashion Terms

- research scientist
- freelance jobs
- textile designer
- textile colorist
- textile stylist
- converter
- production supervisor
- machine operator
- machine technician
- plant engineer
- industrial engineer
- market analyst
- sales representative
- administration
- personnel administrator
- data processing employee

After studying this chapter, you will be able to

- discuss the wide range of careers in the textile industry.
- explain the requirements needed for various textile-related careers.
- discuss professions in textile research and development.
- describe opportunities in textile design and textile production.
- explain careers in textile marketing, sales, and administration.

Millions of people share in our nation's apparel industries. Researchers develop fibers and fabrics. Engineers design machines, and computer experts program systems to mass-produce huge quantities of garments. Workers cut, sew, press, and ship merchandise. Promoters write, photograph, illustrate, display, and model items. These people, and many others, make the apparel industries what they are today.

The textile industry includes every aspect of making, finishing, and selling fibers, yarns, and fabrics. The industry employs artists, chemists, engineers, managers, merchandisers, and many more. It offers jobs at all skill levels. Its constant changes in products and methods keep creating new opportunities.

Salaries in the textile industry are very good for people who can prove their worth. Fringe benefits usually include paid life insurance, medical benefits, retirement plans, paid vacation time, and sick leave. Sometimes tuition refund plans are available for further schooling.

Colleges and technical schools throughout the country offer textile majors in manufacturing, chemistry, engineering, design, and other subjects. Some textile careers require only a high school education and a desire to learn on the job. Most companies run their own training programs. However, advancement is not as high or as fast for those without college degrees. As in most industries, production workers outnumber management.

The textile industry also offers a good choice of where you may want to live. The main textile sales offices are located in New York City. Branch offices are in most other major cities and in foreign capitals. The selling staffs spend much of their time traveling. Most textile manufacturing plants, or mills, are located in small towns. The "textile belt" stretches from New England, down through the Carolinas and Georgia, and across to Texas.

Many, but not all, of the work opportunities in the textile industry are listed in 19-1. Many of the jobs are described in this chapter.

Textile Research and Development

The purpose of research and development, "R & D," is to provide new knowledge, develop new products, and improve old products. Employees in research and development usually work in modern, well-equipped laboratories, such as the one shown in 19-2. They search for facts and then analyze them. They use their findings to make better products. Through research, they try to develop goods that will satisfy the changing wants and needs of consumers, while providing a profit for the company.

The personal requirements to go into research and development include a creative imagination, curiosity, and attention to details. One should like to work alone, as in 19-3, yet have the ability to clearly communicate precise, realistic, and practical results to others. Patience and persistence in working toward a solution are also needed. Sometimes it takes many years to invent a product or to solve a technical problem.

People in textile research and development are employed by fiber manufacturers, textile mills, and private testing laboratories. Government agencies hire chemists and lab technicians to see that textile products on the market meet government standards.

Careers in the Textile Industry	
Accountants	Machine technicians
Administrators	Managers
Advertising and promotion agents	Market analysts
Business planners	Office managers
Card technicians	Personnel administrators
Chemists	Plant engineers
Clerical workers	Product developers
Cloth inspectors	Production supervisors
Colorists	Public relations agents
Computer programmers	Purchasing agents
Converters	Quality control inspectors
Data processors	Research scientists
Designers	Safety experts
Doffers (bobbin changers)	Sales managers
Dyeing supervisors	Sales representatives
Electricians	Screen printers
Engravers	Spinners
Industrial engineers	Stylists
Information specialists	Systems analysts
Knitting machine technicians	Textile testers
Laboratory technicians	Transportation (shipping) specialists
Loom technicians	Warehouse workers
Machine operators	Weavers

19-1 Most of the jobs in the textile industry are listed here. Other specific ones may be created as new technology causes changes in textile industry procedures.

BASF Corporation—Fibers Division

19-2 People who work in research and development try to develop new products or improve existing ones.

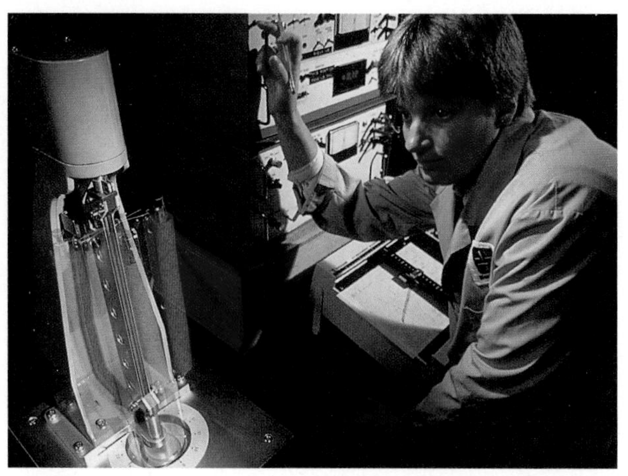

Burlington Industries, Inc.

19-3 The important jobs done by researchers depend on reliable work, often done individually.

Sometimes the researchers must develop the specifications for the standards. A great deal of research is also done at universities. Research positions give graduate students the opportunity to work and gain experience while earning advanced degrees.

Textile Research Scientist

Research and development is done in several different textile areas to satisfy needs for specific end uses. Textile ***research scientists*** develop new synthetic fibers that have certain characteristics. They blend fibers in new and better ways to create desired qualities. They work on different fabric constructions and find new finishes for better fabric performance.

Research and development is technical work, as indicated in 19-4. Researchers do laboratory experiments and study consumer complaints. They keep up with the latest work of their peers through related trade journals. They write reports about their own findings and recommend changes. The work must be done several years before the products will be offered to the retail market.

To be a textile research scientist, a college degree in the sciences is always necessary. Many researchers have degrees in chemistry, chemical engineering, or physics. Many research jobs require advanced degrees. Manufacturing experience in a textile plant during the summers, or as part of a work-study curriculum, is a great asset.

BASF Corporation—Fibers Division

19-4 Textile research scientists have interesting and rewarding jobs with a technical emphasis.

Researchers must enjoy science and like to perform experiments. They must be patient and flexible enough to see a project through from the initial idea to actual installation in a manufacturing plant.

Salaries and fringe benefits are very good for textile research scientists. There are sometimes opportunities to broaden one's career into other areas of company activity such as manufacturing supervision or marketing. Also, personal satisfaction can be very high when complex problems are solved or something new is created. Work hours are usually regular with occasional overtime demands.

Textile research scientists usually belong to professional societies, such as the American Association of Textile Chemists and Colorists. Often they attend seminars and conferences about their subject areas. Expenses for these are usually paid by the companies that employ the scientists.

Textile Laboratory Technician

Textile *laboratory technicians* help conduct research, often working under research scientists. They have high school diplomas and usually technical trade school certificates. Many of the specific procedures they do are repetitive and are learned on the job.

In the lab, technicians set up equipment, write down computations, and help categorize and analyze experiment results. They may duplicate each step of a future manufacturing operation to evaluate its quality and efficiency. They may also test the durability and serviceability of finished fibers, yarns, and fabrics.

Technicians who are *textile testers* work for textile mills or independent testing laboratories. They test new products against the specifications that must be met. They are responsible for checking different fibers, fabrics, or finishes after they have been developed but before they are introduced to the public. They also perform tests during the textile manufacturing process to assure good and uniform quality.

Textile technicians and testers should be able to organize their own work and to work well alone. They should enjoy working with equipment and chemicals. They must be able to follow precise instructions, to do detailed work accurately, and to write thorough reports of their test results.

Textile Design

Textile design employees should have a fine sense of color, a creative imagination, and artistic ability. They also need knowledge of and interest in the fashion field. They should have a fascination for beautiful patterns that are woven, knitted, or printed onto fabrics.

Most jobs in textile design are located in New York City and other fashion centers, where the designing and styling departments of the major textile firms are located. Although most of these positions are offered on a full-time employment basis, it is sometimes possible for experienced people to find ***freelance jobs***. These are short-term jobs done for various firms that end when each assignment is completed. However, because it takes a long time to build up freelance clients, and because employers expect highly professional work on short notice, it is not wise for beginners to consider freelance work.

Textile Designer

A ***textile designer*** creates new patterns and designs, or redesigns existing ones, to be used in the making of fabrics. The designer creates new looks for fabrics for wearing apparel as well as home and commercial furnishings. The fabric designs are usually directed toward a specific end use or market. The designer tries to anticipate coming fashion trends in fabrics. Designers develop a seasonal line of new fabrics for textile manufacturers.

Surface designs of a fabric are applied on top, usually by printing. First the designer sketches the idea. He or she works out one motif, or repeat, of the design on paper or with a computer design program, as in 19-5. Then it is printed out by the computer. A sample can be silk-screened or roller printed to show how the final design will look for mass printing. Then it goes into printing production on a large scale.

Textile designers also create *structural designs*. This is done through knitted patterns or special weaves. Yarns of different textures or colors are combined in interesting ways. Less need for fine drawing ability is required for this, but more knowledge of yarns, technical processes, and machine capabilities is needed. See 19-6.

The use of computers is also widespread for structural design. With older methods, patterns were designed on hand looms or knitting machines in small quantities to show the finished effect in sample swatches. The

Primavision Design System by Cadtex

19-5 Training is needed to effectively use the exciting capabilities of computer design systems.

Fashion Institute of Technology

19-6 This textile student is learning about structural design by weaving a sample of the idea directly on a loom.

designer or a technician then transferred the design into a computer. The computer controlled the loom or knitting machine to produce large quantities of fabric with the design. However, now the entire process can be done by computer. The latest computer systems can be used to create a design and coordinate the design into the structural weave or knitted pattern of the fabric. This is very fast and accurate.

A textile designer must know about new fibers, dyes, and finishes. He or she must understand the latest machinery, as well as new styles and fashion trends. The designer must know how to select, mix, and combine colors. He or she must know how to plan repeats. All processes of textile manufacturing must be understood so the designs that are developed can go into factory production within a specific price category.

Textile designers often get their creative ideas by observing the world around them. They must be sensitive to trends, fads, and current events. They may also look through older fabrics and works of art for inspiration. Many firms have their own fabric libraries for this purpose.

To become a textile designer, you must have education and training after high school. Textile colleges offer four-year degrees in the field. These are partially technical but have strong art emphasis. Courses in art, clothing, and textiles are needed. Some two-year schools of design or technology can prepare you to start out in this career also.

To get a good textile design job, you must develop technical computer skills and textile knowledge to complement your artistic creativity. Competition is keen in the textile design field. There are a limited number of new openings each year. However, the jobs that are available are very rewarding.

Textile Colorist

You may have seen a particular textile design that was offered in several different color combinations. See 19-7. ***Textile colorists*** are the ones who decide which combinations of colors will appeal to customers. They skillfully put those colors into the original textile design by computer or with paints on paper. There might be many changes before final approval is given for production. Then specific instructions are sent to the mill for the manufacturing process.

The position of textile colorist is often an entry-level job for graduates of college or trade school textile design programs. These programs usually include the study of woven

19-7 These four fabrics have the same print but each is done with different color combinations.

Fashion Institute of Technology

19-8 A textile colorist might make up several sample swatches for fabric designs in the same colors. These might be coordinated or combined in apparel outfits or manufacturer's garment lines.

and printed fabrics, color fundamentals, as in 19-8, and creative design principles. A *portfolio* of textile colorings and designs must be submitted when applying for this position. An art or design portfolio is a case of loose, unfolded papers showing a person's creative work.

Being a textile colorist allows a beginner to gain speed and a sense of color by learning to match and paint various color combinations or arrangements on a design created by a textile designer. The colorist is an important member of the designer's team. He or she becomes aware of how colored dyes look on almost any type of fabric. Records of color samples and fabric swatches must be kept. Often research is done of older textiles to get color information about designs used in the past.

Textile colorists must be neat and careful workers. They must be able to follow precise instructions. They must often work at a fast pace to meet deadlines that keep the expensive fabric production on schedule.

Textile colorists who are ambitious and prove their worth may move up to a textile design job. Eventually, they might even fill a managerial position with a textile mill or design studio.

Textile Stylist

Textile stylists have the responsibility of the fabric line from its beginning stages to completion. Textile stylists guide textile designers, as in 19-9, to fulfill consumer desires. They are responsible for most long-range fashion planning of fabric colors, weights, and textures. They sometimes act as a colorist, technician, merchandiser, salesperson, and even designer. They must have an extensive knowledge of the textile industry and a wide range of industry contacts and resources.

Stylists must understand the textile marketplace. They must know what consumers want. They can sometimes even stimulate consumer desires. Successful company sales depend on the stylists' abilities to gauge consumer demand and to stimulate interest in new fabrics.

The stylist is responsible for determining the proper look, or overall concept, of the fabric line. He or she might tell designers and colorists exactly what types of patterns to develop and what fashionable colors to use. The stylist then coordinates the design and

Trevira Polyester—Hoechst Fibers Industries

19-9 Textile stylists coordinate the designs, colors, and end-use projections with those who are creating the fabrics.

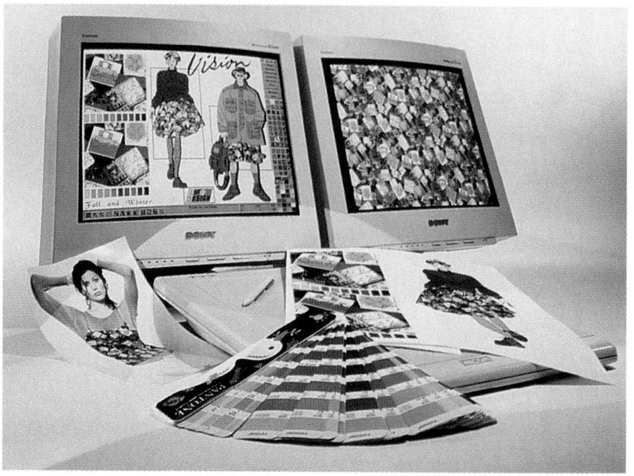

Info Design, Inc.

19-10 Before planning their fabric lines, textile stylists check fashion forecasting information for expected future colors and trends.

color work with the efforts of the production staff at the mill or print plant, often many miles away.

The stylist may supervise one or many designers and colorists in the design studio. In addition, frequent telephone contact with plant employees where the fabric is being produced or printed is needed. Accuracy of the designs and colors must be checked at the mill site to eliminate costly production errors and to assure high sales.

The textile stylist works closely with the company's merchandising department in planning the future fabric lines. Frequent contact must be made with garment manufacturers to learn what kinds of fabrics they will want in order to produce their items. The stylist might also suggest to the manufacturers ways in which particular fabrics can be used.

Stylists also get information on new directions that fashion and color will take from forecasting services, as in 19-10, and fabric editors of trade publications, fashion magazines, or industry associations. The stylist's own good taste and fashion sense are combined with doing research of the marketplace and trade resources. To be a success, this person needs many years of textile industry experience predicting what the public will want to buy more than a year in the future.

Training in the fine arts and textile science is required to be a textile stylist. The varied work requires courses in art, advertising, clothing, and textiles. Creative talent is a must. Retail sales experience with fabrics of ready-to-wear is helpful. Also, experience in a textile mill, to understand production methods and machinery, is valuable.

Textile styling jobs are highly desirable and very competitive. They pay extremely well. The sales and profits of textile firms depend to a great extent on the talents of the stylist. The position is often reached only after many successful years in the textile field. Large textile manufacturing firms may hire several stylists. Each one is responsible for a specific group of fabrics, such as those for children's wear, men's wear, or women's wear.

Assistant Stylist

An *assistant stylist* is a person who is being trained to be a stylist. Assistants set up appointments with customers. They have regular contact with the mills to check on production schedules or problems. They may handle clerical details and correspondence.

Assistant stylist positions are often filled by people who have completed college programs in textile design or textile technology. They also usually have experience as a successful textile colorist or designer.

Textile Production

As already mentioned, first the researchers must develop the fibers, yarns, and fabrics. Next, the designers must create the weaves, knits, and prints in the desired weights and colors. Then the yard goods can be produced. Textile production involves the largest single portion of textile industry employees.

Textile plant operations include opening the bales of fibers, cleaning and straightening the fibers, and spinning the fibers into yarns. Then the yarns are woven, knitted, or matted into fabrics. See 19-11. The textile goods must also be chemically or mechanically finished. Each manufacturing step needs special machinery and trained employees.

Many textile mills run six or seven days a week with three 8-hour work shifts. Shift work schedules often rotate, so employees share the day, evening, and night shifts. They may work six days in a row and have a three-day "weekend" in the middle of the week. That enables them to take care of personal business or enjoy recreation without usual weekend crowds. Also, extra pay is given to those who work the evening and night shifts.

Burlington Industries

19-11 This textile production employee is setting up a circular knitting machine that will produce many yards of knitted fabric.

A textile mill often handles just one part of production—spinning, weaving, knitting, or finishing. The majority of the jobs are located in the Southeast, such as in the Carolinas and Georgia. Others are in the New England states. Many companies have plants in more than one town or city, so employees are subject to transfer if their expertise is needed elsewhere.

Most firms encourage and pay for their management personnel (higher engineers and supervisors) to attend seminars, workshops, and classes to help keep them abreast of current developments in the textile industry, as in 19-12. Otherwise, not much traveling is done by manufacturing employees.

Salaries and fringe benefits are good in textile production jobs. Recognition is given for responsibilities handled well. Working conditions are generally pleasant in modern plants. Many plants even sponsor recreational, social, and athletic activities for employees and their families.

New production jobs are being created as new technology is introduced into the textile industry. New machines and processes are being developed. Computers are playing a growing role in all manufacturing operations. Electronic equipment is used for such tasks as forecasting amounts of raw materials needed, computing how much of each type of fiber or fabric to produce, and giving statistics on quality control.

ITMA (International Textile Machinery Association)

19-12 Many seminars and professional meetings are offered for managers employed by textile firms.

Textile Converter

A key person in textile production is the converter. ***Converters*** are finishing experts. At finishing plants, they decide how various amounts of fabrics should be dyed, printed, or otherwise finished. Converters must be aware of all the needs of customers, so the right finished textiles are made.

Converters calculate the amounts of fabric that will be bought by the firm's customers. They plan the fabric construction and decide on suitable finishes and dyes based on the end uses. Quality standards are calculated for various fabrics depending on how each fabric will be used. Converters also help set prices based on the costs of supplies and production.

Large textile firms employ more than one converter. Each is a specialist in a certain fabric area for a particular market. Each knows what is current in that marketplace and keeps in touch with the customers of those goods. Each must be able to quickly sum up new economic trends relating to his or her particular textile market.

Textile converters must be good at details, figures, and record keeping. They should be well-organized and have excellent communication skills. They must be calm and level-headed when working under pressure.

Assistant Converter

Assistant converters must also be familiar with all the textile production procedures, especially finishing processes. If they perform the duties of assistant well, they can move up to a textile converter's position. They should have completed a two- or four-year textile technology program. College courses are needed in the areas of textile science, fabric construction, mathematics, and textile chemistry. They also should have studied dyeing, color analysis, textile testing, marketing principles, mathematics, and other related areas.

Assistant converters are responsible for keeping accurate production and inventory records. They handle many clerical duties for the textile converter. They also help establish prices based on all the costs involved, plus a margin of profit. This is known as "costing the fabric." Telephone work is necessary in dealing with fiber mills from whom they buy, as well as with customers wanting to know the status of their orders.

Production Supervisor

Many college graduates entering the textile industry begin their careers as ***production supervisors***. They usually have degrees in engineering, textile technology, business, or chemistry. In addition, they can take production management training programs given by the companies that hire them. They coordinate and direct the various mill departments to maintain the highest production and the best quality.

A production supervisor must have qualities of leadership. The ability is needed to motivate others to be loyal to the "team" or firm and give their best efforts toward the finest possible results. He or she must be able to organize work and to solve problems. Supervisors who prove their capabilities can move up to plant management positions or move into sales or development jobs.

Machine Operator

Under the supervisors, there are many workers called ***machine operators***. They operate the machines that do the manufacturing procedures, as shown in 19-13. They do not have college degrees, but they do important, specialized jobs. They generally have a high school education and are trained on the job. However, as the need for skilled workers

Sulzer Ruti, Inc.

19-13 Machine operators in textile manufacturing plants run the production machinery. This woman works in a weaving mill.

increases with computerized, high-tech procedures, more employers are emphasizing formal training in trade or vocational schools. The complicated machines must be handled with a high degree of skill and responsibility.

Production operators must use their minds and their hands. They should have good mechanical aptitude, physical coordination, manual dexterity, and normal vision. However, there are also numerous opportunities for people with disabilities. Also, there is ample job advancement from hourly, paid operator jobs to salaried, managerial jobs for those who work hard and have a good attitude. Most textile production operators are people who enjoy routine tasks and like to do a job alone.

Quality Control Inspector

Quality control inspectors work in all phases of textile production to analyze the quality of the products. They try to find solutions to quality problems when necessary, as shown in 19-14. They carefully examine the fibers, yarns, or fabrics as they come from the extruders, looms, knitting machines, or finishing operations. They check to see that precise standards and specifications are met. Imperfect goods cannot be sold. Thus, inspectors identify production-related problems at different stages of the manufacturing process and write reports based on their findings.

Burlington Industries, Inc.

19-14 When problems in production arise, they are analyzed. Then recommendations are made for solutions. This enables the best possible products to be produced.

People who choose quality control work for a career should have good analytical skills. They should enjoy detail and follow-up work. They should also complete a textile technology or apparel production program.

Machine Technician

Machine technicians keep the complex equipment in good working order. They are often former operators with especially good mechanical skills. They are either trained at the plant or in vocational schools. They do regular maintenance on the machinery. They are also called on to diagnose problems in every type of production machine in the plant. This applies to large gear chains as well as delicate weaving or knitting machine controls. The pay is usually higher than for operators.

Plant Engineer

Plant engineers make sure all environmental systems are operating properly. They are responsible for the heating, air conditioning, electrical, materials handling, and other systems. If these operations are not functioning properly, they must be modified or repaired. Sometimes plant engineers are also consulted by the machine technicians if complex problems occur in production machinery.

Plant engineers must be expert problem solvers. A college degree in engineering is usually required. The job involves a great deal of responsibility. Pay is quite good.

Industrial Engineer

Industrial engineers are cost and efficiency experts. They save companies time and money. They study each operation, as in 19-15, and determine the most efficient and least expensive method of getting it done. They decide what machines are needed for the operations of the mill. They are constantly looking for better ways of doing production jobs without reducing the quality of the final product. They also try to improve the safety of operations.

Industrial engineers must have a complete understanding of the operation of the plant. They must be experts in work methods and job design. They work with plant management and plant engineers to develop the best ways to put new procedures into effect.

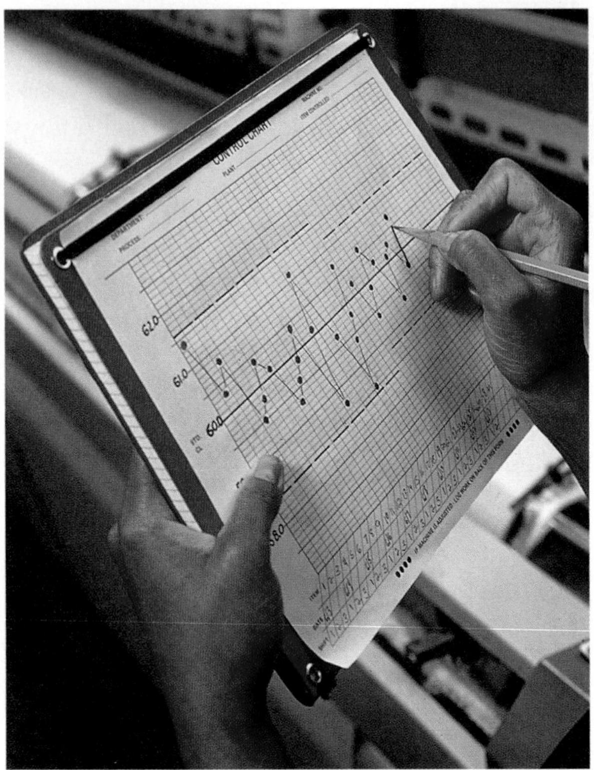

Burlington Industries, Inc.

19-15 Industrial engineers study textile production operations to determine the most efficient methods of doing the tasks.

For this job, companies prefer college graduates with industrial engineering or textile manufacturing degrees. In addition, companies usually provide their own training programs. Industrial engineers are usually people who like scientific or technical work. They like planning and controlling activities.

Textile Marketing and Sales

Marketing is the business of finding or creating a market for the textiles produced. It includes identifying customers, determining the wants and needs of customers, and providing satisfying products at acceptable prices. It also incorporates promotion and distribution at a profit. It helps develop the sales strategy of the firm. Sales refers to the actual exchange of goods from the supplier to the customer for money.

Marketing and salespeople must be fully familiar with their firm's lines, as well as policies and procedures. They must try to satisfy the ever-changing needs and tastes of those who purchase their company's textile products. They must stay on the leading edge of rapid changes in fashion and style.

Market Analyst

Market analysts conduct market research by studying consumer tastes and changing trends. They try to discover future textile needs for all markets. Market analysts are alert to any shift in supply and demand. They keep the research scientists informed of what must be developed. They work closely with their firm's textile stylists. They assist the textile converters in keeping production of the proper finished goods at peak levels. Market research information must be supplied far in advance so the textiles will be ready at the time of demand.

Market analysts help predict the trends of special colors, looks, weights, and textures of fabrics. They coordinate the activities of their textile companies with the rest of the fashion world. They often travel to foreign fashion showings. Market analysts also keep on top of what their competitors are doing in long-range pricing and supply. With a knowledge of general consumer preferences and a great deal of experience, they can make accurate predictions. Their work can be especially exciting if they have been a prime influence behind a successful marketing idea.

Market analysts should have education and training in textiles, business, marketing, economics, psychology, and statistics. They also need to know every aspect of the textile industry.

Textile Sales Representative

Textile ***sales representatives*** are the actual salespeople at various levels of the textile chain. Fiber producers sell to yarn producers or fabric manufacturers. Greige goods manufacturers sell their unfinished fabrics to finishers. Finally, the finished yard goods are sold. To do this, sales representatives show cloth samples, and sometimes finished garments, to apparel manufacturers, as shown in 19-16. They also sell to retail fabric stores and to industrial and household goods producers.

Textile salespeople are the link between their firm and its customers. A continuing relationship is sought with each customer, so the salesperson can be the company's eyes and ears. Salespeople try to advise their managers and market analysts on fashion directions in the marketplace. They stay alert to the activities of competitors, and supply-demand situations. They must give attention to both the needs of the customers and the interests of the company.

Sales personnel are important representatives of their companies. They must be knowledgeable about their products and able to talk intelligently to give reputable advice and direction to each particular customer.

Sales is the most competitive area of the textile industry. It is not the career for everyone. However, for those who have an aptitude for sales, the opportunities and financial rewards can be great. It is a respected and profitable profession for those who are outgoing, yet sincere.

Sales representatives must have personalities that can stimulate desires for their products. They must have poise, self-confidence, and maturity. They must be able to meet and get along with people of all kinds. They need good, convincing communication skills, intelligence, adaptability, and an alertness to current cultural trends. They must be able to make their own appointments and set their own hours.

Textile salespeople must have integrity in order to gain and keep the trust and respect of customers. Honesty in dealing with factors of current sales (quality, dates of delivery, price, etc.) will set the groundwork for future sales to be made. Creativity to make the best of each situation is needed. Sensitivity to respond to the views of others is necessary. Also, proper personal appearance is very important. This means being well-groomed and fashionable rather than sloppy, flashy, or out of style.

Many textile salespeople work in New York City or in corporate offices near their company's production plants. Also, regional or district sales offices are located in other major cities throughout the world. Travel is usually part of a salesperson's life since he or she must be in constant contact with customers and the market.

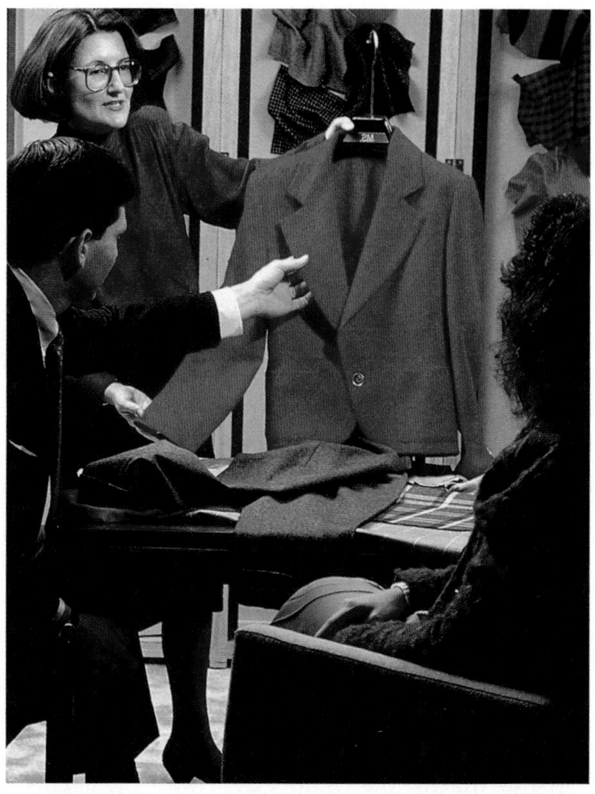

Burlington Industries, Inc.

19-16 With fabric swatches of the company's line displayed on the wall behind her, this textile salesperson is showing examples of how the fabrics can be used in fashionable apparel.

A sales career requires a certain amount of stamina and ability to withstand stress. Salespeople must have a strong desire to sell in this highly competitive business. Sometimes they have to work under pressure, while always keeping high enthusiasm and interest in their lines.

Sales representatives are given earnings and responsibilities in proportion to their performance on the job. For the right hard-working person, this exciting field is both personally and financially rewarding. Pay is either a flat salary or a salary plus a commission on the amount sold. There are opportunities to advance into a marketing manager's job.

Sales Trainee

The entry-level job in textile sales is that of a *sales trainee*, or person learning sales work. Most are hired with a college degree, often in business administration, marketing, the sciences, or liberal arts. They have usually had courses in business, economics, and psychology. Some are textile graduates.

All textile sales personnel are given a period of training by their companies. Some training programs follow a strict schedule, while others are more casual. All of them teach sales trainees about the firms' products and production processes. They cover merchandising techniques and customer needs.

Sometimes sales trainees begin their careers as customer service representatives. Sometimes they are technical service troubleshooters who try to solve problems. This gives them maturity and time to learn about the products before they actually go out to sell them.

Sales trainees perform any duties related to the sales operation. They hang up fabric samples and keep the showroom in shape. They greet customers in the showroom and help present the line. They go out on sales calls to manufacturers with an experienced salesperson. During these calls, they carry the samples, do clerical work related to sales, and assist with the billing, shipping, and handling of orders.

Textile Sales Manager

Sales managers supervise several sales representatives. They might be in charge of the sales of a geographic district or of all the sales of a particular division of the company. They set sales quotas and guide their salespeople to achieve or surpass those quotas.

Sales managers help plan and direct the smooth flow of products from the textile plant to the customer. They must understand every step from research, through design and production, to the final customer. Most successful sales managers have a good sense of fashion and sound judgment about what customers want, as shown in 19-17.

Sales managers must relate easily to people. They must have strong administrative abilities to achieve top sales. The job is demanding and stressful. Success in the job can lead to top management.

Textile sales managers should have a college degree. They may also have an advanced degree, often an M.B.A. (masters of business administration). They should have experience in sales and also understand production. They receive high salaries.

Sulzer Ruti, Inc.

19-17 A good sales manager takes time to meet with customers to help satisfy their important needs.

Textile Advertising and Promotion Agents

Textile *advertising and promotion agents* help boost sales by telling the firm's customers that there is something wonderful to buy! They try to create demand for their textile products. They advertise in apparel trade publications and have displays or exhibits at trade shows. They plan fashion shows or other publicity events to show garments made from the company's fabrics. See 19-18. They prepare instructional materials to show special sewing or handling techniques for their fabrics.

Advertising and promotion agents must have creativity and excellent communication skills. These are needed to plan, develop, and execute campaigns to announce and encourage sales of their textile products. They must meet people easily and have a sincere interest in the company's products. They usually have a college degree in liberal arts, advertising, communications, or journalism.

Textile Administration

Administration is the overall managing of the textile company. Administrators oversee the entire operation. They coordinate the development, manufacturing, and marketing of the fabrics so the company's functions run

Hoechst Celanese/Bendel's

19-18 Olympic figure skating gold medalist Kristi Yamaguchi was hired by advertising and promoting agents to represent the *Celebrate!* acetate fiber through photographs and personal appearances.

smoothly. This enables the company to stay in business with a profit.

Men and women reach management positions in administration after they have acquired maturity and a thorough knowledge of all aspects of the textile business. They are the junior and senior executives all the way up to the president of the firm.

Administrators must have an interest in business and commerce. They must have the ability to make sound decisions, to see how all aspects of the company fit together, and to understand how specific procedures will affect the firm.

Personnel Administrator

The ***personnel administrator*** oversees the people employed by the company. He or she sees that the employees are hired, or fired if necessary. This person also makes sure employees are paid, insured, given benefits, and given retirement compensation.

Personnel administrators must get along well with people. They should have the ability to understand employee problems and to communicate with employees in a friendly and honest manner. They may be in charge of training, work incentives, and award programs. They may work with union grievances, management development programs, insurance claims, continuing education programs, and health and safety programs.

Accounting and Finance Employees

Accounting and *finance employees* keep track of the funds of the company. They keep records of all the money earned and paid out. They manage the company's investments, monitor the cash flow, and deal with financial institutions, such as banks. They also supervise the company's operating statements and financial reports.

Accounting and finance jobs are great for people who enjoy routine and organized work. They are often done alone. Calculators and computers are used. The work is very exacting. Education beyond high school is required. For high-level jobs, a college degree in accounting, finance, or mathematics is desired.

Data Processing Employees

Data processing employees use computers. Computers are used in textile firms to keep track of sales, shipping, billing, inventories, employee records, and payroll. In other capacities, computers are used to monitor textile production processes and to design fabrics.

Data entry employees have basic computer skills since they type routine information into computers, as in 19-19. Computer operators are sometimes high school graduates with technical school training. Computer programmers, who set up the firm's computer systems, usually have college degrees in computer science or math.

Other Administrative Employees

Many other positions in the administration of textile firms are similar in titles and duties to those in other industries. For instance, *office managers* supervise the sales offices or plant sites around the country. They manage equipment and supplies, negotiate with vendors, oversee the company cars, and so on.

Business planners gather data on the company's operations. After analyzing the information, they make recommendations of action to the company's management.

Public relations agents tell the company's story. They "plug" the positive attributes of the company and, if necessary, explain negatives to employees, customers, the community, and the government. They deal with the press when needed.

Purchasing agents are buyers. They may specialize in buying fibers to be made into fabrics. They may be in charge of buying chemicals, dyes, and equipment. They may also buy office equipment, vehicles, machinery, parts, and everything else used by the textile manufacturer. They try to get the best value for each dollar spent. They must be organized, tough, and good with numbers.

19-19 Data processing employees put company-specific information into computer systems for their firms.

Chapter Review

Summary

Millions of people work in the textile industry, in many locations and with good pay and fringe benefits. Textile research and development provides new knowledge, develops new products, and improves old products. Research scientists develop new synthetic fibers, blend fibers in new and better ways, work on different fabric constructions, and find new finishes for better fabric performance. Textile laboratory technicians help research scientists by setting up equipment, writing down computations, and helping to categorize and analyze experiment results.

Textile design employees should have a fine sense of color, a creative imagination, and artistic ability. Textile designers create new surface and structural patterns, or redesign existing ones, for the making of fabrics. Textile colorists develop fashion color design combinations by computer or with paints on paper. Textile stylists have the responsibility of entire fabric lines from the beginning stages to completion. Assistant stylists handle details for textile stylists.

Production at textile mills involves the largest single portion of textile industry employees. Textile converters decide how various amounts of fabrics should be dyed, printed, or otherwise finished for end-use demand. Assistant converters keep accurate production and inventory records, and handle many clerical duties for textile converters. Production supervisors coordinate and direct the various mill departments to maintain the highest production and the best quality. Machine operators do the manufacturing procedures. Quality control inspectors analyze the quality of the products as they go through manufacturing. Machine technicians keep the complex equipment in good working order. Plant engineers make sure all environmental systems are operating properly, and industrial engineers are cost and efficiency experts.

Textile marketing finds or creates markets for the textile products. Market analysts study consumer tastes and changing trends. Sales representatives sell the firm's products to customers. Sales trainees are learning sales work. Textile sales managers supervise several sales representatives in a geographic district or a division of the company. Textile advertising and promotion agents try to create demand for their firm's textile products through advertising, trade show exhibits, special events, and instructional materials.

Textile administration is the overall managing of the company. Accounting and finance employees keep track of the funds of the company. Data processing workers record sales, shipping, billing, inventories, employee records, and payroll on computers. Other administrators include office managers, business planners, public relations agents, and purchasing agents.

Fashion in Review

Write your responses on a separate sheet of paper.

True/False: Write true or false for each of the following statements.

1. Textile-related jobs are available at all skill levels.
2. Most textile firms run their own training programs.
3. In most industries, management personnel outnumber production workers.
4. The "textile belt" stretches from Texas across to California.
5. A good way for beginners to enter the field of textile design is to seek freelance jobs.
6. Fabric designs are usually directed toward a specific end use or market.
7. The position of textile colorist is often an entry-level job for graduates of college or trade school textile design programs.
8. A textile stylist is responsible for determining the proper look, or overall concept, of the company's fabric line.
9. Textile design involves the largest single portion of textile industry employees.
10. Market analysts study consumer tastes and changing trends.

Multiple Choice: Write the letter of the best response to each of the following statements.

11. "R & D" stands for _____.
    A. review and decide
    B. recycle and deliver
    C. recolor and design
    D. research and development
12. Textile technicians and testers should be able to _____.
    A. organize their own work
    B. work well alone
    C. follow precise instructions
    D. All of the above.
13. Textile converters mainly _____.
    A. update or convert production machines so they can produce the latest textile products
    B. decide how various amounts of fabrics should be dyed, printed, or finished
    C. change old production machinery to computerized systems
    D. convert the colors of last year's textile designs to the colors of the upcoming season
14. The person who determines the most efficient and least expensive method of doing each operation of textile production is a _____.
    A. production supervisor
    B. quality control inspector
    C. plant engineer
    D. industrial engineer
15. Which employees work in the administrative part of a company?
    A. Accounting and finance employees.
    B. Office managers.
    C. Purchasing agents.
    D. All of the above.

Fashion in Action

1. Pick a specific job in the textile industry that interests you. Do further study about it in the library or on the Internet, with materials from your guidance counselor, and by talking to people who have that job, if possible. Write a report about the job. Include the qualifications you would need to get the job and what tasks you would perform on the job. Write about the salary range and fringe benefits you could expect and what opportunities there might be for advancement.
2. Try being a textile designer by drawing one motif of a design for a certain type of garment. First jot down several ideas. Then make rough sketches of your ideas. Finally develop your best idea into a finished design. Decide what surface texture the fabric should have and find a fabric swatch with that same texture. Use paints or felt tip pens to show the desired colors. Assemble your papers from each step of the designing process in a booklet. Perhaps the school's art teacher could give comments on the work, as well as the apparel instructor.
3. Make a list of the colleges and technical or vocational schools near you that offer training for the careers described in this chapter. Obtain a catalog from one or more of the schools. Write a short report explaining the specific courses that are required to graduate in a particular textile-related curriculum.
4. Prepare a bulletin board display of textile job opportunities advertised in the local newspaper. What education, training, or skills are required? Are salaries mentioned? Try to call the firm to get more information about the job opportunities.
5. Choose a specific textile product, such as a certain fiber or a fabric with a particular look or end use. Give a sales presentation to the class that includes textile samples and garments, or pictures to support your selling points. Organize your thoughts to speak knowledgeably about the positive aspects of and uses for your textile "line."

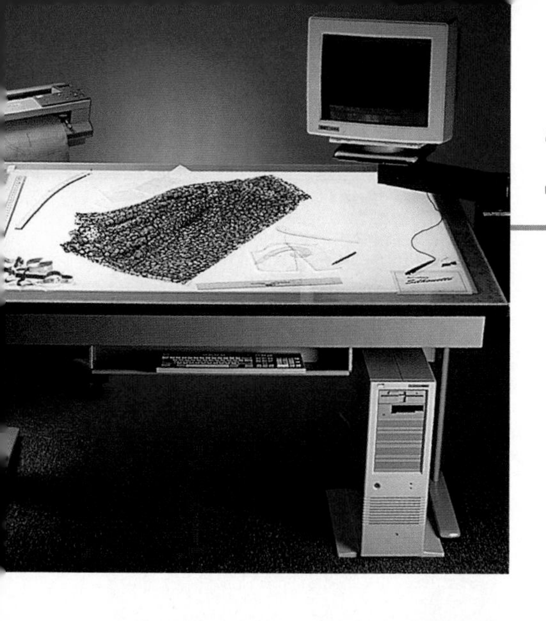

20 Careers in Apparel Design and Production

Fashion Terms

fashion designer
sample maker
value added
pattern maker
pattern grader
assorter
alteration hand
product manager
plant manager
piece goods buyer
costing engineer
distribution
showroom salesperson
traveling sales representative
jobber
C.E.O. (chief executive officer)
division director

After studying this chapter, you will be able to

- describe the work of a fashion designer and the qualifications needed for that career.
- discuss the many types of jobs in apparel manufacturing.
- explain production management careers.
- identify positions in sales and distribution of manufactured apparel merchandise.

Apparel production is the making of fabric into clothing. It is a major industry worldwide. Manufacturing jobs enable new apparel designs to be produced in large quantities and sold to retailers. Many workers are needed with different skills to perform all of these tasks.

Apparel Design

Apparel design is the field of creating new fashionable garments and accessories that have customer appeal. New ideas make the items original and different from others. The success of the manufacturer's business depends on the salability of the designs produced.

Fashion Designer

Fashion designers (also called apparel designers), as in 20-1, create new ideas for garments and accessories. They try to design functional and beautiful items that will be at the leading edge of fashion when they are produced. Designers might specialize in men's, women's, or children's wear. They might concentrate on even more specific areas, such as swimwear, bridal attire, shoes, lingerie, or evening dresses, as in 20-2.

Couture designers usually do two major collection showings per year. Large U.S. manufacturing firms may do four or five seasonal lines during each year. Most lines contain 40 to 75 items.

The greatest number of design employees work for manufacturers who mass-produce low-priced items. These "design stylists," "sketchers/stylists," or "copyists" seldom originate ideas. Instead, they adapt higher-priced designs to meet the price ranges of their customers. They may select the fabrics,

Council of Fashion Designers of America; Roxanne Lowit, Photographer

20-1 Fashion designers, such as Jeffrey Banks and Charlotte Neuville, shown here, sometimes get together for press receptions or award events to benefit the fashion industry.

Fashion Institute of Technology

20-2 Most fashion designers concentrate on a specific type of clothing, such as evening dresses.

coordinate the lines, and oversee other details. They receive lower salaries and less prestige than designers of more exclusive lines. However, entry-level jobs are often available to gain experience in the industry.

Moderately priced apparel is usually produced by medium-sized manufacturers with one or more designers. Each designer has an assistant or a design room staff. Most designers are not well known because their names are not publicized or sewn into the clothes they create. They are strictly important employees working for their firms.

Very high-priced apparel firms employ only the most talented designers. These opportunities are limited, and competition is great. A designer must be recognized as being gifted in the field to fill one of these scarce positions. The salary is extremely high.

Most fashion design jobs are full-time. However, it is possible for experienced designers to work on a freelance basis by doing assignments for several firms when needed. These companies are usually not big enough to have full-time designers of their own. Also, a few designers are self-employed. They may hire sewers to sew the items they sell directly to exclusive clients. Others sell their creations through specialty shops.

Apparel designers work in cities that are fashion centers. Over half of the U.S. designers work in the fast-moving fashion scene of New York City. Most of the rest are in Los Angeles, San Francisco, Chicago, Atlanta, Dallas, Miami, St. Louis, Kansas City, Minneapolis, and Cleveland.

The Work of a Fashion Designer

The idea is the first step in creating an apparel design. Many designers are inspired by swatches of fabrics and forecasting information from textile firms. Others find ideas by doing research in museums, art galleries, and libraries. Some ideas come from world events, from nature, or by observing people on the streets of "fashion forward" cities. Fashion designers are constantly searching for new ideas. They keep up with art and fashion news through trade publications.

As the use of electronic equipment increases, more designers are visualizing their ideas on computers through computer-aided design, or CAD. They can add more or less fullness where desired, work with different color combinations, and see how various trims might look. When the design is just right, it can be saved and printed, or it can be photographed by special equipment to become a slide. The slide, in turn, can be developed into an enlarged fashion photograph.

In smaller or less modernized firms, fashion designers do the sketching by hand. Some prefer this method, since their creativity has flowed onto paper through their pencils or pens for many years, as in 20-3.

Several rough sketches (called croquis), and later completed drawings, are done to show the details of a new design. They indicate the types of fabrics, trims, construction details, and colors the designer has in mind.

20-3 The sweater designer who created this sketch draws directly on paper.

The sketched or computer-generated designs are submitted to top management for approval. Following approval, each design is made into a paper pattern or draped onto a dressmaker's form, usually in muslin fabric. It is cut and sewn. Further details are worked out, and the "prototype" is altered if necessary. Then it is made in the chosen fashionable fabric. It is shown to top management again and becomes a "sample" that merchandisers and salespeople can view and discuss. If the sample is approved as being a worthy part of the line, it will be shown to buyers. Some of the equipment used by fashion designers is shown in 20-4.

Designers must plan and supervise the work of their staff members. In a large firm, they might have one or more assistant designers and sketchers as well as several sample hands. They deal with buyers and fabric salespeople in addition to management, production, and publicity teams.

Designers select fabrics and trims, and they help with the costing of their garments. They work with production and marketing people, and they present their ideas and samples to salespeople and clients at meetings. They often work long, hectic days, especially when a line is being finished for a showing. The work might be intense and tiring, but it can be very rewarding.

One personal reward of design work is the opportunity to express creativity. Another is

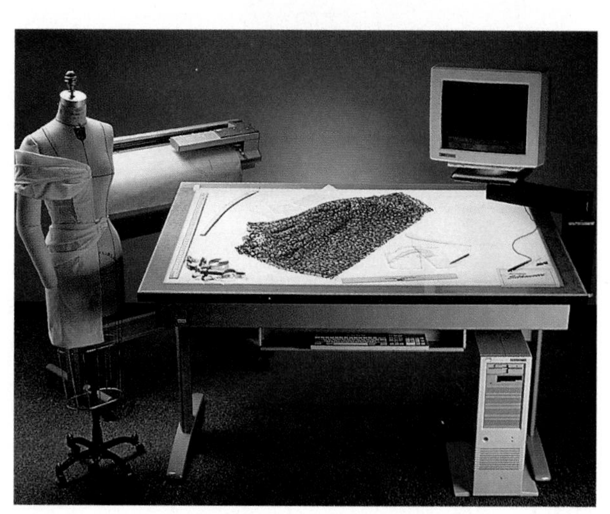

Gerber Garment Technology, Inc.

20-4 Basic equipment used by fashion designers helps them work accurately and efficiently.

the pride that is felt when the products of that creativity are accepted, purchased, and worn by other people.

The financial rewards of design vary a great deal according to experience and talent, the size of the company, and the company's market and location. Salaries are very high for successful top designers.

Qualifications to Be a Fashion Designer

Fashion designers must be imaginative and have a flair for clothing. They should love fashion, fabrics, and beauty. They must be creative artists who can sketch and who have a strong sense of color, line, texture, balance, and proportion. They must be able to generate a constant flow of ideas. They must also have an awareness of changing social and economic trends, so their designs meet the public's demands. Some of this fashion awareness, or flair, is natural aptitude. The rest is gained through experience and education, as in 20-5.

Fashion designers must have technical knowledge of fabrics, trimmings, and fit. Especially important are technical skills in pattern making, draping, and sewing. Designers must understand manufacturing processes and costing, so they can stay within the production capabilities and marketing plans of their firms. They must be able to visualize a finished, three-dimensional garment before it is made. They also must have patience, since the design specifications may have to be revised several times.

Dept. of Consumer Studies, Univ. of Delaware

20-5 Fashion design students study about how well-known designers have achieved their success and fame.

Designers must have enthusiasm, determination, and drive to succeed in this demanding career. They need to be able to work easily and comfortably with others. Flexibility and cooperation are extremely important, especially in a busy and crowded design room. However, designers must also be decisive and believe firmly in their own creativity. They must be able to think on their feet. They must often sell their design ideas to others.

To be an apparel designer, a degree in fashion design is needed from a college or vocational school. Such educational programs include courses in history of costume, drawing, pattern making, draping, and sewing. They also provide knowledge about fabrics and trimmings, the elements and principles of design, and the production and costing of garments. Some have study tours or exchange opportunities with other schools in the United States or abroad. The industry is changing, becoming international in scope and highly computerized. Therefore, the wider and greater your education is, the better your chances are to climb to success.

Assistant Designer

Designers usually first serve apprenticeships as *assistant designers*. This gives them entry-level, on-the-job training in the business. A great deal can be learned while working under experienced designers.

An assistant designer might start by following up on a designer's sketches through draping, as in 20-6, pattern making, sample cutting, or sample making. Clerical duties might involve keeping records of fabric, notions, and trim purchases; making appointments; answering telephones; and running errands.

Later, assistant designers might help the designer select fabrics and trimmings. They might visit retail stores and fashion shows to keep up on new trends. Eventually, they are asked to contribute their own original ideas.

McCall Pattern Company

20-6 Draping is done by pushing fabric around on a dressmaker's form into the desired design. The fabric shape then becomes the basis for the pattern.

They must get along with many people, follow directions accurately, and work well under pressure, often in cramped working conditions.

It is important for an assistant to establish a reputation as a good designer. Then he or she can either move up in the firm or get a job with a different firm as a higher-level designer. Designers tend to change companies often. There is little job security for those who have not become highly recognized.

Opportunities for assistant designers exist in ready-to-wear manufacturing firms or pattern companies in large cities. Sometimes assistant designer positions are available in the summers during college years or as part of work-study programs. When applying, you must present a portfolio of your own design ideas.

Sketching Employees

Exclusive fashion houses sometimes employ *sketchers*. These people make freehand, illustration-quality drawings of the ideas that a designer has draped in fabric onto a dressmaker's form. The work of a sketcher is quite confining, except when asked to meet customers, do promotional work, and assist with presentations of new collections.

Sketching assistants are employed by large manufacturing firms and pattern companies, mainly to record designs. They illustrate the fashion designers' rough sketches in precise, technical detail, as in 20-7. They point out all construction and design features, and add fashion touches. They draw a season's line of samples to be kept with the company records. They "swatch" the sketches by attaching samples of fabrics and trimmings. They also fill out a specification sheet of construction details for each item.

Sketching jobs may include a fair amount of clerical work. Not much original creativity is needed. These jobs are good for people who can draw precise and accurate accounts of other people's ideas at a fast pace.

Sketchers and sketching assistants need professional training and education in fashion design after high school. The programs they study should be strong in the areas of art, design, textiles, and sewing construction. Sketchers should also study draping and pattern making if they plan to move up into fashion design jobs.

Sketchers compete for a limited number of jobs. Hours are often long and irregular. Salaries are generally not very high. However, good experience is gained along with knowledge of how an important design house or firm operates. Also, contacts can be made for future designing or other jobs in the industry.

These jobs require outstanding sketching skills and an extremely high level of fashion

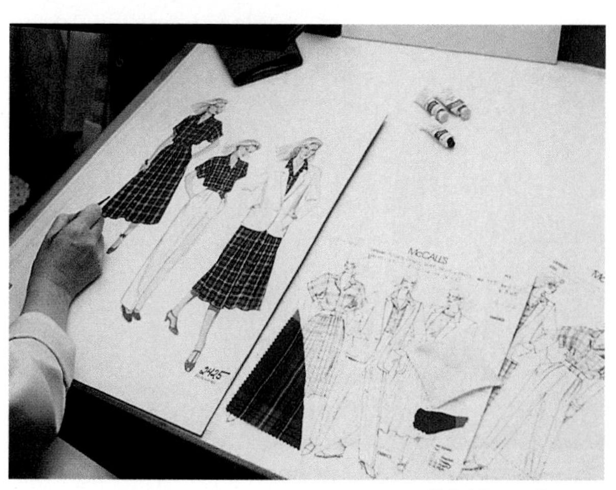

McCall Pattern Company

20-7 A sketching assistant must be able to include all the details of an outfit in a fashion drawing.

sense. When applying, a portfolio of artwork must be presented to the employer. Candidates must look fashionable and be extremely well groomed. They should be articulate, poised, and able to deal with high-level executives.

Sample Maker

A ***sample maker*** (also called a sample hand) sews each designer's sample garment together to make the design a reality. The sample maker proceeds according to the designer's pattern, sketch, and specifications. He or she does all the required machine stitching and hand sewing. This tests the designer's pattern in fabric. It is then put onto a dress form, as in 20-8, and changes or refinements may be made. Later the sample garment is worn by a live model and shown to management.

Sample hands must be skilled in all construction techniques and able to interpret someone else's ideas into garments. In many cases, a factory sewing machine operator has moved up to this position because of skill, hard work, and a good attitude. Sometimes he or she has taken vocational school courses.

Sample making is exacting work. The salary is modest to good. It is higher than those of most other sewing machine operators in the factory. Someone who does outstanding work and pursues more education might even move up to become an assistant designer.

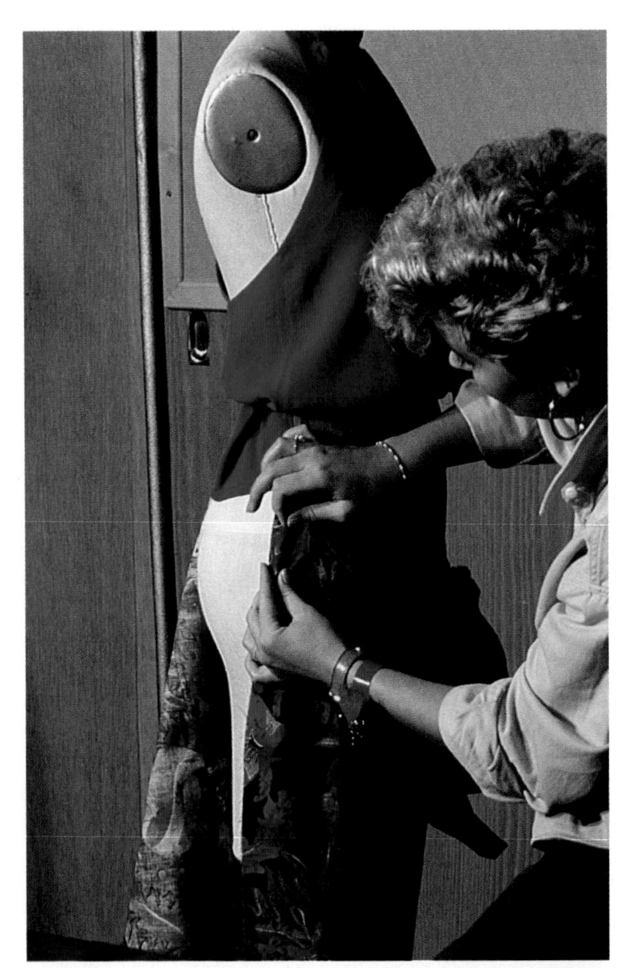

Oregon State University

20-8 A sample garment is sewn, taken apart if needed, and refitted until it is just right.

Apparel Manufacturing

Apparel manufacturing is the production of large quantities of garments. Ready-to-wear apparel is mass-produced along factory assembly lines. Most producers specialize in one type of apparel.

Apparel manufacturing is labor-intensive, requiring many workers. However, worldwide competition is now encouraging fast, flexible, and accurate high-tech production. Old methods are being replaced, requiring trained workers with specialized computer knowledge. In the future, fewer workers with higher skills to operate more advanced technology will be needed.

Apparel production plants are located in small towns and large cities in every part of the United States and in most countries of the world. In fact, many American apparel manufacturers cut garment parts in the United States and then ship them out to other countries for sewing. Eventually, the finished garments are shipped back to the United States and the apparel firms sell the garments to their retail accounts. By doing this, the U.S. company controls the design, materials, and cutting aspects, while the construction work is done in a low-wage country. Duty is paid to the U.S. government only for the value added when the garments re-enter the country. ***Value added*** is the increase in the worth of products as a result of a particular work activity. In this case, it is the amount of work that was done outside the country.

Pattern Maker

A ***pattern maker*** translates an apparel design into pattern pieces that can be used for mass production. The pattern maker cuts a set of heavy manila paper or fiberboard pattern pieces for parts of the garment, as in 20-9. Every detail of the final design is included in this master pattern. It is made in the basic size that the manufacturer uses for its sample garments.

Pattern makers must make patterns that keep fabric yardage down to a minimum. They must be as efficient and precise as possible. They must produce the shape, size, and number of pattern pieces needed to make the garment.

Most apparel manufacturers are now making their patterns with computers. A computer operator selects the particular garment parts that are needed from the multitude of choices programmed into the software. The computer combines the variables and shows a picture of the garment on the monitor. The picture can be changed, and when it is just right, the computer can be programmed to make the master pattern. However, some feel the flair and judgment of a skilled pattern maker cannot be duplicated by a machine.

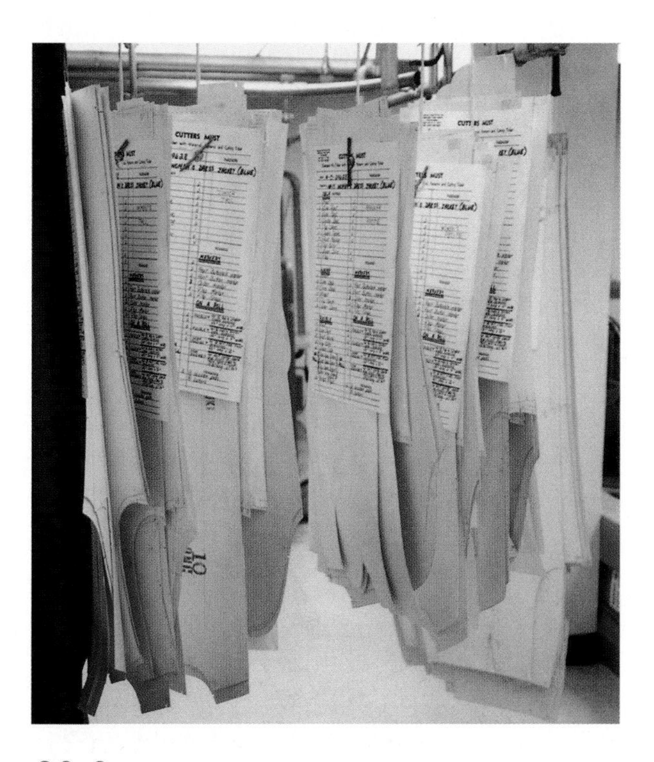

20-9 A set of heavy manila paper patterns is made for each garment design. Here, several are hanging with work-order sheets attached.

Pattern makers need a good background in fitting, flat pattern making, draping, design, and fabrics. They must understand body proportions. They must work well with others. They should have at least two years of vocational education studying pattern making technology. Typical programs offer courses in industrial draping, pattern making, and pattern grading. They also cover the pricing of garments, apparel production processes (construction), and textile science. After becoming proficient at pattern making skills, students should also receive training with pattern making computer systems.

Pattern makers are employed wherever there are apparel manufacturing companies. The salaries for experienced pattern makers are quite good. Assistant pattern makers learn the trade by working closely with experienced pattern makers after finishing their education. Those who are trained to use pattern making computer equipment may earn very high salaries.

Pattern Grader

Working from the master pattern, ***pattern graders*** cut patterns in all of the different sizes produced by the manufacturer. By enlarging or reducing the pattern within a figure type category, all the pattern pieces of the design are made in each size. For instance, the same dress may be produced in misses' sizes 6 to 20. Additions or deletions of exactly the right amounts must be done in the right places to give the same effect as the original design in all sizes.

Pattern grading is highly technical and precise work. It must be done neatly and often at a fast pace under the pressure of production schedules. Pattern makers, or their assistants, often do the grading of the patterns they make. Many large manufacturers now use computers to do the grading work quickly. However, some believe that to keep the feeling of the design, the artistry of the pattern grader is needed.

Pattern graders need skills in drafting, pattern making, and clothing construction. These skills are obtained with at least two years of training at a fashion or technical school. Salaried positions for pattern graders are available mainly with sizeable manufacturers in large cities.

Marker

Markers are employees who figure out how the hard paper pattern pieces can be placed most efficiently for cutting. They trace the pattern pieces in the tightest possible layout, also called a marker. In the ready-to-wear manufacturing industry, almost no fabric is wasted. Computerized plants now have systems that do this electronically.

Spreader

Spreaders lay out the chosen fabric for cutting. They guide bolts of fabric back and forth on a machine that spreads the fabric smooth and straight, layer upon layer. In the past, this job required no higher education since the skills were learned at the plant. However, with electronic equipment, vocational training is needed to operate the new systems. In small or computerized plants with fewer employees, this job is done by cutters.

Cutter

Cutters use power saws or electric cutting machines to carefully cut around the pattern pieces. They cut through all layers in the stack of fabric that is often many inches high. Sometimes they use knives. Shears may be used for just a few layers.

Firms that have modernized may use jets of water or laser beams for cutting out garment parts quickly and accurately. The most modern plants do computer cutting, often without needing a paper marker on the top of the fabric stack. In such a case, the layout arrangement and pattern piece outlines are in the memory of the computer, as shown in 20-10.

Cutters who use power saws or electric cutting machines need physical strength, good manual dexterity, and excellent eyesight. They must take pride in accuracy. They often have some vocational training, but actually learn the job at the plant site from experienced cutters. However, as plants become computerized, highly-trained factory employees are needed. They need education in computer operations and programming, combined with apparel production technology.

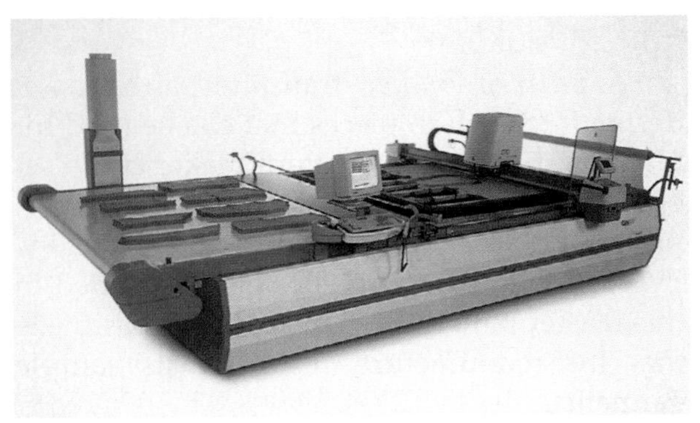

Investronica Systems

20-10 After an operator puts the proper information into the computer system, this equipment does garment cutting automatically.

Assorter

Assorters (also called assemblers) prepare the many cutout parts for sewing if the progressive bundle system is being used. They sort the parts and mark any construction details. They tie garment pieces into bundles of 12 according to color and size before the bundles go to the sewing operators. Later in the production process, assorters bring parts together, such as shirts with their corresponding collars and pockets. Some assorters may also fold and package items after they are finished.

Factories that are mechanized with computer-aided manufacturing (CAM) do not need as many assorters. They have overhead carriers that automatically take garment parts through the operation.

Sewing Machine Operator

Sewing machine operators construct apparel on heavy, fast, industrial power sewing machines. Large numbers of operators work in rows in the factory. For low- and medium-priced ready-to-wear lines, the operators perform just one specific task over and over again. Certain operators sew only seams. Others just put in zippers. Then the work is passed on to another operator who does a different task.

When operators develop more skill and versatility, they can advance to more complicated jobs. They might be assigned to work at specialized double-needle machines or power embroidery machines. They might also be promoted to a job as a sample hand, an inspector, or a training supervisor. Modular work groups also require operators who are skilled in many different sewing tasks.

Sewing machine operators need a basic knowledge of sewing construction. They must be skilled in handling materials and using equipment, as in 20-11. They must have good finger dexterity, coordination, and eyesight. They must be able to do neat, steady, and accurate work at a fast pace in a compact work area. Operators should enjoy doing routine tasks. Also, they cannot be bothered by the loud noise of many machines buzzing at top speed.

Sewing machine operators usually work regular weekday hours. The work can be tiring because of the rapid pace. If they are paid on a piecework basis, the faster they work and the more they finish, the higher their paychecks are. The most complicated procedures, which require more skill, receive a higher rate of piecework pay. Operator's earnings compare well with clerical jobs but may not be as high as factory jobs in other industries. However, pay scales are increasing with the higher skills needed to match the new technology of the industry. Also, sewing machine operators often belong to employee unions.

A high school diploma is recommended to be a sewing machine operator. Trade and vocational schools offer training on power-driven industrial machines. Manufacturing companies give on-the-job training in the specific construction techniques for their garments, as in 20-12. In the future, robotics operations may be perfected to sew parts of garments together automatically. This will replace some operators.

Lands' End Direct Merchants

20-11 Dexterity and skill are needed to operate the many varieties of industrial sewing machines.

Finisher

Finishers are employed mainly by better-quality, higher-priced lines. They hand sew whatever is required to finish garments. They may specialize in one technique, such as attaching fasteners, adding trimmings, putting in hems, or securing linings. There are few opportunities, but the pay is by the hour and may be higher than that of a machine operator.

Hand finishers must have the ability to do rapid, accurate hand sewing. They need to have good eyesight and manual dexterity. No higher education beyond high school is required.

Trimmer and Inspector

Trimmers and *inspectors* are sometimes called cleaners and examiners. Trimmers cut off loose threads and pull out any basting

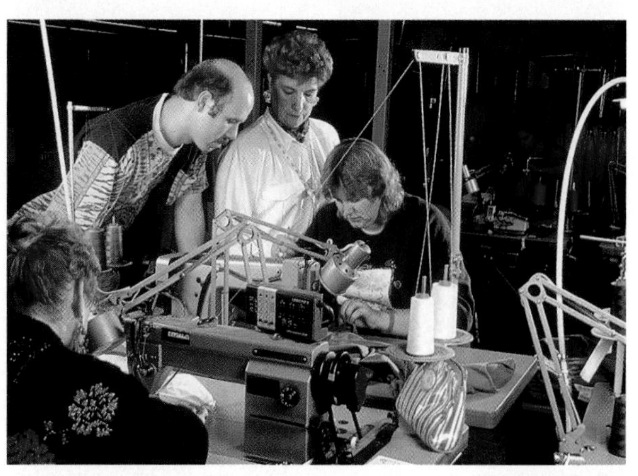

A and Z Industries, Ltd.

20-12 When workers start as new sewing machine operators, or use a machine or procedure for the first time, they are given instruction by their supervisor or other manufacturing experts in the company.

stitches. They also remove lint and spots from finished apparel items. Inspectors examine unfinished garment parts during production, as well as finished garments. They check for construction flaws and imperfections. They pull out the garments that do not meet the standards of the company. They try to prevent items from being seconds or imperfects that would have to be sold cheaply or discarded.

The jobs of trimmer and inspector are combined in small factories. These employees arrange for minor repairs or send the garments to alteration hands in the factory who correct the defects or mistakes.

Alteration Hand

Alteration hands repair defects that have occurred to garments during factory production. They must be able to skillfully perform all basic construction techniques. Experience in clothing construction is needed to handle this job. If merchandise cannot be restored to first quality, the garment is marked as a second. Only extremely bad items are thrown out.

Presser

Pressers flatten seams, iron garment surfaces, and shape garments with steam pressing machines, as shown in 20-13. Pressing is done during construction only in better garments. A final pressing is given to garments in all price ranges at the end of the construction process.

To be a presser, you should have a high school diploma. Vocational school training would be an asset. You must be willing to learn the specifics of the task on the job from a supervisor. You also need a tolerance for steam and heat.

Production Management

Management jobs involve planning, organizing, coordinating, and overseeing the work of others. Managers supervise the various areas of production. They study and prepare reports, attend business meetings, and do computer planning. See 20-14. They guide people and operations so the company can reach its goals. It is demanding work with good compensation.

Most managers have a college background in apparel management, apparel production, or engineering technology. Good math, communication, and problem-solving skills are needed. A combination of production, technical, administrative, and marketing knowledge is required.

A production manager may start as a "management trainee." Maturity, the ability to

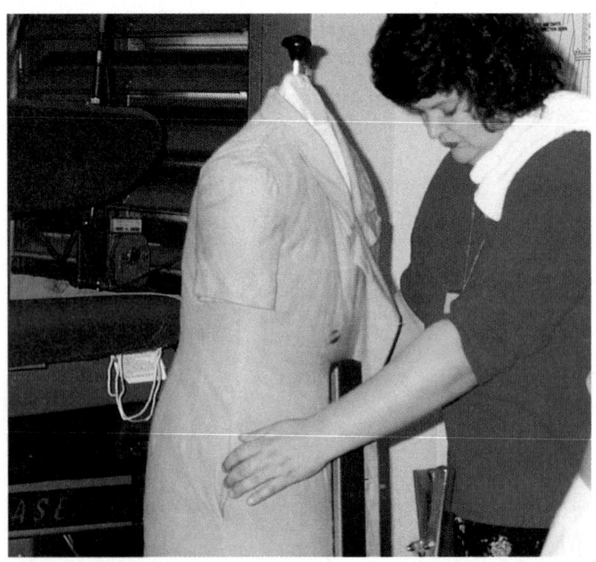

International Fabricare Institute

20-13 Steam pressing machines are made specifically for the types of garments being constructed and needing pressing.

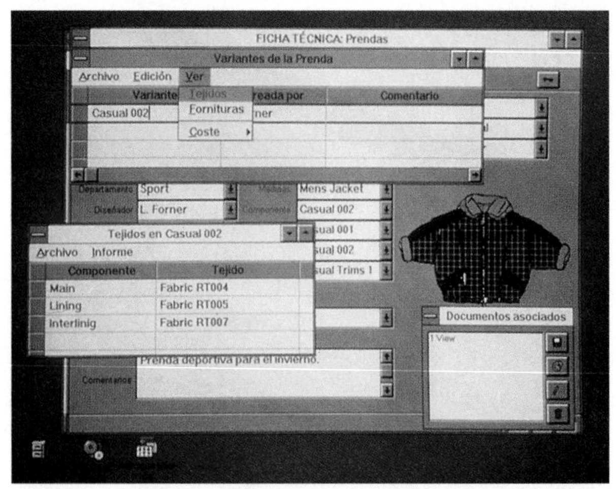

Investronica Systems

20-14 Computerized apparel production planning systems are available in many languages, for use in design rooms and factories throughout the world.

work with varied personalities, and a good sense of organization enable educated employees to advance fairly quickly.

Managers who are hired by large manufacturers may be transferred to several different factory locations over the years. During their careers they may move several times, often to plant sites in small towns. Opportunities for high-paying apparel production management careers are much greater for employees who will consider employment anywhere in the country, or maybe the world!

Product Manager

A ***product manager*** may be in charge of every aspect of one of the company's lines or a specific category of garments within a line. The manager oversees the design, manufacture, selling, and delivery of that line or category of garments.

Product managers must keep up with the latest market trends related to their products. Before production begins, a product manager must decide how the items coordinate with apparel being produced by other firms. He or she must also make sure they fit into the total line of that manufacturing company. Do the colors mix and match with other clothes being sold by the company and other firms this year? Are the products related to the total line of the manufacturer, so the same production machinery can be used to produce them? Can similar sales methods and retail outlets be utilized?

The product manager must also study what the competition is doing. The products must be competitive to be successful. Often the manager must reject a designer's idea because it is not practical or the details of the garment production cost too much. Also, the production schedules must allow delivery dates to be met. The goods must be ready on time and delivered to the retailers who ordered them. All stages of design, manufacture, and distribution must be planned and carried out.

A product manager must be practical and have good business sense. He or she needs education beyond the high school level, usually a college degree. Often the manager has been a successful production supervisor before being promoted to this level. This position has a good salary.

Plant Manager

The ***plant manager*** (also called production manager) is responsible for all operations at the plant. This person is in charge of many employees and has the ultimate responsibility of all production functions. These tasks include estimating production costs, scheduling the flow of work in the factory, and hiring and training new workers. The plant manager must also oversee quality control of the products and supervise all other aspects of production.

It is a complex task to oversee the purchasing, cutting, sewing, pressing, and shipping functions. It is achieved after many dedicated years of working up to this position as a production assistant and supervisor. The plant manager's pay is very high.

Production Assistant

Production assistants do detail work and record keeping for plant managers. They keep track of fabric, trim, and notion samples for the design room. They also assist with the production schedules and keep tabs on the flow of work in the factory. Production assistants are in charge of inventory control. They check on incoming deliveries to the factory and outgoing shipments to customers. They must keep clients informed on the progress of orders, and expedite certain deliveries.

Production assistants get an excellent overview of the entire manufacturing process. They need good math and organizational skills. They must be able to communicate well with many people, especially over the telephone. They should have accuracy, assertiveness, thoroughness, legible handwriting, and a high tolerance for stress. It is best if they have a college degree in apparel production management or fashion merchandising.

Supervisors

Production supervisors oversee and direct the sewing machine operators and other workers on the factory "floor" or workroom. They work in the factory, solving operators' problems. They try to make sure there is a smooth flow of work for the operators at the machines. They should have the ability to motivate others in order to achieve the highest quality and speed of production.

Training supervisors train new operators to do specific tasks or to use specialized machines. They need to know how to do all of the sewing operations in the plant. This usually comes from experience and trade school education in apparel production.

Production supervisors and training supervisors work under a plant manager.

Piece Goods Buyer

Piece goods buyers are purchasing agents. They research and buy the fabrics, trims, and notions that are chosen by the design staff and approved by management. They find the latest information available about textiles and trims. They try to get the highest quality textile products at the lowest possible prices for production of the firm's lines. They also keep tabs on any production problems that might occur with the fabric or trims.

Assistant piece goods buyers help by keeping clear and accurate ordering and inventory records, and swatch files. They assist with the ordering of fabrics, trims, and notions. They follow up on deliveries. Sometimes they must place reorders or consider substitutions. Employees in this entry-level job often accompany and learn from buyers on visits to fabric and trim markets.

Piece goods buyers and their assistants should have an interest in and knowledge of fabrics. They should have good math, communication, and organizational skills. Specialized training in textiles, fashion merchandising, or apparel production management is recommended.

Industrial Engineer

Just as in the textile industry, *industrial engineers* in the apparel manufacturing industry select the most efficient machinery and operational methods. They find ways to cut costs and to improve safety. They analyze techniques to use the least amount of raw materials. They work out "time and motion" studies to establish the piece rate pay for each sewing and production task. They are involved in the physical layout of the plant by designing production lines, storage areas, and workstation progression. They also do production forecasting and planning.

A college degree in industrial engineering or apparel production management is needed for this position. Manufacturers usually supplement that education with their own training programs.

Costing Engineer

Costing engineers determine the overall price of producing each apparel item. They consider the cost of fabrics and notions used, the pay for each construction operation needed to produce it, and all other production fees. They price each separate design in an upcoming line. They may have to travel to various plants to view production operations and consult with plant managers. They sometimes have "costing clerks" who assist them by noting and analyzing the figures on specification sheets.

Quality Control Engineer

Quality control engineers develop specifications for the items that will be manufactured. They are then responsible to see that those standards are met during all of the production phases. They identify quality problems and work with the production staff to correct them. To do this, they may need to travel to many different plant locations. Sometimes checking firms are hired to make sure the quality standards of products really are what they claim to be.

Quality control engineers must have a knowledge of textile technology and garment construction. They must be thorough, well organized, and good with details and follow-up.

Plant Engineer

Just as in the textile industry, apparel production factories have *plant engineers*. These people oversee the physical plants. They maintain the heating, lighting, noise reduction, and other environmental operations.

Sales and Distribution

Manufacturing firms sell their finished goods to retail establishments so consumers can buy them. This is sometimes done through wholesale distributors. However, most production sales are done directly from apparel manufacturers to retailers, as in 20-15.

The Fashion Association / Sigrid Olsen

20-15 Sales employees who work for apparel manufacturers make all necessary arrangements to show their company's garments to retail buyers.

Sales positions offer a great variety of interesting tasks at a fast pace.

Production sales offices and showrooms are located in fashion centers, especially in New York City. They might also be in such cities as Los Angeles, Dallas, Chicago, Miami, Minneapolis, Philadelphia, and St. Louis. Some salespeople represent a giant corporation with many lines. Others sell the goods of several firms.

Distribution is the process of getting the merchandise to the proper locations. For apparel, orders must go from the manufacturer to each of the retail stores in the proper amounts by color, size, and other specifications.

Showroom Salesperson

A ***showroom salesperson*** has an "in house" sales position at the firm's sales offices. This employee presents the line of goods to buyers who visit the manufacturer's showroom. Buyers are told about the newest styling features, fabrics, and colors that are part of the line.

Showroom salespeople have the chance to get to know the buyers of their important, active accounts and to service their needs. One way they help is by suggesting ways retailers can present and display the merchandise in their stores. They may also act as a communication link between the design staff and the retail buyers.

Showroom salespeople must be friendly and outgoing. They must be able to communicate well and follow through on the details of customer service. To do this, they must understand both the manufacturing and retailing of apparel. They must have excellent grooming, dress fashionably, and be poised and confident. They must be able to think quickly and to present a line of merchandise with flair and enthusiasm.

Hours for showroom salespeople are regular much of the time. However, they may be very long during market weeks when much of the buying for each season occurs, 20-16. It is desirable for these employees to have completed a program in fashion merchandising. Such a program would teach them techniques for showing a line, making a presentation, handling objections, and closing a sale. Previous sales experience is also beneficial.

Chicago Apparel Center

20-16 During market weeks, salespeople work long hours showing their lines to many prospective customers.

Other Showroom Sales Employees

Being a *showroom sales trainee* is a good way to break into apparel production sales. Working under experienced salespeople, trainees learn to present the merchandise to buyers in appealing ways. They sometimes model the apparel. They must write up orders accurately and later follow up on shipments. They also deal with buyers over the phone, act as receptionists, schedule appointments, and do filing and other clerical duties.

With experience, a trainee may advance to the position of showroom salesperson and then to *showroom manager*. The manager supervises all of the personnel and activity in the showroom. He or she must make sure everything in the showroom is attractively displayed and that displays are changed regularly. The manager also makes sure that all samples are always in stock for the buyers. He or she handles any buyer's problems that might occur and trains new showroom staff members. An alert showroom manager also keeps the design staff informed of market trends and buyer feedback.

Traveling Sales Representative

Traveling sales representatives make up the "outside" sales force that sells away from the firm's sales office and showroom. They travel around to sell apparel to buyers at department stores and clothing shops.

The representatives are given sample garments and color swatches at sales meetings. The meetings are held at the home office several times a year just before each new line is launched. The product manager shows and describes the garments to the sales "reps." The reps learn how the garments are meant to be worn, how they are made, any special features they have, and what care they need. Each item is given an order number. Any questions about the line are answered. Also, ideas are given to the representatives to help them sell the line.

Traveling sales representatives sell mainly to retail buyers. A great deal of selling is done during market weeks, when retail buyers visit market centers. Sales reps may be responsible for planning exhibits or demonstrations that promote their lines during market weeks. At such shows, salespeople get feedback from the buyers about the wants and needs of the retail stores, as in 20-17. This information is brought back to the product managers so manufacturing can be adjusted accordingly.

Between trade shows, sales reps keep in touch with their accounts (retail stores) and make appointments to sell more items. They inform the buyers about developments, such as price changes, new styles, trends, and fabrics added to the line. They get to know the buyers well and try to expand their markets by getting new accounts. They sometimes offer displays to promote their products, prepare demonstrations about the uses of their products, or present fashion shows at stores. They take orders and follow through on them to make sure that delivery commitments are kept. They must also deal courteously with customer complaints.

Independent sales reps often handle several different lines for many small manufacturers. The manufacturers are not large enough to have their own sales staffs. Independent reps are called ***jobbers***. A jobber in men's apparel might sell a line of shirts from one manufacturer, slacks from another, neckties from another, and sweaters from another manufacturer. Thus, the lines of the companies they represent are not in competition with each other, but the accounts to which they sell are the same.

20-17 Besides taking orders, sales reps can gain valuable feedback from face-to-face meetings with retail buyers.

Jobbers are self-employed but have the same contacts and knowledge as the manufacturers' salespeople. They work on a commission basis. By "repping" many lines, if one doesn't sell well, they still get income from the others. Also, they can drop a bad line and find a replacement as needed.

A traveling sales representative usually works in a certain geographical "territory," or selling area, for the company. He or she travels within that territory, usually by car. Most reps are away from home much of the time. They often work long and unusual hours, especially during peak times. The job is exhausting, but exciting. It is very satisfying for those who enjoy people as well as fashion. With commissions, the financial rewards for successful reps can be great.

Qualifications to Be a Traveling Sales Representative

Sales representatives must be outgoing and likable. They should be enthusiastic and ambitious. They must have a good knowledge of fashion and garments, and have good selling skills. They must also be hardworking and organized, since they usually work on their own.

Sales representatives must be willing to work long and unusual hours. They should like to travel, rather than being in a showroom or office each day. They should not mind being away from home at times. They must also be willing to move to a different territory if the company transfers them.

Sales reps must be truthful to get repeat orders and develop long-standing accounts. They must know and believe in their product lines and be able to speak with confidence about them. They must represent their company in a professional manner while providing their clients with good service.

Most sales representatives are college graduates. They have studied such courses as textiles, economics, psychology, mathematics, merchandising, and computer science.

Sales Manager

A *sales manager* is the boss over several sales representatives in a large firm. He or she usually has been promoted after being a successful, hardworking sales representative. The pay and prestige are high.

Market Research Employees

Market researchers (also called market analysts) study and analyze consumer habits, needs, and wants. They spend time in the marketplace and consult with fashion resources, such as forecasters and textile manufacturers. They access trends of the market to find out what is selling now, what will sell in the future, and why. They find out how long consumers will wear certain items, where they will wear them, and what colors they prefer. They also study what their firm's competitors are doing. Sometimes professional pollsters are hired.

Market researchers discuss their results with product managers. They try to predict the success or failure of a fashion or an apparel item. Their goal is to eliminate mistakes before they are produced.

Sometimes the person doing market research is called a *merchandiser*. A merchandiser determines the direction a manufacturer's line will take for each season based on the market research. In small firms, market research is done by designers or stylists.

Market researchers and merchandisers should have a keen fashion sense and an understanding of styles, colors, and fabrics. They need self-confidence and good communication skills. They also need a knowledge of production processes and excellent contacts with fashion resources.

It is recommended that market researchers or merchandisers have a degree in fashion merchandising, fashion design, or apparel production. They should develop their analytical and organizational skills.

Merchandise control specialists can tell by the rate of merchandise flow how well certain items are selling in various localities. They generally chart sales on computers and draw results from the figures. They might suggest that sales people stress certain items or colors in one part of the country and different ones in other areas.

Jobs in Distribution

Many hourly wage earning jobs are available in distribution. Racks of garments must be moved from place to place. Finished merchandise is taken either to storage areas or to the shipping department.

Employees in the shipping department of an apparel manufacturing firm receive orders from the sales department and fill them. It is important to meet delivery dates in order to satisfy retailers and consumers. Computers are used to keep records of the shipments. Garments are loaded onto trucks for delivery to stores, 20-18.

Employees are needed to pack and ship the merchandise. Most of these jobs do not require education beyond high school. However, these workers must be accurate with details and conscientious about doing a good job. Some of the job titles for these workers include clerk, order picker, checker, packer, and transportation specialist.

20-18 Distribution employees are needed to load finished apparel onto trucks for delivery to retailers.

Top Management

Like other businesses, apparel manufacturing firms have several executive positions. They include members of the board of directors, the president or ***C.E.O. (chief executive officer)***, and several vice presidents. Each vice president is in charge of a certain department, such as sportswear, or segment of the firm, such as overseas operations. There may be several levels of top management.

Division Director

A ***division director*** works above the product managers, plant managers, and sales managers. This person manages the design, production, and sales of an entire line or division of the company. Sometimes he or she is a vice president of the company. In large firms, several division directors may work under a vice president.

A division director must be able to analyze the total scope of situations and make important decisions accurately and with confidence. He or she must be hardworking and able to direct people with pleasant authority. Great respect from others also accompanies the job.

A division director almost always has a college degree. He or she probably has experience in both production and sales. A masters of business administration (M.B.A.) degree is also an asset. The salary for a division director is extremely high.

Administrative Employees

Positions for administrative employees in apparel manufacturing companies are similar to those described for the textile industry. Such positions include personnel (employee relations) staff members, accounting and finance employees, and data processing people. Public relations agents are important, as are tax specialists.

Chapter Review

Summary

Apparel design employees create new fashionable garments and accessories. Fashion designers might do men's, women's, children's wear, or specialize in such areas as swimwear, bridal attire, shoes, or evening dresses. Fashion designers get ideas from fabrics, forecasting information, art events, and world news. Fashion designers must have fashion flair and a knowledge of fabrics and fit. Assistant designers help fashion designers with all details, including clerical work. Sketching employees record designer's ideas or draw and sketch each season's line of samples for the company records. A sample maker sews designers' ideas into sample garments.

Apparel manufacturing is the mass production of garments. A pattern maker translates an apparel design into pattern pieces for factory production. A pattern grader cuts pattern pieces in all of the different sizes to be produced. Markers are employees who figure out how the pattern pieces can be placed most efficiently for cutting. Spreaders lay out bolts of fabric for cutters who use power saws or other machines to cut around the pattern pieces. Assorters put the cutout parts into bundles or onto carriers for sewing. Sewing machine operators construct the garments on industrial sewing machines. Finishers do any needed hand sewing, and trimmers remove loose threads and basting stitches. Inspectors check for construction flaws and imperfections, and alteration hands repair defects. Finally, pressers flatten seams, iron surfaces, and shape the garments.

Production management jobs involve planning, organizing, and coordinating the work of others. Product managers oversee the design, manufacture, selling, and delivery of a line or category of garments. Plant managers are responsible for all employees and functions at a plant site. Production assistants do detail work and record keeping for plant managers. Production supervisors oversee and direct all machine operators, while training supervisors teach specific tasks and the use of specialized machines to workers. Piece goods buyers and their assistants research and buy the fabrics, trims, and notions for apparel production. Plant engineers include industrial engineers to calculate the most efficient machinery and operational methods, costing engineers to determine the price of procuring items, quality control engineers who develop production standards, and plant engineers who oversee the physical plants.

Manufactured apparel is moved to retail locations by sales and distribution employees. Showroom salespeople present goods to buyers who visit manufacturers' showrooms. Other showroom sales employees are showroom sales trainees and showroom managers. Traveling sales reps and jobbers go to retail locations to sell manufacturers' lines, often working long and unusual hours. A sales manager is the boss over the sales reps. Market research employees might be market analysts, merchandisers, or merchandise control specialists. Distribution jobs deal with getting the merchandise to the proper locations.

Top management of apparel manufacturing firms include members of the board of directors, the president or C.E.O., vice presidents, division directors, and others. Administrative jobs are similar to those in textile firms.

Fashion in Review

Write your responses on a separate sheet of paper.

Matching: Write the letter of the general category that matches each specific job.

1. Traveling sales representative.
2. Pattern maker.
3. Sample maker.
4. Piece goods buyer.
5. Sketcher.
6. Cutter.
7. Order picker.
8. Fashion designer.
9. Plant manager.
10. Assorter.

A. apparel design
B. apparel manufacturing
C. production management
D. sales and distribution

True/False: Write true or false for each of the following statements.

11. A design stylist creates original fashions within a certain style category.
12. Assistant designers are employed by manufacturing firms and pattern companies to record designs with precise, technical illustrations.
13. Working from the master pattern, pattern graders cut patterns in all of the different sizes produced by the manufacturer.
14. Spreaders are employees who figure out how the hard paper pattern pieces can be placed most efficiently for cutting.
15. Production supervisors oversee and direct the sewing machine operators and other factory workers.
16. Costing engineers determine the overall price of producing each apparel item.
17. Salespeople who work for apparel manufacturers sell mainly to retail buyers.
18. Showroom sales trainees are also known as jobbers.
19. Most jobs in distribution do not require education beyond high school.
20. A division director tries to divide the work tasks properly among the plant workers.

Fashion in Action

1. Cut out a picture of a garment from a magazine, catalog, or advertisement. In a written report, list the production workers who probably contributed to its creation and completion. Briefly describe what each worker did.
2. Clip fashion drawings (not photos) of three different garments from newspapers or catalogs. Trace each of them and show what changes you would make, to lower their production costs, if you were a design stylist. Have you taken out any seams or trims that would eliminate some sewing procedures? Mount the fashion drawings and your new designs on paper. In an oral report, describe your reasons for the changes you made.
3. Pick a specific job in the apparel manufacturing industry that interests you. Do further study about it in the library, with materials from your guidance counselor, and by talking to people who are in that field. Prepare a written report about the job. Include the qualifications you would need to get the job, the type of company that might hire you, and the tasks you would perform. Estimate the salary range and fringe benefits you could expect and find out what opportunities you might have for advancement.
4. Study the employment want ads of your local newspaper to see if any apparel production jobs are available in your area. If no ads are listed, look in the Yellow Pages under such headings as "Men's Apparel—Whol. & Mfrs." and "Women's Apparel—Whol. & Mfrs." to find the names of manufacturing firms. Check with a few local companies about summer employment during your high school or college years.
5. Design a bulletin board display to show the wide range of career possibilities in apparel design and production.

21 Fashion Merchandising and Other Retail Industry Careers

Fashion Terms

fashion merchandising	stock clerk
retail buyer	head of stock
executive trainee	checkout cashier
retail salesperson	merchandise manager
customer service manager	fashion coordinator
comparison shopper	store manager
personal shopper	branch coordinator

After studying this chapter, you will be able to

- list the career opportunities in fashion merchandising and retailing.
- describe the work of a retail buyer and the qualifications needed for that career.
- explain the duties of persons involved with direct selling and other store operations.
- distinguish between the management positions of merchandise manager, fashion coordinator, store manager, and branch manager.

The giant field of ***fashion merchandising*** involves all of the functions of planning, buying, and selling apparel items. Chart 21-1 gives an overview of merchandising. Large numbers of merchandising jobs exist in the retail industry, which sells goods to individual consumers in exchange for money or credit. Stores do this through direct selling. The indirect selling part of merchandising is promotion, which includes advertising and publicity. Those careers will be discussed in the next chapter.

Many career opportunities exist in the retail industry. People who are interested in textiles and fashion, and who understand the needs and wants of consumers are the most likely to succeed. Employees plan the merchandise that will sell the best. Then they buy it from wholesale sources and promote and sell it to consumers. They must keep a steady flow of merchandise going through their stores as efficiently and profitably as possible.

Retail Job Generalities

Jobs in retailing are available in all geographic locations. Opportunities exist in large and small department stores, chain stores, discount stores, variety stores, and specialty shops. They are also in mail-order houses, catalog stores, manufacturers' outlet stores, and some at-home retail selling companies.

Retailing went through a period of fast growth in the late 1900s. Many types of stores, mail-order businesses, and shopping malls prospered and expanded. After that, the U.S. economy slowed at the same time that the retail industry became oversaturated with too many stores. Some retailers had to file for "Chapter 11" bankruptcy protection. In other

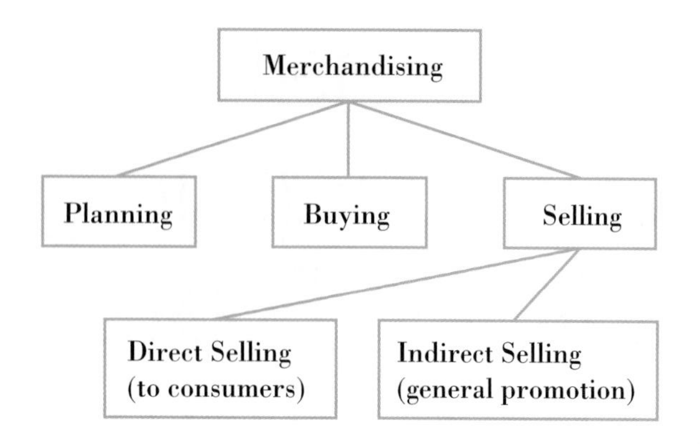

21-1 "Merchandising" is done to some extent by all segments of the textile and apparel industries. However, the retail industry concentrates almost entirely on the merchandising functions. Thus, people often refer to work in the apparel retailing field as employment in "fashion merchandising."

Nautica

21-2 Retailing offers career opportunities for people with many different talents and interests, including designing store fronts and layouts.

words, their debts and management were restructured so they could try to regain their strength. Some other retailers went out of business.

Now the industry is experiencing a reduction to fewer, but larger, retail companies. As a result of this and new technology that has automated many retail tasks, the growth of available retail jobs has slowed. However, many career opportunities still exist. People without any experience must be willing to start at the bottom. Those with education and training can move up more quickly and to higher levels. The opportunities for advancement are good for talented, hardworking people.

Retailing has many levels of employment and degrees of responsibility. The division of job duties varies among different stores. There are jobs buying the merchandise and working with the stock. Other jobs involve promoting the goods, selling the merchandise, overseeing safety and security systems, and designing the stores, as in 21-2.

In small shops, many of the retail duties are combined. A single owner or employee may do all of the planning, buying, receiving, pricing, advertising, displaying, and selling. He or she may also keep accounts and inventory records.

In large stores or chains, specialized tasks are assigned to many different employees. There are also executive and supervisory opportunities, since people are needed to direct the work. Most large retailing firms offer executive training programs for those with college degrees in retail management or merchandising. Some of the best career opportunities are with discount chains and mass merchandisers, although many college graduates would prefer to be associated with upscale fashion stores.

As a general rule, people with jobs in retail merchandising and retailing can expect to work long hours. However, benefits and rewards accompany the effort and time spent. Employees usually receive discounts on personal purchases, and pay is competitive. There is adventure, recognition, and a chance to see the direct results of one's efforts. Often those in retailing get caught up in the excitement of the work and wouldn't be happy doing anything else!

Qualifications for Retail Work

If you want to succeed in retailing, you need to be outgoing and well organized. You should be able to handle figures and details well. You must be able to move and think quickly. Energy and stamina are essential. You must be good at customer relations, value cultural diversity, and able to work under stress. Good leadership abilities and self-confidence help you advance to higher positions. Also, it is very important to have excellent grooming and a sense of fashion.

To prepare for a professional career, you should consider a two- or four-year program in fashion merchandising or retailing. Courses include fashion marketing, sales promotion, fashion buying, merchandise math, and consumer motivation. Other subjects studied are merchandise planning, retail operations, business law, computer science, and small store management. A knowledge of textiles, garment construction, advertising, economics, accounting, as in 21-3, and psychology is valuable for anyone in retailing. Many colleges offer work-study programs that allow you to alternate on-the-job retail work experience with your academic studies.

Experience in any retail store is valuable and transferable. Summer or part-time work in sales, even as a teenager, gives you a taste of this industry and puts experience on your resume.

Once you have training in retailing procedures, you can switch from one store to another in any part of the country. You may even open your own store some day!

One word of warning should be heeded seriously by those who want to pursue retail careers. There are severe employment limitations for anyone who has a record of shoplifting or employee theft. See 21-4. Such information is noted by the security departments of retail organizations. It is kept for many years and checked when employment applications are reviewed. Anyone whose name is on file is considered to be a security risk.

21-3 To move up in the retail industry, it is important to learn the principles of accounting, as well as to have a sound understanding of many other business subjects.

Merchandise Planning and Buying

Planning and buying are key activities of fashion merchandisers. They are done by such employees as retail buyers, resident buying office buyers, buying assistants, and executive trainees.

Retail Buyer

Retail buyers are merchandising employees, as in 21-5, who are responsible for selecting and purchasing goods for their companies. Their goal is to make a profit for the company when the goods are sold. They must estimate future demand for merchandise they buy. The two main types of retail buyers are departmental and classification buyers.

Departmental buyers plan and purchase all the goods for only their own department of a traditional department store. They manage their inventory, pricing, and promotions. They are accountable for sales and profits. They serve as "department manager," overseeing the department's salespeople.

Classification buyers are "central buyers" who specialize in only one category of goods,

Sensormatic

21-4 Shoplifting and employee theft are crimes punishable by the law. People involved in such activities should never be employed in retail work.

In Cooperation with The Fashion Shop

21-5 Besides purchasing merchandise for their departments and working with store employees, retail buyers spend a great deal of time in their offices taking care of important details.

such as infant sleepwear or men's casual trousers. They do the buying for all the stores of one large company. This is more common today. Larger retail organizations have centralized procedures of buying, plus sophisticated distribution of items to their many stores. Sometimes this type of buying is done for a mail-order firm.

Classification buyers work at a central office, often at the corporation's headquarters. They visit major markets, locate new and exciting sources of merchandise, make selections, and place orders. They reorder the goods that sell well. They are not responsible for publicizing or selling the merchandise.

Planning the merchandise involves determining quantities for various styles, colors, sizes, prices, and fashion emphasis. This must be done within a certain budget. It is also timed for quick and profitable sales when consumer demand is high. The store image, as well as activities of competitors, must be watched.

To do purchasing, buyers must locate reliable sources, called "vendors," of goods. They visit manufacturers' showrooms to view and select items, and to agree on prices and delivery dates. Buyers "go to market" several times a year in New York, Los Angeles, Chicago, or other cities. They might also travel to such places as Europe, Asia, or South America to find interesting new items at good prices. Sometimes a resident buyer from the market city may accompany a department store buyer to see goods of manufacturers and wholesalers at the market center. Other times sales representatives call on buyers at their stores or offices.

Merchandise is ordered for the estimated future sales of the stores. Estimated needs are based on current fashion trends and predictions for the socioeconomic and age levels of the customers. Past sales records are also considered. To minimize mistakes, decisions to reorder more of the same merchandise or not are based on the numbers of items sold each day. Electronic data processing equipment is used to help collect this information for buyers. Buyers try to outdo the success of competitive stores as well as their own records from previous years.

Buyers need to make sure the orders from manufacturers are delivered on time and ready to sell, as in 21-6. They authorize payment for the merchandise or issue instructions for its return if defective. Buyers decide what prices to charge for the different items and see that they are properly marked. They decide when and how much to reduce the price of goods that are not selling well. They also help promote, display, and advertise the items.

Buyers instruct the retail salespeople about the new merchandise at meetings. They train employees to sell the goods with the best

Strawbridges

21-6 A buyer must be able to judge if the merchandise received is exactly what was ordered and in condition to be put on the floor for selling.

possible techniques. They supervise the handling of complaints about their merchandise. Sometimes they even do direct selling on the floor to keep in touch with the views of customers.

The roles of retail buyers have been changing. Electronic Quick Response linkages now can reorder retail stock from manufacturers automatically. Also, more private label merchandise is being sold by stores. Therefore, buyers are spending less time on the selling floors of stores. Instead, they are doing more long-range market planning and helping develop the specifications of items to be produced. They are also consulting with the primary sources of merchandise, which sometimes requires international travel.

Buyers' Schedules and Benefits

Buyers must put in long, irregular hours on the job. They may work long days as well as some weekends and holidays. However, there is opportunity for travel to the fashion centers of the world. There is satisfaction when the retailer sells hundreds of items the buyer helped to develop or bought on a hunch. There are also fringe benefits, such as a discount on personal merchandise purchases, paid vacation time, and group insurance.

Buyers are paid salaries rather than hourly wages. As they prove their abilities and assume more responsibilities, their salaries increase. The pay also depends on the size and type of retailer for which the buyer works. However, most buyers will tell you that the paycheck should not be your main goal. First you must love this competitive, stimulating, exacting work making fashion decisions. Top-level buyers develop an enormous range of contacts with fashion writers, manufacturers, fabric houses, designers, and buying offices. A retail company usually picks its top executives from the best buyers.

Qualifications to Be a Buyer

Buyers need a fashion sense and good taste to anticipate what styles and prices will be accepted by their customers. They must keep abreast of fashion trends to know when something has reached its peak and is starting to lose popularity. They must have originality and good ideas on how to promote and sell merchandise. They also need to have good grooming and flair with their personal appearance. They must always be attractively and appropriately dressed.

Buyers need to have technical knowledge about the merchandise as well as a keen business sense. They must be creative and able to organize their work and that of others. They must be outgoing, self-confident, and have a winning personality. They must make sure the personnel of the department work together as a cooperative team. They must also have patience and tact to deal with manufacturers' representatives and customers.

It is often said that buyers need strong feet and a sense of humor! They do need good emotional and physical health and lots of energy. Buyers must be enthusiastic and dedicated to their work. They must be able to remain unruffled when working closely with many people under pressure. They must be fascinated with the business of planning, buying, and selling! People who like this work would be bored in many other types of jobs.

Buyers must be good with figures and details to compute the ongoing sales, margins, markdowns, and inventories. They must have computer skills, as in 21-7, and be able to interpret the trends indicated by numbers on computer printout sheets. They need mature judgment and the ability to respect other people's opinions and points of view. They must be able to take and give directions clearly, respect authority, and demand respect.

Textile/Clothing Technology Corporation

21-7 Almost all jobs in retailing and other fashion careers require good computer skills. Computer use has become universal for merchandise buying careers.

To become a buyer, it is best to have a college degree in merchandising, fashion, or business. Experience, talent, and hard work are also prerequisites. Most buyers are promoted from assistant buyer. Some have been fashion editors for magazines before going into retail work. Some have had responsible jobs for fabric or apparel manufacturers.

Assistant Retail Buyer

Assistant retail buyers help buyers with many different tasks. They help present the merchandise and promotions to the selling staff at sales meetings. They help supervise the sales and stock staffs. They record the sales for the week. They keep track of garments in each style in the department or at all branch stores. With that information, they coordinate transfers of merchandise when needed. They help trace items during shipment, return unsatisfactory items, and place reorders. They keep the books for the buyer and keep track of items ordered, prices, style numbers, and items sold. They help the buyer stay within budget.

As assistant buyers become more experienced, they know the profitable price ranges and merchandise that sells well in their department or classification. They accept responsibility in the buyer's absence. At other times they go to market with the buyer to help select merchandise. In this way, they learn the buying techniques and manufacturing sources of their type of apparel.

Assistant buyers also help with the displays and advertising for the merchandise. They get samples from the manufacturers ahead of time so they can help prepare copy for ads. They continually check ads to see that their products are presented accurately and with the proper mood and image.

Assistant buyers must be organized and have the ability to carry out jobs accurately and quickly. They must be able to communicate well, get along with all kinds of people, and remain calm under pressure.

An assistant buyer is usually a college graduate who has completed the executive training program. Someone with less education must have successful sales and stock experience plus enthusiasm and initiative. An assistant buyer's salary is fair to good.

Resident Buying Office Buyer

Resident buying office buyers help the buyers from their member store do a better job of buying. They research wholesale markets and report on trends through written fashion news bulletins. They arrange for appointments and recommend suppliers. They get samples for store buyers to see before buying decisions are made. Sometimes they go with out-of-town buyers into the market. They make their own buying decisions for their customer stores, if requested. They follow up on deliveries, adjustments, and complaints about merchandise. They also place reorders. They sometimes present merchandise clinics, as in 21-8, or previews of items of the leading resources of goods, as in 21-9.

Frederick Atkins, Inc.

21-8 Resident buying offices keep the buyers of their member stores updated on the latest developments in the apparel marketplace.

Frederick Atkins, Inc.

21-9 Resident buying offices sometimes arrange for fashions to be modeled, so the buyers of their member stores can preview the apparel.

Resident buying office buyers should have excellent communication skills. They must be able to handle work under pressure and manage several tasks at the same time. They must have a highly developed sense of fashion both personally and professionally. These buyers usually work regular weekday office hours. Their offices are located in New York City, Los Angeles, and other fashion centers. Assistant buyers may be hired to help the buyer and to do follow-up tasks.

Executive Trainee

Executive trainees (also called management trainees) receive on-the-job training for potential buyer and management positions. They are interviewed and selected carefully by store managers. They are almost always college graduates. A few may have taken two-year retailing or fashion programs in vocational schools.

To be considered for executive training, candidates need intelligence, leadership abilities, maturity, initiative, and alertness. The applicants must be serious about having retailing as a career. The final selections are usually made by a committee of the store's top managers.

The executive trainee program acquaints the participants with all aspects of merchandising and develops their management skills. They learn about the store's branches, selling departments, and nonselling jobs. The executive trainees spend months in training, learning the business from the ground up. They usually start out selling on the floor. They rotate to every type of department at a fast pace. This acquaints them with the operations and customers of every department. They also do a stint in the stockroom, as in 21-10, and study the store's promotional techniques. Usually they take a trip to the market with buyers and present their own merchandising promotions at their stores.

Executive trainees must be sensitive to conditions around them and able to take constructive criticism and stress. They should be goal-oriented, decisive, and self-confident. They must be enthusiastic, flexible, energetic, and not easily discouraged. They must look like professional fashion experts. They should be able to work with all the store's employees. They should be available for evening, weekend, and holiday work.

21-10 Executive trainees learn about all aspects of retail operations, including stockroom inventory systems.

An executive training position is salaried. However, the salary is low during the training period. An executive trainee who meets the store's high standards of performance may emerge as head of stock or assistant buyer. In several more years, he or she could become a buyer or begin moving up through the management of the company.

Direct Selling

The greatest number of employees in most retail companies are involved with direct selling. They either work under departmental merchandise buyers (previously described) or managers of branch or chain stores.

Retail Salesperson

A ***retail salesperson*** deals directly with the customers. This is usually a beginning job in retailing, but it may be an entire career in itself. It can also be a stepping-stone to higher level merchandising jobs with the store.

Salespeople are valuable employees to the store. They meet the public and represent the store and its image to the outside world. They help customers find what they want.

"Salesclerks" must know every aspect of the merchandise in their department or store, so they can help shoppers make selections. They should be able to show, explain, and recommend merchandise in an enticing way. Top salespeople have the answers to every question about the goods.

Salespeople must make sure all merchandise stays in good condition. Displays must be replenished as merchandise is sold. Salespeople may assist in stock counts and suggest reorders of fast-selling items. They must write out sales slips; compute sales tax; and handle cash, checks, and credit cards according to store policy. They must be able to make change accurately and package the purchases neatly. They may also have to accept returns and refund money.

Retail sales work can be tiring with most of the time spent standing or walking. It is sometimes available part-time or during holidays and summers. It is good experience. For a permanent job, a high school diploma is needed.

Retailers give in-store training to new salespeople, as in 21-11. They give instructions about procedures with sales slips, cash registers, exchanges, and refunds. They familiarize new employees with stock arrangements and teach them how to deal with specific situations that may occur. They suggest what should and should not be said to customers. The instructional program might be short, followed by on-the-job training.

The pay for apparel sales work is generally low to medium. It may be based on an hourly wage, commission (a certain percent of the dollar amount of goods sold by the employee), or on a combination of the two. Most salespeople have to work some weekends, evenings, and holidays. Longer hours may be necessary during busy times, such as the Christmas season, special sales, and inventory time. A big advantage is the 10 percent to 25 percent employee discount on merchandise bought at the store. Other benefits for full-time employees are paid vacation time, sick days, and health insurance.

Symbol Technologies, Inc.

21-11 In order to do their jobs properly, new salespeople are trained according to company procedures. The merchandise is also explained to them.

There are quite a few job opportunities in retail sales because there is a rapid turnover of employees. Jobs are available in almost every city in the United States. The working surroundings for most retail sales jobs are pleasant. Stores are usually well-lit, air-conditioned, and clean.

Qualifications to Be a Retail Salesperson

Salespeople should be neat and attractive, so customers will feel good about buying apparel from them. They should get along with many types of people. Salespeople should be quick to understand what the customer does and does not want. They must always be courteous and pleasant.

Retail salespeople must have enthusiasm for the products they sell. They need good communication skills to describe the merchandise. Basic arithmetic skills are necessary to close the sale. Reliability, honesty, good health, and physical endurance are also required.

Sales work provides an excellent background for almost any higher level job in the apparel industries. The training value of selling experience should not be underestimated. With good performance, a salesperson may advance to assistant buyer or head of stock. To move up fast, or to get to higher positions, a college education is usually needed.

Other Store Operations

There are many other important jobs in retail stores. For instance, people are needed to train new personnel, handle customer problems, and keep tabs on the competition. Employees are needed to do stock control, customer checkout, and office work. Maintenance and security jobs are also available.

Training Supervisor

The *training supervisors* give orientation classes to new salespeople before they start working. They also give updating programs to current salespeople about new equipment or procedures. Some programs might coordinate with the fashion director's information on the season's new colors, styles, terms, and promotional plans. Training supervisors also help plan and run the classes for executive trainees.

The job of training supervisor usually is available only in large department stores. In smaller stores, these duties are combined with another supervisory job. The training job is a combination of office work, teaching, and promotional activities.

This is a salaried position with good pay. A college degree is usually required. Retail experience is a must. A training supervisor must be self-confident, organized, and businesslike. He or she must be able to meet people easily and communicate ideas clearly.

Customer Service Manager

A ***customer service manager*** is sometimes called a "consumer specialist" or "service director." This person serves as an intermediary between the store and its customers. He or she handles complaints and deals with special needs, such as credit purchases, gift wrapping, special orders, or home delivery.

If you are a customer service manager, you investigate problems that consumers are having with the store or its merchandise. You try to solve problems for people, so they continue to like and trust the store. You oversee the returning of items by customers and answer general questions. You see that each salesperson is treating shoppers of all cultural backgrounds fairly and courteously. You take action, if needed, to protect your customers from illegal business practices. You also keep records of all such dealings.

This position is usually gained after much retail experience with the store. Your talent for dealing effectively with people and your abilities at problem solving moves you into this job. Its salary is medium to high.

Alterations Expert

An *alterations expert*, or tailor, is employed by large stores that do not send out their alteration work. This person takes in, lets out, and reshapes garments that do not fit the purchasing customer properly.

An alterations expert must be proficient at all sewing procedures. Such skills are learned from high school or trade school courses, plus lots of sewing experience. A complete understanding of garment fit and adjustment techniques is necessary. This work can be more complicated than sewing apparel from scratch. The pay is low to medium.

Comparison Shopper

Comparison shoppers shop in the store that employs them as well as in competitive stores. Competitive stores are the ones with similar merchandise and the same clientele.

Comparison shoppers compare the merchandise offered by competitive stores with that of their own store. They note the amount of merchandise and how up-to-date it is for the season's new items. They also look at the colors and size ranges offered, and the prices.

Comparison shoppers note store displays and advertisements. They buy merchandise in their store and others to evaluate sales techniques and service provided by salespeople. They also notice the reactions of other consumers.

Comparison shoppers collect facts that are necessary to answer customers' questions and complaints about products, prices, and services among competitive stores. They keep the store management informed, so problems can be anticipated and eliminated, and good programs can be expanded.

A comparison shopper might work for one or several departments of a large store or chain. In a smaller retail outlet, he or she would work for the entire store, probably under the head of stock or assistant buyer. Sometimes this is part-time employment or a temporary job. The pay is fairly low.

There are limited positions available for comparison shoppers. They do not need higher education, but must be able to communicate clearly. They should be organized enough to evaluate their findings and reach meaningful conclusions. Some retailing experience, such as a background in selling and a knowledge of value in merchandise, is desired.

Personal Shopper

A ***personal shopper*** is sometimes called a "fashion consultant." This person has the job of selecting merchandise for customers. This is sometimes done in response to mail or telephone requests. At other times, the personal shopper accompanies customers in the store to offer them fashion advice and to help them select the best items for their needs.

The demand for this type of work is growing. The personal shopper must be familiar with current fashion trends and know the standards of dress for various professions and lifestyles. He or she must have a pleasant and tactful personality as well as the capacity to listen and respond to requests. Education beyond high school may not be required. However, extensive retail experience and a flair for fashion are necessary.

Lands' End Direct Merchants

21-12 The hanging garments shown here are stored according to category and stock number. For mail-order retailing, items can be identified and "pulled" quickly when needed. They are then shipped directly to the customer.

Stock Clerk

Stock clerk, or "merchandise clerical," is an entry-level position for someone who does not have college training. People in this job receive merchandise from the delivery trucks that bring the apparel to the store. They open containers, unpack items, and compare the delivery records with the actual goods that were received.

The condition of the apparel must be checked for any damage or soiling. If there are damaged or lost goods, a report must be filed. New stock items must be entered into the records. Almost all of these "stock control lists" are done on computers. In fact, with the higher efficiency, stock clerk jobs are not increasing in numbers the way they once were.

Stock clerks prepare the merchandise for selling. Price tags are attached. Some items are placed on hangers. The merchandise is then taken to the proper departments as they need it, as in 21-12. Stockroom records are updated again when the goods are delivered to the floor of the store. Stock clerks may also place the items on shelves or racks in the sales area so all displays are neatly filled.

Stock clerks usually help count items during inventory time. They handle returned merchandise by sending it back to the manufacturer or repairing it. Some is put back onto the selling floor at full price, and some is marked down. Records must be adjusted accordingly.

A stock clerk must always know what is on hand and be able to find it. The inventory in the stockroom must be kept in the proper order. Sometimes special orders must be filled. Transfers of merchandise might be made between branches of a chain store. Records of all movements of goods must be updated immediately.

Being a stock clerk is a methodical job. Merchandise must be handled carefully, quickly, and accurately. There may be a lot of lifting, bending, and pushing involved. It may be physically tiring, heavy work. To do it, you must have good health, stamina, and fine eyesight.

Stock clerks usually work regular hours and receive hourly wages. The pay is not very high, but there is usually an employee discount on the purchase of store merchandise. Other benefits include paid vacation time and group insurance.

A high school diploma is usually required for this job. A stock clerk should have a knowledge of mathematics, typing, filing, and basic computer functions. The store gives training in its stockroom procedures, records, and forms. Dependability, a helpful attitude, and legible handwriting all help to do the job well.

Head of Stock

Head of stock is a job with more duties, responsibilities, and pay than that of a stock clerk. This person may be in charge of the stock for one department of a large store or for an entire small store. He or she may have already been a stock clerk or salesperson. This person has proven the ability to get a job done well and quickly. The head of stock works closely with buyers and assistant buyers to make sure the floor is always stocked.

Heads of stock are in charge of and coordinate all the activities of stock clerks. They decide how the inventory should be stored, such as by size, color, or number. They establish work procedures and make sure that damaged or soiled goods are repaired or cleaned. They also make sure that any goods returned by customers are sent back to the manufacturer or put back out onto the selling floor.

Heads of stock help with inventory, supervise the labeling of stock, and process reorders. They make sure the merchandise is put onto the sales floor in an orderly, attractive way. They also keep all of the stock records up-to-date, so the whereabouts of every item is known.

This is an entry-level job for a college graduate. A degree in retailing or merchandising is helpful. With experience and proven abilities, a high school graduate could move into this job.

The head of stock must be well organized and have good record keeping skills. He or she should be tactful and able to motivate others. The head of stock must pay close attention to details. This supervisory job involves both written work and foot work.

The head of stock is salaried. The pay is fair. A store discount is an added benefit. The work might be tiring. If a good job is done, this person might move up to be an assistant buyer.

Checkout Cashier

A ***checkout cashier***, as shown in 21-13, rings up customers' purchases. He or she collects and records customers' payments, and bags or wraps the purchases.

A cashier is usually a high school graduate who has been trained by the store to use the checkout equipment. This might include an electronic cash register, a credit card imprint machine, a tool to remove security tags from merchandise, and other such equipment.

21-13 Checkout cashiers are important to retail stores. They have direct contact with the stores' customers, and they handle the merchandise and money in sales transactions.

Checkout cashiers must be honest and trustworthy. They must be thorough in doing their checkout duties. They should be personable and have a pleasant appearance.

Checkout cashier jobs are readily available in retail stores across the country. A great deal of standing is done, and the pay is quite low. For those who plan to move into other retail jobs, experience as a cashier can be an asset.

Office Workers

Secretaries are needed for the top executives. A secretarial job can give you a broad picture of the store's operations. It can give you a chance to meet important retailing people. You may decide to try direct selling after being a secretary in the office of a retail firm.

Other office positions include *billing agents*, *accountants*, and *office managers*. Resident and corporate buying offices have *buyer's clericals* who answer telephones, schedule appointments, follow up on shipments, and do filing. They may also handle problems with late deliveries or damaged goods. Retail chains have *distributors/planners* who keep track of thousands of units of merchandise through precise, computerized records. They allocate the various items of stock to the many branches that need them.

Office workers should be well organized and good with details. They need to have a good memory and the ability to communicate well with many people, often by telephone. For most of the jobs, an aptitude for working with figures and strong keyboarding skills are needed. Salaries vary according to the educational level, job duties, experience, and expertise of the employee.

Maintenance Workers

Maintenance workers clean regularly as well as paint and repair the store as needed. A knowledge of carpentry and plumbing is recommended. Some workers are electricians. However, for large or specialized jobs, outside tradespeople may be hired. This work is often done at night when the store is closed.

Security Guard

Security guards protect against shoplifters while the store is open and against break-ins when the store is closed. They also work to prevent employee theft. Security guards watch for pickpockets, vandals, and other people doing undesirable activities in the store. They also handle customer safety in emergencies, such as an evacuation in case of fire.

Malls hire security guards to patrol the common areas. Some retail stores have security guards posted near doors. Others have electronic devices that are activated if certain tags pass through the store's doors. Closed-circuit television cameras and mirrors also help security workers do their jobs.

Security guards must be calm, alert, and efficient. Some wear uniforms, and others work in street clothes. They are usually trained in first aid techniques as well as in security work.

Retail Management

Management employees, as pictured in 21-14, are the executives who oversee and coordinate the various parts of a firm. They have college educations and often advanced degrees. They usually have been promoted after being successful, hardworking employees of the company for a long time. The pay and prestige are very high.

Merchandise Manager

A ***merchandise manager*** may be responsible for the total business coordination of several departments or classifications of goods. This person is the direct boss over a group of buyers. For independant department store organizations, as in 21-15, he or she may also oversee the fashion coordinator, sales staff, and stock workers.

Merchandise managers compete with each other for larger sales and profits. They also try to beat their previous records. They search out new and different sources of merchandise. They try to be creative in developing new areas or special departments in the store. They may add glamour or use a new approach in their departments. They encourage their parts of the business to grow and prosper.

If you are a merchandise manager, you have a great deal of responsibility. You act as consultant and teacher by sharing your knowledge and ideas with those under you. You have to organize your duties effectively and work overtime when necessary. You have to gain and keep the respect of employees at all levels in the company. You need patience, diplomacy, and enthusiasm.

Being a merchandise manager is exciting, yet it demands a great deal of time and energy. Educational requirements are high. The prestige and salary are also high.

Strawbridges

21-14 Retail management positions are filled by experienced, knowledgeable people who can plan, organize, and control a firm's operations.

Independent Large Store Organization

- Top management (or store manager)
  - Assistant store manager
  - Merchandise manager
    - Fashion coordinator
    - Buyers
      - Assistant buyers
      - Clerical staff
      - Sales-people
      - Stock workers
  - Operations manager
    - Customer service
    - Maintenance
    - Security
    - Receiving
  - Financial manager
    - Accounting
    - Credit
    - Expense control
  - Promotion manager
    - Advertising
    - Visual merchandising
    - Public relations
  - Personnel manager
    - Hiring
    - Training
    - Employee services

21-15 Although organization charts of different retail companies vary somewhat, the job of merchandise manager is an executive position overseeing goods that are planned, bought, and sold.

Fashion Coordinator

A ***fashion coordinator*** is sometimes called a "fashion director." This person makes sure that all fashion departments of a large retail business are kept updated on the latest trends. He or she advises buyers and merchandise managers about new fashions and assists the promotion departments.

Fashion coordinators tie the merchandise of store departments together to create a fashion whole. When a color or style of clothing is featured in their store, they make sure the right accessories are available, too. They see that the merchandise can be assembled into pleasing wardrobes.

Most fashion coordinators are responsible for assembling and harmonizing the merchandise that goes on display. A theme for each season is suggested to the display manager. Garments and accessories are chosen for window and interior displays, as in 21-16. The publicity department is informed about new fashion trends and products. Direction is given to copywriters and illustrators. Advice is given for radio and television commercials. An overall look and feeling is created for each season.

21-16 A fashion coordinator makes sure that apparel items and accessories available in the store are shown to shoppers by selecting compatible items for store displays.

Fashion coordinators pay attention to what fashion forecasters say. They use that information along with their own tastes, ideas, and fashion instincts to determine which clothes reach the retail stores. In some cases, to get what is wanted, they even make suggestions to designers or manufacturers.

When fashion shows are presented, as in 21-17, the fashion coordinator selects the models, clothing, and accessories. He or she supervises the publicity and preparations for the shows and often acts as commentator, pointing out fashion news. Consumers are shown how a wide range of fashion items can be combined.

Fashion coordinators must cover worldwide fashion centers to get the best advanced information, even before the buyers go to market. They attend trade shows to see what new fabrics and products are being introduced. Information on product research and development gives insight about goods before they become available. The coordinators then develop buying and selling strategies for all component parts of the "total look." The fashion story must then be transmitted to the employees who buy, promote, and sell the merchandise.

The fashion coordinator teaches the buyers and salespeople about new silhouettes, colors, fabrics, styles, and accessories. Sometimes he or she gives talks to community groups about trends and changes in fashion. The job also may incorporate the duties of a training supervisor.

The job of fashion coordinator is time-consuming and demanding. It is glamorous and challenging. It has an interesting variety of duties. There is stiff competition for the few positions. It requires long hours at peak times under considerable pressure. Yet, the excitement of this job is sought after by most aspiring retail people!

Fashion coordinators have a great deal of prestige. Their salaries are high, although they vary among different types and sizes of stores and companies. Fashion coordinators also receive the usual store discounts and other fringe benefits.

In Cooperation with The Fashion Association

21-17 One duty of a fashion coordinator is to supervise all aspects of store-sponsored fashion shows. The audience is informed about the latest style tips, and hopefully the excitement created by the show encourages people to buy items.

Qualifications to Be a Fashion Coordinator

To become a fashion coordinator, you have to be familiar with fashion cycles. You have to understand what affects consumer acceptance or rejection. This means having a highly developed fashion sense. You have to be aware of the style lines that leading designers will be using. You must know the contents of all American and foreign fashion magazines.

Fashion coordinators must be resourceful and flexible. They might spot, for instance, an unusual scarf and visualize an entire store display or promotion around that item. They must be aware of social patterns that change, such as if people are having dinners in restaurants rather than in homes. These general social trends can have a large impact on apparel trends and demand.

Fashion coordinators are expected to be at the important social events to which fashion leaders go. They should participate in a wide range of activities, so they can relate to how their customers live and what the customers want and need. Doing so helps them determine what merchandise to carry.

Fashion coordinators must be able to work tactfully with people. They must be able to schedule their time well. They must be comfortable and confident about speaking to large audiences. They must have poise and good grooming. They must be able to adapt to many situations and to make sound decisions. Good health and energy are needed. A sense of humor is a must!

Fashion coordinators should be enthusiastic, curious, and sensitive. They need imagination, creativity, and initiative. They need a background of successful retail experience. They are always selling, even at the executive level.

To be a fashion coordinator, you need a sound educational background. It is best to have a college degree. You work up to this position after having broad merchandising experience. You may move into this job after being an assistant fashion coordinator or a buyer.

Assistant Fashion Coordinator

An *assistant fashion coordinator* helps a fashion coordinator, mainly by following up on details. He or she sets appointments, makes telephone contacts, books models, and runs errands. The assistant helps put on fashion shows, helps write fashion bulletins, and spends time observing market trends and new looks.

This is a scarce, competitive job available only in large retail firms. Excellent grooming and a keen sense of fashion are required since the assistant must sometimes stand in for the fashion director. Poise, self-confidence, and a good speaking voice are important.

Store Manager

Besides the previously-described jobs that are related to merchandise, store management is a different type of retail career. This is shown on the "operations management" side of the two main tracks for college graduates to move up to retail management positions. See these in 21-18.

The ***store manager*** has probably been named to this position after being a successful assistant store manager or department manager. He or she is in charge of every aspect of the store's operation. The store manager oversees the buying and selling activities as well as the hiring, training, and scheduling of workers. He or she is in charge of promotional activities, the financial accounts, and store security.

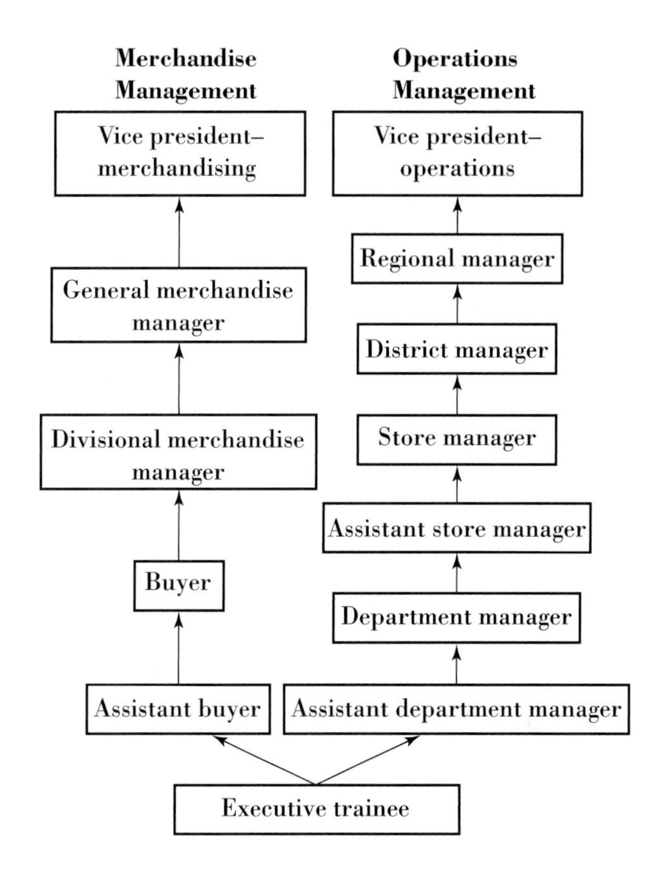

21-18 Retail managers may either move up the merchandise career path or the store operations track, depending on their interests, aptitudes, and skills.

Store managers must sometimes work evenings, weekends, and holidays. However, with dependable and well-trained personnel, the store should run smoothly without the constant presence of the manager. The solving of any large or storewide problem is the responsibility of the manager. Also, credit for a smoothly running operation is given to this person.

Store managers must have initiative, leadership abilities, and lots of energy. They should have a good memory, clear handwriting, and good communication skills. They need an outgoing personality and a fashionable appearance. They must be able to deal with many people in a firm, yet friendly manner. They should have good fashion business sense. They may have to relocate if transferred to stores in different geographic locations.

Other Retail Management

The job of ***branch coordinator*** or district manager is becoming more important and more common in stores as they expand. One store might have several branches, and there has to be at least one person to keep tabs on them as a full-time job. He or she sees that their stock, selling techniques, and general operations coordinate with the main store or central office. This executive position is gained after much retail and management experience.

In centralized retail corporations, a branch coordinator or district manager might be responsible for up to a dozen stores in a geographic area. Then this person might move up to a regional manager, overseeing several districts containing a total of 75 or 80 stores. Regional managers report to the highest store management executive, the vice president of operations.

The merchandise management track moves people up from buyer. These employees might move to divisional merchandise manager (DMM), supervising a group of buyers or a segment of the merchandise the company offers. The next move up would be to general merchandise manager (GMM), responsible for the total retail merchandising of the company. The GMM reports to or has the title of vice president of merchandising.

As in other industries, a retail *personnel director* oversees the hiring of new employees. Sometimes this person must dismiss an employee who has done an unsatisfactory job, is unreliable, or has been dishonest. The personnel director also oversees the fringe benefit programs for the employees.

This position requires a college education in business administration or personnel management. The personnel director has probably not advanced through either the merchandise or operations management chain of jobs. However, he or she knows what tasks each job entails and what qualifications are needed to do those tasks.

Retail firms need dynamic, fashion-oriented executives. The vice presidents and president, like the salespeople, should be thrilled at the idea of selling. They must be creative and appreciate creativity in others. They must know how to be selective. They must be able to make decisions based on a thorough knowledge of the latest business and fashion trends. They must know what resources are available and how to use them. They must know and understand the firm's customers, products, personnel, promotional programs, and all other details.

Chapter Review

Summary

Fashion merchandising involves planning, buying, and selling of apparel items. Jobs are available in different types of retail businesses in all geographic locations. Being outgoing, well organized, and detail-oriented are qualifications for retail work.

Retail buyers may do planning and buying exclusively. Departmental buyers oversee the goods for their own department, while classification buyers specialize in only one category of goods for all the stores of one company. Buyers put in long, irregular hours and receive good salaries and benefits. Buyers must have fashion and business sense, as well as technical knowledge about merchandise. Assistant retail buyers help buyers with all the details of the job. Resident buying office buyers are in market centers to help the buyers of their member retailers. Executive trainees are college graduates learning to do merchandising jobs.

The greatest number of retail employees do direct selling. Retail salespeople help customers find what they want. Retail salespeople should be neat and attractive, get along well with people, be quick to understand customer wants, and always be courteous and pleasant.

Other store operations jobs include training supervisors who give orientation classes to new salespeople, customer service managers who serve as intermediaries between the store and its customers, and alterations experts who remake garments to fit customers. Comparison shoppers analyze merchandise and services of their own and competitive stores. Personal shoppers select requested merchandise for customers. Stock clerks receive, unpack, check, record, prepare, and move goods for selling. The head of stock is in charge of the stock for one department of a large store or for an entire small store. Checkout cashiers ring up customer purchases, record payments, and bag the items. Office workers include secretaries, billing agents, accountants, office managers, buyer's clericals, distributors/planners, and others. Additionally, maintenance workers keep stores clean and in good repair, while security guards protect against theft and safety hazards.

Just as for other industries, managers oversee and coordinate the various parts of retail firms. Merchandise managers coordinate the business of several departments or classifications of goods.

Fashion coordinators keep all fashion departments of large retail businesses updated on the latest trends. This prestigious job requires a deep understanding of fashions and consumer acceptance of products. Assistant fashion coordinators follow up on details for fashion coordinators. Store managers are in charge of every aspect of their store's operation. Other retail managers include branch coordinator, district manager, divisional and general merchandise managers, personnel directors, and others.

Fashion in Review

Write your responses on a separate sheet of paper.

True/False: Write true or false for each of the following statements.

1. Fashion merchandising is a giant field that involves all of the functions of planning, buying, and selling apparel items.
2. In direct selling, goods are sold to consumers in exchange for money or credit.
3. Retailing is currently experiencing a period of fast growth with many new companies that are prospering and expanding.
4. Retail buyers spend most of their time on the selling floor of the store in order to find out what consumers want.
5. The position of retail buyer is an entry-level job requiring only a high school degree.
6. A classification buyer does the buying of a specific merchandise category for all the chain stores of one particular company.
7. A customer service manager deals with special customer needs, complaints, and problems.

8. Comparison shoppers shop in the store that employs them as well as in competitive stores.
9. A personal shopper selects merchandise for customers.
10. Maintenance workers guard against shoplifting, employee theft, and vandalism.
11. A merchandise manager is responsible for the total business coordination and all functions of several departments of a store.
12. A branch coordinator updates all fashion departments on the latest trends and ties the merchandise of the departments together to create a fashion whole.

Short Answer: Write the correct response to each of the following items.

13. List five duties of retail salespeople.
14. List five duties of stock clerks.
15. List five duties of fashion coordinators.

Fashion in Action

1. Make an appointment to have a short meeting with a manager in a local retail store. Before the meeting, prepare a list of questions about the manager's educational background, previous retail experience, and job duties. Share what you learn with the rest of the class in an oral report.
2. Visit the personnel office of a large department store. Ask for a list of all the different kinds of jobs in the store. Use that information as the basis for a bulletin board display.
3. Pretend to be a retail buyer for a specific department of a local department store. Write a report describing the department and your duties. Clip pictures from catalogs or magazines of some of the items you would buy for your department to sell. Describe the colors and sizes of the merchandise. Explain why you selected those particular items. Tell how you would increase sales on items that were not selling well.
4. Check the classified ads in your local newspaper for job openings in retailing. Clip the ads and divide them into specific job categories, such as salesperson, checkout cashier, and stock clerk. Follow up on three of the ads to find out about the pay and any qualifications needed to get the jobs. Prepare an oral report for the class on your findings.
5. Interview three experienced apparel salespeople from different local stores. Ask how they approach and greet customers. What help do they offer to the customers? How do they stay informed about the merchandise they sell? What training have they had for their jobs? Do they have to abide by employee dress codes? Compare your findings with those of your classmates in a class discussion.

22 Careers in Fashion Promotion

Fashion Terms

account executive
art director
graphic designer
advertising director
display designer
fashion illustrator
fashion model
fashion photographer
photo stylist
fashion writer
press kit
copywriter
editor
audio-visual work
publicity
public relations agent

After studying this chapter, you will be able to

- discuss careers in fashion advertising and display.
- explain the work of fashion illustrators, models, and photographers.
- describe employment in fashion journalism and audio-visual work.
- discuss the field of fashion publicity.

Fashion promotion is the indirect selling part of merchandising. This area of employment has unlimited, diverse career possibilities. Some are glamorous and exciting. Most are creative and competitive. All are challenging.

A firm can have the best products in the world, but if they are not sold, there is no profit. Thus, a market demand must be created. Indirect selling helps create a market through promotion, such as advertising and publicity.

Indirect selling is aimed at the general buying public as a whole. It does not deal individually or personally with consumers as retailing does. People in indirect selling jobs show, draw, model, photograph, write about, and talk about fashions to attract the attention of the general public. Sometimes promotional activities are designed to bring people into a store or mall, as in 22-1.

Indirect selling is done by a network of supporting businesses to the apparel industries. The network includes advertising agencies and various print and broadcast media that spread the news of fashion trends.

Fashion Advertising

People in advertising try to attract and inform audiences in the hope of selling products. To succeed, they must know the merchandise and its outlets. They need to be familiar with various approaches to reach the audiences of different products. They must formulate and follow complete advertising programs for products.

Account Executive

An advertising agency employee in charge of handling specific advertising accounts is called an ***account executive***. Account executives work closely with merchandising and marketing people from the businesses they represent. Some advertising agencies have a fashion expert or advisor on their staff, while

Manufacturers Outlet Mall (MOM), D'Angelo Photography

22-1 Entertainment is one part of a promotional plan to attract customers into this mall.

other agencies are devoted entirely to apparel. Their accounts are usually medium to large textile or apparel manufacturers and small to medium retailers. Most large retailers have their own advertising departments.

Often ad agency account executives coordinate entire corporate advertising campaigns. These consist of a series of ads or various promotional techniques to create a special impact. Usually a campaign has a theme with all advertising activities corresponding to that theme. The advertising is aimed at reaching as many customers of the manufacturing or retail business as possible.

Successful account executives continue to represent the same businesses year after year. They must be innovative and up-to-date on the fashion news of those businesses. They must be aware of consumer desires. They must identify target audiences. They must know where to place the ads to create the greatest impact for the particular firms they are representing.

Account executives relate a client's message to the creative staff of the advertising agency. They spell out exactly what should be told through print or broadcast. They specify the feeling the advertising should portray, such as the image of the retail store being represented. Then the creative staff takes over.

Art Director

Art directors are sometimes called advertising designers. They conceptualize the ads for newspapers, magazines, direct-mail flyers, radio, television, signs, and outdoor media, as in 22-2. They try to design the best advertising for the budgeted price. Besides advertisements, they also design "collateral materials." These include brochures, annual reports, packaging, hangtags, logos, trademarks, and other corporate image projects. The collateral materials often accompany products for sale. They usually contain important information, sometimes required by law, while they catch the customer's eye to help sell the product.

Most of the career opportunities for art directors in the fashion industries are with ad agencies and retail stores. Only a few art directors are employed by buying offices, textile producers, apparel manufacturers, trade organizations, and magazines.

There is stiff competition in advertising design. Job titles and responsibilities differ depending on the size of the agency or firm, and on the types of projects to be done. Salaries, fees, or commissions vary widely.

W.L. Gore & Associates, Inc.

22-2 The concept for this advertisement was suggested by an art director who wanted to convey a certain message and feeling.

Creative work must often be done under the pressure of a deadline. There may be long hours and limited vacations.

To become an art director, education is needed after high school in a college or vocational school. Courses include basic design, drawing, painting, lettering, photography, typography, advertising, and promotion. Audio-visual skills are also learned. Most schools have typesetting and printing facilities. Many also have well-equipped radio and television labs, as shown in 22-3.

Studies in computer science are essential for advertising graduates, too. Computer graphics and computerized typesetting are becoming widespread in the advertising field.

Graphic Designer

Graphic designers come up with what is needed for advertisements and collateral materials based on the ideas of the art director, as in 22-4. When they have clearly illustrated the ideas and gained the approval of the art director, agreement must be obtained from the clients. Finally, the finished work is prepared by the graphic designer or some junior members of the staff.

A graphic designer must have a degree in art or advertising design. At some colleges, a bachelor of fine arts degree can be obtained with a packaging design major. This degree has regular art, design, and advertising courses. In addition, other classes are offered in packaging materials, such as glass, plastic, paper, and industrial materials.

Other Advertising Design Employees

Entry-level jobs that can lead to graphic design positions include those of layout artists and paste-up/mechanical artists. These people do the finished art to prepare the work for reproduction. In some agencies, these jobs are combined and done by one person.

Layout artists design the layouts for ads, usually under the supervision of a graphic designer or art director. They specify the typefaces and do "comp" renderings that indicate what the finished ads will look like when they are printed.

Paste-up/mechanical artists put together the elements of the layout, such as the type, line drawings, and photographs. They do the arrangement of art and copy based on directions given by those above them in the organization.

People seeking positions such as these should have a degree in advertising design or illustration. They should have a portfolio showing their precision and accuracy in doing advertising layouts. They need to have the ability to work quickly under pressure to meet deadlines. They must be good at details and have the ability to follow instructions and take criticism. They must do neat, accurate, precise, and thorough work.

Fashion Institute of Technology

22-3 Audio-visual labs give students "hands on" experience with equipment they might use in promotional jobs after graduation.

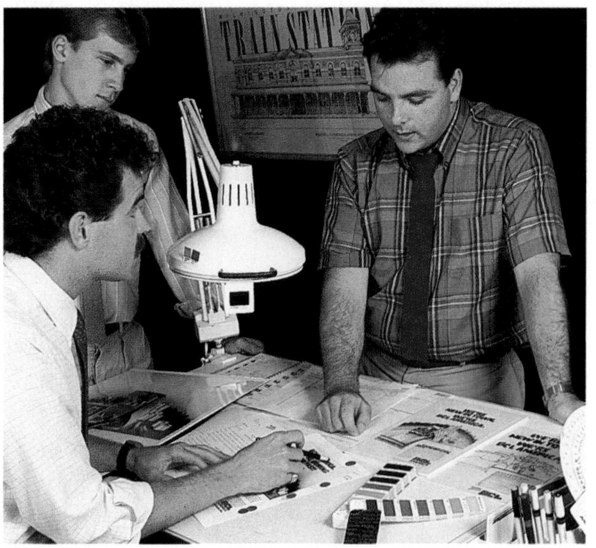

Jones Kent & Associates

22-4 Graphic designers work with other advertising employees to create the most effective advertisements possible.

Advertising Director

Advertising directors work for retail stores and publications. They oversee all the advertising activities. People with similar responsibilities in manufacturing firms are usually called "marketing specialists."

Some advertising directors work with account executives at advertising agencies. Others supervise their companies' own ad departments. They direct the planning of advertising and promotional campaigns. This includes the preparation of print and broadcast advertisements, as well as the creation of selling aids, labels, signs, and packaging.

The advertising director of a retail store coordinates the design and distribution of all the store's newspaper ads, catalogs, and other direct-mail pieces. He or she also helps plan promotions, keeps track of each buyer's advertising budget, and supervises the art department. This position requires both education and experience. Experience may be gained either at an advertising agency or by working up through the store's advertising or fashion promotion department.

The advertising director of a publication is in charge of selling and arranging ad space in the publication. He or she interacts a great deal with the account executives at advertising agencies. This person also works with the advertising directors of individual firms to coordinate their advertising in the publication.

Advertising directors must have good organizational skills and administrative abilities to see that deadlines are met. They need self-confidence, persuasiveness, and the ability to communicate ideas. They must be imaginative, alert, ambitious, and able to deal with creative people. They need enthusiasm and physical stamina. They must have the ability to originate and develop ideas that are sound and practical yet new and unusual. They must be capable of doing market research, media analysis, and mass communications planning.

Advertising directors are usually college graduates. They have taken courses in advertising, psychology, fashion, business, writing, English, printing, photography, and art. Advertising directors earn high salaries, but work hard for their pay and status.

Fashion Display

A *display* is a visual presentation of merchandise. It might be in the window of a store or in an interior location, as shown in 22-5.

Whatever form it takes, the display is intended to show customers what merchandise is available and how items can be combined and accessorized. Each display should be eye-catching to encourage consumers to select and buy merchandise.

Display Designer

Display designers, sometimes called display artists, create displays. They usually gear their work toward special events or promotional activities, as shown in 22-6. They often emphasize seasonal and holiday themes, such as summer vacation, back-to-school, and New Year's Eve. Sometimes displays stress a new color, fashion trend, or cultural event.

Display designers must study and understand lighting techniques, accessorizing, and

JC Penney

22-5 Interior and window displays for fashion items, including home fashions, are planned and put together by display managers and designers.

International Linen Promotion Commission

22-6 Designers created this special display to spotlight linen garments at a Paris festival.

the use of props. They need training in the use of design elements and principles, and in window trimming. They must have a good imagination. A knowledge of carpentry and sewing is helpful. They should also be familiar with the related areas of photography, lettering, painting, fashion, and merchandising.

Some vocational schools offer specific programs in display design. Education is usually followed by an apprenticeship in the actual use of lighting, mannequins, and props under a display designer or display manager.

Display Manager

Display managers often work for retail stores. In different companies they may be the same as, or work under, "visual presentation directors." They consult with the buyers, public relations agents, creative artists, and photographers to plan and arrange displays. They prepare rough sketches, outline ideas, and make blueprints or models for the display staff to follow. They sometimes arrange for television monitors to be placed in particular locations and select the videotapes for customers to view.

The display manager should have a thorough knowledge of retail markets. He or she decides which trends and items to promote in displays, and what feeling the displays should have.

A college degree is preferred for this position. Small stores and businesses that do not have their own display staff hire display agencies to do this work when needed.

Other Fashion Promotion Careers

Other forms of promotion go hand in hand with advertising and display to announce and help sell the latest fashions. They include fashion illustration, modeling, photography, journalism, and audio-visual work.

Fashion Illustrator

Fashion illustrators draw garments that have been designed and produced by others. Complete costumes are usually shown with fashionable accessories. Illustrators try to show the good points of apparel to promote and sell the fashions.

Fashion illustrators are employed mainly by retail stores, pattern companies, and advertising agencies. Some are hired by fashion magazines and other publications. Design or display studios, buying offices, and textile and apparel manufacturers also use fashion illustrators.

When illustration is done for a pattern company, emphasis is given to seams and to trimming details of the apparel. The intent is to give the sewer an idea of the construction required. For fashion magazines or trade publications, the illustrator might point out trends or garment features being described by the fashion writers. For retailing and advertising uses, exotic background touches might be added. The illustrator should be able to catch the attention of viewers and tempt them to buy what is shown.

Illustrators who work for buying offices draw the latest fashions to show them to their member retailers. Those who are employed by textile firms illustrate fashionable uses for the line of textiles. In manufacturing firms, the drawings done by illustrators often record a season's line.

Competition for illustration jobs is keen. Most opportunities are located in big cities. Illustration jobs are usually paid by salary. The amount of pay depends on the person's talent as well as on the size or type of firm. The hours are usually regular.

Good illustrators who become well known often decide to be independent, freelance artists. They do this after gaining a great deal of on-the-job knowledge and making many good contacts. Freelance work can be more profitable than staff work. It is usually more challenging and satisfying, but less steady. A freelancer may have many or few assignments and must complete them as requested within set deadlines.

Qualifications to Be an Illustrator

To be a fashion illustrator, you must be artistic and have a knowledge of fabrics and fashions. You need flair, initiative, and determination. You must be able to work rapidly under the pressure of deadlines. You should keep up with current art and fashion news through trade publications, magazines, art and costume exhibits, fashion shows, and displays.

To succeed in this field, your natural artistic talent should be further developed at a university or trade school. In these programs, art courses concentrate on drawing the human figure, as shown in 22-7. There is also training in advertising design, all art media, and the use of art studio equipment. Many students develop their own distinctive styles of illustrating.

When hunting for a fashion illustration job, a neat portfolio containing examples of your best work is needed. Your illustrations should be of interest to the type of employer you contact. Your ability will be judged on the basis of your presentation.

Fashion Institute of Technology

22-7 Students in fashion illustration courses often develop their skills by drawing live models.

Fashion Model

Fashion models wear garments and accessories to try to show them off. Modeling is a combination of the fashion, advertising, and performing worlds. Models stand, turn, and walk to demonstrate the features of clothing. This might be done at restaurants, department stores, seminars, trade shows, and conventions.

Some models show companies' lines in the showrooms of designers or manufacturing firms. This is called *mannequin work*. These models usually work full-time, regular daily hours for the firm. They receive salaries and company benefits. Extra part-time models are employed for several weeks during the retail buying seasons.

Runway work is done in front of live audiences, as in 22-8. These models appear in fashion shows, usually walking and turning on a runway.

Work done in front of cameras is called *photography work*. Photographic models pose for pictures used in press releases from manufacturers and other firms, as in 22-9. Their photographs may appear in fashion magazines, trade publications, newspapers, advertisements, and pattern catalogs. They are also in mail-order catalogs and direct-mail flyers, as well as on garment packaging and billboards.

Modeling is competitive and demanding work. It can be glamorous. However, employment is often unpredictable with irregular earnings. Runway and photographic models are paid on a fee-per-hour or per-day basis. Fees increase as the model gains experience and name recognition.

Most models are in their twenties, but there is a growing demand for models of all ages. It is more acceptable now to have older and heavier models represent the way people really look. Some very young models are needed to model children's clothes, as in 22-10.

Because of the desire for youth, a modeling career may only last a few years. For female models, the average working time is about 10 years. For male models, it is about 20 years. Models should realize that they may have to go into another line of work in the future. Modeling can be a stepping-stone to other apparel careers. Also, part-time modeling can supplement the regular income from another job.

Qualifications to Be a Model

To go into modeling, you need to have good health and perfect grooming. You need fashion consciousness and above average appearance. You must have physical stamina as well as plenty of determination and patience. You should have poise, style, and a flair to move effectively in the clothes you wear. You should also have a pleasant, outgoing personality. A model has to "sparkle," look confident, and have good posture.

The Fashion Association / Marithe & Francois Girbaud

22-8 These models are doing runway work in a fashion show for a live audience.

JC Penney

22-9 Photographs such as this one are used to show consumers the types of merchandise that a retailer offers for sale.

Kmart

22-10 Although most fashion models are in their twenties, some models are needed from infancy to old age.

Photographic models must be photogenic. They must convey pleasure, surprise, or other feelings for long hours under bright lights and often in hard-to-hold poses. They must be responsive to direction and able to move with spirit and grace.

For modeling, your figure or physique should be tall and thin. It helps to be well proportioned and long-waisted, and to have long legs as in 22-11. A slim figure makes clothes look good and is what our society considers to be "ideal." Also, photos make the figure look heavier than it really is.

Models are hired to fit clothing to be shown. The smaller sizes of a fashion line are usually used. Most female models should fit a size 8 or 10. Male models should usually be a size 40 or 41 long.

Some people strive too hard to become a model because this career seems so glamorous. In reality, there is a great deal of pressure to maintain a certain physical image. There is also job insecurity and an irregularity of assignments. Drug use and eating disorders sometimes develop, which are extremely damaging. These conditions cause dehydration, malnutrition, brittle hair and nails, poor teeth, dark circles under the eyes, and harm to all other parts of the body. These habits harm the chances to be a model, rather than helping it, and can cause death.

To maintain fitness, models should get plenty of sleep, eat balanced meals, and exercise regularly. They must be ready for a job at any time. They must be able to adapt to erratic scheduling. The career demands self-discipline and sometimes the sacrificing of personal pleasures.

To become a professional model, you may want to go to an accredited modeling school. You will learn the techniques of modeling, such as how to stand, pose, and move properly. You will study posture, speech, hairstyles, and makeup. Additional training or experience in dance, drama, art, fashion design, and retail sales could be helpful.

Modeling school courses do not guarantee you a modeling career. Make sure the school has a job placement service before you decide to attend. Thoroughly read and understand any contracts you are asked to sign.

JC Penney

22-11 A long-waisted body with long legs is desirable for models to show most fashions the best.

To register with a modeling employment agency, you will need several enlarged, unretouched photos that show you in various poses. One should be full-length, one close-up, and one smiling. You will fill out an application form listing your name, address, age, height without shoes, weight, and body measurements. Often modeling agencies also train models.

Most models are employed in major cities. Although New York City has the greatest number, jobs are also available in most other cities. Large businesses almost always hire models through agencies. However, you may apply directly to advertising agencies, newspapers, retail stores, or apparel manufacturers. In each case, include photographs and a resume listing training, experience, and personal specifications.

Fashion Photographer

Fashion photographers take pictures that show fashionable clothes and accessories looking their best. They also try to express moods through the settings and compositions of their photos. They use creative props and interesting backgrounds. Photographers may take pictures of fashion apparel on live models or still shots of fashion merchandise.

When using a model, a fashion photographer often tries to make the picture look like the model is in motion. The model moves and poses in front of the camera. Many pictures are taken. After printing, the best one is chosen to be used.

Some fashion photographers shoot motion footage on film or videotape. Fashion films and tapes are used for television ads, educational purposes, and other promotional viewing. Often they play on monitors in retail stores to update consumers and to help sell merchandise.

Photography is considered to be one of the most flexible forms of artistic expression in fashion communication. Photographers are associated with photo studios, advertising agencies, publications, and large stores. Their work may require some travel. They may work as salaried employees or as freelancers. Only top talent receives top pay.

A fashion photographer should have an interest in all art forms, fashion trends, and people. Besides talent and imagination, photographers must have sound technical training. They must understand lighting techniques and how to use professional cameras and other film-related equipment, as in 22-12. They must be able to use both black-and-white and color film effectively. A knowledge of darkroom procedures is also necessary.

Fashion photographers need trade school training after high school. A college degree is usually not necessary. They must learn the technical aspects of photography. They should also study fashion display and advertising design. A portfolio must be prepared to show prospective employers the types and quality of work the photographer is capable of doing.

Other Fashion Photography Employees

Assistant photographers are apprentices getting on-the-job experience under professional photographers. They test the lighting, take sample photos, and help prepare sets and props for backgrounds. They may clean the studio and make minor repairs to equipment. When they have proven their technical abilities, they also work with the darkroom equipment.

Photo stylists work in photography studios, in advertising agencies, or as freelancers. They book models, accessorize apparel, obtain props, pin up hems, iron garments, and pick up and return merchandise. They may work long hours during heavy work periods. They must understand both fashion and photography. They need enthusiasm, stamina, flexibility, resourcefulness, a high tolerance for stress, and a strong sense of style and color.

Fashion Institute of Technology

22-12 Professional photographers must know how to use many types of cameras, lighting techniques, props, and other equipment to get expert results.

Fashion Writer

Fashion journalists pass along textile and apparel information through the mass media. ***Fashion writers*** work mainly for newspapers and fashion magazines. Sometimes full-length books are written on topics, such as wardrobe planning, colors for individuals, or general fashion and apparel information like this textbook.

A fashion reporter for a newspaper may write a daily or weekly column, or periodic feature stories. He or she may write about fashions seen at important social and cultural events. Some fashion writers do freelance work part-time from home. For monthly magazines, full-time writers usually do more than one article. These often are accompanied by several fashion photos or illustrations.

Press kits from manufacturers and advertisers are sometimes used as sources of information for fashion writers. They include photos along with descriptions or short articles about a company's latest apparel. Otherwise, fashion writers plan, research, and write the articles from scratch. They gather material by conducting personal interviews, making phone calls, and attending fashion events. Some fashion news items come over the wire services.

A writer often covers a specific subject area, such as textiles, accessories, coats, knitwear, or certain designers. Sometimes the writer serves as a technical consultant to other writers or staff members, to photographers, and to advertising people. Sometimes the writer must travel to get a firsthand look at the latest fashions during the semiannual Fashion Press Weeks. He or she must keep in touch with key people in fashion, textile production, and apparel manufacturing.

Qualifications to Be a Fashion Writer

Fashion writers must be creative. They must write precisely while being under the pressure of deadlines. The latest interesting and useful information must be researched and explained to the proper audience. The information must always be presented clearly and thoroughly.

Members of the "fashion press" must have a flair for writing and a keen sense of fashion. They must keep up with changes in the industry and have the ability to spot newsworthy trends or feature material. Enthusiasm is needed. Writers must be able to organize their time and work well alone. They must also be outgoing to arrange and conduct interviews. Writers must be able to think and act quickly to obtain as much information as possible from each interview.

To become a fashion writer, you should have a college degree in journalism, combined with merchandising and apparel courses. Advertising courses are also helpful. Good computer skills are required.

If you want to be a fashion writer, you should read fashion publications to see the styles of other writers. When you apply to publications that allot space to fashion, take samples of your writing with you. Become familiar with a newspaper or magazine before applying for a job or submitting writing samples. By doing this, you can point out your specific qualifications that blend well with that publication.

Beginning salaries in fashion journalism are moderate, with increases for education, experience, and ability. Fringe benefits vary among employers. Working hours may be irregular, depending on deadlines and events to be covered. The work is exciting and offers challenges and variety. However, lots of competition exists for a limited number of jobs. Most opportunities are located in large cities.

Copywriter

Copywriters compose the messages that describe items that are being promoted in ads and catalogs. They write the information as editorial text or as blurbs that accompany photographs or illustrations, as in 22-13. "Copy" includes the words, lettering, and numbers that appear in type.

22-13 Copywriters are needed to write the descriptions that describe merchandise in mail-order catalogs.

Copywriting is done for the promotional materials of manufacturers. It is done for the brochures and direct-mail advertisements of retail stores. It is done for mail-order catalogs and apparel trade publications. Often, apparel items must be accurately described with as few words as possible. Copywriters work from essential information that is given to them, such as garment and fabric descriptions, prices, sizes, or stores where available.

Copywriters should have creativity, maturity, and the ability to translate ideas into words quickly and under pressure. They must be thorough, flexible, and able to spot trends and identify resources. They also need good keyboarding skills.

A copywriter should have a degree in advertising or journalism, with a fashion emphasis. Also, trade schools offer training in advertising and communications. When applying for a job, a copywriter should have a portfolio with examples of his or her written copy.

Editor

Editors usually supervise fashion writers and copywriters. They might be in charge of one or more departments of a publication, such as fabrics or accessories. They should have extensive experience in their fields of expertise. They are administrators as well as journalists. They set policies and give out assignments. They sometimes supervise photography sessions. They are responsible for all information being accurately and creatively presented to the correct readership audience. Eventually, they might become editor-in-chief of a fashion magazine.

The activities of fashion editors vary according to the importance that each publication and its readers attach to fashion information. For large general publications, the fashion editor is assisted by a fashion staff that might include photographers, illustrators, and writers. The editor is the publication's authority on all aspects of fashion and apparel. He or she interprets fashion news, influences the acceptance of fashions, and advises manufacturers about what fashions the readers want to buy.

Editors must have imagination, integrity, vision, and administrative skills. They should be well-traveled, have had good cultural experiences, and be confident and alert at all social events. Besides flair and proven ability, this person must have a business mind, know the apparel industries, and be well organized. Jobs as editors are hard work. However, they have high prestige and outstanding pay.

Audio-Visual Work

Apparel-related ***audio-visual work*** involves radio, television, and multimedia presentations. These jobs involve planning programs, writing scripts, getting props, and producing or being in the presentations. The productions might be commercials, promotional fashion tapes, TV shopping programs, or entertainment talk shows with fashion themes. They could be interviews with fashion experts or coverage of designer collection runway shows for television viewing. The productions might be educational, such as consumer programs or sewing shows. They could be aimed at training people, such as instructing or updating retail store employees.

Audio-visual workers must have a knowledge of fashion as well as writing, speaking, and dramatic skills. They need to write descriptive, informational dialogue quickly and accurately, as well as present ideas clearly and simply. They must be confident and outgoing while sounding natural and sincere with a pleasing voice and appearance.

For this detailed work, you must have good organizational skills. You will have frequent deadlines, long rehearsals, and tight and irregular schedules. Instinctive timing is necessary, as well as the ability to think and act professionally in any situation. You must be able to meet and get along with people under all circumstances. You will have to maintain poise and self-confidence, even under pressure. You must always be up-to-date on new fashion trends.

A degree in audio-visual communications is recommended for this work. Courses in writing, speech, and drama are needed. You may also enter this field after gaining fashion, advertising, or journalism expertise. Employment might be with a broadcast station, advertising agency, marketing firm, or video production company.

These competitive jobs are challenging, have lots of variety, and may offer some travel. There is interesting contact with people. Every day is different. The limited positions go to those who have ambition and the ability to convince others. Sometimes people "make" jobs or opportunities that were unrecognized by others.

Fashion Publicity

Publicity is time or space given without charge by various media because the message is considered to be newsworthy. Publicity differs from advertising, which is paid promotion. However, this "free advertising" also serves to enhance sales appeal, just like other forms of promotion.

Public Relations Agent

Public relations agents, or publicists, tell the story of a firm, or its products or services, through various media. They strive to get editorial mention and photographs in publications, "plugs" on broadcasts, and favorable remarks in public speeches. They speak about products at meetings, conferences, and conventions. They give demonstrations and public presentations, such as fashion shows. Sometimes they give training courses for retail salespeople and buyers. They also try to get endorsements from, or tie-ins with, schools and universities, sport teams, or special events.

Public relations employees prepare and send out press releases to the media, as in 22-14. They produce and present filmstrips, posters, movies, transparencies, booklets, and promotional/educational kits. They publish and distribute newsletters, reports, and bulletins for teachers, researchers, store buyers, students, and consumers. Public relations agents also serve as technical consultants to writers, photographers, and artists. They study the market to seek new product uses. They try to present their products better than the competition.

Creative people are hired by public relations firms or public relations departments of manufacturers, retail stores, and trade associations. Public relations representatives of fabric houses, pattern companies, sewing machine manufacturers, and other apparel-related firms travel around the country. They are constantly alert for opportunities to mention and promote their companies' products. They also relate consumer needs and desires back to their firms.

Clotilde, Inc.

22-14 A public relations agent might send this photo along with supportive copy to city newspapers if the sewing expert shown is going to visit there. Then, hopefully, people will want to attend the expert's seminars and buy her books.

Public relations agents must have lots of imagination and good writing skills. They must have a thorough understanding of their products and know how to find or create news value to promote them. They must be able to anticipate and predict trends. They must be familiar with all kinds of advertising media and selling techniques.

Public relations agents must be convincing, yet tactful. They should have confidence and drive. They must know how to schedule their time wisely. They should not be shy about speaking in front of groups, and should know how to use visual aids. They should have social poise and a pleasing voice and appearance. They should enjoy some traveling.

Most public relations employees have college degrees. The courses are similar to those offered in other advertising and promotion programs. A portfolio of publicity campaigns is needed when applying for a public relations job. Pay is high once experience and success are achieved.

Chapter Review

Summary

Fashion promotion helps create market demand among the buying public. Advertising tries to sell specific products by attracting and informing audiences. Account executives with advertising agencies handle specific accounts. An art director conceptualizes ads to put into various media. Graphic designers create the ads based on the ideas of the art director. Other advertising design employees include layout artists and paste-up/mechanical artists.

Fashion display is a visual presentation of merchandise. Display designers create displays based on special events, promotional activities, holidays, or seasons. Display managers often work for retail stores, planning and arranging displays showing the latest merchandise.

Fashion illustrators draw garments that have been designed and produced by others. Illustrators must be artistic and know about fabrics and fashions. Modeling combines fashion, advertising, and performing to show off garments. Models might do mannequin, runway, or photography work. For modeling, you need good health, grooming, posture, and appearance.

Fashion photographers take pictures that show clothes and accessories looking their best. Assistant photographers do photo preparation and testing work, while photo stylists book models and gather merchandise and props for photo shoots.

Fashion writers work mainly for newspapers and fashion magazines, sometimes using press kits for information. Members of the fashion press must prepare concise, creative copy under the pressure of deadlines. Copywriters compose descriptive messages used in ads, catalogs, and other promotional materials. Editors supervise fashion writers and copywriters. Audio-visual workers plan programs, write scripts, get props, and produce or appear in commercials, promotional fashion tapes, TV programs, or other presentations.

Fashion publicity is free promotion that is considered to be newsworthy. Public relations agents, or publicists, strive to get comments and photos in publications, on broadcasts, and in public speeches.

Fashion in Review

Write your responses on a separate sheet of paper.

Matching: Match the general categories to the specific job titles.

1. Account executive.
2. Graphic designer.
3. Display designer.
4. Art director.
5. Public relations agent.

A. fashion advertising
B. fashion display
C. fashion publicity

True/False: Write true or false for each of the following statements.

6. Indirect selling is aimed at the general public as a whole rather than dealing individually or personally with consumers.
7. Account executives with advertising agencies keep the financial books.
8. Advertising directors oversee all the advertising activities for their companies.
9. A display manager decides which trends and items to promote in displays and what feeling the displays should give.
10. Fashion illustrators draw garments that have been designed and produced by others.
11. "Runway work" is the term for modeling a manufacturer's line of clothes in a showroom.
12. Most female models should fit a size 4 or 6.
13. Editors compose the message that describes items that are being promoted in ads and catalogs.
14. Employment in audio-visual work involves radio, television, and multimedia presentations.
15. Public relations agents are constantly alert for opportunities to mention and promote their companies' products.

Fashion in Action

1. Write a fashion article for a newspaper or magazine. Choose a new fashion or accessory item that interests you. Talk with manufacturers and retailers about it. See if any information is available in the library or on the Internet about its popularity now or in the past. Inform your audience about it in your article, enthusiastically telling about its fashion appeal.
2. Choose three fashion articles from newspapers. Read them and write a brief report about what each one says. Then tell which one you liked best and explain why.
3. Design a newspaper advertisement for an apparel product that you like. Begin by gathering information about it. Clip a photograph or illustration from a magazine, newspaper, or catalog. Paste it onto a piece of heavy paper or cardboard. Add lettering for headlines as well as some advertising copy. Direct the feeling of it toward a certain audience you are trying to reach.
4. Pick a specific job in fashion promotion that interests you. Do further study about it in the library or on the Internet, with materials from your guidance counselor, and by talking to people who are in that field. Prepare an oral report about the job. Include the qualifications you would need to get the job, the type of company that might hire you, and the tasks you would perform. Estimate the salary range and fringe benefits you could expect, and find out what opportunities you might have for advancement.
5. Ask local modeling schools or agencies about their programs and policies. What modeling courses do they offer? What are the costs and requirements? What job placement efforts do they make for their models? Prepare a written report about your findings.

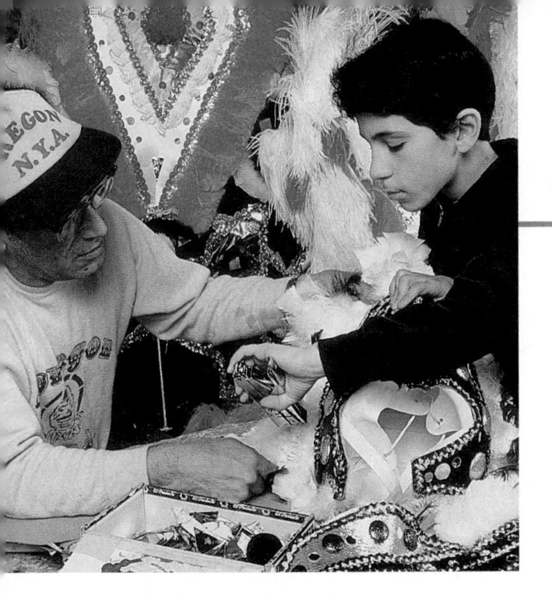

23 Other Careers and Entrepreneurial Opportunities

Fashion Terms

county extension agent
educational representative
home sewing industry
costume curator
theatrical costuming
entrepreneur
sole proprietorship
partnership
corporation
business plan
cottage industry
consignment
trading company
freelancing
consulting

After studying this chapter, you will be able to

- discuss careers in education and extension.
- list and describe careers in the home sewing industry.
- describe museum and library work as well as theatrical costuming.
- discuss careers in the clothing care field.
- explain how you could be an entrepreneur in an apparel-related business.

Many apparel-related careers fall outside the areas of the textile industry, apparel design and production, retailing, and fashion promotion. They have various duties and require a number of different personal qualifications. One of these careers may be especially suited to you.

Apparel Educators

Apparel educators give instruction in clothing and merchandising classes, extension work, and adult and consumer education courses. All these teaching jobs need people who are knowledgeable and have various amounts of education and work experience.

Classroom Teacher

Classroom teachers in the apparel area of Family and Consumer Sciences (FACS) are in charge of clothing classes at the junior and senior high school levels. They teach about textiles, fashion, grooming, clothing selection, consumer education, apparel care, clothing careers, and sewing construction. Teachers in vocational schools give training that can lead the students directly to gainful employment. They teach courses in commercial clothing construction, clothing alteration and repair, pattern making, modeling, art, and retailing skills.

Trade schools that offer programs in fashion design, illustration, retailing, apparel production, photography, and other fashion subjects need qualified teachers. Colleges and universities also need apparel instructors and professors. They teach courses in textiles, apparel design, market research, merchandising, family and consumer sciences education, fashion journalism, advertising, and others.

Teachers are professionals who have many roles. Besides instructing classes, apparel teachers demonstrate products and procedures. They must make purchasing decisions about textbooks, sewing machines, mannequins, and other equipment and supplies. When fashion concepts are taught, they might assume the role of designer, stylist, or fashion coordinator. The teacher has the opportunity to spread knowledge and enthusiasm to others, as in 23-1.

Teaching provides a routine with variety. Instructional plans follow a certain master curriculum. However, as fashions change and different students enroll in the classes, the specific content of the courses vary. Teachers learn continually as they read and study to stay up-to-date. They have a fair amount of flexibility and freedom in their work. They are needed in every state. They can work in small towns or large cities.

Most teaching jobs have good fringe benefits. Low group insurance rates and good retirement programs are common. Generous vacation times are scattered throughout the year. Teachers who have school-age children have time to spend with them. The formal teaching day may end in mid-afternoon, but teachers often have work correcting papers or preparing lessons during other hours.

College instructors have shorter teaching hours, but they have other duties in professional areas and are assigned to be advisors to certain students. Sometimes they conduct research or write for professional publications.

The pay for teachers is good, but not high. However, the great amount of vacation time is a bonus. Teachers can use that time to conduct personal business, enjoy hobbies, travel, or pursue further studies.

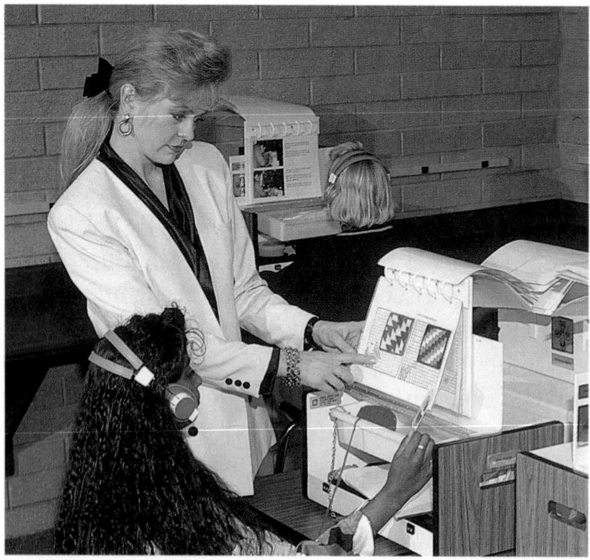

Chec Systems

23-1 Teachers share their knowledge with others in interesting, informative ways.

Qualifications to Be a Classroom Teacher

To be a teacher, you must like to work with people and help them. You need good communication skills to explain facts, give directions, demonstrate procedures, answer questions, and discuss ideas. You should be flexible, fair, and patient. You should be able to gracefully accept and give criticism.

Clothing teachers in junior and senior high school need at least college bachelor's degrees. Many have master's degrees and, therefore, receive higher salaries. Teachers in vocational schools should have education beyond high school. However, their work experience and job expertise are often more important. To teach at the college level, a master's degree is required. A doctorate is preferred and is rewarded with a higher salary and better opportunities for advancement.

If you think you might be interested in teaching, get involved now in school organizations such as Family, Career, and Community Leaders of America (FCCLA). This group highlights career opportunities and stresses leadership. Another helpful organization might be the Future Teachers of America (FTA). Test your teaching ability by doing volunteer work with young people. Evaluate how you feel and react in teaching-related activities.

County Extension Agent

County extension agents are hired and paid by state land grant universities under the U.S. Department of Agriculture. They serve both urban and rural areas. They teach apparel, family, and consumer subjects to groups such as 4-H clubs, community organizations, and individual families. They work informally in communities and in people's homes. Sometimes they hold classes or workshops. They help develop programs based on the needs and desires of the people in their counties.

Extension agents speak to groups, attend meetings, and plan and put on demonstrations. They write newspaper columns and prepare booklets and educational materials.

They prepare information for radio and television programs, and they often appear on the programs. They also spend lots of time traveling within their counties.

County extension agents' salaries are similar to those of teachers. They usually have longer work hours and shorter vacations than high school teachers. They have a great deal of freedom to plan and work on their own. They usually get deep satisfaction from working closely with the people in their communities.

Qualifications to Be a County Extension Agent

Teaching experience is excellent preparation for this work. Extension agents should be organized and able to plan their own work and the way they use their time. They should want to improve people's knowledge, skills, and lives. They must get along well with other people but also like to work alone. They must be resourceful, imaginative, and good at solving problems.

To be an extension agent, you should have the ability to present material in a clear and interesting way. You should have a pleasant personality and poise in making public appearances. You must have a genuine interest in family and consumer sciences and a willingness to learn and teach.

County extension agents must have college degrees in family and consumer sciences, with some courses in education. They need experience in the practical use of the skills they teach. They should be knowledgeable in communications, advertising, psychology, and art.

State and Federal Extension Work

Extension work is also done at state and federal levels. There is often a clothing specialist at the state level. This person is a member of a university staff and may teach some courses there. He or she also travels around the state to train and help county agents.

State and federal clothing specialists determine the needs and interests of people and coordinate all of the clothing programs. They prepare educational materials, as in 23-2, and give presentations. Another area of work is to analyze research data and report the findings.

23-2 Extension clothing specialists sometimes prepare materials for distribution to those who need the information.

Sometimes extension services must be coordinated with other public agencies.

Advanced degrees are needed for state and national extension careers. Clothing specialists must be able to communicate well through all media. They must meet, teach, and work with people easily. They must also be able to work independently and be organized with their tasks.

Adult Education

Adult education courses are usually held at night since most of the people who take the courses work during the day. Teaching these courses is often a part-time or extra job. It is usually done at a school or community center.

An apparel-related adult education teacher might instruct adults about pattern alteration, basic construction, tailoring, or sewing with new fabrics. He or she might teach clothing care, clothing repair and alteration, consumer skills, or other subjects.

Adult education teachers usually have developed the skills they teach through work experience. Some are high school or vocational teachers during the regular workday.

Consumer Education

Consumer education combines teaching with business. It is a full-time professional position with manufacturing firms. It can sometimes be done part-time with retail stores. Employees who do this work are often called ***educational representatives***.

Educational representatives promote their firms' products by teaching about them. They instruct dealers and consumers about sewing machines, patterns, notions, textiles, laundry equipment, or other products. They teach about their products at in-store classes and demonstrations, as in 23-3. They give presentations at trade shows, special seminars, and other events. They are important links between their companies and consumers.

Educational representatives for manufacturers assist dealers by helping with store events. They prepare the retail staff for special sales, sewing programs, and fashion shows. They answer consumers' questions.

When they are not out in the field, educational representatives work in the consumer education offices of their firms. There they plan the informative programs that are presented around the country. They also prepare educational and promotional materials about the use and care of their products.

A consumer educator employed by a retail store might teach sewing classes to the store's clientele, as in 23-4. Specialty classes might be taught before holidays. For instance, the sewing of Christmas gifts might be taught in a fabric store in November. In a yarn store, one might teach knitting, crocheting, or macramé to promote the store's products. The educators often have to prepare instructional leaflets and make samples to show, as in 23-5. Pay is either a flat fee per session or a percentage of the total fees collected. Sometimes the educator is given all the fees since the class brings customers into the store to buy supplies for the course.

Consumer educators should be well-groomed and well-mannered. They should have poise and a pleasing voice and personality. They should be creative, have good personal taste, and cooperate well with others. They need good communication skills, self-confidence in front of large groups, and loyalty to their firms.

Fabrications

23-3 Educational representatives from manufacturers often give demonstrations of their companies' products. This helps to sell products as well as to inform consumers about the use of the products.

Fabrications

23-4 Consumer educators at retail fabric stores hold classes to teach customers how to sew.

Nancy's Notions

23-5 Educators are more effective if they show samples that illustrate the points they are teaching.

The pay varies greatly according to job responsibilities and personal qualifications. A college degree in textiles and clothing, fashion, or a related field is needed to work for a manufacturer. A high level of skill may be the only requirement to teach at a store.

The Home Sewing Industry

Business firms in the ***home sewing industry*** deal with nonindustrial sewing machines, notions, fabrics, and patterns. There are books, magazines, radio and television shows, advertisements, and videotapes aimed at home sewers. Some mail-order firms cater to home sewers and professional dressmakers. They sell specialized items through magazine ads, direct-mail campaigns, and catalogs. All the businesses in the home sewing field need fashion-oriented personnel.

Employees of commercial pattern companies design, produce, package, and sell the patterns that are purchased by home sewers. These people combine their artistic talent and technical expertise as a team to create the patterns. Many of the duties of the jobs are similar to the related positions in apparel manufacturing and fashion promotion. The educational and personnel requirements are also similar. The office staff and top management jobs for all firms have similar duties and personal qualifications, too.

Commercial Pattern Development

The marketing department of major pattern companies collects consumer statistics. *Fashion information directors* gather and interpret the latest fashion information about silhouettes, colors, fabrics, and accessories. *Merchandising directors* figure out what the company's customers will want in patterns.

With this information in hand, *designers* create fashionable garments within the many pattern categories offered. They might specialize in such design areas as dresses, sportswear, or children's patterns. *Pattern makers* make heavy paper patterns for the parts of the new designs, as shown in 23-6. Then *garment makers* construct the muslin prototype and fashion fabric samples of each design. Finally, *fitting models* try on the samples to check fit, drape, and movability. They model for the design staff and members of management.

Design directors supervise the designers, pattern makers, and garment makers in producing the new pattern designs. Design

The McCall Pattern Company

23-6 Pattern makers work out every detail of each design in heavy paper pattern pieces.

directors are in charge of the initial designs. They also check the fit and approve the final garments.

Simultaneously, *fabric editors* obtain samples of the latest fabrics. They catalog, display, and supervise the samples in the company's fabric library. They also order fabrics and help "swatch" patterns for dressmaking. This means they recommend fabrics to be used for the sample of the design.

Accessory editors research and obtain the latest available styles in shoes, scarves, buttons, and belts, as in 23-7. They organize the items in the accessories room. The accessories are used to create finished ensembles for illustration and photography.

Pattern graders make larger and smaller versions of the pattern pieces for each design, with the help of a computer. *Checkers* look over patterns to see if notches line up and if facings match necklines or other garment edges. They also check to see that other cutting and sewing markings have been properly included.

Pattern Guide Sheet and Envelope Production

Technical writers create clear sewing directions that must be easy to read and follow. These writers must have journalism skills as well as knowledge of sewing construction, fabrics, notions, and patterns. They also need computer skills, since some standardized "how to" copy is programmed into computers. *Diagram artists* sketch the technical drawings to accompany the written directions of the guide sheet.

The McCall Pattern Company

23-7 This pattern company accessory editor is coordinating a belt with the design of a new sewing pattern.

Illustrators make a finished fashion drawing of each design. Garments are shown in the fabric selected by the designer. The artwork appears in the company's counter catalog and on the pattern envelope. *Markers* calculate pattern layouts and fabric requirements for the designs. These are done on computers, and the best versions are printed on the guide sheets.

Finally, *layout designers* mount the guide sheet copy and diagrams on layout sheets for printing. They also prepare the layouts for pattern envelopes. *Printing plant employees* produce the finished pattern pieces, guide sheets, and envelopes. Special machines fold the pattern pieces and insert them into the envelopes. The counter catalogs are also printed and bound at the printing plant.

Pattern Sales and Promotion

Sales and promotion employees include home office staff members and field staff people who call on accounts in geographic territories. *Retail coordinators* are fashion and promotional liaisons between pattern companies and retail fabric stores. They provide the retail stores with promotional help. *Educational representatives* prepare education materials and teaching aids such as booklets, posters, and filmstrips. They must have top construction skills, since they answer questions about sewing problems that customers ask them to solve.

Fabric Sales

Salespeople in fabric stores, as shown in 23-8, have similar duties to other retail salespeople. However, they must have a thorough understanding of fibers, fabric construction, finishes, care of textiles, and sewing procedures. They must be able to interpret information from pattern envelopes in order to advise customers. Today, some fabric stores are placing more emphasis on home decorating fabrics than apparel fabrics.

Subscription fabric club personnel work at a central mail-order office and warehouse. They do not deal face-to-face with customers. They buy and stock quantities of fabrics. They cut

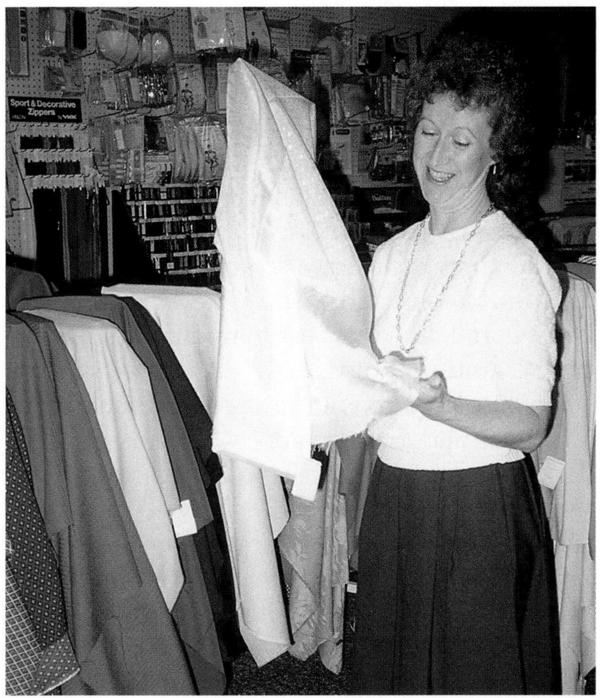
Fabrications

23-8 Retail salespeople who work in fabric stores give assistance to home sewing customers.

swatches and prepare periodic mailings to send out to their members. They design advertisements and place them in home sewing publications. They send bills and receive the yearly subscription fees.

Subscription fabric club personnel measure and cut the fabrics that are ordered. They also collect from their shelves the patterns, notions, and sewing tools that have been ordered, and send them to the customers. They receive merchandise returns from members who are not satisfied. They must be organized and good with details.

At-home fabric sales representatives sell fabrics from displays in their homes. They offer personal service, design and fabric advice, and one-stop shopping in a quiet atmosphere. Sometimes they put together total wardrobe packages of fabrics, patterns, and notions.

Video Demonstration Work

Home sewing videotapes are available with information on wardrobe planning, pattern alterations, and sewing techniques. Some give step-by-step directions in various sewing procedures for particular fabrics, garments, or notions. They go into more detail than most pattern guide sheet instructions. Some videotapes demonstrate beginning sewing skills. Others give instruction in advanced techniques and tailoring.

Sewing videotapes offer flexibility of use. Viewers can watch them whenever they have time. They also have the option of stopping and reviewing portions when needed.

Videotapes may be offered for rent or purchase through fabric stores, mail-order firms, schools, 4-H groups, or extension agents. Some can be viewed in retail stores.

To do video work, preproduction planning is very important. The right equipment and supplies must be on hand. Knowledge and experience in electronics and communications are needed. People appearing on the tapes must be experts in the subjects being discussed. They should seem pleasant and look fashionable. They also need self-confidence and good communication skills.

Textile and Clothing Historians

Fabrics and apparel of the past represent the living patterns of our ancestors. Such antique textiles and clothing give a special richness to the fine art collections of museums and libraries.

Costume curators as well as historians, scientists, and conservators deal with fabrics and apparel of the past. They locate, identify, and determine the age of textiles, apparel, and accessories from past cultures. Scanning electron microscopes are often used to determine the age and condition of fibers and stitching threads.

Costume curators and conservators carefully restore old textiles and garments. They repair broken and frayed areas. They also remove soil that could prematurely age and damage the fabrics, as in 23-9. Dust, dirt, and insects are harmful to old textiles.

Costume curators and conservators record their findings, as in 23-10. Then they store the clothing and textiles in places with proper temperature and humidity. They use "archival" acid-free paper products for storage. Curators also make sure the items are kept in darkness or under low light levels. Ultraviolet rays are screened out to minimize fading.

Costume curators and conservators work in museums to develop and care for historic costume collections. They prepare exhibits that display the historical garments as accurately as possible. They must know every detail of the apparel of past eras and other cultures. They often deal with inaugural gowns, wedding gowns, and other ceremonial garments because these special ones have been saved. Everyday clothes from long ago are quite rare, but they often tell a great deal more about what life was like in former times than special garments do.

Costume scholars also work for large city and university libraries. They collect and catalog old and new fashion drawings, clippings, slides, photos, films, and books. They often give lectures with slides depicting historic costumes and their influences on modern fashions and fabrics. They have access to detailed information that can date, describe, authenticate, and classify old apparel, as in 23-11. They help people find specific information by researching what has been written about various aspects of clothing. Sometimes they assist the wardrobe designer of an historic film.

23-9 Wet cleaning of antique silk damask fabric is done by trained conservators using special equipment.

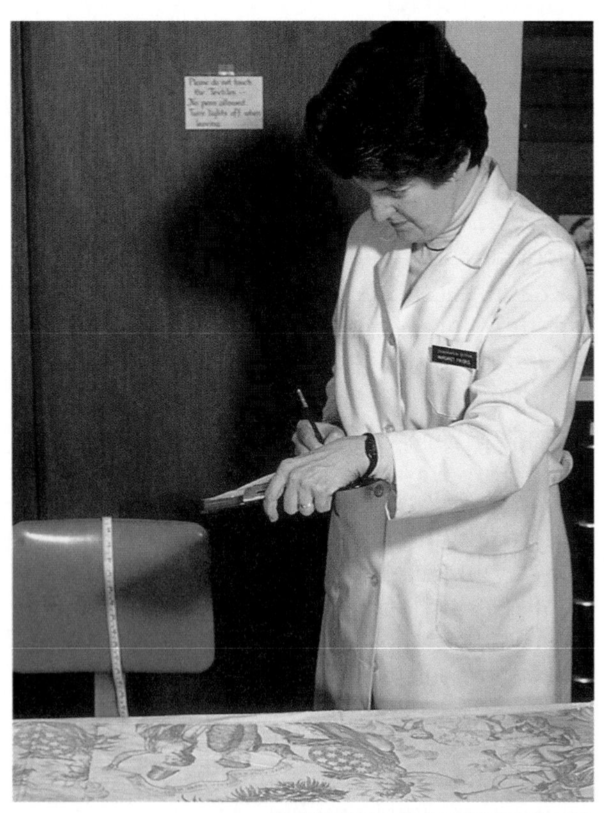

The Henry Francis Du Pont Winterthur Museum

23-10 After examining an historic textile, a "condition report" is written by the conservator.

Dept. of Consumer Studies, University of Delaware

23-11 The age and origin of historic garments can be determined by costume scholars.

Costume curators and conservators generally have college degrees in science, textiles and clothing, history, or art history. Many have advanced degrees. Apprenticeship training or graduation from an art conservation training program is recommended for this work. Also, one must have patience to do a job carefully. For library work, a background in library science is recommended. Interests in textiles, science, history, and fashion are needed.

Work with historic textiles and costumes can be stimulating and personally rewarding. The hours are sometimes long and are often spent doing independent work. The pay is medium. Professional esteem may be very high. Those with experience and expertise might also be instructors for others, such as apprentices going into this career.

John McGrail, National Geographic Society

23-12 Although theatrical costumers do not always create flamboyant costumes, such as this bright parade outfit, their work often involves interesting designs.

Theatrical Costumers

Theatrical costuming is done for opera, ballet, stage plays, circuses, movies, advertisements, television productions, and parades, as in 23-12. A costumer might work for a theater company, a movie studio, a costume shop, or a television network. Sometimes outside costume designers are hired to work with the wardrobing staff for certain shows.

Theatrical costumers work from scripts to learn about the types of activities for which outfits are needed. They should know the culture and time period being represented, the income level of the characters, and the mood or desired effect. Designers must create complete costumes that give the right look under various lighting conditions. They have to work within certain limitations, such as the size of the stage or screen. They have to coordinate the costumes with the props. All of this must be done within a specified budget.

There are several levels of responsibility and pay in theatrical costuming. A beginner may start as a *wardrobe helper*, sometimes called a "costume technician." Wardrobe helpers organize the costumes and accessories by character and scene. They do shopping and other footwork to collect everything that is needed. They help with research to make sure the designs are authentic. They help the actors and actresses dress for the production and care for the wardrobes afterward. They also do repairs on the costumes before, during, and after performances. All members of the wardrobe crew must work together as a team.

With experience, a wardrobe helper might eventually work up to become a *wardrobe designer*. This position as head of the costume department is rewarded by very high pay and great self-esteem. Recognition is given in the credit lines of a movie or the program of a stage production. However, as in other types of fashion designing, a very small percentage of people make it to the top. There are important contributions and enjoyable careers for those at all levels of theatrical costuming.

Theatrical costumers should have great flair and creative imagination. They must be able to work with emotional and artistic personalities. They need a thorough knowledge of lighting, staging, and special techniques. They also need a solid background in art, design, and history.

To do theatrical costuming, you would need fashion school training and an apprenticeship.

You would also need a wide range of skills, including sketching, pattern making, draping, and sewing. You might even have to glue or staple once in a while, as in 23-13. Membership in the United Service Artists of America would be desirable and is often required.

Clothing Care

Textile care and maintenance is a leading service industry. Most of these businesses are small and privately owned. There are many jobs in commercial laundries and dry cleaning establishments. Sometimes several jobs are combined for one person to do.

A high school diploma is needed for careers in clothing care. Many require training or experience under skilled workers in the field. Some vocational schools and the International Fabricare Institute train people in basic fabric knowledge, spot removal, and the use of clothing care equipment, as in 23-14.

In most clothing care jobs, you must have good eyesight and be able to work well with your hands. The work must be accurate, and you must be able to concentrate well as tasks are repeated. You should have a thorough knowledge of textiles and a good attitude to maintain high standards.

Dry Cleaning Businesses

With dry cleaners, a *checker* receives and returns the garments. This is a sales job with lots of customer contact. Checkers represent their companies to the public. They also handle payment, minor complaints, and telephone calls. They should have a neat appearance, a pleasing voice, and patience. They should be courteous, like people, and have knowledge of clothing and cash register procedures.

A *marker* puts identity tags on the clothes and sorts them. The garments are inspected for stains, damage, and items left in pockets. Trims or buttons are removed, if necessary, for the cleaning process. Garments are classified according to fiber content, color, and type. This determines how each will be cleaned. Some items are marked for special handling.

A *spotter* treats stains with proper spot removal methods. The fabric of the garment must not be damaged. He or she must understand fibers, finishes, dyes, and stain removal. The spotter should have an interest in fashion and some knowledge of clothing construction.

John McGrail, National Geographic Society

23-13 A proficiency at many different sewing and craft skills is needed to assemble professional looking costumes.

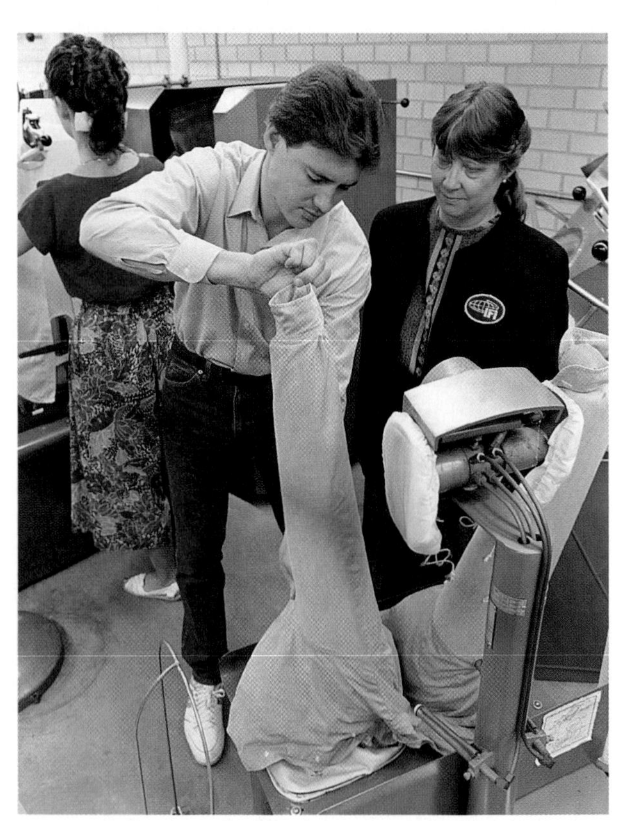

International Fabricare Institute

23-14 Specialized training is available for those who want to enter clothing care careers.

A *dry cleaner* is responsible for cleaning the garments. The garments are weighed to determine the poundage of each load. Then they are cleaned in either petroleum or synthetic solvents. The solvents must be kept pure and clear. Specialized cleaning equipment is used. Some clothes are tumble dried, and some are dried in the air.

To be a dry cleaner, you must have physical strength and mechanical ability. An understanding of chemistry and mathematics is helpful. You may start out as a helper and work up to this job.

A *washing operator* does laundry with wetwashing and drying equipment. The proper water level, temperature, and amounts of detergents and chemical agents must be selected for each load. Each load must be balanced to prevent equipment damage. To do this job, you must have physical strength and mechanical ability.

A *finisher* restores the cleaned garments to their original appearance and shape using various kinds of finishing equipment. Finishers soften fabrics and remove wrinkles by using steam and various types of presses in combination with hand irons. Skill is needed to handle the equipment on the many different garment designs and fabrics. The garments are then hung on hangers or folded neatly.

A *sewer* does any needed repairs or alterations. He or she might restitch seams and attach linings that have become loose. A sewer may also replace buttons, secure shoulder pads and trims, and put in hems. A person in this position must have complete knowledge of sewing construction.

An *inspector* examines the apparel after the cleaning and finishing processes. Garments are returned to the proper departments for correction if they do not meet the plant standards. Memos to the customers are sometimes attached regarding special garments. To be an inspector, you would have to understand the expected standards as well as the dry cleaning and laundry operations. You would need to have good judgment as well as tact.

An *assembler* gathers together all the pieces of a customer's order. He or she matches invoice descriptions and numbers with garment tags. An assembler's job must be done carefully and accurately so no items are lost or misplaced.

A *wrapper* (sometimes called a "bagger") packages the finished apparel to maintain the quality of cleaning or finishing. This person places garment bags over dry-cleaned items and wraps laundered clothes. He or she then attaches invoices to the packages that are placed on the rack ready to be returned to customers.

Commercial Laundries

A few jobs are available in self-service laundries. *Attendants* make change, sell laundry products, and assist customers with laundry problems and the use of machines. They should have a general knowledge of fibers, fabrics, and clothing. Attendants must be courteous and look clean and neat. They may have to work some evenings, holidays, and weekends.

Institutions, such as hospitals and hotels, employ *laundry specialists* to work in their on-site laundries. They maintain the linens, such as uniforms, aprons, sheets, towels, tablecloths, and napkins. Also, some outside laundry businesses specialize in commercial work and clean only items from institutions and large businesses. Laundry specialists need formal training or experience in an apparel care establishment.

Linen Supply Service

Linen supply services own linens and rent them to institutions and businesses. These types of firms have pick-up, laundering, and delivery services. Panel truck drivers are needed, as well as laundry equipment operators and maintenance workers. Office employees schedule pickups and deliveries, purchase new and replacement linens, and handle the financial accounts.

Entrepreneurs

Entrepreneurs are people who start their own businesses. More and more people want to work for themselves. As owners, they assume the risk and management of their enterprises. A business might be run from home, an office, or a store. It may focus on retail sales, provide a service, or manufacture a product, as in 23-15. The business might be set up as a ***sole proprietorship***, with only one owner. It could be a ***partnership***, with two or more owners. The business could also be a ***corporation***, which is a separate legal entity that gives the owner certain protections. A lawyer may be needed to set up the company.

Entrepreneurs should look at all the factors involved in starting and running a business. They should take business courses and specialized education classes to get all the preparation possible. They should have work experience in another firm that deals with the same type of product or service. They should analyze such questions as those in 23-16 before starting an entrepreneurial endeavor.

An entrepreneur should develop a ***business plan*** that defines the idea (purpose), operations, and financial forecast of the proposed company. The business plan will help the entrepreneur crystallize his or her thinking about the venture. Also, it is needed in order to borrow money from lending institutions.

Entrepreneurs should be able to buy supplies, keep financial records, and manage people. They need to set up business bank accounts and have business cards and letterhead stationery printed. They need to get special insurance and must understand taxes and government regulations. See 23-17. Information and guidance might be available from the local chamber of commerce or small business association.

Fashion Institute of Technology; John Senzer, Photographer

23-15 As an entrepreneur, you might want to design, make, and sell stuffed animals or other types of toys.

Self-employment is a big responsibility. Discipline is needed to schedule work and meet deadlines. Long hours must often be worked, and concerns about work usually linger after hours. Employees sometimes have to be hired and trained, and work must be delegated. Expenses must be kept down, while efforts are made to increase profits. Great satisfaction is received from a business venture that becomes a success. However, many small businesses fail each year because of poor planning, poor business skills, and lack of money for a solid start.

A Home-Based Business

More and more people are earning some or all their income from working at home. Some people use knitting machines to produce sweaters, or monogramming machines to

Some Questions to Ask Before Starting a Business

Does the business fill a gap or real need in the market?

Is the timing right, so the demand is growing?

Do I have enough education and experience in this type of work?

What are my strengths and weaknesses in this type of work?

Where can I get the extra expertise I need (courses, workshops, consultants, etc.)?

Do I have helpful industry contacts?

What is my target market and the image I want to project?

What are my promotional and pricing strategies?

Who are my competitors, and what are their strong and weak points?

Where will the business be located, and why should it be there?

How and where will I get the capital I need to finance the venture?

Are there organizations that I should join (chamber of commerce, trade associations, etc.)?

Am I willing to put in lots of hard work, possibly at the expense of my personal time?

23-16 These questions are important to answer as a person considers starting a business.

Regulations to Check Before Starting a Business

Licenses or Permits—State or local licenses are often needed to make or sell certain items or to provide certain services. Your local Economic Development Office or Department of Public Health should be able to help you with this.

Zoning Ordinances—In almost all communities, certain areas are designated for residential use (for homes), and other areas are designated for commercial use (for businesses). Most businesses cannot be located in a residential zone. Also, commercial zones are often further restricted to certain types of businesses, such as for offices, light industry, or heavy manufacturing. For zoning information, check with your local planning commission.

Sales Tax—Many states have a sales tax on certain categories of items that are sold directly to consumers. If your product or service is within one of the categories, you must collect the tax from your customers and pay it to the state. Check with your County Clerk's Office about sales tax procedures.

Labeling—If you plan to manufacture apparel items, even on a small scale, certain information must be included on permanent, sewn-in labels. The Federal Trade Commission can provide you with specific details about this.

Registering a Business Name—The name of your business can be registered if no one has already registered it for their use. This protects the name for your use only. Your County Clerk's Office can get you started on this procedure.

23-17 Local, state, and federal laws affect businesses. Owners must find out about official regulations and obey them or face penalties.

personalize jackets. Some write fashion articles for publications. Some specialize in making seasonal craft items, like the woman in 23-18.

A home handiwork business is called a ***cottage industry***. The labor force of a cottage industry consists of family units working in their homes with their own equipment. It can be just one person working at home or many people who are hired to work individually in their own homes. This reduces overhead and provides flexibility.

The Du Pont Company

23-18 Craft items made at home can be sold through various retail outlets in one locale. They may also be distributed nationally by specialty stores or mail-order houses.

Selling might be done directly from your home. Also, products might be made at home and sold on ***consignment***. In that case, you would place your goods for sale in a store. You would be paid a percentage of the retail price if and when the goods are sold. You might also consider renting a booth at fairs or shows to sell your products directly to consumers.

You may be able to sell products made at home to regular retail outlets. To do this, you must make appointments with buyers to meet at their convenience. You must make sales presentations, pack and unpack samples, and fill out order forms. This is time-consuming, often frustrating, and sometimes rewarding. You might even pay commissions to sales representatives to sell your products in various parts of the country. As a business grows, sales outlets and methods change. In all cases, you must supply items of consistent quality in the amounts needed after the orders have been taken.

Most of the time, home-earned income is lower than that earned from employment with an outside firm. However, there are many advantages. Home-based businesses allow you to combine family responsibilities with income enhancement. Income is earned without losing the flexibility and personal rewards that being at home offers.

There are some tax advantages related to home-based businesses. Check with an accountant or tax expert about home expenses that may be allocated to the business. Publications by the Internal Revenue Service or Small Business Administration are also helpful. Keep careful records of business-related travel, legal and professional fees, supplies, business publications, educational expenses, and so on.

Some trade schools and colleges offer courses to prepare people to have home-based businesses. Some have continuing education classes in the evenings for adults already out in the workforce. Students prepare financial, marketing, and feasibility studies to see if their business ideas are sound. They look at the realities of taxes, insurance, patents, liabilities, and investment capital. They also consider their personal values and lifestyles, to evaluate the effects of a home-based business on family members.

Owning Your Own Retail Store

Great amounts of textile and apparel products are distributed by specialty stores. Some are privately owned and successful for many years. However, apparel shop entrepreneurship is risky and has a high failure rate. Lack of working capital and incompetent management are often the reasons for failure. There is sometimes too much optimism combined with too little experience and knowledge. To open your own retail shop, you first need a sound background in the business.

As a general rule, if a store can get through the first two years, the chances for survival are good. The rewards with a successful shop are many. It can be wonderful to be your own boss. You can decide where the shop is located, what image it has, and how it is run.

There are various ways to establish ownership of a retail shop. You can buy a business that is already in operation or buy into a partnership in it. This takes some negotiating after the assets and profitability of the store have been evaluated. Another way is to start a retail business from scratch. A store facility must be rented or bought, merchandise must be ordered, advertising must announce the store, and employees may have to be hired. Another way to have your own retail establishment is to buy a franchise. This has a lower rate of failure, since the parent firm gives some guidance and protection. In all cases, a capital loan will probably have to be acquired from a bank or other source for start-up costs.

Small shop owners often employ only one or two others to help operate the store. They perform all the retail functions of buying, stock control, promotion, and direct selling. The owner must locate sources, price the goods, and offer credit and personal services to customers. He or she probably does all the paperwork, arranges displays, and possibly even sweeps the floor! There may be low or uneven income for the owner until the store becomes well-established with satisfied customers.

Owners of retail shops must know their target market and their competition. They must carry goods that meet a need or are in demand. They must cater to their market segment of customers with the proper merchandise and image. They should have good relations with people in the community.

Independent store owners need expertise in many areas, as outlined in 23-19. They have to be able to maintain and analyze business records. They must be able to interpret financial data in order to improve business performance. They have to update their inventory control systems when needed. They must know how to develop seasonal merchandise plans. They must know how to control costs and build sales, so the "bottom line" shows a profit.

A Dressmaking or Tailoring Shop

Professional *dressmakers* and *tailors* are expert sewers who do custom sewing, alterations, and repairs for others. Some people do the work in their homes. Others have storefront operations.

Custom sewers make garments for individual clients. Sometimes they specialize by doing only wedding gowns or tailored suits. They help the customer choose a flattering style and fabric. They take the person's measurements. They construct the garment, check fit, and include all finishing touches.

Areas of Expertise Needed to Own a Small Retail Shop

- How to maximize your buying dollars to get appropriate quantities, assortments, and styles for your target customers
- How to merchandise your store according to the "fashion calendar"
- How to maintain proper inventory levels and overcome slow or nondeliveries of merchandise
- How to calculate open-to-buy and stock-to-sales ratios
- How to build and keep good credit and financial health
- How to establish and maintain good relationships with suppliers
- How to maximize an image of high customer service

23-19 These areas of expertise, and others, are needed for owners of apparel retail stores. They can be gained through reading, attending courses and seminars, and having work experience in similar businesses.

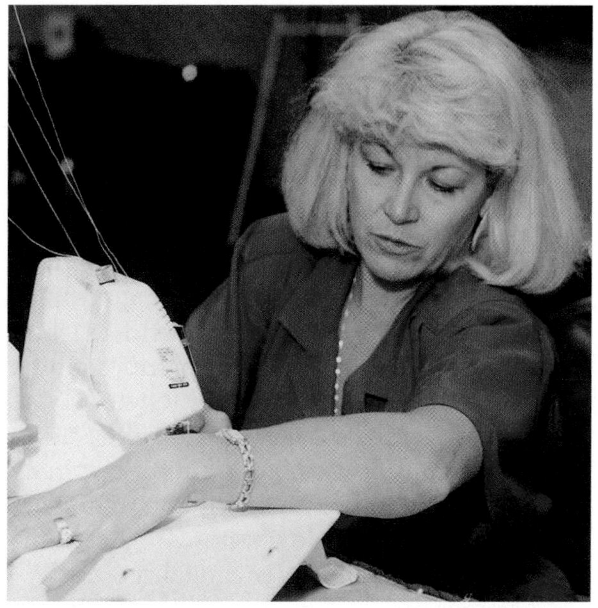

Bernina of America, Inc.

23-20 To have a successful dressmaking or tailoring shop, you must take your work seriously to do an expert job on each project.

Dressmaking is much more than following a printed pattern. Many dressmakers and tailors build a clientele on their ability to create a garment according to customer specifications. Sometimes they make apparel with just a sketch or photograph to follow. To do this, they must understand form, proportion, fit, and color. They must also be good at pattern making and draping.

Dressmakers and tailors must have a keen interest in clothes and enjoy sewing, as in 23-20. They must be experts at all the skills of construction. They should know professional techniques and shortcuts. With experience, they develop speed and accuracy. Good eye-hand coordination is needed, as well as finger dexterity and a fashion sense.

Dressmakers and tailors must have a knowledge of what types of fabrics are suited to various garment designs. They must be neat, accurate, and patient. They must be tactful with people and work well under pressure.

Dressmakers and tailors sometimes master this trade through apprenticeships. However, vocational or trade school courses in fashion design, pattern making, construction, and textiles are helpful. The best preparation is to study textiles and clothing in college with business courses taken as electives. Sometimes extension clothing specialists or adult education courses offer "sewing for profit" workshops.

Special books and magazines are also available. They tell you the pitfalls to avoid, the best ways to begin, and how to make your enterprise grow. They give helpful advice and keep you up-to-date. Your local fabric store should carry them or know the addresses so you can subscribe. You might also want to join associations for dressmakers and tailors. One example is the Custom Tailors and Designers Association of America, described online at www.ctda.com.

A dressmaking or tailoring shop can be started with a relatively small investment. It has good potential for growth. With so many people employed outside the home, apparel services that offer quality work and proper fit are in demand. Dressmakers and alterations experts may also do work for department stores, specialty clothing shops, and dry cleaning establishments.

An Apparel Production Business

Despite the existence of many giant garment firms, the apparel manufacturing industry has an abundance of small, privately-owned businesses. It is possible for enterprising and creative people to set up businesses and prosper. Some capital is needed for office space and to purchase fabric. The cutting and sewing can be sent out to contractors.

To succeed, small apparel manufacturers must produce styles that will find acceptance. They have as good a chance to create popular designs as large firms. In fact, they often have the advantage of being able to act quickly to respond to sudden fashion shifts. They buy smaller amounts of fabrics and produce only a few designs in small lots. They often set the fashion pace for the industry because they create more "fashion-forward" or "future-inspired" designs. Profits can be high with the right ingenuity. However, a single season with a bad line can wipe out a small, under-capitalized firm.

Apparel production entrepreneurs should become personally acquainted with store buyers, fashion editors, suppliers, and contractors. Personal appearances in stores that carry the products can help promote the line. Such visits also give the entrepreneurs customer feedback for improvements or changes. "Plugs" in fashion articles can be a big boost.

Small business centers and trade schools offer courses in how to start up apparel production businesses. Students are often people with design talent and experience with large firms who want to go out on their own. In the courses, students learn how to develop a sales force, manage personnel, and work with contractors. They also learn various finance methods relating to sales and production. Workshops are taught by industry leaders and consultants, with guest speakers in specific fields of expertise.

At the contracting end of the business, capital is required for machinery, rent, utilities, and employees' wages. Sometimes there is demand for a specialized type of production, such as sewing appliqués onto garments. In such a case, a "niche" can be filled very successfully by a manufacturing entrepreneur.

A Trading Company

A ***trading company*** is an import/export business. Trading companies contract for, or buy, merchandise from other countries. They arrange for shipments of goods to the United States, as in 23-21. They sell the merchandise to domestic retailers who, in turn, sell it to consumers. Most trading companies specialize in a particular product category, such as silk apparel from Hong Kong or hand-knit items from Scotland.

National Cotton Council

23-21 Goods that trading companies have bought or sold in other parts of the world are shipped into and out of the United States on container ships.

A trading company may be run by one or two people who have some warehouse space and a small office. To start this type of a business, product line expertise is needed. Contacts are needed with overseas suppliers and domestic retail accounts. A great deal of traveling has to be done. The costs of the travel are business expenses.

A Mail-Order or Internet Business

Many people find a market through *mail-order* or Web site sales for the apparel-related products they produce. Such products include specialized patterns, supplies, manufactured garments, kits to be made into apparel, and "how-to" books for home sewers. Mail-order businesses advertise in publications that are read by consumers who can use what they have to offer. Sometimes they send out advertising materials to a targeted mailing list of names. However, the postage and other costs for this add up quickly. Other times they supply their products to an existing mail-order house or Internet site that reaches their target audience. In that case, they get about 40 percent of the retail selling price.

Inventory planning for a mail-order or Internet sales business can be tricky. There is no sure way of knowing how many orders will be received. You could be swamped with orders you can't fill fast enough, or, if you build up a large inventory anticipating heavy orders that don't materialize, you could be stuck with unsold goods.

A small mail-order business can be run from home. Basement or garage space can be used as a warehouse. Additional space and equipment are needed to produce whatever is being sold. Packing materials and office supplies are also required. A post office box is usually rented and used as the address for the business.

Mail-order is a relatively simple way to start a business. It requires organizational abilities and a willingness to work hard. It also requires a talent for selecting products that people will want badly enough to fill out a form and send a check or make a call. A toll-free phone number may be needed. The business can be started on a part-time basis until it grows into a full-time job. The owner can dress as desired and work whatever hours are convenient.

Products sold through mail-order and Internet marketing should be fairly lightweight and easy to package. They should be easy to describe technically and creatively in words. The market for the products must be identifiable and easy to reach. The prices must attract customers while giving you a realistic profit for your time and effort.

Owners of mail-order and Internet businesses must obey Federal Trade Commission and post office regulations. Products and prices cannot be misrepresented in advertisements. Items must have proper labeling. All products must be safe for consumers to use. Also all orders must be filled within 30 days of receipt. If this cannot be done, the customers must be contacted. They must be given the opportunity to cancel the order with a refund or to accept later shipment.

Freelancing and Consulting

Freelancing is the selling of expert skills. ***Consulting*** is the selling of expert ideas and advice. Both are service businesses that offer help to companies or consumers. They sell expertise rather than products. A great deal of successful experience is needed to establish credentials as an expert.

Freelancing and consulting can be done to serve any of the apparel industry segments. (Previous chapters have mentioned freelance work in fashion design, illustration, and photography.) These positions can also serve consumers. For instance, *color analysis consultants* find the best colors for specific clients. They guide clients in the color coordination of their apparel and cosmetics, as in 23-22. *Wardrobe consultants* show consumers how to combine fashion items, and help them plan and manage their wardrobes and purchases. They are especially helpful in establishing "personal image" programs for busy career people.

Freelancing and consulting jobs generally have few inventory requirements. Advertising materials are usually distributed to announce and promote the service. The work can be done full-time or part-time. Sometimes it is started during free time from other employment until enough contacts have been made to make it a steady career.

Evening courses offer instruction in establishing freelancing or consulting businesses. They teach people how to promote themselves and their talents, develop client lists, and minimize financial risks. Methods of building personally and financially rewarding businesses are discussed.

BeautiControl

23-22 Color analysis consultants are trained to determine what colors are the most flattering for individuals' apparel and cosmetics.

Chapter Review

Summary

There are fashion careers outside of textiles, apparel, retailing, and promotion, such as apparel educators. Classroom teachers instruct students about fashion, grooming, clothing selection, sewing, and other apparel subjects. Teachers should have good communication skills, flexibility, patience, and the desire to help people. County extension agents, with state and land grant universities, develop educational programs based on the needs and desires of people in their counties. Extension agents should be organized, resourceful, and work well independently. State and federal clothing specialists coordinate apparel extension work.

Adult education teachers teach courses in the evenings, usually at schools or community centers, to local residents who sign up. Consumer educational representatives teach about and promote products of manufacturing firms for whom they work.

The home sewing industry involves nonindustrial sewing machines, patterns, fabrics, and notions. Employees in commercial pattern development include fashion information directors, merchandising directors, designers, pattern makers, garment makers, fitting models, design directors, fabric editors, accessory editors, pattern graders, and checkers. Pattern guide sheet and envelope production workers include technical writers, diagram artists, illustrators, markers, layout designers, and printing plant employees. Coordinators with retail stores and educational representatives who prepare teaching aids are sales and promotion employees. The home sewing industry also has retail fabric salespeople and video demonstration employees.

Costume curators, as well as historians, scientists, and conservators, locate and identify old apparel, record their findings, and do repairs and restoration work. Theatrical costumers work for theatre companies, movie studios, costume shops, and television networks as wardrobe helpers.

Clothing care is a service industry with jobs in commercial laundries and dry cleaning firms. Workers for dry cleaners include checker, marker, spotter, dry cleaner, washing operator, finisher, sewer, inspector, assembler, and wrapper/bagger. Commercial laundries have attendants and laundry specialists. Linen supply services own linens that they pick up, launder, and deliver to institutions and businesses.

Entrepreneurs are self-employed, and they assume the risk and management of their enterprises. A home-based business can be a cottage industry, sell items on consignment, or earn money from home another way. Owning your own apparel retail store has a high risk of failure, but many specialty stores are independently owned. A dressmaking or tailoring shop does custom sewing, alterations, and repairs for others. An apparel production business can be started with good designs, with the cutting sewing sent out to contractors.

A trading company buys merchandise from other countries and sells it to domestic retailers. A mail-order or Internet business takes orders via mail, telephone, or a Web site and ships items to the customers. Freelancing and consulting are service businesses that sell expert skills or expert ideas and advice to those who pay for the service.

Fashion in Review

Write your responses on a separate sheet of paper.

True/False: Write true or false for each of the following statements.

1. Work experience and job expertise are more important than higher academic degrees for teachers in vocational schools.
2. County extension agents work only in rural areas.
3. Technical writers create clear sewing directions for commercial patterns.
4. Subscription fabric clubs are mail-order operations that sell fabrics and other sewing supplies to consumers.

5. Video demonstration work is sometimes done for a home sewing audience.
6. Costume curators locate, identify, and determine the age of textiles, apparel, and accessories from past cultures.
7. Wardrobe helpers are consultants who do color analysis and "personal image" wardrobe planning.
8. Linen supply services own linens and rent them to institutions and businesses.
9. Entrepreneurs are people who start their own businesses and assume the risk and management of their enterprises.
10. Sole proprietorships are small manufacturing firms that make and sell shoe parts in the footwear industry.
11. Selling something by consignment is when you are paid to sign over the ownership of your product idea to a manufacturing firm.
12. Small apparel manufacturers often have the advantage over large firms to create popular designs because they can act quickly to respond to sudden fashion shifts.

Multiple Choice: Write the letter of the best response to each of the following statements.

13. Which of the following is NOT involved in commercial pattern development?
    A. Design director.
    B. Fabric editor.
    C. Accessory editor.
    D. Wardrobe designer.
14. Which of the following is NOT involved in the dry cleaning business?
    A. Finisher.
    B. Retail coordinator.
    C. Inspector.
    D. Assembler.
15. Trading companies _____.
    A. employ extension agents to expand their trading territories
    B. are import-export businesses
    C. are firms that trade (barter) products or services with other companies rather than deal in cash or credit
    D. take in merchandise to sell on consignment

Fashion in Action

1. Pick one of the jobs described in this chapter. Do further study about it in the library, on the Internet, with materials from your guidance counselor, and/or by talking to people who are in that field. Prepare an oral report about the job. Include the qualifications you would need to get the job and the type of company for which you would work. Describe what tasks you would perform on the job, the salary range and fringe benefits you could expect, and what opportunities you might have for advancement.
2. Try being a clothing teacher. Prepare a plan for a 10-minute lesson to be taught to a clothing class. Study the subject and prepare any charts, handouts, or samples that would be needed. Then do the actual teaching to your class.
3. Try being an educational representative. Pick an apparel-related product to demonstrate. Give an educational talk and demonstration to your class. Be prepared to answer audience questions about the product.
4. Using an old commercial pattern that will no longer be used, take the contents out of the envelope. Mount pattern pieces, guide sheet pages, and the envelope on poster board as a visual aid for an oral report. Describe the jobs of at least six of the employees who contributed to the design, production, packaging, and selling of the pattern. Ask classmates if they can describe the jobs of any other employees who contributed to the finished pattern.
5. Develop a plan to start an apparel-related mail-order business. Describe the product and explain how you would produce it. What supplies would you need? How would you package it to be sent to customers? Create a sample advertisement. Identify your target market and the publications in which you would place your ad. Check with the post office about how you would sign up for a post office box and what the charges would be. Also check on postage rates for the size and weight of the packages you would be shipping. Write a report about your findings.

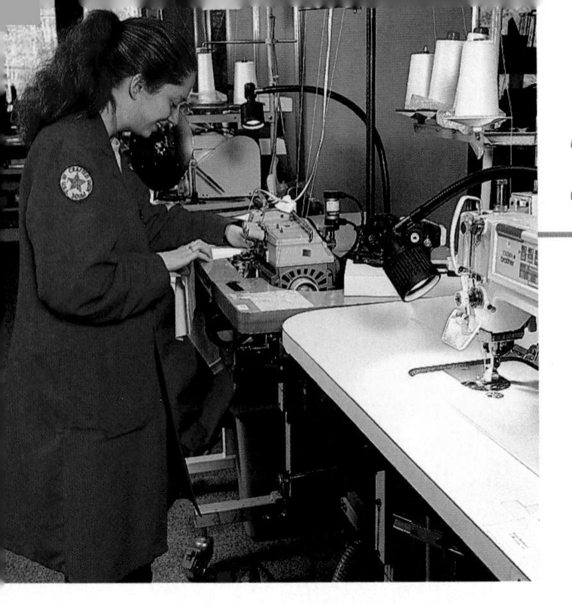

24 A Career for You

Fashion Terms

salary	employment agency
fringe benefits	resume
aptitude	cover letter
extracurricular	job interview
work-study program	references
placement office	application form
transcript	mentor
networking	

After studying this chapter, you will be able to
- evaluate careers that might interest you.
- explain how to find out about yourself, specific careers, and educational resources.
- discuss how to prepare a resume and how to apply and interview for jobs.
- explain what is needed to become a success.

After studying the apparel-related careers described in the previous chapters, it is now time for you to think about choosing a career. With planning, preparation, and dedication, you should be able to find a type of work that suits you. You should be able to enjoy the work and achieve success in it.

Vast Opportunities

Fashion is everywhere! Those who know the business and have talent, ambition, and ability can succeed in a wide variety of jobs.

Fashion offers some of the world's most fascinating, challenging, high-paying, and exciting careers, as in 24-1. Maybe you have dreamed of becoming the president of a large store chain, a famous designer, or a top fashion model. Only a few people can achieve, or really want, these positions. They are obtained with lots of hard work, strong ambition, personal sacrifices, and skill.

Good fashion jobs are not easy to get. Thus, why do so many people seek employment in this field? With such variety of opportunities, there is something for everyone. Many like the excitement of being part of a dynamic and fast-moving world. With creative ideas and good technical skills, hardworking employees can move ahead and gain recognition.

Varying Rewards

The personal and financial rewards of fashion careers are as varied as the jobs themselves. Some require worldwide travel. Others permit you to stay near home. Some pay top dollar, while others provide a meager livelihood. Some apparel employees are salaried. Some are paid commissions, and others are paid by the hour. Taxes and other deductions are taken out of employees' paychecks. Also, some jobs have better fringe benefits and employee protection than others.

A ***salary*** is a fixed amount of pay, usually received once or twice a month, for services performed on the job. A salary is determined by the company's need for a person's services, education, and work experience. Most jobs requiring college degrees are salaried. They are considered professional positions.

Italian Trade Commission/Gianni Versace

24-1 The fashion world offers glamorous jobs such as modeling.

A salaried employee does not receive overtime pay for extra hours worked. On the other hand, pay is not deducted if the employee takes a long lunch break or leaves work early for a personal reason. Thus, a salary is the amount of money paid for a particular person to do a particular job.

Some employees receive an *hourly wage*. Their wage is the amount of pay for each hour they spend doing the job. "Wage-role" people usually punch in and out on a time clock. If they work overtime, they are paid an extra amount. In fact, overtime is usually paid at one-and-a-half times the regular hourly wage. Work on Sundays and holidays is sometimes paid at two times the regular rate. On the other hand, wage earners who work fewer hours than normal during a pay period are paid only for the time they worked. Wage earning jobs usually do not require higher education.

A *commission* is a percentage of the dollar amount of products sold by an employee. Commissions serve as incentives for sales employees to sell as much as possible. The more a person sells, the more he or she earns. Often a small "base" salary or hourly wage is paid, with a commission added above it.

Fringe benefits are additional inclusions, such as disability insurance, life insurance, medical programs, and pension plans. They may also include paid vacation time, sick leave, bonus payments, and discount privileges for merchandise purchased. Perquisites ("perks") are extras, such as personal parking spaces or first-class airline tickets on business trips.

Employee protection is provided by government legislation, labor unions, and private agencies. The Equal Employment Opportunity Act, Fair Labor Standards Act, and Equal Pay Act are a few of the laws that protect employees. These laws are explained in 24-2.

In general, apparel industry occupations offer less job security than many other businesses. Employees tend to change jobs frequently because of the changeability and seasonal nature of much of the industry. Few jobs offer a typical nine-to-five workday. Production schedules or other deadlines must be met at peak times. Also, retail stores must be staffed when other businesses are closed. Freelancers, consultants, and other self-employed people must provide their own benefits, such as health insurance and retirement accounts.

Laws That Provide Employee Protection

Equal Employment Opportunity Act—Prohibits employers from discriminating against any employee or job applicant because of race, color, religion, national origin, sex, physical or mental disability, or age.

Fair Labor Standards Act—Sets the national minimum wage and overtime pay practices.

Equal Pay Act—Requires that men and women be paid the same amount for doing the same jobs under the same conditions.

24-2 The intent of these laws is to provide fairness to those being employed.

Choosing a Career

Preparation for your career should include careful planning and study. Your career decision is very important. It will affect your self-concept and self-esteem. It will affect how others view you. It will determine where you will live, what hours you will have for work and relaxation, and how much money you will make. It will also influence what friends and social activities you will have. Your career choice will affect all aspects of your adult life.

You will probably work for 40 years or more, which is a major portion of your life. It is your duty to yourself to choose a career carefully. This will ensure that your work years will be exciting and satisfying. Don't assume that a job has to be dull or routine. The most satisfying career for you will have work activities that you enjoy and will give you the lifestyle that you seek. It can be a source of continued learning and pleasure from the day you begin.

To choose a career that is right for you, first get to know yourself. Then learn as much as possible about various fields of work. With an understanding of your own interests and abilities, and information about various career paths, you can make the best possible career choices. You can set your personal goals and aim for them. You can get the education and develop the skills needed for entry-level jobs in the area that seems most satisfying to you.

Find out About Yourself

A sincere self-evaluation is needed to determine the kind of work you can do and would like to do. One way to do this is to take tests that determine your aptitudes. Your ***aptitudes*** are your natural talents. They can help you predict your suitability to different jobs. Aptitude tests are often given in schools. They can be arranged through your school vocational counselor. The results indicate the types of work in which you would probably excel.

For instance, suppose you had an aptitude for writing. You could probably be successful in the fields of fashion or retail advertising, fashion journalism, public relations, and audiovisual work. With an aptitude for science, you might develop new textile fibers and finishes or do product planning. With a natural talent for working with people, you might want a career in sales, education, or public relations.

A person with a well-proportioned body and a photogenic face might choose to be a model, as in 24-3. Creative or artistically talented people do well in design, display, advertising, photography, and illustration. Some people have aptitudes for logical thinking and working with numbers. They may be successful in managing retail units, planning factory production, or doing market research.

Begin to pay attention to your job-related interests. What are your likes and dislikes? Do you like to work alone or with others? Do you like to travel or work at the same place every day? Would you want to live in a large city, in a small town, or near a university? Do you like to do scientific experiments? Do you like to sew, draw, sell, or write?

Do your interests seem to revolve around products, people, or data? Do you like organized activities, such as office work? Do you

J. Jill

24-3 This woman has natural attributes that make it possible for her to be a fashion model.

prefer repeated activities, such as factory work? Do you enjoy a variety of changing activities, such as sales or administrative work? Do you like calm times, or do you handle pressure well?

You will make interesting discoveries as you "take stock" of yourself. Think about your specific accomplishments. Take the time to think seriously about times, events, or jobs that may have provided you with good learning experiences. Make a chart, like the sample chart in 24-4, listing your accomplishments in school, at work, and during leisure activities.

Next, think about the skills or abilities you used to achieve your accomplishments. What did you like or dislike in each situation? Fill in your chart with skills and interests that correspond to the accomplishments. This will give a picture of you as a working individual so far, even if you haven't had actual jobs.

Find out About Careers

To make wise and realistic choices concerning your future vocation, you must also investigate specific careers. You should learn about the job possibilities that exist within different career fields. You should understand the personal qualifications and educational requirements needed for the jobs that interest you. You must plan ahead to find the most meaningful career for *you*.

Read books that describe various careers. Review sections of this book that describe individual jobs that especially interest you. Use your school's guidance or vocational counselor as a resource. Check what information is available in the library, on the Internet, or on file at nearby career centers. Go through the list shown in 24-5 as you study the pros and cons of various careers.

Talk with people in the careers that interest you. Ask about jobs, the needed education and training, and what advancement you could expect. Check with your state employment office. Write for information from trade associations and organizations in the fields you are considering.

Finally, try some jobs to test your career interests. Work experience while you are still in school can help you decide what you might

Self-Evaluation		
Accomplishments	**Skills**	**Interests**
Got highest grade in summer computer word processing course at nearby community college.	Ability to type and use computer in productive ways.	Liked challenge and achievement of learning more uses of computers.
Scout bazaar fund-raiser: In charge of receiving goods and distributing them to appropriate selling areas.	Ability to organize and carry out plan while cooperating with others. Ability to direct others and communicate clearly.	Liked planning floor layout and working with people. Disliked record keeping.
Painted neighbor's house during spring break of junior year.	Ability to organize my time and stick to a task until finished.	Liked the money and independent schedule. Liked sense of achievement when finished. Disliked outdoor labor.
Clerk at local grocery/deli during summer vacation.	Learned to maintain stock on shelves. Learned to operate cash register.	Liked dealing with customers. Disliked filling shelves unattractively due to space limitations. Would like to be more creative.
Varsity soccer team won conference championship.	Ability to work well with others to achieve a common goal.	Liked team spirit and competition.

24-4 As you have more jobs and experiences, your lists of accomplishments and skills will grow. Your interests will become more clearly defined.

want to do. Perhaps you could find a job during evenings, weekends, or summers.

Find out About the Needed Preparation

Some apparel industry jobs need no advanced education beyond high school. Some require vocational or technical training. *Certificate courses* in trade schools take one, two, or three years to complete. Two-year college or trade school *associate degrees* are available in such fields as merchandising, fashion design, apparel production management, and apparel production engineering.

Four-year college *bachelor degrees* (bachelor of arts, bachelor of science) are also available in apparel-related curriculums. Majors include textile science, fashion design, merchandising, and other textiles and clothing specializations. These are offered by colleges of art or family and consumer sciences within universities.

A broader education is received at colleges than at trade schools. Some say this offers more career choices once the graduates are out in the working world. More complicated careers may require advanced degrees. For instance, a *masters degree* or a *doctorate* may be needed to conduct research, as in 24-6.

Admission requirements for most colleges specify a high school diploma or its equivalent. Some require certain scores on SAT or ACT standardized tests. One or the other of these tests is usually taken during a student's junior or senior year of high school. Some colleges require a written essay, an interview, and a personal recommendation. Trade schools may require either an art portfolio or a home exam for art and design majors. A home exam enables those with talent and an interest in fashion to be accepted without a portfolio.

The more creative and challenging career positions, and those with potential for career advancement, are all highly competitive. Higher education is desirable to land such jobs. The prospects of advancement depend on one's skills and personality, in addition to one's educational background. For instance, if you have sewn for years and then study fashion merchandising, your sewing skills may put you ahead when you are employed.

Strive Toward a Goal

Achieving success is easier when your efforts are focused on a goal. Therefore, it is to your advantage to set a career goal. However, beware of being too specific. Remember that a

Aspects of Various Careers

To choose a career, know the following aspects of jobs in that career path:

- duties and responsibilities
- training requirements
- compensation (pay, fringe benefits, etc.)
- location
- working conditions
- job security
- advancement opportunities
- status and prestige

24-5 When you choose a career, consider the entire career path. You may have to take a job or two that you don't enjoy in order to reach challenging positions with good pay.

U.S. Borax & Chemical Corporation

24-6 Research and development of products, such as textile fibers or laundry cleaning agents, often requires postgraduate study for advanced degrees.

fashion career could open up anywhere along the apparel channel of distribution, from raw materials to the final consumer purchase.

Employers often consider an applicant's ***extracurricular*** activities, such as clubs and sports, as well as academic records when they hire personnel. Activities, especially those with leadership roles, can prepare students to deal with challenges in work situations. To succeed in the working world, employers know that people must have "life skills" as well as training in technical areas.

To find out about schools that offer training and education in fashion, consult your school guidance counselor or a librarian. Several handbooks are available that index schools in many ways. They describe each school's curriculum, size, costs, student body, housing, and financial aid programs. They also give the address of each school's admissions office.

You might want to talk about schools with people who work in apparel careers that interest you. They can tell you which have the best reputations and the best results in placing their graduates into good jobs.

Pursue the Needed Education and Training

After deciding which schools interest you, contact the admissions office of each one. You will be sent a catalog and any other information you need to know about the school and how to apply.

Some universities that are not near fashion centers have *reciprocal agreements* with schools in such cities as New York, Paris, Rome, and London. Students go there for a semester, year, or some other recommended length of time. They see and learn about the actual workings of most aspects of the apparel industries. They receive credits that transfer to their original institutions.

Many schools offer ***work-study programs***, 24-7. They are often called co-op programs at lower levels of education and internships at higher levels. In some programs, students are placed in temporary jobs that alternate with their semesters of academic studies. In others, students work two days a week while taking a reduced course load, or spend a summer on a particular job. They gain valuable work experience, learn about different apparel jobs, and make important contacts in the industry.

Textile/Clothing Technology Corporation

24-7 Work-study programs allow students to combine education with related work experience. For instance, apparel production management students might spend a summer at this industry technology center.

Work-study programs make the transition from the classroom to a full-time career position easier. The realities and responsibilities of a job are already recognized.

For those without degrees of higher education, clothing-related businesses often have *on-the-job training programs*. A new employee must learn how to do a particular job. He or she might spend a day, a week, or longer working with an experienced employee before being asked to work independently.

Some companies have *apprentice programs*. The apprentice spends many months learning all aspects of a trade by practical experience under skilled workers.

Landing That Job

The first step in career development is to choose a field of employment that interests you and for which you are qualified. The next step is to enter the field by getting a job. It may not be easy to find the job of your choice. You must take this task seriously and give it your best effort. It can mean the difference between a rewarding lifetime career and an uninteresting job that you must go to each day just to make a living.

Job Hunting

Job hunting can take several different paths. First, let as many people as possible know that you are looking for a job. Ask your friends if there are job openings where they work. Stores and factories sometimes post job openings in a window, on a bulletin board, or near the entrance as in 24-8. Accept help and advice from your parents and their friends. The guidance or vocational counselor of your school may also have business and industry contacts. Follow up promptly on all leads.

Most institutions of learning have a ***placement office*** that arranges contacts for jobs. They want their graduates to find work because that makes them look good which, in turn, attracts future students. Take full advantage of their services. They may help you prepare a letter of application and a resume.

24-8 Retail stores sometimes put a sign at their entrance to entice shoppers to apply for a sales job while they are in the store.

They will provide records ***(transcripts)*** of your academic grades. They will also help you select portfolio items that show your skills if you are applying for design or illustration jobs.

A professional-looking *portfolio* should contain a collection of your best and most creative work. It should show a balanced representation of your current styles of art and abilities. Your work should be mounted neatly. It should be in a folder for protection. Each piece should have your name on it.

To locate job openings, check with the offices of trade associations as well as professional and technical societies. Read the *classified ads* in newspapers and trade publications. You could even put a "situation wanted" ad telling about yourself in some publications. Describe the type of job you are seeking and list your qualifications. The general news section of the paper might even help you. If there is an article about a new or expanding business in your chosen field, go apply for a job!

Contact the *personnel offices* of large industrial firms and department stores. Pleasantly ask about any possible job openings. Tell them about your interests and abilities. Try to make a strong, personal impression so they will remember you. If no jobs are available, ask if you could fill out an application for them to keep on hand in case something opens up. Follow this with a letter thanking them for allowing you to fill out the application. Also include your resume and a request for an interview. Each time you make a polite contact with such an office, you are spreading the word of your availability. You are also gaining experience and confidence.

An important aspect of job hunting is networking. ***Networking*** is the exchange of information or services among an interconnected group of people. Lines of communication flow throughout such a network of friends, relatives, and old school acquaintances who are already employed in the industry. When looking for a job within a certain field of work, call any acquaintances you have in that field. Ask your parents to tell their friends. Eventually, your name and availability will circulate throughout the industry, and you may be called for a job by someone you have never met.

Employment Agencies

You can also look for a job through ***employment agencies***. There are two kinds: public and private. Public employment agencies are run by the state or federal government. They are supported by tax money. They do not charge a fee for their services.

Private employment agencies place employees with firms who ask them to find people to fill specific positions. They charge a fee for their services. The total fee is paid by the employer or sometimes the job seeker. Other times the fee is split between the job seeker and the employer. In all cases, read the contract carefully and be sure you understand all parts of it before you sign it.

Before going to an employment agency, make an appointment by telephone. A chat on the phone can tell you if the agency handles the types of jobs you want. When you go to your appointment, take several copies of your resume with you.

Self-Employment

The option of being an *entrepreneur* might also be considered. To start your own business, seek information from small business associations and your local chamber of commerce. Another source of knowledgeable people is the Service Corps of Retired Executives (SCORE).

To start an entrepreneurial venture, you must have a product or service that is in demand. You should have knowledge, skills, and lots of drive. Start-up money (capital) is often needed to buy supplies, build up inventory, and hire employees. It is usually smart to start small and enlarge the business gradually.

As mentioned in Chapter 23, starting your own business is risky. Many new ventures fail. However, under the right conditions, and with lots of hard work, you could have a successful company of your own.

Preparing a Resume

A ***resume*** is a brief, written account of your qualifications for a job. It gives personal, educational, and vocational information about you. It acquaints a prospective employer with your qualifications, experiences, goals, and interests. A resume is a means of advertising yourself.

A neat, clean resume is very important. It can bring interest and attention to you, and persuade an employer to grant you an interview. A sample resume is shown in 24-9.

Be sure to be very honest about all the information in your resume. Never misrepresent your experience or work history. Focus on your strong points that are related to the job.

The resume should include your name, address, and telephone number at the top. List an occupational goal that is in keeping with the job for which you are applying. Tell what education you have, listing the most recent first. Include general dates, degrees, and honors. List extracurricular activities and special achievements in school.

Indicate in your resume any work experience that you have had. List the most recent first and describe the rest in reverse order. Volunteer work is equally important. Also list your special skills. Special skills might include your speed of typing, skills with machines or instruments, or proficiency in a foreign language.

Your resume must be neatly typed, photocopied, or professionally printed. It should not be handwritten. It should be well spaced with proper margins. Try to keep all of the information on one page. The resume should not have any spelling or grammatical errors. Ensure this by having someone proofread it for you or do a computer spell check. When mailing it to apply for a job, always send a cover letter of application with it.

Writing a Cover Letter

A ***cover letter*** is a short business letter sent with your resume. The purposes of the letter are to express your interest in a job and to persuade a potential employer to read your resume and grant you an interview.

The cover letter should be positive and polite. It should introduce you in a way that makes you stand out above the others who want the job. It should tell why you are interested in the job.

The cover letter should make it clear why you are qualified. Try not to limit yourself to a single job title if you have the education and skills for several different types of jobs. If possible, omit any reference to salary until you talk face-to-face with a hiring interviewer.

John W. Gordon
2119 Cedar Drive
Hillsboro, Oregon 97799
Phone: (503) 555-1234
E-mail: jwgordon@hills.net

OCCUPATIONAL GOAL:	Retail Department Manager	
EDUCATION:	Portland Technical College Retail Management Course; Upper quarter of class.	2003-2006
	North County High School Hillsboro, Oregon General curriculum; Football captain; Athletic Leadership Award.	2003 Graduate
WORK EXPERIENCE:	Apparel Stock Clerk Tallmore Discount Store Reedville, Oregon Recorded inventory; Stocked shelves; Transported incoming apparel to proper areas.	June, 2002 to September, 2003
	Sports Activities Counselor Camp Tockovista Batterson, Oregon Organized and lead sports activities for resident campers, ages 8-14.	Summers of 2000 and 2001
	Yard work, pet care, and snow shoveling.	1998 and 1999
SKILLS AND INTERESTS:	Sports, working with others, keyboarding 40 words per minute, computer use experience, physical strength, record-keeping skills.	
	REFERENCES AVAILABLE ON REQUEST	

24-9 A resume should be brief, truthful and positive, and neatly typed.

Also, omit sending a photograph unless that is specifically requested.

To make a good impression, the cover letter must be neatly typed. It must not have any spelling, grammar, or typographical errors. It must not be a photocopy. Each letter sent must be an original. A computer is great for this job. A sample cover letter is shown in 24-10.

Send a letter and resume only to places where you are really interested in working. Try to find out the name of the person who is doing the hiring. Address your letter to that person rather than just sending it to the company in general. If you are answering a newspaper ad, follow the specific instructions that are given. In all cases, enclose your resume and ask for an interview.

If you do not receive a response within a week or 10 days, follow up with a pleasant telephone call. Ask if your resume was received. You may state that you definitely are interested in working for the company and feel that you could do a very good job. Maybe then you will be given an appointment for an interview!

The Interview

A ***job interview*** is a face-to-face meeting between you and the person who hires employees for a company. It is the time when you try to present yourself as the best candidate

2119 Cedar Drive
Hillsboro, Oregon 97799
April 20, 20xx

Mr. Fredrick Ashton, Personnel Manager
Kroley's Department Store
2006 East Augusta Road
Portland, Oregon 97790

Dear Mr. Ashton:

I am eager to apply for the Retail Department Manager's job in your store that was advertised in this morning's *News Journal*. I believe my education and experience qualify me well for the opening.

Enclosed is my resume. I will be receiving an associate degree in Retail Management next month. I could begin working immediately after that.

The field of retailing has always seemed exciting and challenging to me. I am enthusiastic about starting my lifetime career with Kroley's Department Store. Please call me at (503) 555-1234 to arrange an interview. You may also e-mail me at jwgordon@hills.net. I look forward to hearing from you and talking with you in the near future.

Thank you for your kind consideration in this matter.

Sincerely,

John W. Gordon

John W. Gordon

24-10 Your cover letters might be similar to this one. Each one must be typed individually since the content must be geared to a specific job.

for a particular job. It is also the time to find out what the company has to offer you.

Getting Ready

Preparing yourself for the interview is extremely important for a successful outcome. Try to be ready for all interview questions by practicing the answers. As shown in 24-11, you may be asked about your goals, career plans, previous jobs, school background, and special skills. You may have to explain why you are interested in that particular company. You may have to state what you plan to be doing one, five, and ten years in the future. Think about your answers in advance. Practice talking about your background, experience, professional interests, and future plans. Try this with a friend or family member, or alone in front of a mirror. Take note of your "body language" since that also tells a lot about you.

Learn all you can about the firm ahead of time. Find out what the company makes or sells, what the customers are like, and if there are branch operations. You can do this by talking to friends who work there or by doing some research. Large firms may have annual reports or other descriptive materials on file at the library. If you are interviewing at a store, browse at its various branches ahead of time. If

Questions Often Asked at Job Interviews

About work experience:

What was your most recent job?

What duties and responsibilities did you have?

Tell me some things you have liked about your previous jobs.

Describe the most/least interesting work you have done.

What are some of the things that a job has to offer in order to give you satisfaction?

Why did you leave your last job?

About education and training:

What were your average grades in college (high school)?

What subjects did you enjoy the most/least?

Did you participate in extracurricular activities? Which ones?

About your interests:

Why have you chosen this career field?

Why do you want to work for this firm?

What would you like to be doing one/five/ten years from now?

Do you want to manage others? Create? Do detail work? Be outdoors? Be at a desk? Have regular or irregular hours?

Do you usually prefer to work alone or with lots of people?

What are your feelings about being relocated (transferred)?

What do you feel your assets/liabilities are as a person?

What are your most-liked/least-liked leisure-time activities?

24-11 These and other questions might be asked during job interviews. They should be answered with well thought-out answers.

the firm makes products, familiarize yourself with them at stores that stock them.

Decide why you are suitable for the job. Review your qualifications and experiences. Be mindful of your special skills and talents. Be rested when you go to the interview so your thinking is sharp. You will then be ready mentally and physically.

Take time to dress properly for the interview. In most cases, you should dress as you would dress for the job. A basic tailored suit is often good, even if more casual garments might be worn after you are employed. Avoid casual, dressy, or gaudy extremes. Cleanliness and good grooming show that you take pride in yourself. This is very important, especially in apparel businesses. The interviewer will judge you on your appearance and behavior as well as your professional qualifications.

Before you go to the interview, write down the company's address. Also write down the time of your interview and the name of the person you are to see. Have these details with you on a small note pad so you don't have to rely on your memory. If you need a portfolio, have it in perfect order. Take an extra copy of your resume. Also take a small pad of paper, a pencil or pen, and your Social Security card. If you do not have a Social Security number, apply immediately at the closest office of the Social Security Administration.

Take the names, addresses, and phone numbers of at least three ***references*** with you. These are people who know you well enough to recommend you for a job. They can vouch for your work attitude and character. They must be adults, but not relatives. The best choices are former bosses, teachers, or your counselor. If necessary, you might list your religious leader, neighbors, or family friends. Ask these people for permission before you give their names as references.

Allow extra time to get to the interview in case you have traffic problems or difficulty finding the address. You will then arrive relaxed and in plenty of time. Showing up late for an interview is not acceptable. However, phone the interviewer's office if you must be delayed.

Attend the interview alone. Do not take a friend or relative. You must show that you are independent and responsible.

Applications and Tests

You may be asked to fill out an employment application form when you arrive for an interview. An ***application form*** requests the applicant's personal, academic, and employment information, 24-12. Some of the questions may

THE Southwest CENTER

DEPARTMENT STORE DIVISION

APPLICATION FOR EMPLOYMENT

NAME (LAST — FIRST — MIDDLE) | SOCIAL SECURITY NO. | DATE

PLEASE MENTION ANY OTHER NAME UNDER WHICH YOU HAVE WORKED OR BEEN EDUCATED

PRESENT ADDRESS (NO. AND STREET) (CITY OR TOWN) (STATE) (ZIP) | PHONE NUMBER

WHAT HAS PROMPTED YOUR APPLICATION TO OUR COMPANY?

POSITIONS DESIRED 1. 2. | MINIMUM SALARY REQUIRED | DATE AVAILABLE FOR WORK

	YES	NO		
HAVE YOU EVER PREVIOUSLY APPLIED FOR EMPLOYMENT WITH THIS COMPANY?			WHERE	WHEN
HAVE YOU EVER BEEN PREVIOUSLY EMPLOYED BY THIS COMPANY (INCLUDING THE GLOVES DIVISION)?			WHERE	WHEN
IF YOU ARE NOT A CITIZEN OF THE UNITED STATES DOES YOUR VISA OR IMMIGRATON STATUS PREVENT LAWFUL EMPLOYMENT?				
DO YOU HAVE ANY PHYSICAL DEFECTS OR CHRONIC AILMENTS THAT WOULD LIMIT YOUR PERFORMANCE IN THE POSITIONS STATED ABOVE?			DESCRIBE	
HAVE YOU EVER BEEN CONVICTED OF SHOPLIFTING, THEFT OR OTHER FELONIES?			DESCRIBE	

EDUCATION

NAME AND ADDRESS OF SCHOOL	DATES ATTENDED		MAJOR STUDY	DEGREE RECEIVED	GRADE AVERAGE
	FROM	TO			
HIGH SCHOOL ADDRESS					
COLLEGE OR UNIVERSITY ADDRESS					
BUSINESS OR TECHNICAL ADDRESS					
OTHER ADDRESS					

HOBBIES, SPORTS, SCHOOL AND OTHER ACTIVITIES WHICH MAY HAVE CONTRIBUTED TO YOUR JOB SKILLS | TECHNICAL OR BUSINESS SKILLS INCLUDING SPECIAL COURSES AND TRAINING

MILITARY SERVICE

IF YOU HAVE SERVED IN THE U.S. ARMED SERVICES, WHAT BRANCH?

LENGTH OF TIME SPENT IN MILITARY SERVICE | HIGHEST RANK ACHIEVED

DESCRIBE DUTIES AND TRAINING

PLEASE COMPLETE OTHER SIDE OF APPLICATION

EMPLOYMENT RECORD

START WITH PRESENT OR MOST RECENT EMPLOYER PLEA

NAME OF COMPANY
ADDRESS
TITLE OR POSITION
DUTIES
REASON FOR TERMINATION

NAME OF COMPANY
ADDRESS
TITLE OR POSITION
DUTIES
REASON FOR TERMINATION

NAME OF COMPANY
ADDRESS
TITLE OR POSITION
DUTIES
REASON FOR TERMINATION

NAME OF COMPANY
ADDRESS
TITLE OR POSITION
DUTIES
REASON FOR TERMINATION

REFERENCES

PLEASE DO NOT INCLUDE FRIENDS OR RELATIVE EDUCATION FOR MORE THAN ONE YEAR.

NAME | POS

PERMISSION IS GRANTED THE S.W. CENTER TO INVESTIGATE MY PERSONAL HISTORY AND SOLICIT STATEMENTS FROM ANY PERSON OR ORGANIZATION WITH WHICH I HAVE EVER BEEN ASSOCIATED. IN CONSIDERATION OF THE RECEIPT OF THIS APPLICATION BY THE S.W. CENTER, I RELEASE SAID COMPANY AND ALL PERSONS OR ORGANIZATIONS FROM ANY LIABILITY ARISING FROM SUCH STATEMENTS, THEIR SOLICITATION OR USE.

FAILURE TO PROVIDE COMPLETE AND ACCURATE INFORMATION ON THIS APPLICATION WILL CONSTITUTE GROUNDS FOR IMMEDIATE DISMISSAL. I ALSO UNDERSTAND THAT I MUST PRESENT PROOF OF DATE OF BIRTH WITHIN 30 DAYS FOLLOWING EMPLOYMENT.

I AGREE THAT IF I AM EMPLOYED I WILL ACCEPT THE TERMS OF BENEFIT PLANS FOR WHICH I BECOME ELIGIBLE AND THAT I WILL FOLLOW THE RULES AND POLICY PROVISIONS DESCRIBED IN THE EMPLOYEE HANDBOOK.

SIGNATURE OF APPLICANT

IT IS THE POLICY OF THE S.W. CENTER TO PROVIDE EQUAL EMPLOYMENT OPPORTUNITY TO ALL APPLICANTS AND EMPLOYEES WITHOUT REGARD TO THEIR RACE, COLOR, RELIGION, SEX, AGE, NATIONAL ORIGIN OR HANDICAPS.

F603-8 (7/86)

24-12 This application form for employment with a retail firm is typical of those used by most large companies.

duplicate information already given in your resume. However, the company may require you to complete an application form anyway.

Read through the entire form first so you can sort out the answers in your head. Then fill out the form completely and neatly. A messy or illegible application may imply to the company that you are a careless worker. Follow the directions exactly. If you do not understand a question, ask for an explanation instead of filling it in incorrectly. Refer to an extra copy of your resume for dates of previous jobs and other information you need for the application.

Automated job applications are now being used by some companies. They electronically record, format, sort, and screen employment applications. People looking for retail jobs, or shoppers who may be interested, can be directed by signs to stop at kiosks in stores that are equipped with computer terminals and telephone connections. This systemizes the hiring of hourly employees for retailers and provides easy access to the process for you and others who are looking for a job.

Before being offered a job, you may also be asked to take one or more tests. Some are *performance tests*. They test keyboarding or computer skills that are needed for the job. Others are *written tests*. They test aptitude for the job and general knowledge. Don't be frightened. Everyone else who applies there has to take them, too. Just do your best.

During the Interview

An average interview with the personnel director of a firm usually lasts 15 to 30 minutes. Try to relax. When you introduce yourself, call the interviewer by name if possible. Look him or her in the eye and shake hands pleasantly and firmly. When asked to sit, be straight, but relaxed, in the chair. Good posture makes a good impression. Do not smoke or chew gum. Try to avoid all nervous mannerisms. Also, avoid putting items on the interviewer's desk unless you are asked to do so. Put your portfolio or other materials next to your chair until you are asked to present them.

Let the interviewer lead the conversation. Be courteous and alert. Clearly answer all questions in a pleasant tone of voice with complete, informative sentences. Answer honestly; don't just say what you think the interviewer wants to hear. After all, you want a job that suits you, not the interviewer. Sell yourself by being quietly confident about your qualifications. Loud self-applause will not win you the job, nor will small talk about personal interests. Emphasize all information that relates to the job. Indicate your flexibility and willingness to learn. Try to be businesslike and positive in your speaking.

Some interviewers discuss your portfolio with you to learn about your abilities and ideas. Others might look at it without any discussion or comments. It is not recommended that you leave your portfolio with an employer. Instead, arrange to bring it back another time so others in the firm can see it and evaluate your work.

Ask any necessary questions at appropriate times during the interview. These may be about the duties and responsibilities of the job. You may also want to know more about the background and operations of the company. These questions give you information about the firm while you are giving information about yourself. However, avoid taking notes until after the interview is finished.

If the job appeals to you and there has been no mention of salary by the end of the interview, you might ask what the pay is. In turn, you might be asked what you think you are worth! It is, therefore, wise to know the average salary range for specific jobs.

The company's policy on salary review might come up naturally if pay is discussed. That policy defines how often employees are considered for raises. A job at one company may start at a lower salary than another, but get faster pay raises to allow for higher earnings sooner.

Do not ask about vacations, lunch breaks, or fringe benefits until all other aspects of the job have been discussed. Someone will fill you in on those details if you are offered a job. When they ask to hire you, you should ask all of the questions that are still unanswered in your mind.

If you are offered a job during the interview, and you are sure you want it, give a definite acceptance with thanks. Find out the starting date, time, to whom you should report, and where. You will probably have to fill out personnel and payroll forms.

If you are not sure about accepting a job offer, thank the employer and ask if you can take some time to think about it. Clarify any questions about the job and specify when you will call back with your answer. Also, jot that date and time down so you won't forget! If you decline, express your appreciation graciously. You may be interested in considering employment with that firm some time in the future.

At the end of the interview, stand up and shake hands with the interviewer. Thank him or her for taking the time to meet with you. Follow each interview with a mailed thank-you letter, like the sample one in 24-13.

Keep a record of which employers expect you to call them and which ones will call you about their hiring decisions. If you do not hear from a company within a week or so, call the interviewer's office to thank him or her again for seeing you. At this time, you may also ask if anyone has been hired for the job. Show your interest, but do not be pushy. Sometimes the person who shows the most interest is the one who ends up with the job.

A number of job rejections can be discouraging. However, you learn a great deal from your early interviews and do better as you gain experience. You may get an exciting job offer when you least expect it. Actively pursue all leads that interest you. Follow up in an aggressive, but businesslike, manner. Go after what you want.

For Future Jobs

If you apply for jobs that suit your interests, aptitudes, abilities, and education, you will always be on the right track. If you are also careful in planning the resume, cover letter, and interview, you will most likely find the

2119 Cedar Drive
Hillsboro, Oregon 97799
May 30, 20xx

Mr. Fredrick Ashton, Personnel Manager
Kroley's Department Store
2006 East Augusta Road
Portland, Oregon 97790

Dear Mr. Ashton:

Thank you for giving me the opportunity of interviewing with you yesterday. It was a pleasure to meet you and to hear about the operations of Kroley's Department Store.

My interest in working for your firm is high, and I would appreciate being able to prove my worth in a top-notch retail firm such as yours.

Please call me at (503) 555-1234 to discuss this with me further. As shown on my employment application form, I can also be reached via e-mail at jwgordon@hills.net.

I look forward to hearing from you soon.

Sincerely,

John W. Gordon

John W. Gordon

24-13 Sending a thank-you letter to follow up on a job interview will show that you are polite and interested in the job.

employment you seek. Little details can put you above the others. You will benefit from the experience of each try for a job, even if you are not hired. You are gaining confidence and maturity needed for just the right opportunity!

One last hint to improve your lifetime job-seeking skills: keep a file of your employment activities. Include in it photocopies of all letters submitted for jobs and notes about what follow-up was done. Include copies of your resume as it is updated through the years. List the names, addresses, and telephone numbers of your employers and the dates of your employment with them. Also write down job descriptions, salary information, and the names of reference people. This will become a ready reference file for moving ahead along your career path when opportunities present themselves.

Becoming a Success

Your first job may determine your pattern of achievement and advancement throughout your working life. It calls for good manners, a pleasing appearance, and your best actions.

Develop Positive Personal Traits

To become a success, work on developing the following positive personal traits. If perfected, they will become automatic habits, and you will be on your way to the top.

Drive is the energy and persistence to accomplish goals. It causes you to have the best attitude and to put forth the most effort toward your job.

Self-esteem is a deep conviction of your own worth. With honesty, high ethics, and hard work, you will always be proud of your actions. Keep a positive attitude about yourself and your abilities.

Reliability is dependability. It means you can be trusted to do as you say you will do. People are confident that you will not let them down.

A *sense of responsibility* is the ability to always account for your own behavior and decisions. You accept blame for your mistakes and learn from them. (No one is perfect!) On the other hand, you are entitled to praise for your accomplishments.

Dedication is a devotion of your time and enthusiasm to the job and to the firm. Be willing to learn new skills. Accept new ideas. Try your best.

Awareness is a knowledge of what is happening. Observe the job, the people, and other operations so you always know what is going on. You may learn many valuable details about your work and the entire operation.

Cheerfulness is the tendency to look for the good side of every situation. Be optimistic. Have a positive attitude and good expectations for yourself and others. This makes everyone feel better, as in 24-14.

Cooperation allows you to get along with others. Be considerate, pleasant, and concerned about those around you. That creates cooperation and makes the work easier for everyone. Pitch in and help coworkers when you can without interrupting your own work schedule. You can then ask them for help when you need it. Experienced employees have wonderful amounts of information and advice. Use them as resources and appreciate their experience.

Other Important Factors

Besides the personal traits just mentioned, a few steps can be taken to further fashion careers. They may also take time and effort on your part. They include developing communication skills, finding a mentor, and building positive responses to pressure. They also include developing decision-making and

Lands' End Direct Merchants

24-14 A cheerful attitude about one's work helps that person become a success and helps those around that person feel good.

leadership skills, as well as presenting a professional image.

Develop Effective Communication Skills

These give you the ability to receive and transmit information. You need reading skills to understand directions and other written information. You need writing skills so you can give directions and messages to others. Learning to listen well and speak clearly is also very important. Computer skills are "a must" to communicate effectively. Most jobs rely heavily on the accurate exchange of information.

Find a Mentor

A ***mentor*** is a faithful counselor who values your abilities and potential. He or she guides you along your career path. For instance, a supervisor can watch your progress and give you new assignments that will help your career. This personal relationship is especially important in larger, more impersonal firms.

Develop a Positive Response to Pressure

This is easier by nature for some people than for others. It is essential for many jobs in the apparel industries. Producing good, creative results under pressure is a valuable ability. Meeting deadlines and filling sales quotas are types of pressure. Some people enjoy this challenge and are "turned on" to perform their best. On the other hand, pressure makes other people nervous and unable to perform well.

Develop Decision-Making Skills

Decision making is the determining of a course of action from all the options. It is the process of choosing what seems to be the best of several alternatives. It requires an understanding of the situation and may involve some risk. It also involves accepting the responsibility for the decision after it is made.

Develop Leadership Skills

Leading others involves motivating and managing them. This means getting members of a group to work together to further the group's interests. A skillful leader gives employees the opportunity to attain individual goals and rewards while they are performing well for the company.

Present a Professional Image

By presenting a businesslike image of yourself through your appearance and actions, you can further your career. When you look and feel professional, you will perform and be viewed as a successful member of your profession.

You Can Do It!

Success is obtained with hard work, strong ambition, and diverse skills. The people who end up in the highest jobs are those who continue to learn and grow. They are the ones who enthusiastically look for and accept new challenges. Often, someone with less talent but more traits for success will work harder and achieve more than others. Sincere enthusiasm for the work is one of the first requirements for success, 24-15.

Whatever career you choose, you must give it your best effort. Apparel-related jobs require a neat appearance, good grooming, and good health. You need energy, pride, a cooperative manner, and a sense of good taste. The responsibility of confidentiality must also be taken seriously, so new fashion designs, company policies, or research discoveries are not told to the wrong people. To get ahead, you must follow instructions accurately, take suggestions and criticism well, and finish all tasks promptly. You must be willing to do more than you are asked. Enthusiasm, perseverance, and ambition are important traits for success. It is no small order, but you can do it!

24-15 Pleasure and satisfaction result when you enthusiastically give your best effort toward your job.

Chapter Review

Summary

Fashion is everywhere and offers vast job opportunities. Fashion careers give varying personal and financial rewards, depending on the jobs.

Do careful planning and study to choose your career. First, get to know yourself with a self-evaluation of your aptitudes and interests. Then investigate fields of employment that match you. Find out about the preparations for careers that interest you. Set a career goal and strive toward it. Pursue the needed education and training, including internships, work-study, or apprentice programs.

Do serious job hunting to get a job in your chosen field by using placement offices and networking and contacting personnel offices. You might use the help of employment agencies. Also, self-employment might be an option.

Carefully prepare a resume to give to prospective employers. Send it with a cover letter to the correct person at each firm that interest you. For job interviews, be ready to answer questions, fill out application forms, and take performance and written tests. Try to be relaxed and honest during interviews. As you go through your career, keep a file of your employment activities to use for reference.

To become a success, use good manners, a pleasing appearance, and your best actions. Develop positive personal traits, such as drive, self-esteem, reliability, and a sense of responsibility, dedication, awareness, cheerfulness, and cooperation. Also, find a mentor, present a professional image, and develop effective communication skills, a positive response to pressure, decision-making skills, and leadership skills. Continue to learn and grow through your career to gain the success you deserve.

Fashion in Review

Write your responses on a separate sheet of paper.

True/False: Write true or false for each of the following statements.

1. Most administrators are paid by commissions rather than salaries or hourly wages.
2. In general, apparel industry jobs offer more occupational security than many other businesses.
3. Your aptitudes can help you predict your suitability to different jobs.
4. Two-year colleges and trade schools offer associate degrees.
5. Employers often consider extracurricular activities as well as academic records when they hire personnel.
6. Schools and businesses often have work-study programs that allow students to gain work experience while taking classes.
7. Private employment agencies do not charge a fee for their services.
8. Volunteer work should not be mentioned in a resume.
9. It is wise to specify salary requirements in a cover letter to prevent misunderstandings.
10. A good way to relax and keep your breath fresh during a job interview is to quietly chew gum.

Short Answer: Write the correct response to each of the following statements.

11. Name three laws that protect employees.
12. List five ways your career decision will affect your life.
13. What should you take with you to a job interview?
14. List five personal traits that people need to achieve success.
15. How does a mentor affect a person's career?

Fashion in Action

1. With a classmate, conduct a mock interview in front of the class for a specific position. Practice job-related questions and answers beforehand.
2. Look through the classified ads in a local newspaper for apparel-related job openings. Clip them out and mount them on a bulletin board. Discuss the variety of career opportunities available in your area from entry-level jobs to top executive positions.
3. Prepare your own resume, listing your education, honors and activities, work experience, skills, and interests. Then write a cover letter to an imaginary company asking for an interview for a specific job that might really suit you.
4. In the counselor's office or reference section of the library, look through handbooks that list trade schools and colleges. Find five schools that offer courses to prepare students for apparel careers. In a written report, compare their curriculums, sizes, costs, settings, and other specifics.
5. Do research on the Americans With Disabilities Act (ADA). Give an oral report to the class telling when law was enacted, what it provides for employees with disabilities, and how companies must comply. Relate it to the *Laws That Provide Employee Protection* in Figure 24-2 on page 421.

Glossary

Comprehensive Glossary of Fashion and Apparel Terms as Used in the Context of this Textbook

(For descriptions of specific fabrics, consult the Glossary of Popular Apparel Fabrics on pages 164-168 of this textbook.)

A

A.A.M.A. American Apparel Manufacturers Association. (4)

abrasion resistance. Ability of a fiber or fabric to withstand surface wear and rubbing. (8)

absorbent. Having the ability to take in moisture. (8)

accented neutral color scheme. Color plan that combines white, black, or gray with a bright accent. (10)

accessories. Articles such as belts, hats, jewelry, shoes, gloves, and scarves added to complete or enhance an outfit of apparel. (2, 14)

accessory editor. Commercial pattern company employee who obtains the latest accessories to be used in creating finished ensembles for illustration and photography. (23)

account executive. Advertising agency employee responsible for handling specific clients' accounts. (22)

accountant. Employee who records and summarizes business transactions and reports the results. (19, 21)

accounting and finance. Department of a firm that keeps track of the funds of the company. (19)

acetate. Manufactured fiber made from cellulose acetate. (8)

acrylic. Manufactured fiber with wool-like qualities. (8)

added visual texture. Surface design printed onto fabrics. (11)

administration. Overall managing of a company or other organization. (19)

adornment. Decoration or ornamentation. (1)

adult education. Educational courses for adults, usually offered in the evenings. (23)

advertising. Paid promotional message by an identified sponsor. (6, 7, 15, 22)

advertising and promotion agents. Employees who try to create a demand for the firms products through advertising and promotion. (19, 22)

advertising designer. *See* art director. (22)

advertising director. Employee who oversees all the advertising activities of the organization. (22)

agile manufacturing. A data capture system of information, production, and delivery of custom garments to individual consumers. (6)

A-line. Dress that is narrow (fitted) at the shoulders, has no waistline seam, and becomes wider at the hemline. An A-line skirt is fitted at the waist and has extra width on each side at the hem. (2)

alteration hand. Apparel manufacturing employee who repairs defects that have occurred to garments during factory production. (20)

alterations expert. One who takes in, lets out, and reshapes standard-sized garments so they fit the customers who buy them. *Also* called a tailor. (21)

American Fashion Critics Awards. *See* Coty Awards and Cutty Sark Awards. (4)

American Fiber Manufacturers Association, Inc. Trade organization for producers of manufactured fibers in the United States. (8)

analogous color scheme. Plan using adjacent or related colors on the color wheel. (10)

anchor stores. Stores that provide the attraction needed to draw customers to shopping malls. (7)

angora. Fiber from the Angora rabbit. (8)

anidex. Manufactured elastic fiber with resistance to chemicals, sunlight, and heat. (8)

antistatic. Chemical finish applied to fabrics to prevent the buildup of static electricity so garments will not cling to the body of the wearer. (9)

apparel. General term that includes men's, women's, and children's clothing. (2)

apparel industries. Businesses that center around textiles, garment manufacturing, and retailing. (4)

apparel manufacturing industry. Group of firms that design and construct garments. (6, 20)

apparel marts. Buildings or complexes that house permanent showrooms of apparel manufacturers. (6)

application form. Paper requesting personal, academic, and employment information to be filled out by a job applicant. (24)

applied lines. *See* decorative lines. (11)

appliqué. Design made separately and then sewn onto fabric or a garment; a cutout decoration. (14, 17)

apprentice program. Long-term on-the-job training learned from an experienced skilled employee. (24)

aptitude. One's natural talent or suitability to some type of activity. (24)

aramid. Manufactured fiber that is lightweight, tough, and resistant to flames and chemicals. (8)

armchair shopping. Gathering product and availability information at home through catalogs and public media, often as a planning step before doing the actual shopping. (15)

art director. Advertising agency employee who conceptualizes ads for various media and designs collateral materials. *Also* called an advertising designer. (22)

artificial suede. Nonwoven polyurethane/ polyester fabric that looks and acts like real suede. (9)

as ready. Expression used by manufacturers referring to agreements to deliver merchandise to retailers when it is ready, rather than by a specified date. (7)

ascot. Broad scarf that is looped over itself and worn at the neck. (3)

assembler. Dry-cleaning firm employee who gathers together all the pieces of a customer's order after they have been cleaned or laundered. *Also* see assorter. (23)

assets. Best qualities or parts of a person's body that should be emphasized or accentuated by apparel. *Also,* a person's financial resources. (12)

assistant. Someone who helps or aids another, especially in a particular job or set of duties. (19)

associate degree. Two-year college or trade school degree. (24)

assorter. Apparel manufacturing employee who sorts and prepares the cut garment parts to go through the production sewing assembly line. *Also* called an assembler. (20, 23)

asymmetrical. Different on one side from the other when divided by a center line. A garment design in which the right side is not the same as the left side. *Also,* informal balance. (3, 12)

at-home fabric sales. Job of selling home sewing fabrics and notions from one's home. (23)

attitudes. One's feelings about or reactions to people, objects, or ideas as formed from the person's values. (1)

audio-visual work. Employment involving radio, television, and multimedia presentations. (22)

automatic dryer. Laundry appliance that dries fabric items with heat and a tumbling action. (18)

automatic fund transfer. Transfer of money done electronically by computer from a person's bank account to a store's account when a purchase is made. (16)

avocation. Secondary employment or hobby done for pleasure. (24)

avant-garde clothes. Daring and wild designs that are unconventional and startling. (2)

B

Bachelor's degree. Bachelor of Arts or Bachelor of Science degree from a four-year college or university. (24)

back fullness silhouette. Recurring fashion style with extra fullness at the back only. (2)

bagger. *See* wrapper. (23)

balance. Principle of design that implies equilibrium or steadiness among the parts of a design. (12)

balanced plaid. Design of crossing lines and spaces that are the same going out in both the lengthwise and crosswise directions. *Also* called even plaid. (9)

bale. Large bundle or closely pressed and bound goods, such as cotton or other fibers. (8)

bargain. Favorable purchase; a high value of merchandise in exchange for a relatively small amount of money. (16)

basic apparel. Garments that are worn most often and are the core of a person's wardrobe. (14)

basic stock. Store merchandise that is constantly in demand. It is stocked continuously on an ongoing basis. (7)

basket weave. Variation of the plain weave with two or more filling yarns passing over and under the same number of warp yarns. (9)

bast fibers. Strong, woody fibers that lie in bundles just under the bark in the stems of various plants. (8)

bateau neckline. *See* boat neckline. (2)

batik. Form of resist dyeing in which wax is used to cover the area where dye is not wanted. (9)

batting. Soft bulky mat of fibers, usually sold in sheets or rolls, used for warm interlinings and stuffings. (9)

batwing sleeve. Kimono sleeve style that is very low and loose at the underarm with hardly any curve between the side waistline and the sleeve bottom. (3)

beauty. Quality that gives pleasure to the senses and creates a positive emotional reaction in the viewer. (1)

beetling. Mechanical finishing process for cotton or linen fabrics that pounds them flatter, which gives a harder surface with increased sheen. (9)

bell silhouette. Recurring style with fullness at the bottom. (2)

benchmarking. The continuous process of measuring a company's products, services, and practices against those of other companies that are extremely good. (6)

bias grain. Diagonal grain of fabric, with true bias being at a 45 degree angle to the lengthwise and crosswise grains. (9)

billing agent. Employee in charge of keeping track of amounts due and sending out bills for those amounts. (21)

biodegradable. Ability to be broken down into natural waste products that do not harm the environment. (18)

blazer. *See* sport coat. (3)

bleaches. Chlorine or oxygen type laundry products used to clean, whiten, and brighten clothes. (18)

bleaching. Chemical process that removes color, impurities, or spots from fibers or fabrics during fabric finishing or garment laundering. (9, 18)

blend. Yarn made by spinning together two or more different fibers (usually staple fibers). (8)

block printing. Printing by hand using carved wooden or linoleum blocks instead of screens or rollers. (9)

blouson. Dress style with blousy fullness above the waist, usually with a fitted skirt and a belt. (3)

boat neckline. Neckline that is high at the front (and usually back) and is wide on the sides, going straight across from shoulder to shoulder. *Also* called bateau neckline. (3)

Bobbin Show. Yearly trade convention of the A.A.M.A. (4)

bodice. Apparel area above a waistline seam, usually closely fitted. (2)

body build. Total human structural form established by relationships or proportions of body parts. (12)

bolls. *See* cotton bolls. (8)

bonding. Process of laminating two or more materials together with adhesives, plastics, or cohesion (self-bonding). (9)

border print. Design pattern that forms a distinct border, usually along one or both selvage edges of a fabric. (9)

boutique. Specialty shop or a section within a department store devoted to unusual merchandise that is often presented in a nontraditional manner. (4, 7)

box loom. Loom using two or more shuttles for weaving fabrics with filling yarns that differ in fiber type, color, twist, or size that produces patterns such as plaids and ginghams. (9)

braiding. Intertwining three or more strands to form a regular diagonal pattern down the length of the resulting cord. *Also* called plaiting. (9)

branch coordinator. Retail executive who coordinates several branch stores of a company. (21)

branch stores. Additional stores operated from a flagship store and with the same name and merchandise. (7)

brand name. A trade name that identifies the product and/or its manufacturer. (16)

bridge jewelry. Jewelry made to look like fine jewelry, with a less expensive price. (14)

bridge lines. Designer ready-to-wear apparel priced between better and couture categories. (4, 14)

brushing. Finishing process in which rotating brushes raise a nap surface on fabrics. *Also* called napping. (9)

builders. Ingredients in laundry products that inactivate hard water minerals. (18)

business plan. Written plan that defines the idea (purpose), operations, and financial forecast of a proposed entrepreneurial venture. (23)

business planner. Employee who gathers data on the company's operations, analyzes the information, and makes recommendations of action to the company's management. (19)

business-to-business ("B2B") e-commerce. The transacting of business between companies on the Internet. (6)

button-down collar. Collar style with points that button to the shirt. (3)

buyer. *See* retail buyer. (21)

buyers clerical. Retail employee who answers telephones, schedules appointments, follows up on shipments, does filing, and handles other matters to help a retail buyer. (21)

buying. Exchange of money or credit for goods or services. (15)

C

caftan. Long, flowing, robelike garment. (3)

calender printing. *See* roller printing. (9)

calendering. Mechanical finishing process in which fabric is passed between heated rollers under pressure to produce special effects such as high luster, glazing, or embossing. (9)

camel's hair. Fashionable specialty wool from the two-humped Bactrian camel. (8)

cap sleeve. Very short sleeve, like a sleeveless armhole at the underarm and a short kimono sleeve going out from the shoulder. *Also* called French sleeve. (3)

cape. Cloak that hangs from the neck and shoulders and has no sleeves. (3)

capital. Financial worth or accumulated investment cash needed to start, expand, or run a business. (5, 6, 23, 24)

cardigan. Sweater that opens down the front. (3)

carding. Cleaning and straightening staple fibers using a machine with fine wire teeth to form a continuous, untwisted strand called a sliver. (8)

career. Field of employment (vocation) in which a person usually progresses through assignments that give consecutive levels of achievement. (24)

cash loan. Money borrowed from a bank, credit union, or finance company that is repaid with interest according to a written agreement. (16)

cash purchase. Transaction in which goods or services are bought by paying the full amount with cash, check, or automatic fund transfer. (16)

cashmere. Luxurious specialty wool from the cashmere goat. (8)

catalog showrooms. Retail outlets that display items available through their catalogs. (7)

cellulosic fibers. Fibers composed of or derived from cellulose from plants, such as cotton and linen. (8)

central buying office. Buying office of a chain store operation responsible for selecting and purchasing the merchandise handled by those stores. (7)

C.E.O. Chief executive officer of a firm. (20)

certificate courses. Trade school training programs that take one to three years to complete. (24)

chain stores. Groups of stores that are owned, managed, and controlled by a central office without an actual main or flagship store. (7)

Chambre Syndicale. Trade association for the haute couture houses of Paris. (4)

channel of distribution. Route that goods and services take from the original source, through all middle people, to the ultimate user. (7)

charge slip. Paper for later payment, signed when a credit purchase is made, that tells the details of the transaction. (16)

checker. Commercial pattern company employee who looks over the pattern pieces to check that all cutting and sewing markings are included and correct. *Also,* a dry-cleaning firm employee who receives and returns garments. (23)

checkout cashier. Retail store employee who collects and records customers' payments and bags the merchandise after finalizing the sales. (21)

chemical finishes. Finishes that become part of the fabrics through chemical reactions with the fibers. (9)

chemical spinning. Forming manufactured fiber yarns by hardening and solidifying filaments coming from the spinneret. (8)

chemise. *See* shift. (3)

chevron. Pattern having the shape of a "V." (11)

circular. Very full skirt that forms a circle when laid flat. (3)

class market. The few consumers who buy high fashion clothing. (2)

classic. Item that continues to be popular and acceptable over a long period of time even though fashions change. (2)

classification buyer. Central office retail buyer of only one type of goods for all the stores of one large company. (21)

classified ads. Ads in newspapers and trade publications that are divided into classes or categories; source of lists of job openings. (24)

classroom teacher. Instructor who teaches in a school classroom. (23)

closures. Zippers, buttons, snaps, hooks and eyes, Velcro®, or any other fasteners that enable the wearer to get into and out of garments. (3)

clothing specialist. Extension agent, usually at the state level, who spreads clothing knowledge. (23)

coat. Warm, weatherproof garment that is worn over regular clothing. (3)

coatdress. Heavy dress that looks like a coat but is worn as the main garment rather than over another garment. (3)

C.O.D. Cash on delivery. This requires payment for goods in cash when they are received by the purchaser. (16)

cold water detergents. Laundry products designed to clean fabrics in cold water. (18)

collateral materials. Promotional materials other than actual advertisements, such as brochures, annual reports, packaging, hangtags, logos and trademarks, and corporate image projects. (22)

collection. Manufacturer's or designer's line (total group) of designs or creations for a specific season. (4, 5)

color. Element of design; hue. (10)

color analysis. A study to analyze a person's skin tone and to determine what colors are most flattering for that person to wear. (10)

color analysis consultant. Person with the job of doing color analyses. (23)

color schemes. Plans for harmonious combinations of colors. (10)

color wheel. Circle, with colors shown, used as a guide to study how to choose and combine colors. (10)

colorfast. Term designating that the color in a fabric will not fade or change with laundering, dry cleaning, or time and use. (9)

combination yarn. A ply yarn composed of two or more yarns that differ in fiber composition, content, and/or twist level, or composed of both spun staple and filament yarns. (8)

combing. Step in cotton and wool (worsted) processing that mechanically straightens the fibers and extracts foreign matter and short or tangled fibers. It produces a stronger, more even, finer, and smoother yarn. (8)

commission. Method of payment based on a percentage of the dollar amount of products sold by an employed person. (21, 24)

commissionaires. Agents in foreign countries who are hired by retailers to help with the buying from those countries. (7)

communication. The giving and receiving of verbal and nonverbal messages (such as through one's appearance). (13)

comparison shopper. Employee who shops in the store that employs him or her and in competitive stores to examine the merchandise, prices, and customer services. (21)

comparison shopping. Comparing the qualities and prices of similar items in different stores before buying. (16)

complementary color scheme. Plan using hues across from each other on the color wheel. (10)

completion date. Date designated on a purchase order by a retailer to a manufacturer, after which the order is subject to cancellation. (7)

composite garment. Garment made by a combination of the tailored and draped method. (2)

compressive shrinkage. Finishing process resulting in a controlled standard of fabric shrinkage of not more than one percent. Trademarked as Sanforizing® (9)

computer-aided design (CAD). The use of electronic equipment to combine and visualize design ideas and to make patterns and prepare them for cutting. (6)

computer-aided manufacturing (CAM). Computer-aided manufacturing; utilizes electronics for the production of apparel. (6)

computer imaging. CAD function which enables a three-dimensional figure to be shown on a computer monitor and turned so it can be seen from all sides and angles. (6)

computer-integrated manufacturing (CIM). Computer-integrated manufacturing that combines CAD, CAM, robotics, and company information systems to approach "hands-off" production. (4)

computer shopping. *See* telecommunication shopping. (7)

conformity. Act of obeying or agreeing with some given standard or authority. (1)

conservator. *See* costume curator. (23)

consignment. Placing merchandise for sale in a store and being paid a percentage of the retail price if and when the merchandise is sold. (23)

consulting. The selling of a person's expert ideas and advice as a service business. (23)

consumer. Someone who buys and uses goods. (2, 15)

consumer aids. Educational information put out by manufacturers, retail firms, or trade associations to inform consumers about their products. (15)

consumer education. The combining of teaching with business, often done for a manufacturing firm or a retail store. (23)

consumer specialist. *See* customer service manager. (21)

contractor. Manufacturer who does any or all the cutting, sewing, and finishing work for other apparel producers under a contractual arrangement. (6)

converters. Companies or individuals who do textile finishing to satisfy market demand and end uses. (19)

convertible collar. Collar style that can be worn buttoned at the neck or open to form a "V" shape with a lapel. (3)

cool colors. Hues, such as green, blue, and violet, that serve as reminders of water or the sky. (10)

cooperative advertising. Advertising done jointly by a manufacturer and retailer with the costs shared. (6, 7)

copy. The words, lettering, and numbers that appear in type as part of a written work. *Also,* a duplicate of someone else's work, such as a garment design. (2, 22)

copywriter. Person who composes the messages that describe items that are being promoted in ads, catalogs, magazines, etc. (22)

cord. The result of twisting together a number of ply yarns. (8)

cords. Jeans made of corduroy fabric. (3)

corporation. Business set up as a separate legal entity, giving the owner certain protection. (23)

cosmology. Ancient Chinese science that deals with the order of the universe and describes yin and yang traits. (13)

costing. Procedure done to figure the expenses of producing something. (6)

costing engineer. Apparel manufacturing employee who determines the overall price of producing each apparel item. (20)

costume curator. Fabric and apparel historian usually employed by a museum or library. (23)

costume jewelry. Fairly inexpensive jewelry, often of unusual materials or plated with metals, and sometimes containing artificial stones. (14)

costume technician. *See* wardrobe helper. (23)

cottage industry. Home handiwork business with family units (or a network of people) working in their homes with their own equipment. (23)

cotton. Natural fiber obtained from the boll of the cotton plant. (8)

cotton belt. Band of southern U.S. states where cotton is grown. (8)

cotton bolls. Pods of cotton plants where cotton fibers grow. (8)

Cotton Incorporated. Trade organization for cotton growers. (8)

Coty Awards. Another name for the American Fashion Critics Awards presented annually from 1943 to 1978 for the most creative and outstanding designers of women's wear, accessories, and menswear. They were replaced by the Cutty Sark Awards. (4)

Coty Hall of Fame. Honor bestowed on designers who won a Coty Award three different years. (4)

county extension agent. Educator who teaches about and demonstrates various topics and skills (including those related to fashion) to community groups and individuals. (23)

course. Row of loops or stitches running across a knit fabric. (9)

couturier. Designer who creates original, individually designed high fashions and usually owns the fashion house. (2, 4)

cover letter. Short business letter sent with a resume to express interest in interviewing for a job. (24)

cowl. Neckline style that is draped with flowing folds. (3)

crabbing. Process of heating wool fabric under tension in a hot liquid, then cooling it under tension to give it dimensional stability. (9)

Crafted with Pride in U.S.A. Council, Inc. Unified coalition of textile/apparel industry companies and organizations that encourages consumers to buy apparel "Made in the U.S.A." (6)

craze. Passing love for a new fashion that is accompanied by a display of emotion or crowd excitement. (2)

crease resistant. Term denoting a fabric chemically treated to resist and recover from wrinkling. (9)

credit card. Plastic card that establishes the holder's ability to charge goods and services at participating businesses. (16)

credit limit. Maximum financial amount a person may have outstanding on a charge or other credit account. (16)

credit purchase. A promise to pay for goods or services in a certain, specified way at a later date. (16)

credit rating. Evaluation of the financial standing of a person or business based on past records of debt repayment, financial status, etc. (16)

crimp. Curl in fibers that looks like wavy lines or coil springs under a microscope; gives fibers elasticity and resiliency. (8)

croquis. First rough sketch of a garment design. (5)

cross dyeing. Method of dyeing blend or combination fabrics to two or more shades by using dyes with different affinities for the various fibers. (9)

crosswise grain. Direction of the filling yarns across the fabric from selvage edge to selvage edge. (9)

cuff. Band at the bottom of the sleeve, pant leg, or other area. (3)

cuff links. Ornamental accessories that fasten the wrist of shirt sleeves by passing through buttonholes on French cuffs. (14)

culottes. Pants that are made to look like a skirt. (3)

curved lines. Elongated marks that are rounded or somewhat circular. (11)

custom-designed. Apparel created specifically for a particular person with special fit, design, and fabric. (2)

customer service manager. Retail employee in charge of handling customer complaints and returns as well as special needs such as gift wrapping, home delivery, and special orders. (21)

custom-made. Apparel sewn for a particular person who has ordered it, usually after seeing a sample garment, sketch, or picture. *See* made-to-order. (2)

custom patterns. Patterns made to fit individual measurements. (5)

cutter. Apparel manufacturing employee who cuts out fabric garment parts for production. *Also,* the machine that does the cutting. (20)

cutting. In fabric finishing, a process of cutting the ribs of floating filling yarns with razor-sharp discs to create a cut pile surface. In apparel production, the cutting out of garment parts before sewing. (6, 9)

Cutty Sark Awards. Another name for the American Fashion Critics Awards presented from 1979 to 1988 to designers voted to be the most creative and outstanding by fashion editors of newspapers and magazines. They replaced the Coty Awards. (4)

cylinder printing. *See* roller printing. (9)

D

Daily News Record (DNR). Trade publication of the textile and menswear industries. (4)

dart. Short, tapered, stitched area that enables a garment to fit a figure. (2)

dart loom. Weaving machine with darts that shoot across with individual lengths of yarn. (9)

data processing. Computer use for administrative tracking of sales, shipping, billing, inventories, employee records, payroll, etc. (19)

debit card. Card used for electronic transfer of money between a store and bank to pay for purchases. (16)

décolleté. Low neckline, often with bare shoulders. (3)

decorative lines. Applied lines created by adding details to the surface of clothing. (11)

denier. Term to describe filament thickness or diameter. Higher numbers indicate thicker yarns. (8)

department manager. Person responsible for a particular selling area in a retail store. (21)

department store. Retail establishment that offers a large variety of many types of merchandise placed in appropriate departments. (7, 15)

departmental buyer. Traditional retail buyer who plans, purchases, and manages all the goods for only one department of a store. (21)

design. A specific arrangement of parts, form, color, fabric, line, and texture to create a fashion style concept. Also, the plan used to put an idea together. (7, 21)

design director. Home sewing pattern company employee who supervises the designers, pattern makers, and garment makers for new designs. (23)

designer. Person who creates new versions of concepts or styles in fabrics, apparel, accessories, etc. (4, 22, 23)

designer label. Designer's trademark or logo on a garment that usually gives status to the wearer. (16)

designer patterns. Patterns for replicas of designer fashions offered to home sewers by home sewing pattern companies. (4, 5)

detergents. Laundry products made synthetically from chemicals that suspend and hold dirt and grease away from clothes during washing. (18)

diagonal lines. Elongated marks that slant rather than being vertical or horizontal. (11)

diagram artist. Someone who does technical drawings, such as the precise sketches that accompany the written directions of a commercial pattern guidesheet. (23)

diaper. Basic garment for infants of folded cloth or other absorbent material drawn up between the legs and fastened near the waist. (17)

directional print. Fabric design with an up and down direction. (9)

direct-mail marketing. *See* mail-order retailing. (7)

direct printing. *See* roller printing. (9)

direct selling. The exchange of merchandise to individual consumers in return for money or credit. (7)

dirndl. Skirt style that is gathered only slightly for minimum fullness. (3)

discharge printing. Method of adding the desired design to a fabric by bleaching out color from the cloth with rollers in a determined pattern. *Also* called extract printing. (9)

discount store. Retail outlet that sells merchandise at consistently low prices, usually in a simple building with low overhead, and utilizes mass retailing methods with few customer services. (7)

disinfectants. Sanitizers that are sometimes used in laundering to control or eliminate infections by reducing or killing microorganisms. (18)

display. Visual presentation of merchandise or ideas. (22)

display designer. Craftsperson who creates displays. *Also* called a display artist. (22)

display manager. Employee in charge of visual displays at a retail store or other business. (22)

distribution. Process of getting merchandise to the proper locations. (20)

distributor/planner. Retail chain employee who keeps track of merchandise and allocates various stock items to the branches that need them. (21)

division director. Management employee in charge of all aspects (including design, production, and sales) of an entire division of a company. (20)

dobby attachment. Loom attachment that permits the weaving of geometric figures. (9)

dolman sleeve. Kimono type of sleeve with a lowered underarm. (3)

domestic production. The manufacturing of goods in one's own country. (6)

double-breasted. A front closure that laps over and has two vertical rows of buttons. (3)

double knit. Fabric produced on a weft knitting machine with two sets of needles and yarns, knitting two fabrics as one. (9)

double ticketing. Garment sizing that combines two or more specific sizes into general categories such as small, medium, and large. (16)

doupion silk. Silk from two or more silkworms who spin their cocoon together. It has irregular, thick-thin filaments that produce a slubbed effect and is used in making shantung fabric. (8)

down. The light, fluffy feather undercoating of geese and ducks that is an extremely effective insulator. (8)

draped garment. Garment that is wrapped or hung on the human body and has characteristic folds of soft fabric. (2)

drawing. Process of pulling or stretching laps, slivers, rovings, or continuous filament tow to align and arrange the fibers for more length, strength, and uniformity. (8)

dress code. Written or unwritten rules of what should or should not be worn by a group of people. (1)

dressmaker. Expert sewer who does custom sewing, alterations, and clothing repairs for others. *Also* called a tailor. (23)

drip-drying. Method of drying laundered items by hanging them up dripping wet without squeezing or wringing on nonrusting hangers or racks. (18)

dropped shoulder. Sleeve style with a horizontal seam going around the upper part of the arm. (3)

dry cleaner. Employee in a dry-cleaning firm who does the actual cleaning of garments with specialized equipment and solvents. (23)

dry cleaning. Process of cleaning textile items with nonwater liquid solvents or absorbent compounds. (18)

dry spinning. Process in which a solution of manufactured fiber-forming substance is extruded into a heated air chamber to evaporate the solvent, leaving a solid filament. (8)

durable finish. Fabric finish that lasts through several launderings or dry cleanings, but loses its effectiveness over a period of time. (9)

dyeing. Method of giving color to a fiber, yarn, fabric, or garment with either natural or synthetic dyes. (9)

E

easy-to-sew patterns. Designs offered by commercial pattern companies that are simple to cut out and make. (5)

editor. Journalist who supervises writers (reporters) and copywriters. (22)

educational representative. Consumer educator who promotes the products of a firm by teaching about them and their uses. (23)

educator. Person who teaches or instructs others. (23)

Egyptian cotton. Fine, long-staple cotton fiber with a smooth, silklike texture. (8)

elasticity. Stretchability, or the ability of a strained material to recover its original size and shape immediately after stretching. (8)

electronic data interchange (EDI). Computer linkages between companies to automatically communicate inventory levels, purchase needs, and other information throughout the supply chain. (6)

electronic retailing. A form of in-home shopping using a computer or interactive cable television. (7)

elements of design. Building blocks of design that include color, shape, line, and texture. (11)

emboss. To give a permanent raised and indented design to a fabric using a calendering process with engraved, heated rollers. (9)

emphasis. Principle of design that uses a concentration of interest in a particular part or area of a design. (12)

empire (om-peer). Dress with a high waistline. (3)

employee protection. Safeguards provided to workers by labor unions, private agencies, and government legislation. (24)

employment agency. Public or privately run company that tries to match available jobs with qualified job applicants. (24)

empower. To give official authority and autonomy to employees to make their own work-related decisions. (6)

entrepreneur. Person who starts his or her own business and who assumes the risk and management of the enterprise. (23)

enzymes. Proteins that speed up chemical reactions. In laundry products, they help break down certain soils and stains into simpler forms that can be removed more easily. (18)

epaulet sleeve. *See* saddle sleeve. (3)

ergonomics. Human engineering that matches human performance to the tasks being done, the equipment used, and the environment. (6)

e-tailing. Selling to consumers electronically via the Internet; Web-based retailing. (7)

even plaid. *See* balanced plaid. (9)

executive trainee. Retail employee who receives on-the-job training for potential buyer and management positions. *Also* called management trainee. (21)

exports. Commercial products sent out of a country to other countries. (5, 6)

extenders. Less expensive garments and accessories that can be mixed and matched with the basic apparel of a wardrobe to multiply the number of outfits. *Also* called multipliers. (14)

extension agent. Educator who teaches and demonstrates clothing and other knowledge and skills to community groups and individuals. (17, 23)

extract printing. *See* discharge printing. (9)

extracurricular. Activities not falling within the regular academic scope of the curriculum, such as a school's athletic programs and clubs. (24)

extrude. To force or push out. Liquid substances are extruded through spinnerets to make manufactured fibers. (8)

F

fabric. Cloth made from textile fibers or yarns by weaving, knitting, etc. (9, 16)

fabric editor. Commercial pattern company employee who obtains samples of the latest fabrics for the firm's fabric library and designs. (23)

fabric finishes. Sizings that restore the original body to fabrics. *Also* see finishing. (9)

fabric softeners. Laundry products that give softness and fluffiness to washable fabrics and control static cling. (18)

factory outlet. Store owned by a manufacturer who sells company products to the public at reduced prices. (7)

fad. Temporary, passing fashion or item that has great appeal to many people for a short period of time, then dies out quickly. (2)

fashion. Type of clothing that is widely popular at any given time. (2)

fashion conscious. State of being aware of and wanting new fashionable items, usually for self-expression and peer approval. (15)

fashion consultant. *See* personal shopper. (21)

fashion coordinator. Retail executive who coordinates the merchandise of various departments, organizes in-store fashion promotions, and keeps employees updated on the latest fashion trends. *Also* called a fashion director. (21)

fashion cycle. Periodic popularity, disappearance, and later reappearance of specific styles or general shapes; a cycle of the rise, popularization, and decline of particular styles. (2)

fashion designer. Person who creates new ideas for garments and accessories. (4, 20)

fashion forward. Phrase implying the leading edge of fashion trends. (6, 23)

Fashion Group. Organization of fashion professionals who are committed to promoting the American fashion industry. (4)

fashion houses. Couture firms, each with a designer who creates original, individually designed fashions. (4)

fashion illustrator. Promotion artist who makes drawings of garments that have been designed and produced by others. (22)

fashion information director. Commercial pattern company employee who gathers and interprets latest fashion information. (23)

fashion journalist. Person who passes along textile and apparel information through the mass media. (22)

fashion leaders. Innovative, trend-setting people who have the status and credibility to introduce and popularize new styles. (2)

fashion merchandising. The planning, buying, promoting, and selling of apparel and other fashion merchandise to meet customer demands as to price, quantity, quality, style, and timing. (21)

fashion merchandising services. Firms that can be hired by retailers to provide merchandise and market information. (7, 21)

fashion model. Person who wears garments and accessories to show them. (22)

fashion photographer. Person who takes pictures that show fashionable apparel and accessories looking their best. (22)

fashion piracy. The stealing of design ideas, or the use of a design without the consent of the originator. (2, 4)

fashion show. Planned presentation of a group of styles, often as part of the promotion of a season's new merchandise. *Also* called a fashion showing. (6, 21)

fashion trend. Direction in which fashion is moving. (2)

fashion writer. Person who writes about fashion for newspapers, magazines, or books. (22)

Federal Trade Commission (FTC). Governing body over textile and apparel legislation and labeling. (15)

felting. Process of tangling, shrinking, and matting fibers (often wool) to form a nonwoven fabric. (9)

fiber. Long, thin, hairlike natural or manufactured substance that is the basic unit of textile products. (8)

fiber dyeing. The dyeing of fibers before they are spun into yarns. (8)

fiberfill. Manufactured fibers, often of polyester, especially engineered for use as filling material for warm insulation. (8)

fiberglass. Fibers with excellent insulation qualities and fire resistance, not used for apparel because of their heavy weight and low abrasion resistance. (8)

figure. Shape of a girl's or woman's body. (12)

filament. Long, fine, continuous thread found naturally as silk and extruded as manufactured fibers. (8)

filling knits. *See* weft knits. (9)

filling yarns. Crosswise yarns running from selvage to selvage at right angles to the warp yarns in a woven fabric. (9)

film. Thin sheet, usually of vinyl or urethane, sometimes used as a coating over fabrics. (9)

fine jewelry. Expensive jewelry of fine metals and often containing precious or semiprecious stones. (14)

finisher. Apparel manufacturing employee with better quality, higher-priced lines who does whatever hand sewing is needed to finish garments. *Also,* a dry-cleaning firm employee who shapes, irons, and packages clothes that have been cleaned or laundered. (20)

finishes. The processes through which fibers, yarns, and fabrics are passed to improve their appearance, feel, and/or performance in preparation for their end uses. (9)

first impression. What people think of someone (a feeling or reaction) when they first see or meet him or her. (13)

fit. How tight or loose a garment is on the person who is wearing it. (2)

fitted. Garment or garment part that is shaped to follow the lines of the body, such as a fitted dress or a fitted sleeve. (2)

fitting model. Person employed by an apparel manufacturing firm or pattern company who tries on garment samples to check fit, drape, and movability and to show the samples to design staff members and management. *See* mannequin work. (22, 23)

flagship store. Original "parent" store that gives direction and merchandise to its branch stores. (7)

flame resistant. Term denoting a finish that prevents fabric from supporting or spreading a flame and causes the fabric to be self-extinguishing when removed from the source of ignition. (5, 9)

Flammable Fabrics Act. Law that specifies burning standards for household textiles and apparel. (15)

flaps. Decorative fabric pieces that fall down over the openings of pockets. (3)

flared. Garment such as a skirt or pants that widens near the bottom and has some fullness at the hem. (3)

flat drying. Method of drying laundered garments, usually used for wool, knit, and leather items such as gloves and sweaters. Items are laid on a flat, absorbent surface away from direct heat and shaped to their original dimensions. (18)

flax. Plant from which the natural fiber linen is obtained. (8)

fleece. The hair of sheep from which the natural fiber wool comes. (8)

float. In weaving, the portion of a yarn that extends over two or more perpendicular yarns to form certain designs. *Also,* in a knit, a portion of yarn that extends for some length without being knitted in. (9)

flocking. Method of cloth ornamentation in which a glue substance is put onto a fabric in a pattern, with finely chopped fibers sprinkled on to produce a design with texture. (9)

ford. Style or design that is produced at the same time by many different manufacturers at many different prices. (6)

forecasting services. Consultants that foresee the colors, textures, and silhouettes to predict coming fashion trends. (5)

formal balance. Equilibrium created in a design with symmetrical parts, such as design details being the same on each side of a center line. (12)

franchise. Business arrangement in which a firm grants a retailer the right to use a famous or established name and trademarked merchandise in return for a certain amount of money. (4, 7)

freelance jobs. Independent work of an individual who does short-term jobs for various firms that end when each assignment is completed. (19)

freelancing. The selling of expert skills. (23)

French sleeve. *See* cap sleeve. (3)

fringe benefits. Extra compensation other than pay, such as vacation time, insurance, sick leave, and pension plans. (19, 24)

full-fashioned. Knits produced on a flat knitting machine that have been shaped by adding or reducing stitches. (9)

full skirt. Skirt that is pleated or gathered for fullness. (3)

fulling. Mechanical finishing process for wool fabrics that makes them stronger and denser by tightening the weave with controlled shrinkage. (9)

Fur Products Labeling Act. Law that protects consumers against deceptive information on labels and in advertisements of garments of or containing fur. (15)

fusible web. A sheet of binder fibers that can act as an adhesive because its softening point is relatively low. (9)

G

garment. Any major article of clothing. (2)

garment district. Area within a fashion city where garment businesses are located. (4)

garment dyeing. The dyeing of constructed garments by apparel manufacturers to fill retail orders for requested colors. (9)

garment industry. Another name for the apparel manufacturing industry. (4)

garment maker. Person whose occupation is sewing, such as constructing muslin prototypes and fashion fabric samples of designs for a commercial pattern company. *Also* called tailor or dressmaker. (23)

garment parts. The sleeves, cuffs, collar, waistband, and other components that make up a complete garment. (2, 3)

gathers. Fullness in a garment created by pulling fabric together. (3)

gauchos. Pants that end just below the knee with legs like wide tubes. (3)

gauge. Number of stitches or loops per inch in a knitted fabric. (9)

generic name. An identification for each family of manufactured fibers grouped by similar chemical composition. (8)

glass fiber. *See* fiberglass. (8)

gored. Skirt style with panels formed by vertical seamlines. (3)

gown (infant). Full-length garment with sleeves and a drawstring closing at the bottom. (17)

gradation. A gradual increase or decrease of similar design elements used to create rhythm in a design. *Also* called progression. (12)

grading. Scientific process of making garment patterns into larger or smaller sizes. (5)

grain. The direction of the lengthwise and crosswise yarns or threads in a woven fabric. (9)

graphic designer. Advertising artist. (22)

greige goods. Fabrics that are just off the loom or knitting machine in an unfinished state. (5)

grippers. Heavy-duty types of snaps. (17)

grooming. A person's cleanliness and neatness of nails, hair, teeth, and body. (13)

growth features. Attributes of garments that allow them to be "expanded" as children grow. (17)

guide sheets. Illustrated directions for all cutting and sewing steps that are included in commercial patterns. (5)

gusset. Wedge-shaped piece of fabric added to give more ease of movement at a kimono sleeve underarm or other area of a garment. (3)

H

hackle. The combing of flax fibers to straighten and clean them. (8)

halter. Brief garment worn on the upper body, usually in hot weather. (3)

hand. The way fabric feels to the touch. (9)

hangtags. Detachable "signs," usually affixed to the outside of garments, that provide promotional information. (15)

hardening a substance. Removal method for such stains as candle wax and gum from textile products. (18)

harem pants. Flared pants that are gathered in at the ankles. (3)

harmony. Pleasing visual unity of a design created by a tasteful relationship among all parts within the whole. (12)

haute couture. Firms in the high fashion industry of Paris (or elsewhere) whose designers create original, individually designed fashions. (2, 4)

head of stock. Retail store employee responsible for all the stock functions of a certain department or store and for coordinating the activities of stock clerks. (21)

heat setting. Process of heating and stretching a fabric of manufactured fibers to bring it to its correct width and length. (9)

heat transfer printing. Method of printing fabric by transferring the design from preprinted paper by contact heat. (9)

Hemline Index. Theory that links the rise and fall of fashion hemlines to the rise and fall of stock market indexes. (2)

herringbone pattern. Broken twill weave in which the wale changes direction at regular intervals to produce a zigzag effect. (9)

high fashion. The very latest or newest fashion, usually of fine quality and beautiful fabric and therefore expensive. It often seems extreme and unusual. *Also* called high style. (2)

high sudsing detergents. All-purpose laundry detergents. (18)

hip-huggers. Pants with their top lower than the regular waistline. (3)

home sewing industry. Businesses that deal with the production and selling of nonindustrial sewing machines, notions, retail fabrics, and patterns. (23)

hood. Head covering that is attached at the garment neckline. (3)

horizontal lines. Elongated marks that go from side to side like the horizon. (11)

horseshoe neckline. Neckline that is high at the back neck, but goes down like a "U" in front. (3)

hosiery. Stockings, including pantyhose, tights, and all other socks. (14)

hourly wage. Employee payment based on a predetermined amount of pay for each hour spent doing the job. (24)

house boutique. Small retail shop owned by a couturier that sells items with the couturiers label. (4)

hue. The name given to a color. (10)

I

identification. Process of establishing or describing who someone is or what something does. (1)

illustrator. *See* fashion illustrator. (22)

image. Visual representation projected by a person, business, or organization that forms the mental picture of that person or group as seen by others. (13)

imports. Goods that come into the country from foreign sources. (5, 6, 7, 16)

impulse buying. Making purchases that are sudden and not carefully thought out. (15)

independent sales rep. Self-employed salesperson who handles several different lines for many small manufacturers. (20)

indirect selling. Nonpersonal promotion aimed at a large general audience. (7)

individuality. The quality that distinguishes one person from another; self-expression. (1)

industrial engineer. Cost and efficiency expert who coordinates people, material, equipment, space, and energy to save time and money for his or her firm. (19, 20)

Industrial Revolution. The time, roughly between the late 1700s and mid-1800s, when the hand crafting economy changed to a machine manufacturing economy. (4)

infant gown. *See* gown (infant). (17)

infant kimono. *See* kimono (infant). (17)

informal balance. Equilibrium in a design created with an asymmetrical arrangement in which design details are divided unequally from the center. (12)

ink jet printing. Printing onto paper or fabric with colored droplets from computer driven micronozzles. (9)

innovation. The creative, forward-thinking introduction of new ideas. (5)

inseam. Seam on the inside of pant legs from crotch to bottom of hem. (3)

inside shop. Apparel firm that does all stages of garment production itself, from fabric purchasing to the distribution of finished garments. (6)

inspector. Apparel manufacturing employee who checks garment parts during production, as well as finished garments, for flaws and imperfections. *Also* called an examiner. *Also,* a dry-cleaning firm employee who examines apparel after the cleaning and finishing processes. (20, 23)

installment plan. Credit arrangement by which a down payment is made toward a specific large purchase, and a contract specifies the periodic payments and finance charges. (16)

intensity. The brightness or dullness of a color; color purity. (10)

interfacings. Fabric pieces between the outer cloth and lining or facing of a garment, usually to give support and extra strength. (16)

interlock knit. Lightweight and stretchy double knit fabric with a smooth surface on both sides and a very fine lengthwise rib. (9)

intermediate hues. Colors made by combining equal amounts of adjoining primary and secondary hues. Examples are blue-violet, yellow-green, and red-orange. *Also* called tertiary hues. (10)

inventory. Supply of goods to be sold. *Also,* an itemized list of what one has, such as a wardrobe inventory. (7, 21)

inventory control. *See* stock control. (7, 21)

investment dressing. Having several good quality garments that will last a long time and not go out of style. (14)

ironing. Process of using an iron to remove wrinkles from damp, washable clothing with heat and pressure. (18)

ironing board. Flat, padded, cloth-covered surface, usually on legs, that is used for ironing. (18)

irregulars. Articles of merchandise with slight imperfections or defects that are sold to consumers at reduced prices. (2)

J

jabot (ja-bow). Ruffled or lace trimming effect on the front of men's or women's shirts on the collar or going down from the neckline. (3)

jacket. Short coat. (3)

Jacquard loom. Machine that weaves large and intricate designs with a series of programmed punch cards. (9)

jagged lines. Lines that change direction abruptly and with sharp points like zigzags. (11)

jeans. Sturdy pants, usually made of denim. (3)

jewel neckline. Plain, rounded neckline that encircles the base of the neck. *Also* called round neckline. (3)

jewelry. Ornamental apparel decoration, usually made of various metals and sometimes containing gemstones. (14)

job. Particular assignment of work to be done with specific duties, roles, or functions. (24)

job interview. Face-to-face meeting between a job applicant and the person who hires employees for a company. (24)

jobber. Independent sales rep who handles several different lines for many small manufacturers. (20)

joint venture. Partnership of two firms for advantages, such as a domestic firm and a foreign producer for production and sales overseas. (6)

jumper. Low-necked, one-piece garment with shoulder straps worn over a blouse or sweater. (3)

jumpsuit. Garment with a bodice attached to pants. (3)

K

kimono (infant). Full-length garment with sleeves that is closed in the front or back with snaps or ties. (17)

kimono sleeve. Sleeve that is a continuous extension out from a garment's armhole area with no seamline connecting it to the bodice. (3)

knickers. Pants that end just below the knee where they are gathered to a band or strap. (3)

knitting. Method of fabric construction done by looping yarns together. (9)

knock-off. Copy of another, usually higher priced, garment. (2)

L

label. Small piece of ribbon or cloth permanently attached to a garment that provides information, most of which is required by law. (15)

laboratory technician. Aide who helps researchers conduct their work. (19)

lamb's wool. The soft, fine fiber from young sheep up to seven months old. (8)

laminated. Fabric layers joined with an adhesive. (9)

lap. A continuous wide sheet of fibers, such as cotton. (8)

lapel. Pointed part of the front neckline of a garment below the collar that is folded back with the collar and looks like a continuation of the collar going down from a "V" notch along the outer edge. (3, 16)

lapel pin. Pin worn in the buttonhole on the lapel of a suit. (14)

laser cutter. A device that cuts out garment parts with an intense, powerful beam of light that quickly vaporizes the fabric. (6)

laundering. Washing apparel or other textile items with water and laundry products. (18)

laundry specialist. Person employed by an institution such as a hospital or hotel to work in an on-site laundry. (23)

layaway purchase. Deferred purchase arrangement in which a store puts an item away for a certain length of time when the intended purchaser makes a deposit toward buying it in full at a future date. (16)

layette. Assembled minimal needs of clothing and textile goods for a newborn infant. (17)

layout artist (layout designer). Advertising design employee who renders layouts for ads. *Also,* a commercial pattern company employee who mounts the guidesheet copy and diagrams on layout sheets for printing. (22, 23)

lead time. Amount of time between when a customer places an order for merchandise and the desired delivery date. (6)

leased department. Department within a retail store operated by an outside firm. (7)

lengthwise grain. The direction the warp yarns run in a fabric, parallel to the selvages. (9)

leno weave. Fabric construction which produces an open effect using crossing pairs of warp yarns. (9)

liabilities. Physical drawbacks that should be minimized or camouflaged by apparel. (12)

licensing. Arrangement whereby a manufacturer is given the exclusive right to produce and market goods that bear the famous name of someone who, in return, receives a percentage of wholesale sales. (4)

lifestyle. Way of life that is made up of the activities a person does. (13)

line. Element of design that is a distinct, elongated mark as if drawn by a pencil or pen. *Also,* a collection of styles and designs that will be produced and sold as a firm's new selections for a given season. (6, 11, 18)

line drying. Method of drying laundered items by hanging them on a clothesline, usually outside. (18)

linen. Natural fiber obtained from the stalk of the flax plant. (8)

linen supply service. Company that rents linens, such as sheets, towels, and tablecloths, to institutions and businesses. (23)

lingerie. Feminine undergarments and nightwear. (3)

lining. Inner layer of fabric sewn inside a garment. (3, 16)

lint. Cotton fiber after it has been ginned. Also, any fuzz, fine ravelings, or short fibers of yarn or fabric. (8)

logo. Symbol that represents a person, firm, or organization. (4)

loom. Machine for weaving fabric. (9)

loss leader. Item priced so low that the retail outlet makes little or no profit on it but uses it to attract shoppers into the store. (7)

low quality. Items of only fair standards of construction, materials, and design. (16)

low sudsing detergents. Laundry detergents containing sudsing control agents, especially recommended for front loading, tumbler-type washers that are "suds sensitive." (18)

lowered waistline style. Dress with a long torso and a waistline seam down toward the hips. (3)

M

machine operator. Production worker who operates a machine that does a manufacturing procedure. (19, 20)

machine technician. Manufacturing plant employee who keeps the production equipment (machinery) in good working order. (19)

made-to-order. *See* custom-made. (2)

mail-order retailing. Direct-mail marketing done by selling to consumers through a catalog. (7, 20, 23)

maintenance worker. Employee who cleans, paints, makes repairs, and otherwise keeps the physical structures of a business in good condition. (21)

management. Supervisory employees who plan, organize, coordinate, and oversee the work of lower-level employees. (20)

management trainee. *See* executive trainee. (21)

mannequin. A lifelike figure used to display clothing. *Also* the French term for a human fashion model. (7)

mannequin work. Live fashion modeling, especially when showing a manufacturers line in a showroom. (22)

manufactured fibers. Fibers that are produced artificially from substances such as cellulose, petroleum, and chemicals. (8)

margin. Difference between the retail selling price and the wholesale cost of merchandise; the amount of money made. (7)

markdown. Retail price reduction that is made in hopes of selling certain merchandise, but which lowers profits. (7)

markdown money. Funds to retailers from manufacturers to compensate for losses when the selling prices of goods must be reduced. (7)

marker. Long piece of paper that has a drawing of the layout of the pattern pieces for fabric cutting in garment manufacturing; an employee who makes such layouts. *Also see* sorter. (6, 20, 23)

market. Group of potential customers. *Also,* the place or area (city or mart) in which buyers and sellers congregate. (6, 7, 19, 20)

market analyst. Employee who conducts market research by studying consumer tastes and changing trends to predict future demand for specific goods. *Also* called market researcher. (19, 20)

market weeks. Periods of time when retail store buyers come to look at collections shown by manufacturers at their showrooms. (6, 7)

marketing. The business of planning, pricing, promoting, and distributing goods or services for a profit. *Also,* finding or creating a market for specific goods or services. (5, 19)

marketing specialist. Advertising director of a manufacturing company. (19)

markup. Monetary amount added to the cost (billed price) of goods to create the retail selling price. (7)

mass market. The bulk of average people who buy medium-to low-priced, mass-produced, ready-to-wear garments. (2)

mass retailing. Large stores that sell volumes of merchandise at low prices, with a minimum of customer service for shoppers. (7)

maternity fashions. Apparel for pregnant women. (17)

mechanical finishes. Fabric finishes that are applied mechanically rather than chemically. (9)

mechanical spinning. Method of pulling (drawing) and twisting staple fibers together to obtain continuous lengths of yarns. (8)

medium quality. Items of reliably good construction, materials, and design. (16)

melt spinning. Process in which a solution of manufactured fiber-forming substance is extruded into air or other gas, or into a suitable liquid, where it is cooled and solidified. (8)

mentor. Faithful counselor who values a person's abilities and potential and who guides that person's career. (24)

mercerization. Caustic soda finishing treatment for cellulosic textiles to increase the luster, strength, absorbency, and dyeability of the fibers. (9)

merchandise clerical. *See* stock clerk. (21)

merchandise control specialist. Employee who charts the rate of sales of particular items in various localities and suggests what merchandise should be stressed where. (20)

merchandise manager. Retail executive responsible for the total business coordination of several departments or classification of goods. *Also,* a commercial pattern company employee who figures out what the company's customers will want in patterns. (21)

merchandiser. Employee who determines the direction a manufacturer's line will take each season based on market research. (20)

merchandising. The process through which products are designed, developed, and promoted to the point of sale. It involves varying degrees of planning, buying, advertising, and selling. (6)

merchandising director. Executive who heads up the merchandising activities of a firm and tries to determine what the firm's customers will want. (23)

metallic fiber. Metal, often combined with plastic or a different fiber, that has decorative uses in apparel. (8)

microdenier fibers. Extremely thin filament manufactured fibers that are soft, luxurious, and drapable. (8)

microelectronics. Computer-related procedures and technology. (5)

mildew-resistant. Able to resist mildew as a result of a chemical fabric finish. (9)

mill. Textile production plant that spins fibers into yarns and/or manufactures fabrics from yarns. (5)

mineral fiber. Fiber taken from a mineral source. (8)

mix-and-match. Apparel items that can be combined with many other items or garments to create different outfits. (14, 17)

modacrylic. Manufactured fiber of modified acrylic synthetic polymer that is resistant to flames and most chemicals. (8)

model. *See* fashion model. (22)

modesty. The covering of a person's body according to the code of decency of that person's society. (1)

modular manufacturing. Flexible, highly productive apparel production method where employees are divided into independent module work groups that sort out problems and agree on their own work assignments and schedules. (6)

mohair. Long, smooth, silky hair from the Angora goat. (8)

money conscious. Awareness of the cost of items of merchandise and of one's personal monetary resources. (15)

monochromatic color scheme. Plan that uses different tints, shades, and intensities of one color. (10)

monofilament yarn. Any single filament of a textile fiber, usually of a high denier. (8)

moth resistant. Able to resist moths and carpet beetles as a result of chemical finishing. (9)

motif. One unit of a design that is usually repeated. (11)

multifilament yarn. Yarn consisting of many continuous filaments or strands twisted together. (8)

multipliers. *See* extenders. (14)

multisize patterns. Home sewing patterns with several sizes printed together on the same pieces. (5)

N

nap. A layer of fiber ends raised from a fabric surface. It appears different when viewed from different directions. (9)

napping. *See* brushing. (9)

natural fibers. Fibers made from natural sources, the most common of which are cotton, linen, wool, and silk. (8)

need. Something necessary for a person's continued existence or survival. (14)

needle punch. To mechanically interlock fibers with a needle loom to make a nonwoven fabric characterized by regularly placed punched holes. (9)

needle trades. Term referring to the garment manufacturing or apparel industry. (4)

networking. The exchange of information or services among an interconnected group of people. (24)

neutrals. Black, white, and gray rather than true hues. (10)

niche retailing. Dividing the total consumer market into narrow target markets with specific tastes or lifestyles. (7)

nightgown. Loose garment worn in bed. (3)

nightshirt. Nightgown that resembles a shirt. (3)

non-cellulosic fibers. Category of manufactured fibers made from chemical compounds called polymers. *Also* known as synthetic fibers. (8)

nonwoven fabric. Assembly of textile fibers held together by mechanical interlocking in a random web or mat, by fusing, or by bonding. (9)

NRF. National Retail Federation, the trade association of leading American retailers. (7)

nylon. The first fiber made totally from chemicals. A very strong manufactured fiber. (8)

O

occupation. Career or vocation. (24)

odd-figure pricing. The retail pricing of merchandise a few cents less than the next higher dollar to make the merchandise psychologically seem less expensive. Examples are $2.99 and $19.98. (7)

odd lots. Incomplete assortments of goods, such as overruns or discontinued items, bought by retailers at reduced prices and sold at low retail prices. (7)

office manager. Administrative employee who supervises a particular office for a company. (19, 21)

off-price discounters. Stores that carry well-known brand name merchandise at lower than normal prices. (7)

offshore production. Manufacturing that is done overseas. (6)

offshore sourcing. The wholesale buying of goods from overseas producers. (7)

olefin. Manufactured fiber, sometimes called polypropylene, that is lightweight and nonabsorbent with good wicking power. (8)

open-to-buy (OTB). The amount of merchandise (in dollars or units) that retail buyers are permitted to order for their store, department, or apparel category during a specified time period. (7)

opposition. Rhythm created in a design when lines meet to form right angles. (12)

outlet stores. Retail stores associated with manufacturers that sell overruns and seconds. (7)

outside shop. Apparel manufacturing firm that handles everything but the sewing and sometimes the cutting; it utilizes contractors to do those steps of production. (6)

overall print. A printed design that is the same across the entire fabric. (9)

overdrawn. An account that has checks written against it for more money than it contains. (16)

overruns. Extra first-quality apparel items that were produced by a manufacturer but not ordered by retailers. (2)

P

package priced. Each garment, or package of items, has its individual price marked on it for customers to read. (7)

packaging. The covering wrapper or container in which some merchandise is placed. (15)

padding. Bulky, stuffed material placed inside a garment to give added shape. (16)

pajamas. Loose, usually two-piece, lightweight garment worn while lounging or sleeping. (3)

palazzo. Pants that are flared from the waist and very full at the bottom. (3)

pants (slacks, trousers). Garments that cover each leg separately. (3)

parka. Heavy winter jacket with a hood and a warm fuzzy lining. (3)

partnership. Business structure with two or more owners. (23)

paste-up/mechanical artist. Advertising employee who puts together the elements of a layout, such as the type, line drawings, and photographs. (22)

pattern grader. Apparel manufacturing or commercial pattern company employee who makes patterns in all the different sizes to be produced. (20, 23)

pattern maker. Apparel manufacturing employee who translates an apparel design into pattern pieces that can be used for mass production. (20, 23)

peer group pressure. Force that makes people want to be like their friends. (1)

performance test. Test of skills, such as shorthand, typing, or computer use, often required when a person applies for a job. (24)

Permanent Care Labeling Rule. Law that requires manufacturers to attach clear and complete permanent care labels to apparel items. (15)

permanent finish. Fabric finish that lasts the life of the garment. (9)

permanent press. Term describing a finish on a garment to retain its smooth appearance after laundering, especially if tumble dried. (9)

personal selling. Direct selling, usually done through parties held in homes. (7)

personal shopper. Retail employee who selects merchandise for customers in response to mail or telephone requests, or who accompanies shoppers in the store to help them select merchandise. *Also* called a fashion consultant. (21)

personal style. A person's pleasing appearance that is achieved when design is used to the best advantage for that individual. (12)

personality. The total characteristics that distinguish an individual, especially his or her behavioral and emotional tendencies of thoughts, feelings, and actions. (1, 13)

personnel administrator. Administrator in a firm who oversees the people employed there, especially in regard to hiring, firing, and receiving pay and benefits. *Also* called a personnel director. (19, 21)

personnel office. Office of a business firm that deals with employee concerns. (19)

photo stylist. Photography employee who books models, accessorizes apparel being photographed, obtains props, pins up hems, irons garments, and picks up and returns merchandise used in pictures. (22)

photographer. *See* fashion photographer. (22)

photography work. Fashion modeling done in front of cameras. (22)

physique. Shape of a boy's or man's body. (12)

piece dyeing. The dyeing of yard goods in fabric form after weaving or knitting rather than as fibers, yarns, or garments. (9)

piece goods buyer. Purchasing agent who researches and buys the fabrics, trims, and notions for an apparel manufacturing firm. (20)

piecework system. Manufacturing procedure in which each specific task is done by a different person along an assembly line. (6)

pile. Fabric surface effect with tufts, loops, or other projecting yarns. (9)

pill. To accumulate little balls of fibers on the surface of a spun yarn fabric, usually caused by rubbing and wearing. (8)

Pima. Specific cotton fibers of high quality and with naturally long fibers. (8)

pinsonic thermal joining. The use of ultrasonic vibrations to quilt fabrics together with a series of "welds" in a chosen design. (9)

placement office. Office within a school in which people specialize in arranging contacts for jobs for graduating students. (24)

placket. Decorative strip of fabric over a sleeve vent, closure, or fastener. (3)

plain weave. Simplest and most common fabric weave in which each filling yarn passes successively over and under each warp yarn, alternating each row. (9)

plaiting. *See* braiding. (9)

plant engineer. Production plant employee who keeps all environmental systems, such as heating, air conditioning, electrical, and material handling, operating properly. (19, 20)

plant manager. Manufacturing management employee who is responsible for all operations at a plant site. *Also* called a production manager. (20)

pleats. Structured folds of cloth that give fullness in a garment. (3)

ply. One of the strands in a ply yarn. (8)

ply yarn. Yarns of two or more single yarns twisted together for extra strength, added bulk, or unusual effects. (8)

pockets. Built-in "envelopes" in a garment to hold items. (3, 16)

polybenzimidazole. A manufactured fiber that does not burn, melt, or drip. (8)

polyester. Manufactured fiber made by melt-spinning and known for its outstanding wrinkle resistance and easy care qualities. The most widely used textile fiber. (8)

polymers. Chainlike structures of molecules from which many manufactured fibers are made. (8)

polypropylene. *See* olefin. (8)

poncho. Unshaped, blanket-like outer garment with a slit or hole in the middle so it can be slipped over the head. (3)

portfolio. A case of loose, unfolded art or design papers showing a person's creative work. (19, 22, 24)

preschooler. Young child between the ages of 2½ and 5 years. (17)

preshrinking. Procedure done with heat and moisture to prevent fabric shrinkage of more than three percent in either direction. (9)

press kit. Promotional packet of photos and news copy about a manufacturer's line distributed to the media in hopes of receiving good publicity. (22)

presser. Apparel manufacturing employee who irons and shapes garments after, and sometimes during, production. (20)

pressing. Placing a heated iron on fabric and then lifting, rather than using a gliding motion as in ironing. (18)

pressing cloth. Piece of fabric placed between an iron and an item being pressed to prevent shine or other damage to garment fabrics. (18)

prêt à porter (pret-ah-por-tay'). The French ready-to-wear apparel ("pret") industry. (4)

price markets. Categories into which merchandise is placed according to its retail selling price, often described as high, moderate, or low. (2)

primary hues. The three basic colors of red, yellow, and blue. (10)

princess. Dress style with seamlines going up and down the entire length and no horizontal waistline seam. (3)

principles of design. Rules of balance, proportion, emphasis, and rhythm concerning how the design elements should be used. (12)

printing. Process for adding color, pattern, or design to the surface of fabrics. (9)

priority. Order of importance. (14)

private label. Merchandise produced specifically by or for a retailer with a special trademark or brand name owned by the store or group of stores. (7)

product development director. Management employee who establishes the specifications for products to be made by or for the firm. (7)

product manager. Production management employee who is in charge of every aspect of a product line or a specific category of items within a line. (20)

production assistant. Manufacturing employee who does detail work and record keeping for the plant manager. (20)

production manager. *See* plant manager. (20)

production supervisor. Manufacturing plant employee who coordinates and directs various operations to maintain the highest production and the best quality. (19, 20)

productivity. A measure of how efficiently or effectively resources such as labor supply, machines, and materials are used. (6)

profession. Vocation or category of employment. (24)

profit. The incoming money that is left over in a business after all the outgoing costs have been deducted. (6)

progression. *See* gradation. (12)

progressive bundle system. Piecework apparel manufacturing in which cut garment parts are packaged into bundles of dozens to go through the sewing operations. (6)

promotion. Indirect selling to encourage public acceptance through advertising, displays, exhibits, and publicity. *Also,* any advertising or merchandising effort done nationally or locally to improve the sale of items. (2, 7)

proportion. Principle of design concerning the spatial relationship of the parts in a design to each other and to the whole. *Also* called scale. (12)

protective clothing. Apparel that provides physical safeguards to the human body, preventing harm from the climate and the environment. (1)

protein fibers. Natural fibers of animal origin, such as wool, silk, and specialty hair fibers. (8)

prototype. The original correct version of something; the first full-scale trial garment of a new design. (5)

public relations agent. Publicist who tells the story of a firm or its products or services through various media or in person. (19, 22)

publicity. Free promotion such as nonpaid messages to the public about a company's merchandise, activities, or services. (7, 22)

pullover. A garment, such as a sweater, that slips over the head when put on or taken off. (3)

purchase order. Document written by a buyer that authorizes a seller to deliver certain goods at specified prices. (7)

purchasing agent. Employee who does the buying of supplies and equipment for a company. (19)

pure silk. Silk fabric with no metallic weightings and no more than 10 percent by weight of dyes or finishing materials. (Black silk may have up to 15 percent.) (8)

pure wool. Fabrics that are made from all-wool fibers or yarns that have never been used before. (8)

purl knit. Knitted fabric with pronounced horizontal (crosswise) ridges with superior stretch and recovery in both directions. (9)

Q

quality control. Employees and activities that analyze the quality of manufactured products and solve problems when necessary so salable goods are produced. (19, 20)

quality control engineer. Production management employee who develops specifications for items that will be manufactured. (20)

quality control laboratories. Testing labs that evaluate the performance of merchandise according to established standards. (20)

Quick Response (QR). Business strategy that ties together all parts of the U.S. textile-apparel-retail pipeline as one unified industry rather than as individual segments. (6)

quilted fabrics. Three-layer fabrics with batting in the middle, usually held together by machine stitching. (9)

quotas. Limitations established by a government on quantities of certain categories of goods that can enter a country during an established time span. (6)

R

radial arrangement. Rhythm in a design created by lines emerging from a central point like rays. (12)

rag trade. Old, insiders term for the garment manufacturing industry. (4)

raglan. Sleeve style with a shaped seam in the garment originating from the underarm. (3)

ramie. Bast fiber nicknamed "China grass" that is sometimes used as a linen or cotton substitute. (8)

raschel knit. Patterned or intricate knit fabric made on a raschel knitting machine. (9)

ravel. The unweaving at the cut raw edges of fabric as threads come loose. (9)

rayon. The first commercial manufactured fiber, made from regenerated cellulose. (8)

ready-to-wear (RTW). Apparel that is mass-produced in factories according to standard sizes as opposed to being custom-made. (2, 4, 6)

reciprocal agreement. A cooperative pairing of services, such as a school in a fashion center offering parts of its program for credit to students from a university away from the fashion scene. (24)

recycled wool. Wool fibers from previously made wool fibers. (8)

reeling. Process of unwinding silk filaments from softened cocoons and then winding the filaments onto reels. (8)

references. People who know a person well enough to recommend him or her for a job. (24)

reinforcements. Strengthening materials or supports. (16)

renewable finish. Fabric finish that is temporary but can be replaced or reapplied. (9)

repetition. Method of creating rhythm in a design by repeating lines, shapes, colors, or textures. (12)

research and development (R & D). Activities done to provide new knowledge, develop new products, and improve old products. (19)

research scientist. Technical employee, usually with an advanced degree, who works in laboratories to discover, develop, and improve products. (19)

resident buying office. Business organization located in a market center that serves client retailers with market information and buying help. (7, 21)

resiliency. Ability of a fiber or fabric to spring back to its original condition, size, and shape when crushed or wrinkled. (8)

resin. Waxy substance added to fabrics to give them more body, shape, or tension and wrinkle resistance. (9)

resist dyeing. Fabric coloring procedure in which areas that are not to be colored are restricted from the dye while areas to be colored are left exposed to the dye. (9)

resources. The money, time, and skills a person has available to make wardrobe improvements or carry out other plans. *Also,* a retailer's term for suppliers, or vendors of wholesale merchandise. (14)

response time. The amount of time it takes a manufacturer to produce and deliver merchandise after it has been ordered. (6)

resume. Brief, written account of a person's qualifications for a job. (24)

retail. The selling of merchandise directly to consumers through a store or other outlet. (2, 7)

retail buyer. Merchandising employee responsible for selecting and purchasing goods for a retail company, with the goal of making a profit for the company when the goods are sold. (21)

retail coordinator. Commercial pattern company employee who is a fashion and promotion liaison between the pattern company and retail fabric stores. (23)

retail salesperson. Employee who deals directly with customers by selling merchandise, preparing sales checks, and receiving payments for the sales. (21)

retting. The soaking of bundles of flax stalks in water to allow bacterial action to loosen the outside flax fibers from the rotting woody center stem. (8)

revolving charge account. Credit account that may be paid in full or in monthly installments with added finance charges but may not exceed a total predetermined credit limit. (16)

rhythm. Principle of design concerned with the pleasing arrangement of the design elements so a feeling of continuity or easy movement of the observer's eye is produced. (12)

rib knit. Double knit fabric with pronounced vertical (lengthwise) ridges and great crosswise stretch. (9)

rib weave. A plain weave with a corded effect created by using coarser yarns in one direction and regular yarns in the other direction. (9)

robe. Long, loose garment worn over nightwear. (3)

robotics. Mechanically accomplished tasks done by automated equipment. (5)

roller printing. The application of designs to fabric using a machine with a series of engraved metal rollers around a large padded cylinder from which one color of dye paste is applied at a time. *Also* called direct, calender, or cylinder printing. (9)

rotary screen printing. A combination of roller and screen printing in which dye is pushed through a perforated cylindrical screen to apply each color. (9)

round neckline. *See* jewel neckline. (3)

roving. The condensed sliver that has been twisted slightly and drawn as an intermediate step in spinning between the original sliver and finished yarn. (8)

rubber. Stretchy, waterproof substance with some uses in apparel. (8)

rubbing. Mechanical finishing technique that uses friction to give a soft surface to fabrics. (9)

runway work. Modeling, usually on a long platform, in front of a live audience. (22)

S

saddle sleeve. A particular raglan sleeve design. *Also* called epaulet sleeve or strap-shoulder sleeve. (3)

safety features. Attributes of garments that reduce the chances of injury or other hazardous occurrences. (17)

salary. Fixed amount of pay, usually received once or twice a month, for an employee to perform the duties of a particular job. (24)

sale. Reduction in retail price in order to clear out goods and/or bring in more customers. (7)

sales. The exchange of goods from the supplier to the customer for money. *Also,* retail promotional events with merchandise offered at lowered prices. (7)

sales manager. Person in charge of a certain segment of sales responsibility and the boss of sales representatives. (19, 20)

sales person. Retail employee who deals directly with customers by selling merchandise, preparing sales checks, recording sales, and receiving payment from customers. (19)

sales representative. Employee who sells the products of a company to the next level of user in an industry's chain. (19)

sales resistance. Self-control to avoid making unplanned purchases while shopping. (15)

sales trainee. Person learning to do sales work; often an entry-level job. (19)

sample. The model or trial garment made up exactly as it will look when sold. (6)

sample hand (sample maker). Skilled sewer who makes apparel design samples. (6, 20)

samplings. Small quantities of garments placed in retail stores to get indications of consumer reactions to them. (6)

Sanforizing®. *See* compressive shrinkage. (9)

saran. Fiber suited to household and industrial uses because of its low absorbency, high resiliency, and resistance to chemicals and weather. (8)

sateen. Satin weave fabric with floats running in the filling direction giving lower luster than satin. (9)

satin. Satin weave fabric with floats running in the warp direction giving high luster. (9)

satin weave. Basic fabric weave that has long yarn floats on the surface giving a smooth, lustrous appearance. (9)

scale. Proportion or size relationship of the parts in a design to each other and to the whole. (12)

scoop. Neckline style that is lowered and round in front. (3)

scour. To wash raw wool to remove sand, dirt, and the natural oil lanolin. (8)

scraping. Stain removal method done with a dull knife or a spoon. (18)

screen printing. Printing method similar to stenciling in which the coloring matter (dye paste) is forced through untreated areas of a fabric screen onto the material being printed. (9)

scutching. Step in linen processing during which rollers crush the flax stalks to complete the separation of the soft fibers from the harsh, woody parts. (8)

seam. A stitched line that joins two garment parts together. (2, 16)

seamstress. Person whose occupation is sewing, such as constructing muslin prototypes and fashion fabric samples of designs for a commercial pattern company. *Also* called tailor or dressmaker. (23)

seasonless clothes. Apparel that can be worn during most of the year, such as lightweight woolens, knits, or corduroy garments. (14)

secondary hues. The colors of orange, green, and violet made by mixing equal amounts of two primary hues together. (10)

seconds. Apparel items that are soiled, have flaws, are missing trim, or have other defects. (2)

secretary. Office worker who handles correspondence and manages other work for an executive. (21)

security guard. Employee who protects records, persons, and property against theft or harm. (21)

self-help features. Attributes of garments that enable children, as well as seniors or people with disabilities, to dress themselves. (17)

selvage. Strong lengthwise edges of fabric that do not ravel, formed when the filling yarns turn to go back the other direction. (9)

sericin. Gluelike substance released by the silkworm when spinning the cocoon that holds the cocoon together. (8)

sericulture. Science of raising silkworms from moths to the cultivation of silk fibers. (8)

service director. *See* customer service manager. (21)

set-in sleeve. Sleeve that is stitched to the garment around the regular armhole. (3)

Seventh Avenue. New York City's "Fashion Avenue" along which designers' showrooms and apparel industry company offices are located. (4)

sewer. Dry-cleaning firm employee who repairs garments; a person who sews. (23)

sewing machine operator. *See* machine operator. (20)

shade. A darkened color made by adding black to a hue. (10)

shape. Element of design signifying the silhouette, or overall outline, of a garment or other item. (11)

sheath. Dress that hangs from the shoulders and has inward shaping at the waist, but no waistline seam. (3)

shearing. Mechanical finishing procedure where projecting fibers are cut or trimmed from the face of fabric. *Also*, cutting off the fleece from sheep. (8, 9)

shift (chemise). Straight, loose-fitting dress that has no inward shaping at the waist and no waistline seam. (3)

shirt sleeve measurement. The length from the back center base of the neck, across the shoulder, and down the arm around the bent elbow to the wristbone. (16)

shirtwaist. Dress style that is like a dress-length, semifitted, tailored shirt with a belt or sash at the waist. (3)

shoplifting. The stealing of merchandise from a store. (15)

shoppertainment. Providing consumers with entertainment value while they shop, for a fun and memorable experience. (7)

showroom manager. Employee who supervises the personnel and activity in a firm's showroom. (20)

showroom salesperson. Employee who makes "in house" sales presentations of a line of goods to buyers who visit the manufacturer's showroom. (20)

shrinkage control. Chemical finish used to give a fabric dimensional stability. (9)

shuttle. Device that pulls filling yarns back and forth from edge to edge in weaving. (9)

shuttleless loom. Weaving device that carries the filling yarns by steel bands attached to wheels or other method without a shuttle. (9)

signature line. A collection of apparel or commercial patterns that has the endorsement of a celebrity. (5)

silhouette. The shape of a clothing style shown by its outer lines. (2)

silk. Natural fiber obtained from cocoons spun by silkworms. (8)

silk screening. *See* screen printing. (9)

singeing. Mechanical finishing procedure of burning off protruding fibers from yarn or fabric to give a smooth, uniform surface. (9)

single knit. Stretchy knit fabric constructed on a single-needle, weft knitting machine. (9)

single-breasted. A front closure that laps over and is held shut with one row of buttons. (3)

sizing. Solution applied to fill up the spaces between yarns in fabric to increase the fabrics weight, body, and luster. (9)

sketcher. Fashion house employee who makes illustration-quality drawings of the ideas that a designer has draped in fabric onto a dressmaker's form. (20)

sketching assistant. Employee of an apparel manufacturing firm or pattern company who records a seasons line of designs in precise technical illustrations for the company's records. (20)

skort. Shorts combined with a skirt panel across the front. (3)

slacks. *See* pants. (3)

sleeve board. Miniature ironing board used for pressing small areas such as sleeve seams. (18)

sleeveless. Garment design that has no sleeves. (3)

sliver. Continuous ropelike strand of loosely assembled staple fibers without twist. (8)

sloper. A basic pattern in a certain size, or to particular body measurements, used as the basis for creating fashion patterns. (5)

smart cards. Similar to credit cards, these contain a microchip that can store information such as body scanning measurements or financial balances. (7, 16)

soaking. Stain removal method to eliminate spots from washable fabrics. (18)

soaps. Biodegradable bar or granule cleaning products made mostly from natural fats and lye. (18)

social saturation. A point reached by a fashion when it is overused, is no longer novel or exclusive, and becomes dull and boring. (2)

soil release finish. Chemical finish for fabrics that eases the removal of soil and stains. (9)

sole proprietorship. Business with only one owner. (23)

solution dyeing. Process of adding color to a fiber solution before it is extruded. This gives a clear, rich color with high colorfastness since the pigment is a part of the fiber. (9)

sorter. Dry-cleaning firm employee who puts identity tags on incoming clothes, sorts them, inspects for stains, checks pockets, and classifies them according to cleaning procedures required. *Also* called a marker. (23)

sourcing. The seeking of vendors or producers of desired goods. (21)

spandex. Manufactured fiber that is strong, lightweight, and durable with great elasticity. (8)

specialty store. Retail outlet that handles a specific kind of merchandise or one category of goods, such as shoes, bridal attire, or children's apparel. (7)

specialty wools. Fibers from the hair of goats, camels, or other animals. (8)

spinneret. Metal disc containing many tiny holes through which liquid fiber-forming solutions are forced to form manufactured filaments. (8)

spinning. Process or processes used in the production of yarns from staple fibers or the extrusion of manufactured filaments. (8)

split-complementary color scheme. The use of one color with the two colors on each side of its complement on the color wheel. (10)

sponging. The dampening and drying of wool fabrics to prevent them from shrinking with later dry cleaning. *Also*, a method of removing stains from fabrics. (9)

sport coat. A classic, suit-type jacket. *Also* called a blazer. (3)

spotter. Dry-cleaning firm employee who treats stains. (23)

spreaders. Machines that hold bolts of fabric and roll back and forth to spread yard goods onto long tables in high stacks for garment cutting. *Also*, apparel manufacturing employees who lay out fabric for cutting. (6, 20)

spun silk. Fabric that is made from staple silk fibers or waste silk. (8)

spun yarns. Yarns made by spinning (mechanically pulling and twisting) staple fibers together into a continuous length. (8)

stain-resistance. Ability to resist spills and other soiling because of chemical finishing that makes the fibers less absorbent. (8, 9)

standards. The criteria set by authorities who judge products to verify certain levels of quality. (15)

staple fibers. Short fibers of various lengths from natural sources or cut lengths of manufactured fibers. *Also*, in retailing, an item that is always in demand and always kept in stock. (8)

starches. Sizings that restore body and crispness to fabrics that have become limp from laundering and wear. (18)

status. One's position or rank in comparison to others. (1)

stock clerk. Retail employee who receives and marks merchandise, moves it to the selling floor, maintains inventory records, and otherwise handles merchandise inventory. *Also* called a merchandise clerical. (21)

stock collar. Collar that is an imitation of an ascot. (3)

stock control. The receiving, storing, and distributing of merchandise in a retail store. *Also* called inventory control. (7)

stock dyeing. The dyeing of natural fibers in staple form. (9)

stock keeping unit (SKU). The smallest unit for which sales and stock records are kept for inventory control and retail identification systems. (7)

store manager. Retail executive responsible for the profitable operation of an entire store and all its functions. (21)

store ownership groups. Corporations formed by individual stores joining together to share information, expertise, and centralized volume buying. (7)

store policies. Rules of a retail establishment, often for the safety and protection of shoppers. (15)

Stores Magazine. Trade publication of the NRF. (7)

straight. Skirt that goes down straight from the hipline with no added fullness and a slim silhouette. *Also*, pants that are the same width at the bottom as at the knee. (3)

straight lines. Elongated marks that are not curved or jagged. (11)

strap-shoulder sleeve. *See* saddle sleeve. (3)

structural design. Texture or interest created when fabrics are manufactured. (19)

structural lines. Constructed lines in clothing (seams, darts, pleats, tucks, and garment edges) that are formed when a garment is made. (11)

structural texture. Surface interest created when fabrics are manufactured. (11)

style. A particular design, shape, or type of apparel item distinguished by the particular characteristics that make it unique. *Also*, possessing a characteristic or distinctly "smart" way of doing things. (2, 3)

stylish. Wearing or using the styles that are currently popular and fashionable. (2)

stylist. Person who redesigns existing garments rather than creating new fashion designs. *Also*, one who advises about styles in apparel or other categories of goods. (6)

subscription fabric club. Mail-order fabric sales organization. (23)

sundress. Dress for hot weather with a skirt attached to a brief upper garment. (3)

surface designs. Designs applied on the right side of fabrics, usually by printing. (19)

surfactants. Surface active agents in detergents that reduce the surface tension of water, help loosen and remove soil, and suspend the soil in the water until it is drained. (18)

surplice. Having a diagonally overlapping neckline or closing. (11)

sweater. A knitted (or crocheted) covering for the upper body. (3)

sweater knit. A loosely knitted stretchy fabric made with large yarns. (9)

sweatshops. Manufacturing plants that pay less than fair wages, use child labor, do not recongnize overtime worked, have unclean conditions, or violate workers' rights in other ways. (6)

symmetrical. Containing formal balance, or design details that are identical on both sides and an equal distance from the center. (12)

synthetic fibers. Textile products manufactured totally from chemicals. (8)

T

tabs. Decorative fabric pieces that go out from the edge of pockets or other areas of a garment. (3)

tailor. *See* alterations expert and dressmaker, garment maker, and seamstress. (21, 23)

tailor system. Manufacturing system in which all sewing tasks for a garment are done by a single operator. (6)

tailored garment. Garment that is made by first cutting garment pieces and then sewing them together to fit the shape of the body. (2)

tapered. A silhouette, such as for pants, that narrows near the bottom or hem. (3)

target market. The block of consumers a company wants as customers, and toward whom it directs its marketing strategy. (7)

technical writer. Person who authors technical material, such as the sewing directions on a commercial pattern guidesheet. (23)

technology. Scientific discovery or modification of products and methods, usually advanced through research and development. (5, 7)

telecommunication retailing. Offering shopping via communication devices, such as televisions and computers. (7)

television retailing. *See* telecommunication retailing. (7)

temporary finish. Fabric finish that lasts until washing or dry cleaning. (9)

tent dress. Large, billowy dress that hangs loosely from the shoulders. (3)

tentering. Drying and stretching fabric to a specified finished width under tension. (9)

tertiary hues. *See* intermediate hues. (10)

textile belt. Area of the United States stretching from New England down through the Carolinas and Georgia across to Texas where most domestic textile manufacturing is done. (8)

Textile/Clothing Technology Corporation [(TC)2]. An industry-wide, not-for-profit, textile/apparel coalition that researches high-tech innovations in apparel production processes and helps the industry implement them. (6)

textile colorist. Textile industry employee who decides which color combinations in textile designs are most likely to appeal to customers. (19)

textile converter. Firm that buys or handles greige goods for finishing. *Also*, an individual textile production employee who decides what textiles should be made or how various amounts of fabrics should be dyed, printed, or otherwise finished. (5, 19)

textile designer. Textile industry employee who creates new patterns and designs, or redesigns existing ones, to be used in the making of fabrics. (19)

Textile Fiber Products Identification Act. Legislation assigning the Federal Trade Commission to organize, name generically, and define the products of the manufactured fiber industry, and to require labeling of textile products with percentages of fibers present. (15)

textile industry. Business firms that deal with any or all processes needed to produce fibers, yarns, and finished fabrics. (5, 19)

textile stylist. Textile industry employee who is responsible for a fabric line from its beginning stages to completion, including long-range fashion planning of fabric colors, weights, and textures. (19)

textile tester. Technician with a textile mill or an independent testing laboratory who tests new products against specifications. (19)

texture. Element of design concerned with the surface quality of goods, or how they feel and look. (11)

texturing. Process of crimping, looping, or coiling manufactured filaments by using chemicals, heat, or special machinery. This increases bulk, resilience, and softness, or creates a different surface texture. (8)

theatrical costuming. Job of finding or creating appropriate apparel for opera, ballet, stage plays, circuses, movies, advertisements, television productions, and parades. (23)

thirty-day charge account. Consumer credit arrangement whereby the bill must be paid in full within 30 days of the billing date. There is no extra finance charge on the bill. (16)

throwing. The twisting together of several silk filaments to form a silk yarn suitable for weaving. (8)

tie dyeing. A form of resist dyeing using tight folds to form barriers to the dye where desired. (9)

tint. A lightened color made by adding white to a hue. (10)

toddler. A young child, usually between 1 and $2^1/_2$ years of age, who is actively moving or walking. (17)

Tokyo Collection. Association organized to introduce Japanese designers to buyers, the public, and the press. (4)

Tommy Awards. Statuettes awarded in the United States for the best printed fabrics. (4)

top management. A firm's highest executives. (20, 21)

tow. A large, continuous strand of manufactured fiber filaments in loose, ropelike form before being cut into staple. (8)

trade associations. Groups that promote or further the interests of a certain industry or trade. (4)

trade deficit. Condition that exists when the amount of imports is greater than the amount of exports. (5)

trade publications. Magazines, newspapers, and books that deal specifically with a certain industry or trade. (4)

trademark. A product's own brand name, often represented by a logo, that identifies it as belonging to a particular seller or manufacturer. (16)

trading company. Import-export business that contracts for, trades, or buys merchandise from other countries. (23)

trainee. Person learning to do a particular job. (19)

training supervisor. Manufacturing plant employee who trains machine operators to do specific tasks or to use specialized machines. *Also*, a retail store employee who gives orientation classes to new salespeople and updating programs to current salespeople. (20, 21)

transcript. Complete, official, written record of a person's educational courses and grades received. (24)

transition. Fluid rhythm created when a curved line leads the eye over an angle. (12)

travel wardrobe. Apparel a person takes on a trip. (17)

traveling sales representative. Member of a firm's "outside" sales force who sells the company's products away from the firms sales office and showroom. *Also* called sales reps. (20)

triacetate. Manufactured cellulosic fiber with a crisp hand and heat-setting qualities for permanent pleating. (8)

triad color scheme. Plan that uses three colors equidistant on the color wheel. (10)

tricot. Drapable, warp knit fabric that does not run. (9)

trimmer. Apparel manufacturing employee who cuts off or pulls out loose threads and removes lint and spots from finished apparel items. *Also* called a cleaner. (20)

trimmings. Various types of ornamentation added to garments. (16)

trousers. *See* pants. (3)

true bias. Grainline that runs at a 45 degree angle, or halfway between the lengthwise and crosswise grains of a fabric. (9)

trunk measurement. The length of the body from shoulder to crotch. (17)

trunk show. A limited amount of time during which a collection of a designer's or producer's samples are brought into a store, and orders are taken directly from customers for later delivery. (6)

Truth-in-Lending Law. Federal statute requiring grantors of consumer credit to reveal the true cost of credit in uniform, easy-to-understand terms. (16)

tubular silhouette. A recurring style that is slim and straight from top to bottom. (2)

tunic. A hip-length or longer blouse or shirt that extends down over pants or a skirt and is often belted. (3)

tussah. Fabric woven from wild silk fibers. It has an uneven, coarse texture and is usually tan in color. (8)

twill weave. A basic fabric weave characterized by diagonal wales produced by a series of staggered floats. (9)

U

umbrella skirt. Skirt style with many narrow gores that open and close as the wearer moves. (3)

undertone. Subdued trace of a color seen through another color or modifying the other color. Every skin color has an undertone of either yellow (warm) or blue (cool). (10)

uneven plaid. A design of crossing lines and spaces that are different in one or both directions. (9)

uniform. Outfit or article of clothing that is specific to everyone in a certain group of people. (1)

unit production system (UPS). Computerized piecework apparel manufacturing system in which the cut pieces of a garment are hung (loaded) together on an overhead product carrier that moves them through the line. (6)

UPC (universal product code). Standardized bar codes that are printed onto machine-readable merchandise tickets for electronic scanning. When scanned, all needed information about inventory or retail transactions is automatically entered into the computer system. (7)

V

value. The lightness or darkness of a color between almost white to almost black. *Also*, the degree of worth or measure of benefit of something, such as highest quality of garment materials, construction, and fashion for the lowest price. (10, 16)

value added. The increase in worth of products as a result of a particular work activity. (2)

values. Ideas and beliefs that are important to people. (1)

variant. A manufactured fiber modified slightly (within its generic group) during production resulting in a change in the properties of the fiber. (8)

variety store. Retail outlet with a wide assortment of lower-priced merchandise displayed on open counters. (7)

vat dyeing. Coloring fabrics (usually of cellulosic fibers) with dyes that become water insoluble and are very colorfast. (9)

vendor. One who sells, specifically a manufacturer or wholesaler, from whom a retailer buys goods. (21)

vent. An opening that goes up into a sleeve from the opening of the cuff. (3)

vertical lines. Elongated marks that go up and down. (11)

vest. Sleeveless, close-fitting, jacket-like garment that covers the chest and back. (3)

video demonstration work. Employment in the production of videotapes such as those that give sewing demonstrations or fashion reports. (23)

video merchandising. The use of videos in retail stores to show new fashion trends, promote merchandise, and build customer traffic. (7)

vinyon. Manufactured fiber used most often in bonded and nonwoven fabrics for industrial clothing. (8)

virgin wool. Wool fibers that have never been used before. *Also* called pure wool or 100% wool. (8)

visual merchandising. Presenting goods in an attractive and attention-getting manner though displays, exhibits, and special promotional activities. (7)

visual presentation director. *See* display manager. (22)

vocation. *See* career. (24)

W

waistband. Band of fabric that fastens together at the waistline of a garment. (3, 16)

wale. In woven fabrics, one of a series of ribs or cords running in a particular direction. In knitted fabrics, a column of loops lying lengthwise in the fabric. (9)

want. Desire for something that gives satisfaction but is not absolutely needed. (14)

wardrobe. The apparel a person owns, including all garments and accessories. (2)

wardrobe consultant. Person who shows consumers how to combine fashion items to enhance personal or professional images and helps them plan and manage their wardrobes and purchases. (23)

wardrobe designer. The head of theatrical costuming for an entertainment production organization. (23)

wardrobe helper. Theatrical worker who helps to obtain, make, and organize costumes and accessories by character and scene and to help the actors dress for the production. *Also* called a costume technician. (23)

wardrobe inventory. Itemized list of apparel, including accessories, that a person has. (14)

wardrobe plan. A "blueprint" of action to be taken to update or complete a person's wardrobe in the best way. (14)

warm colors. Hues, such as red, orange, and yellow, that appear to be hot like the sun or fire. (10)

warp knits. Fabrics made on flat knitting machines using many yarns and needles, with loops interlocking in the lengthwise direction. (9)

warp yarns. Yarns that run lengthwise (parallel to the selvage) in woven fabrics. (9)

wash load. The clothes put into a laundry machine to be washed together. (18)

washer agitation. The mechanical action that helps to loosen and remove soils during laundering. (18)

washing operator. Employee who does laundry with wet-washing and drying equipment. (23)

water repellent. Renewable finish that enables fabrics to shed some water, but cannot resist heavy rain or make fabrics completely waterproof. (9)

water softener. Chemical agent or mechanical device that neutralizes or removes the mineral ions found in hard water. (18)

waterproof. Finish that prevents water from passing through a fabric. (9)

wearing ease. The room in a garment, beyond the actual body measurements, which the wearer needs to move comfortably in the garment. (16)

weaving. Procedure of interlacing two sets of yarns placed at right angles to each other, usually done on a loom. (9)

Web-based retailing. Selling to consumers electronically via the Internet; "e-tailing". (7)

weft knits. Fabrics knit with one continuous strand of yarn going crosswise. *Also* called filling knits. (9)

weft yarns. *See* filling yarns. (9)

weighted silks. The addition of salts from tin, lead, or iron to silk fabrics to make them heavier and more drapable. (8)

weskit. *See* vest. (3)

wet spinning. Process in which a solution of manufactured fiber-forming substance is extruded into a chemical bath that regenerates and solidifies the polymer. (8)

wholesale. The selling of goods in large lots to retailers. (2)

wicking. The dispersing or spreading of moisture or liquid through a given area, such as pulling body moisture to the surface of a fabric where it can evaporate. (8)

wild silk. Uneven, rugged silk fibers from uncultivated silkworms. *See* tussah. (8)

window shopping. Gathering fashion or other merchandise information by looking in stores, or at their display windows, without making actual purchases. (15)

Women's Wear Daily (WWD). Trade newspaper covering all aspects of the women's fashion industries. (4)

woof yarns. *See* filling yarns. (9)

wool. Natural fiber obtained from the fleece of sheep. (8)

Wool Bureau, Inc. Trade association of United States wool growers. (8)

Wool Products Labeling Act. Law enforced by the Federal Trade Commission that requires the label on a wool product to give the percent of fiber content and the source. (15)

woolen yarn. Wool fabrics made from shorter fibers by the woolen system that are relatively dense and have soft, fuzzy surfaces. (8)

work-study program. A co-op or internship arrangement whereby an academic institution (school) places students in temporary jobs that relate to and alternate with their studies. (24)

worsted yarn. Wool fabrics made from longer combed fibers by the worsted system that have tight, smooth surfaces. (8)

wrap style. Garments that wrap around the body and overlap at the side-back or side-front. (3)

wrapper. Dry-cleaning firm employee who packages the finished apparel to maintain the quality of cleaning and finishing. *Also* called a bagger. (23)

written test. Test of general knowledge and aptitudes, often required when a person applies for a job. (24)

Y

yang. In Chinese cosmology, the active, rugged elements of personality with large physical characteristics. (13)

yard goods. Manufactured lengths of fabrics. (5, 9)

yarn. A continuous, usually twisted, strand of fibers suitable for weaving, knitting, or other processing into fabrics. (8)

yarn dyeing. The dyeing of yarns before they are woven or knitted into fabrics. (9)

yin. In Chinese cosmology, the passive, timid, delicate elements of personality with small physical characteristics. (13)

yoke. A band or shaped piece, usually at the shoulders or hips, to give shape and support to the garment below it. (3)

Index

G

H

L

M

N

O

P

Q

R

S

T

U

V

W

Y

Z